Engineering Tools for Environmental Risk Management – 3

Engineering Tools for Environmental Risk Management – 3

Site Assessment and Monitoring Tools

Editors

Katalin Gruiz

Department of Applied Biotechnology and Food Science, Budapest University of Technology and Economics, Budapest, Hungary

Tamás Meggyes

Berlin, Germany

Éva Fenyvesi

Cyclolab, Budapest, Hungary

CRC Press
Taylor & Francis Group
Boca Raton London New York

CRC Press is an imprint of the
Taylor & Francis Group, an **informa** business

A BALKEMA BOOK

Published by: CRC Press/Balkema
P.O. Box 11320, 2301 EH Leiden, The Netherlands
e-mail: Pub.NL@taylorandfrancis.com
www.crcpress.com – www.taylorandfrancis.com

First issued in paperback 2020

CRC Press/Balkema is an imprint of the Taylor & Francis Group, an informa business
© 2016 by Taylor & Francis Group LLC

Typeset by MPS Limited, Chennai, India

No claim to original U.S. Government works

ISBN 13: 978-0-367-57426-0 (pbk)
ISBN 13: 978-1-138-00156-5 (hbk)

British Library Cataloging in Publication Data

A catalogue record for this book is available from the British Library

Library of Congress Cataloging-in-Publication Data

Names: Gruiz, Katalin, editor. | Meggyes, T. (Tamás), editor. | Fenyvesi,
Éva, editor.
Title: Site assessment and monitoring tools / editors, Katalin Gruiz,
Department of Applied Biotechnology and Food Science, Budapest University
of Technology and Economics, Budapest, Hungary, Tamás Meggyes, Berlin,
Germany, Éva Fenyvesi, Cyclolab, Budapest, Hungary.
Description: London, UK : CRC Press/Balkema is an imprint of the Taylor &
Francis Group, an Informa business, [2016] | Series: Engineering tools for
environmental risk management ; volume 3 | Includes bibliographical
references and index.
Identifiers: LCCN 2016028941 (print) | LCCN 2016037662 (ebook) | ISBN
9781138001565 (hardcover : alk. paper) | ISBN 9781315778761 (eBook PDF) |
ISBN 9781315778761 (ebook)
Subjects: LCSH: Pollution–Measurement. | Environmental monitoring. |
Environmental risk assessment. | Environmental indicators.
Classification: LCC TD193 .S57 2016 (print) | LCC TD193 (ebook) | DDC
628.028/7–dc23
LC record available at https://lccn.loc.gov/2016028941

Visit the Taylor & Francis Web site at
http://www.taylorandfrancis.com
and the CRC Press Web site at
http://www.crcpress.com

Table of contents

4 *In-situ* and real-time measurements for effective soil and contaminated site management   **245**

K. GRUIZ, É. FENYVESI, M. MOLNÁR, V. FEIGL, E. VASZITA & M. TOLNER

Preface

This is the third volume of the five-volume book series "Engineering Tools for Environmental Risk Management". The book series deals with the following topics:

1. Environmental deterioration and pollution, management of environmental problems
2. Environmental toxicology – a tool for managing chemical substances and contaminated environment
3. Assessment and monitoring tools, risk assessment
4. Risk reduction measures and technologies
5. Case studies for demonstration of the application of engineering tools

The authors aim to describe interactions and options in risk management by

– providing a broad scientific overview of the environment, its human uses and the associated local, regional and global environmental problems;
– interpreting the holistic approach used in solving environmental protection issues;
– striking a balance between nature's needs and engineering capabilities;
– understanding interactions between regulation, management and engineering;
– obtaining information about novel technologies and innovative engineering tools.

This third volume provides an overview on the basic principles, concepts, practices and tools of environmental monitoring and contaminated site assessment. The volume focuses on those engineering tools – such as the in situ and real-time measurement methods – that enable integrated site assessment and decision making and ensure an efficient control of the environment. Some topics supporting sustainable land use and efficient environmental management are listed below:

– Efficient management and regulation of contaminated land and the environment;
– Early warning and environmental monitoring;
– Assessment of contaminated land: the best practices;
– Environmental sampling;
– Risk characterization and contaminated matrix assessment;
– Integrated application of physical, chemical, biological, ecological and (eco)toxicological characterization methods;
– Direct toxicity assessment (DTA) and decision making;

- Online analyzers, electrodes and biosensors for assessment and monitoring of waters;
- *In situ* and real-time measurement tools for soil and contaminated sites;
- Rapid on-site methods and contaminant and toxicity assessment kits;
- Engineering tools from omics technologies, microsensors to heavy machinery;
- Dynamic characterization of subsurface soil and groundwater using membrane interface probes, optical and X-ray fluorescence and ELCAD wastewater characterization;
- Geochemical modeling: methods and applications;
- Environmental assessment using cyclodextrins.

Abbreviations

2,4D	–	2,4-dichlorophenoxyacetic acid
AAS	–	atomic absorption spectrometry
ACE	–	acetylthiocholine
AChE	–	acetylcholinesterase
ADCP	–	acoustic doppler current profiler
ADI	–	acceptable daily intake
ADS	–	accelerated diffusion sampler
AE	–	alkyl ethoxilate
AHH	–	aryl hydrocarbon hydroxylase
Ag-ddtc	–	silver diethyldithiocarbamate
AMS	–	advanced monitoring systems
ANAIS	–	autonomous nutrient analyzer *in situ*
ANOVA	–	ANalysis of VAriance
APE	–	alkylphenol ethoxylate
APNA	–	autonomous profiling nutrient analyzer
AS	–	acid sulfate
AWACSS	–	Automated Water Analyzer Computer Supported System
B(a)P	–	benzo(a)pyrene
BioMEMS	–	biological micro-electro-mechanical systems (sensors)
BL-MTB	–	bioluminescent magnetotactic bacteria
BOD	–	biological oxygen demand
BOD5	–	5-day BOD test
BTEX	–	aromatic hydrocarbons of benzene-toluene-ethylbenzene-xylenes
CA	–	cluster analysis
CABERNET	–	Concerted Action on Brownfield and Economic Regeneration Network
CBEM	–	community based environmental monitoring network
CHARM	–	channel relocatable mooring
CHCs	–	chlorinated hydrocarbons
CD	–	cyclodextrin
CDF	–	cumulative distribution function
CENS	–	Center for Embedded Networked Sensing
CL:AIRE	–	Contaminated Land: Applications in Real Environments
CLARINET	–	Contaminated Land Rehabilitation Network for Environmental Technologies

CLEA	–	contaminated land exposure assessment
CMBCD	–	carboxymethyl beta-cyclodextrin
CMT	–	condition monitoring technologies
COD	–	chemical oxygen demand
CPAHs	–	carcinogenic PAHs
CPT	–	cone penetrometer testing
CRDS	–	cavity ring-down spectroscopy
CSV	–	cathodic stripping voltammetry
CTD	–	Conductivity-Temperature-Depth recorder
CTM	–	continuous toxicity monitor
CWA	–	Clean Water Act
CYP1A	–	Cytochrome P450 1A
CZT	–	cadmium, zinc, telluride detector
D	–	dose
DAS	–	data acquisition system
DCS	–	distributed control system
DDD	–	dichlorodiphenyldichloroethane
DCE	–	dichloroethylene
DDE	–	dichlorodiphenylchloroethane
DDT	–	dichlorodiphenyltrichloroethane
DELCD	–	dry electrolytic conductivity detector
DGGE	–	denaturing gradient gel-electrophoresis
DGT	–	diffusive gradients technology
DNA	–	deoxyribonucleic acid
DNAPL	–	dense non-aqueous phase liquid
DNAzymes	–	catalytic DNAs
DO	–	dissolved oxygen
DOAS	–	differential optical absorption spectroscopy
DoD	–	U.S. Department of Defense
DoE	–	U.S. Department of Energy
DPC	–	1,5-diphenylcarbazide reagent
DPD	–	n,n–diethyl–p–phenylenediamine sulfate reagent
DST	–	decision support tools
DTA	–	direct toxicity assessment
dtk	–	droptestkits
DTNB	–	5,5'- dithio-bis (2-nitrobenzoic acid)
EDA	–	exploratory data analysis
EDCs	–	endocrine disrupting chemicals
EDXRF	–	energy-dispersive X-ray fluorescence analyzers
EF	–	enrichment factors
EFNMR	–	Earth's field NMR
ELCAD	–	electrolyte cathode discharge spectrometry
ELISA	–	enzyme linked immunosorbent assay
EO	–	Earth Observation Portal
EPA	–	Unites States Environmental Protection Agency
EQC	–	environmental quality criteria
ERA	–	environmental risk assessment
ESEB	–	Environmental Stewardship Equipment Bank

EsM$_{20}$	–	sample mass causing 20% inhibition
ESP	–	effective sample proportion
ESR	–	electron spin resonance
ETV	–	environmental technology verification
EW	–	early warning
EWES	–	evanescent wave excitation system
EWS	–	early-warning systems
EWI	–	evanescent wave immunosensors
EWOP	–	evanescent wave fiber optic immunosensors
FAME	–	fatty acid methyl esters
FAST	–	flow assay and sensing system
FFD	–	fuel fluorescence detector
FIA	–	flow injection analysis
FIAM	–	free-ion activity model
FID	–	flame ionization detector
FISH	–	fluorescence *in situ* hybridization
FMO	–	flavine-dependent monooxygenase
FOCS	–	fiber optic chemical sensors
FTIR	–	Fourier transform infrared spectroscopy
GAW	–	Global Atmosphere Watch
GC	–	gas chromatography
GC/MS	–	gas chromatograph equipped with mass spectrometry detector
GEMS	–	World Weather Watch – Global Environment Monitoring System
GEOSS	–	Global Earth Observation System of Systems
GFP	–	green fluorescent protein
GHS	–	globally harmonized warning system
GLRL	–	green light red light device
GLRS	–	grating light reflection spectroscopy
GMR	–	geomagnetic resonance
GnRH	–	gonadotropin-releasing hormone
GOOS	–	Global Ocean Observing System
GPR	–	ground penetrating radars
GPS	–	Global Positioning Systems
GRP	–	geogenic radon potential
HABs	–	hazardous algal blooms
HANAA	–	handheld advanced nucleic acid analyzer
HAP	–	high-altitude platform
HDPE	–	high density polyethylene
HPACD	–	hydroxypropyl alpha-cyclodextrin
HPBCD	–	hydroxypropyl beta-cyclodextrin
HPGCD	–	hydroxypropyl gamma-cyclodextrin
HRGB	–	Hungarian regional geochemical background
HRP	–	horseradish peroxidase
HSR	–	hyperspectral imaging
HRSC	–	high-resolution site characterization
HSI	–	hyperspectral irradiometer
HSPs	–	heat shock proteins
HSR	–	hyperspectral

IBI	–	biotic integrity index
ICP	–	inductively coupled plasma spectrometry
ICS	–	industrial control system
IM	–	inductive modem
IMM	–	inductive modem module
IMS	–	ion mobility sensors
IQR	–	inter-quartile range
IR	–	infrared spectroscopy
ISCO	–	*in situ* chemical oxidation
ISCR	–	*in situ* chemical reduction
ISE	–	ion selective electrodes
ITC	–	index of trophic completeness
K$_{ow}$	–	octanol-water partition coefficient
LAS	–	linear alkylbenzene sulphonate
LDPE	–	low density polyethylene
LOEsM	–	lowest effect mass
LIB/LIBS	–	laser-induced breakdown spectroscopy
LIDAR	–	light detection and ranging
LIF	–	laser-induced fluorescence
LNAPL	–	light non-aqueous phase liquid
LOC	–	lab-on-a-chip
LOD	–	limit of detection
MAD	–	median averaged deviation
MAO	–	monoamine oxidase
MBs	–	molecular beacons, oligonucleotide hybridization probes
MEMS	–	microelectromechanical systems
MFO	–	cytochrome P-450-dependent monooxygenase
MIA	–	magnetic immunoassay
MIP	–	membrane interface probe
MNA	–	monitored natural attenuation
MOSEAN	–	multi-disciplinary ocean sensors for environmental analyses and networks
MPEIA	–	magnetic particle enzyme immunoassay
MRI	–	magnetic resonance imaging
MRI	–	multiplexer radiometer irradiometer
m-RNA	–	messenger ribonucleic acid
MRS	–	magnetic resonance sounding
MS	–	mass spectrometry
MSR	–	multispectral radiometers
MT	–	metallothioneins
μTAS	–	micro total analysis systems
MWD	–	Mining Waste Directive
MXR	–	multixenobiotic resistance protein
NADH	–	nicotinamide adenine dinucleotide
NAPL	–	Non-Aqueous Phase Liquid
NDIR	–	nondispersive infrared (detector)
NDVI	–	normalized difference vegetation index
NETL	–	National Energy Technology Laboratory

NIR	–	near infrared spectroscopy
NMR	–	Nuclear Magnetic Resonance
NOAA STAR	–	NOAA Center of Satellite Applications and Research
NOEsM	–	no effect mass
NOPP	–	National Oceanographic Partnership Program
NPDES	–	National Pollutant Discharge Elimination System
NRC	–	National Research Council
NT	–	needle trap
NTREE	–	National Round Table on the Environment and the Economy's National Brownfields Redevelopment Strategy
O/C	–	organophosphate and carbamate type insecticides, pesticides
ORP	–	oxidation reduction potential
O-SCOPE	–	ocean-systems for chemical, optical, and physical experiments
OUR	–	oxygen uptake rate
P	–	toxicitiy of the pore water
PAH	–	polycyclic aromatic hydrocarbon
PAN	–	1-(2-pyridylazo)-2-naphthol) indicator
PAOs	–	phosphorus accumulating organisms
PAR	–	photosynthetically active radiation
PBDE	–	polybrominated diphenyl ether
PCA	–	principal component analysis
PCAPS	–	passive capillary samplers
PCBs	–	polychlorinated biphenyls
PCE	–	perchloroethylene
PCP	–	pentachlorophenol
PCPT	–	piezocone, an electric CPT cone
PCR	–	polymerase chain reaction
PDMS	–	polydimethylsiloxane
PEC	–	predicted environmental concentration
PeCOD	–	photo-electro chemical oxygen demand technology
PH	–	percussion hammer
PID	–	photoionization detector
PMRSE	–	passive magnetic resonance subsurface exploration
PNEC	–	predicted no effect concentration
PLC	–	programmable (logic) controller
POP	–	persistent organic pollutants
PRI	–	photochemical reflectance index
PXRF	–	Field-portable X-ray fluorescence measuring devices
Q	–	questions
qPCR	–	quantitative PCR
QSAR	–	Quantitative Structure Activity Relationship
RA	–	risk assessment
RAMEB	–	randomly methylated beta-cyclodextrin
RAP	–	site remedial action plan
RAPD	–	random amplified polymorphic DNA expression
R/B	–	respiration to biomass rate
RCR	–	risk characterization ratio
RDO	–	rigged dissolved oxygen

RDX	–	Research Department explosive, cyclotrimethylenetrinitramine, IUPAC
name:		hexahydro-1,2,5-trinitro-1,3,5-triazine
RESCUE	–	Regeneration of European Sites in Cities and Urban Environments
RFLP	–	restriction-fragment-length polymorphism
RIANA	–	river analyzer
RMSE	–	root mean square error
RNA	–	ribonucleic acid
RNA-Seq	–	RNA sequencing
RODTOX	–	rapid oxygen demand for toxicity assessment
ROS	–	reactive oxygen species
ROST	–	rapid optical screening tool
RPAS	–	remote piloted aircraft systems
rRNA	–	ribosomal RNA
RRR	–	risk reduction rate
RT	–	reverse transcriptase
S	–	toxicity of the saturated soil or whole sediment
SAR	–	synthetic aperture radar
SARS	–	severe acute respiratory syndrome
SAW	–	surface acoustic wave sensors
SCADA	–	Supervisory Control and Acquisition of DAta
SCAPS–LIF	–	site characterization and analysis penetrometer system
SDD	–	silicon drift detector
SEAS	–	spectrophotometric elemental analysis system
SEPA	–	Swedish Environmental Protection Agency
SERS	–	surface-enhanced Raman scattering
SIA	–	sequential injection analysis
SiPIN	–	silicon PIN diode detector
SIR	–	substrate-induced respiration
SMAP	–	soil moisture active/passive
SPADNS	–	4,5 dihydroxyl-3-(p-sulfophenylazo)-2,7-naphthalene-disulfonic acid-Na salt reagent
SCPT/SCPTU	–	seismic cone penetration testing
SPE	–	solid-phase extraction
SPME	–	solid-phase microextraction
SPR	–	surface plasmon resonance
SPT	–	standard penetration testing
SQC	–	soil quality criteria
SRB	–	sulfate reducing bacteria
SRT	–	solids retention time
STT	–	soil testing triad/triplet
TBI	–	Trent biotic index
TCD	–	thermal conductivity detector
TCE	–	trichloroethylene
TDL	–	tunable diode laser
TDR	–	time domain reflectometry
TDT	–	time domain transmissometry

TDT	–	time delay transmission
TIN	–	triangular irregular network
TMS	–	toroidal ion trap mass spectrometer
TNT	–	trinitrotoluene
TOC	–	total organic carbon
TarGOST	–	tar specific green optical screening tool
T-RFLP	–	terminal restriction fragment length polymorphism
TSA	–	time series analysis
TSS	–	total suspended solids
U	–	UNKNOWN responses
UAS	–	unmanned air system
UAV	–	unmanned air vehicle
UNEP	–	United Nations Environment Programme
UV	–	ultraviolet light
UVOST	–	ultraviolet optical screening tool
Vtg	–	vitellogenin
VNTR	–	variable number of tandem repeat (marker-based bioassay)
VOC	–	volatile organic compounds
WCMC	–	World Conservation Monitoring Centre
WET	–	whole effluent toxicity
WET sensor	–	sensor for water content (W), electrical conductivity (E), and temperature (T)
WFD	–	Water Framework Directive
WHO	–	World Health Organization
WMO	–	World Meteorological Organization
WSN	–	wireless sensor network
XRF	–	X-ray fluorescence

About the editors

Katalin Gruiz graduated in chemical engineering at Budapest University of Technology and Economics in 1975, received her doctorate in bioengineering and her Ph.D. in environmental engineering. Her main fields of activities are: teaching, consulting, research and development of engineering tools for risk-based environmental management, development and use of innovative technologies such as special environmental toxicity assays, integrated monitoring methods, biological and ecological remediation technologies for soils and waters, both for regulatory and engineering purposes. Prof. Gruiz has published 35 papers, 25 book chapters, more than hundred conference papers, edited 6 books and a special journal edition. She has coordinated a number of Hungarian research projects and participated in European ones. Gruiz is a member of the REACH Risk Assessment Committee of the European Chemicals Agency. She is a full time associate professor at Budapest University of Technology and Economics and heads the research group of Environmental Microbiology and Biotechnology.

Tamás Meggyes is a research co-ordinator specialising in research and book projects in environmental engineering. His work focuses on fluid mechanics, hydraulic transport of solids, jet devices, landfill engineering, groundwater remediation, tailings facilities and risk-based environmental management. He contributed to and organised several international conferences and national and European integrated research projects in Hungary, Germany, United Kingdom and USA. Tamás Meggyes was Europe editor of the Land Contamination and Reclamation journal in the UK and a reviewer of several environmental journals. He was invited by the EU as an expert evaluator to assess research applications and by Samarco Mining Company, Brazil, as a tailings management expert. In 2007, he was named Visiting Professor of Built Environment Sustainability at the University of Wolverhampton,

UK. He has published 130 papers including eleven books and holds a doctor's title in fluid mechanics and a Ph.D. degree in landfill engineering from Miskolc University, Hungary.

Éva Fenyvesi is a senior scientist, a founding member of CycloLab Cyclodextrin Research and Development Ltd. She graduated as chemist and received her Ph.D. in chemical technology at Eötvös University of Natural Sciences, Budapest. She is experienced in the preparation and application of cyclodextrin polymers, in environmental application of cyclodextrins and in gas chromatography. She participated in several national and international research projects, in the development of various environmental technologies applying cyclodextrins. She is author or co-author of over 50 scientific papers, 3 chapters in monographs, over 50 conference presentations and 14 patents. She is an editor of the Cyclodextrin News, the monthly periodical on cyclodextrins.

Chapter 1

Integrated and efficient characterization of contaminated sites

K. Gruiz

Department of Applied Biotechnology and Food Science, Budapest University of Technology and Economics, Budapest, Hungary

ABSTRACT

Volume 1 (Gruiz *et al.*, 2014) of this book series discussed the problems of contaminated land and Volume 2 (Gruiz *et al.*, 2015) dealt with the tools used to assess the adverse effects of environmental contaminants. This third volume looks into the concepts of contaminated land assessment, site investigation and innovative engineering tools for the management of contaminated water and soil.

This Chapter 1 describes and assesses those sites already contaminated to prevent deterioration or further damage and specifies target quality and the extent of risk reduction. Chapter 2 continues this topic and concerns early warning and soft monitoring for the general observation of the environment.

Management of contaminated and derelict land, and pollutant spills requires the integrated knowledge and tools of geotechnics, chemical engineering, agriculture, mining, land remediation and waste treatment. First, the problems must be explored, then the sites investigated, results evaluated, environmental and human risks assessed, socioeconomic impacts characterized and the best risk reduction measures selected. Stakeholders must be involved in the entire process. Experience gained from the management of chemicals, wastes, air, water and agricultural soils should also be applied to contaminated land.

1 INTRODUCTION

The risk model of a contaminated site and the aim of assessment determine the assessment strategy. An optimal tool battery is needed to assess

(i)   The type of pollution and the contaminants;
(ii)   Occurrence, concentration, effects, environmental transport, and fate of contaminants;
(iii)   The state of endangered or contaminated environmental compartments and
(iv)   The receptors, i.e. (potential) users of (suspected) contaminated land – including both the ecosystem and humans.

The variables of contaminated site assessment are summarized in Figure 1.1.

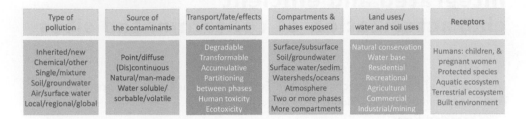

Type of pollution	Source of the contaminants	Transport/fate/effects of contaminants	Compartments & phases exposed	Land uses/ water and soil uses	Receptors
Inherited/new Chemical/other Single/mixture Soil/groundwater Air/surface water Local/regional/global	Point/diffuse (Dis)continuous Natural/man-made Water soluble/ sorbable/volatile	Degradable Transformable Accumulative Partitioning between phases Human toxicity Ecotoxicity	Surface/subsurface Soil/groundwater Surface water/sedim. Watersheds/oceans Atmosphere Two or more phases More compartments	Natural conservation Water base Residential Recreational Agricultural Commercial Industrial/mining	Humans: children, & pregnant women Protected species Aquatic ecosystem Terrestrial ecosystem Built environment

Figure 1.1 Variables of contaminated site assessment.

As a preamble, the terms used are clarified here. The terms land, site and soil, as used in this book, have a very similar meaning, as virtually throughout the technical literature (see also Chapter 3, Volume 1 of this book series, Gruiz *et al.*, 2014).

The term 'land' can be broadly interpreted and may refer to an area, a site, ground, or soil. Land is defined by the online Dictionary (Dictionary.com, 2015) as follows:

Land is any part of the earth's surface not covered by a body of water; the part of the earth's surface occupied by continents and islands; an area of ground, with reference to its nature (e.g. arable land); with specific boundaries, with specific use (e.g. agricultural, urban). From the point of view of ownership, land is "any part of the earth's surface that can be owned as a property, and everything annexed to it, whether by nature or by the human hand".

In this book, the term 'land' is used in a generic sense: a large area with all its environmental compartments, including their uses and users.

Site has a less versatile meaning, but still carries uncertainties. It may refer to the position or location of a town, a building, and its environment, and it may be an exact plot of ground on which anything is, has been, or is to be located (Dictionary.com, 2015). In this book we use the term 'site' for areas with specific coordinates and boundaries, mainly for contaminated or deteriorated sites. Soil and groundwater are typical compartments at a site. *On-site* means taking place or located at the site, whereas *off-site* means away from the site; some activities such as sample analysis, soil treatment, etc. may differ, depending on whether it is performed on- or off-site.

Soil is the top layer of the earth's surface. It consists of rock and mineral particles mixed with organic matter. The term soil is also used in the sense of a place favorable for plant growth and as the habitat of an ecosystem. Soil is inhabited by billions of 'invisible' microorganisms and by visible plants and animals. The terrestrial ecosystem's services are attributed to soil. Subsurface waters (groundwater and deep- or fossil waters) are also considered part of the soil. Soil is a three-phase system of gas, liquid and solid, with special forms of soil air, soil moisture, absorbed, capillary and other interstitial waters, water bodies of confined or unconfined aquifers or frozen water in permafrost regions, as well as soil particles of different size and structure, including base rock.

Other terms often used are *site assessment* and *site investigation*. They are more or less interchangeable, some literature and legislation prefer assessment, others investigation. Their key feature is the collection of information (historical information, databases and measured data), evaluation, and interpretation of these data in a study.

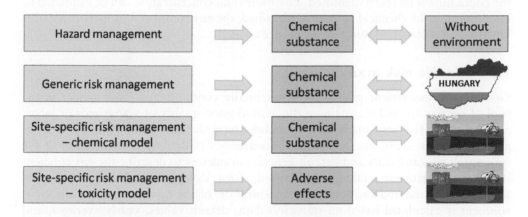

Figure 1.2 Hazard-based, generic and site-specific risk management of contaminants.

A study usually comprises a description, the site's conceptual model, collected and measured data, calculated risk values compared to no risk or acceptable risk values (see Figure 1.2), evaluation of the site from a socioeconomic point of view, and the options for further risk management.

As shown by Figure 1.2, environmental pollutants and contaminated land can be handled based on:

– Hazard management of hazardous agents, especially chemical substances;
– Generic risk management of hazardous agents and larger spatial units;
– Problem- or site-specific environmental management.

1.1 Hazard management of chemicals

Hazardous agents are discharged into the environment in smaller or larger quantities during their production, use and in their waste phase. This is the origin of all anthropogenic environmental pollution. Hazard management of chemicals, products, and wastes with known components is based on the intrinsic material properties of these substances. In addition to the material properties, environmental hazard also depends on their environmental fate and behavior. Typical management activities underpinning the use of the hazard-based approach, – i.e. independently from the location and the users – are the regulation (authorization, restriction, etc.) of pesticides, biocides and industrial chemicals, as well as classification and labeling of chemical products (e.g. the generic risk and safety information written on the packaging of a hazardous product). The hazard-based thinking excludes site-specific considerations and does not *per se* include the produced or used amount of the chemical substance. Hazard-based legislative measures do not deal with individual site characteristics, special environmental conditions, non-average human and ecosystem sensitivities, land uses, or special human behavior. These issues are handled by problem-specific or site-specific risk management tools. Nevertheless, information collected for hazard assessment is also essential for site-specific risk assessments. Hazard information can only be used when the contaminant has been identified. Environmental concentration can be estimated if, in addition to the chemical substance identified, the environmental properties are also known, and a transport and fate model are created.

1.2 Generic risk management

Generic risk assessment related to a potentially contaminated environment is the tool of prevention and maintaining the desired good quality of watersheds or larger areas. Generic risk assessment uses data about hazards (flammability, toxicity, mutagenicity, reprotoxicity, etc.) and fate characteristics (volatility, degradability) of the (potential) contaminants and default generic parameters to describe the environment (for example Hungary, Danube watershed, Toka Valley, etc.). The quantity most probably released and the resulting concentration of the contaminants in the environment is calculated based on statistical data, default values, yearly averages, and GIS-based watershed maps. These are values which are true for the entire area but do not necessarily hold for small sites. Generic risk assessment is the basic tool of environmental regulation and generic preventive measures. Unfortunately, it is not suitable for the management of small sites. Proper risk mitigation tools for regional or watershed-scale are:

– Prevention by reduction, restriction or ban;
– Elimination of sources and the remediation of contaminated land to protect surface waters.

In cases when life phases (production, use, and waste) of contaminating agents are independent of each other, site-specific risk assessment and risk reduction provide the only risk management option. Examples are abandoned sites with unknown contaminants, areas contaminated with chemicals produced and used in the past but currently banned, or simply sites with multiple uses in the past.

1.3 Site-specific risk management

Management of potentially or *de facto* **contaminated land** needs targeted risk assessment, which specifically deals with the predicted or the existing adverse impact. The assessment of an adversely impacted site can be approached from the angles of (i) the current (or potential) contaminants or (ii) the manifested (or potential) adverse effects already identified.

1.3.1 The chemical model – based on the contaminant concentration

The contaminants are the starting point of the chemical models. Their transport, fate, and effects are characterized on a known site based on the contaminants' measured or calculated concentrations. The exposure parameters are determined according to land uses and the identified local receptors, from which the adverse effects and the risks can be predicted. Transport and fate models enable calculation of the concentration of the chemical substance at different locations and times from the identified source location. Exposure, i.e. the predicted environmental concentration (PEC), is responsible for the adverse effects when humans and the ecosystem meet the contaminated environment. The determination of the chemical substance's distribution and the mapping of the environmental concentration on the site, in the neighborhood or in the future, must be based on a transport and fate model, even if sampling and analysis enable concrete site characterization. The validation of the model by measurements is the next necessary step. The information necessary for decision making and managing the problem is derived from the primary data by modeling and mapping. It explains why an environmental concentration at a site is always a predicted value. In those cases where the conceptual model has not been established and refined yet, and the contaminants and their sources have not been identified, sampling and analysis cannot be performed (since one does not know where sampling should take place and what should be analyzed) and what statistical data and historical information must be used (see also tiers of site assessment).

On establishing the conceptual model of the site and identifying the contaminants, sources and transport pathways, the risk reduction can also be planned.

1.3.2   The 'direct toxicity' model – based on measured adverse effects

When an *adverse impact* observed or predicted from early warning signals is the starting point of the management procedure, two alternatives are possible:

–   If the type, location and strength of the adverse effects are known (e.g. an impact map is available), one can trace back to the origin of the impact and identify the possible sources and causes. This approach is applicable for single-point sources when the transport pathways are obvious (runoff, leachates, floods, etc.) and when sensitive and high-resolution maps are available e.g. from monitoring of the biota by remote sensing.

–   As another alternative, management can be based on the type and extent of the measured adverse effects, but – contrary to the chemical model – without trying to identify the contaminants. In such cases, the targeted (acceptable) effect level is compared to the measured or predicted adverse effects, and the scale of toxicity reduction is proportional to the ratio of the actually measured to the targeted (acceptable) effect level (often the 'no effect' level). Typical application of this approach is where the problem is caused by too many unknown contaminants such as wastewaters, solid wastes, abandoned industrial or storage sites, or illegal waste disposal sites. The approach based on toxic risk can also be applied when the adverse effect is heavily influenced by environmental and contextual issues (presence of chemical species with different effects, high soil sorption capacity, low mobility and bioavailability of the contaminant), and the simple chemical model would not work. The assessment results reflect actual toxicity which includes the enhancing or buffering/compensating effect of the environment (contaminant mobility or bioavailability). The subsequent interventions seek to reduce toxicity by natural attenuation or stabilization of soil contaminants or by controlled flow parameters of leachates, runoff and wastewaters.

The tiered assessment approach, with direct toxicity assessment in the first step, may represent an economically feasible solution in the above cases, given that the negative samples will be excluded immediately after the first tier, and expensive chemical analyses will only be applied for those showing unacceptable toxicity. Since biosensors are becoming increasingly available and widespread, bioindicator-based *in situ* assessment and decision making is becoming more common as the basis for risk management.

The task of risk management becomes rather complex when discharge and pollution are separated in space and time. The use of plant nutrients and pesticides, the management of the resulting diffuse pollution on a watershed scale, and the non-traceable use of import products are typical examples of non-identifiable diffuse sources. Two approaches to manage this type of problem are (i) watershed-scale risk assessment of individual chemical species and (ii) complex chemical, ecological, and/or toxicological monitoring of watersheds. Monitoring results can validate the generic risk assessment outcome. In today's practice, the two approaches are not integrated into one common management methodologies.

Efficient site investigation applies a tiered tool battery. The assessment procedure can be controlled by decisions taken at points suitably arranged within the flowchart. This helps to achieve optimal efficiency in terms of time and cost. A properly designed flowchart ensures that the result of a tier specifies the next one and that iterative circles prevent superfluous data from being collected. Optimized tiering enables the exclusion of negative cases as early as possible. Figure 1.3 shows the stepwise site investigation (1 through 3 with the lists of data to be collected) and the risk assessment and decision making tiers based on the results of site investigation.

The flowchart and the assessment test battery vary according to circumstances, and the following must be clarified before developing the concept and planning the investigation:

- Is the contaminant unknown or has it already been identified?
- Is there a single contaminant or a mixture of contaminants?
- Which environmental compartments and phases are affected?
- Is the source, the transport pathway, or the target of the pollution process on the site itself?
- Do the contaminants stem from point or diffuse sources?
- Is the land suitable for present or future land uses? Is land use typically industrial, agricultural, residential, or recreational? Are, or were, the subsurface and surface waters used as drinking water?
- Who are the land users who may act as receptors, e.g. the aquatic ecosystem, terrestrial ecosystem, and humans?
- Have food chains been contaminated?
- Is the spatial extension local, regional or global?
- In terms of temporal extension, is the contaminated site an old, inherited site or a new one? Can it be contaminated in the future?
- What is the aim of the assessment?

According to their aim, the typical site assessment activities are:

- Environmental survey: recording the actual status of the environment;
- Time-related assessment: recording the changes in the environment;

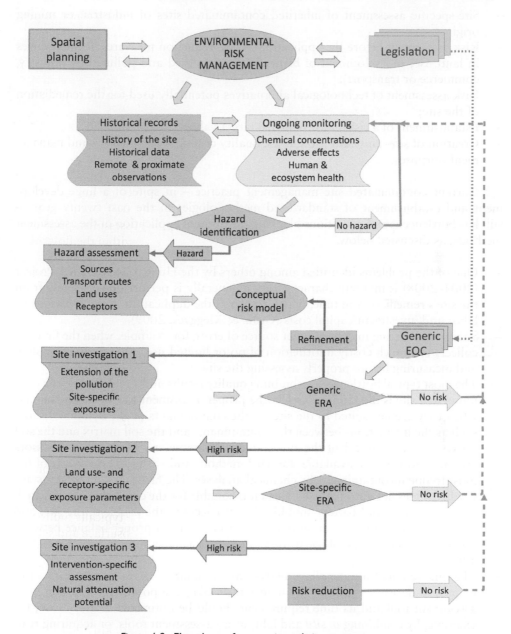

Figure 1.3 Flowchart of contaminated site management.

- Compliance monitoring in order to maintain a certain targeted quality through environmental management;
- Site-specific risk assessment of new industrial, agricultural, residential or recreational activities to support spatial planning and prevent and control contamination and risk;
- Site-specific risk assessment of accidental or regular discharges;

- Site-specific assessment of inherited contaminated sites of industrial or mining origin;
- Risk assessment before the application of risk reduction measures or the changes in land use, redevelopment of brownfields (sites used and polluted by industry, commerce or transport);
- Risk assessment of technological alternatives potentially used for the remediation of the site;
- Establishment of a risk-based remedial target value;
- Creation of site- and land-use-specific quality criteria for regulatory and management purposes.

Current contaminated site management practice – in spite of a huge development and establishment of standardized methodologies in the past twenty years – still has bottlenecks and conditions blocking the optimal application of the assessment methods, as discussed below.

- One of the problems identified among others by the Hungarian MOKKA Project (2004–2008) is that site characterization generally is performed separately from the site's remediation in time, which results in the duplication of the assessment's labor and investment/capital costs (Gruiz & Meggyes, 2009).
- Suboptimal tiering may be another source of error, for example, when the first tier collects too much costly information of no or limited use, e.g. starting sampling and measuring before properly assessing the site.
- The most typical bottleneck in obtaining quality results and a real picture of a contaminated site is the failure to select the proper assessment tools. This is the case when they are not suited to the site, to the contaminants or the dynamic factors such as the interactions between the contaminants and the soil matrix and the soil ecosystem or the effect of the environmental parameters. Most of the assessors are satisfied with the available standard methods and with the assessment of the most frequent contaminants by chemical analyses. The fact that a method is standardized does not automatically mean it is suitable for the case being investigated. The assessment tool battery should take into account the problem, the site, and the management target. An optimum method ensures a proper balance between the physico-chemical, biological and ecotoxicological methods and combines their advantages by synergy.
- The time required for sampling, analysis and evaluation of the results is a barrier to proper decision making. The longer they take, the poorer the decision. The assessment tool and its time requirement should be optimized in every case, for example, by combining *in situ* and laboratory assessment tools, or acquiring real-time data for immediate decision making and intervention.
- Planners and decision makers are required to know the available tools and their applicability, as well as the problem and the site in detail to create the optimal tool battery. However, because the problem and the site are not known at the beginning of the process, a dynamic and iterative concept has to be applied, which results in a gradual, cost- and time-effective progress. It requires the integration of the risk management tasks of (i) site investigation, (ii) risk assessment, and (iii) selection of the best risk reduction tool (which is cost-, time- and eco-efficient).

- A dynamic site assessment tool can provide the assessor with information not only about the current condition/status and risks posed by the site, but also about the site's own risk mitigation potential. *In situ* measurements enable a rapid and expedient mapping of the contaminants. Mapping of adverse effects may help to identify sources, transport pathways and the environmental compartments exposed. The integration of chemical analysis and biological/toxicological tests may characterize adverse effects, mobility, bioavailability, biodegradation, and bioaccumulation of the pollutants. Natural risk-reducing processes (e.g. contaminant biodegradation) can also be identified, measured, or confirmed by the combination of contaminant analysis and toxicity tests.

Appropriate *handling of uncertainties* is a basic task in the course of contaminated site management. The main uncertainties to be managed are:

- Environmental heterogeneities;
- Poor quality data and information;
- Lack of information about site history, geochemical and hydrogeological status, land uses, etc.;
- Inaccurate conceptual model of the site;
- Lack of statistics-based sampling;
- Shortcomings of assessment plan;
- Inadequacy of analytical and testing methodologies: type, sensitivity and accuracy of the methods, and sample preparation;
- Non-optimal tiering of the assessment.

The creation of the optimal assessment tool battery for the management of contaminated sites can be supported by effective grouping of assessment. The applied chemical, biological, or ecosystem level models approach the real environment, but still remain at a certain distance from reflecting the actual contaminated site and are loaded with uncertainties (see Volume 1 of this book series, Figure 11.14, Gruiz *et al.*, 2014 and Volume 2, Figure 1.6, Gruiz *et al.*, 2015).

2 EFFICIENT MANAGEMENT AND REGULATION OF CONTAMINATED LAND

Contaminated land management has been developed for many years and is currently regulated by a range of laws, regulations, standards and protocols around the world. This section gives an overview of the current state of the relevant regulations and the development toward sustainability. Since contaminated land consists of soil, water and air, they should be managed in an integrated way, instead of the currently applied separate procedures.

2.1 Current legislations and management practices

In the US, contaminated site management is regulated by the 'Comprehensive Environmental Response, Compensation and Liability Act' (CERCLA, 2011), also known as 'Superfund' for abandoned hazardous waste sites and by its amendment, the 'Superfund Amendments and Reauthorization Act' (SARA, 1986). Some other acts are closely related such as the Small Business Liability Relief and Brownfields Revitalization Act (2002), the Oil Pollution Act (1990), Resource Conservation and Recovery Act (RCRA, 1976), RCRA Corrective Action Program (2003), and laws, regulations, and policies pertaining to underground storage tanks (UST, 2005).

Similar to the US, in most European countries, national clean-up projects have been performed and comprehensive regulations and management methodologies established. Unfortunately, no uniform regulatory framework has been set up for the whole of Europe, as it is discussed in Volume 1 of this book series (Gruiz *et al.*, 2014). Current European contaminated land policies are not part of a general land policy and management framework, which would ensure sustainable land use and improve soil quality throughout Europe.

Remediation of contaminated soil needs a similar approach and technologies as the maintenance of good quality of soil in general. Emerging *in situ* technologies utilizing the natural risk-reducing capacity of the soil ecosystems represent a major advancement. What distinguishes soil quality management from contaminated soil management?

–   Scale of divergence from the target quality.
–   Time scale: soil deterioration is generally caused by a long-term adverse impact and loss of balance, while contamination in most cases is a short-term action.
–   Prevention of soil deterioration should be a continuous effort in order to maintain good soil quality and services. Soil remediation is a one-off task, generally performed in connection with changes in land use or land ownership.
–   Urgency from the point of view of future land use. Natural maintenance of soil quality means that the soil ecosystem is able to spontaneously and rapidly compensate for processes unbalancing equilibrium. Soil quality management in general keeps soil functions and services compensating for soil deteriorating effects, resetting the original equilibrium state or adapting to new conditions. Remediation of a soil whose characteristics are far from the target means directing soil processes back to the original or a new equilibrium over the long term.
–   Healthy soil needs continuous monitoring and quality maintenance. Contaminated soil requires soil remediation. Deteriorated soil is typically amended i.e. its structure, composition and functions are improved by additives. Derelict land should be rehabilitated i.e. returned to its healthy equilibrium and functions. Extremely deteriorated soil must be remedied similarly to contaminated soil. After remediation it needs continuous quality management (see also Chapter 3, Volume 1 of this book series, Gruiz, 2014).

Land/soil management in general covers the following:

–   Definition of the land and soil type;
–   Definition of land use;
–   Setting the necessary (target) quality;
–   Monitoring of the quality of the land based on physicochemical, ecological, toxicological quality indicators;

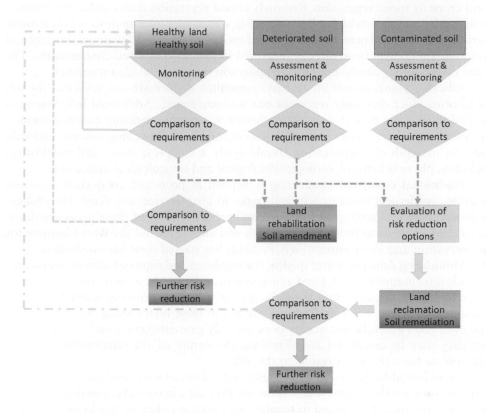

Figure 1.4 Management cycle of contaminated or deteriorated land.

– Evaluation of the monitoring results;
– Comparison of targeted and monitored indicator values;
– Identification and planning of measures to maintain soil quality over the long term (prevention of contamination, supporting ecosystem in self-recovery, or applying a remediation technology if efficiency is lacking) or reducing the risk by adjusting land use to soil quality (see Figure 1.4).

These steps are similar to those applied in watershed management or in the management of hazardous chemicals but their implementation faces soil- and land-specific difficulties and uncertainties.

Let us take the example of monitoring. Soil monitoring and sampling needs high-level conceptual planning and a problem-specific sampling strategy as well as an adequate handling of uncertainties on grounds of extremely high spatial variability. Some land uses are advantageous because the land cover, for example, crops, forests, pastures, can be monitored either on the field or remotely, and soil quality indicators can be derived from vegetation-related indicators. Thus changes can be identified and their time-course evaluated. Hyperspectral remote sensing, for example, is an effective tool for the detection and characterization of chemical contaminants in agricultural food crops or forest vegetation. Remotely sensed vegetation data can be extrapolated to invisible (non-detectable) soil characteristics. Remote direct analysis is only feasible during non-vegetated periods. One can extrapolate to the plants' material/nutrient and contaminant content from directly measured soil components and contaminants. The same methodology can be used for mapping soil type and soil characteristics.

Indicator plants grown on an easy-to-monitor surface are not sufficient for soil monitoring since they only represent one warning signal. Additional information is needed from other indicators: soil geochemistry, structure, organic material content, presence, and characteristics of soil microorganisms, soil-dwelling animals and wild life. An in-depth characterization should apply a complex, integrated monitoring, including physicochemical, ecological/biological and toxicological indicators.

Traditional ecological monitoring is applied in protected areas such as nature reserves, wilderness areas or national parks to characterize and detect the changes in biodiversity and ecological functions and services. Most of these areas are monitored worldwide using individual protocols and indicators and the World Institute for Conservation and Environment (WICE, 2012) has started their harmonization.

Monitoring drinking water quality is a regulated and rigorous activity and it may provide information for soil. Land as a source of water can be monitored and managed more or less in the same way since the goal is to ensure drinking water for human consumption. Deep subsurface waters are not in close relationship with the surface and even good groundwater quality does not fully guarantee good surface soil quality, but they may be closely related. Knowing the nature of the relationship, one can extrapolate from the water results to the soil.

Uncultivated land, abandoned mining and industrial sites, and abandoned waste disposal sites are the riskiest areas; however, they are almost never monitored. Urban land and soil are not subjected to regular investigation either, except hygienic control in special cases.

Contaminant transport pathways, and not only the soil itself, must be identified and monitored to obtain a complete picture. Monitoring should cover the contaminants derived from the atmosphere – transported by wind (dust) or precipitation (contaminated rain) – from the water and sediments (transported by floods) and from other solid-phase compartments as leachates of dissolved contaminants or eroded solid matter (both transported by water).

The US Superfund program (2015) and the European research and development projects dealing with contaminated land management (CARACAS, CLARINET, NICOLE, RACE, SNOWMAN, etc.) have collected a wealth of experience and expertise, laid down major guidelines and utilized them in contaminated land legislation, risk management, and remediation. The basic principles and developed methodologies helped establish uniform thinking in contaminated land management and brownfield clean-up. Risk-based management and remediation were the key points in establishing the common bases for decisions, i.e. to quantify risk. Some of the major projects were:

– Concerted Action on Risk Assessment for Contaminated Land (CARACAS, 1996–1998);
– Contaminated Land Rehabilitation Network for Environmental Technologies in Europe (CLARINET, 1998–2002);
– Network for Industry Contaminated in Europe (NICOLE, 1996–1999).

These development projects established the principles and methods of site-specific risk management. Many sampling and analytical as well as modeling issues were clarified. The risk assessment methodologies, the principles and methods for the creation of site-specific environmental quality criteria, and remedial target values were established. Decision making as the focal point of environmental management was thoroughly analyzed, several decision support tools (DST) were developed from knowledge bases to expert systems, and a number of alternative engineering tools for site characterization, risk assessment, monitoring and remediation were established. The necessity of an integrated approach was emphasized and scientists, researchers, consultants, government experts, owners of contaminated land and engineers were involved in the conceptual and methodological developments and in decision making (Spira *et al.*, 2014).

2.2 Trends in contaminated land management

In the beginning, contaminated site management and brownfield redevelopment were merely aimed to find a technologically and economically efficient solution to the problem. The most cost-effective remedial technology is usually understood in terms of environmental and economic efficiency, i.e. the one with the lowest specific cost among all feasible options. The economic assessment was later supplemented with the expected benefits and characterized by a cost–benefit analysis where the benefit from future land uses and social components were also considered. Initially the risk-based approach chiefly included local-level human risk and costs of the technology application, but later on ecological risks and broader socioeconomic considerations, life cycle assessment and the inclusion of global aspects (scale of contribution to global environmental problems = global footprints) resulted in a move toward sustainability.

Ecological efficiency and sustainability issues (green technologies, etc.) came to the fore in the 2000s, but a fully holistic approach is still awaited. This is understandable because today's scientific knowledge does not enable harmonizing and balancing the seemingly opposing interests of humans and ecosystems, artificial and natural, eco-friendly and economic, environmental protection and social security, and many other issues that are controversial at the current level of our knowledge and management practice. However, development will show that the controversial issues of today will just form two sides of the same coin because mankind's interest cannot be in conflict with the global ecosystem over the long term.

One step toward a more complete soil management may be that soil contamination – considered to be one type of soil deterioration – is managed together with other types of soil deterioration, taking into account soils' regeneration potential and available technical tools to support the natural regeneration process. The risk posed by soil contaminants can be further refined and interpreted based on the information provided by dynamic monitoring of soil deterioration and regeneration capability, plus the information on current and future land uses. Another feasible approach may be the corrective maintenance of soil quality by amending soil with the missing structural and nutritional components, thus maintaining terrestrial habitats and soil services over the long term. A combination of traditional (agricultural, geotechnological, mining, etc.) soil technologies, new knowledge about contaminated soil remediation, and general sustainability requirements can provide soil management solutions applicable overall (see Section 2.4).

2.3   Contaminated land – contaminated soil, water and air

The management of contaminated land addresses the (i) atmosphere, (ii) surface waters with their sediments, and (iii) soil and groundwater, but it is not integrated into one general regulation and management concept. In spite of the current knowledge on the inseparability of the environmental compartments from each other – supported also by legal tools such as the Integrated Pollution Prevention and Control (IPPC, 2008) and the legislation for licensing and permitting in many countries – there is still no uniform methodology for land and soil management. A solution is needed to ensure the maintenance of healthy soil quality and valuable land use as well as the continuous quality increase of deteriorated and contaminated land and soil.

The political systems and finances of countries influence their environmental legislation with the effect that air and water have high priority and enjoy intensive protection and chemicals are strictly regulated worldwide. Soil is far behind in this respect: it is treated as if it were not an integral part of the environment, and its essential functions are ignored. It is still being looked at as an issue for agriculture or building ground, and soil microbiota often fail to be considered when a remedial target or other quality criterion is being established. Only human exposure, primarily via water and food is taken into account.

Waters, and surface waters in particular, enjoy better perception in terms of management because of the environmental and health risks of contaminated waters are noticed at a fairly early stage. As a result, similar concepts and regulations have been developed in most parts of the world. Receptors of contaminated land, including the terrestrial ecosystem and interacting humans, have also been less studied, and the available methods are less uniform compared to water and aquatic ecosystems. However, water cannot be properly managed without considering adjacent land: contaminated land endangers surface waters. A more holistic approach is therefore necessary for sustainable management of the environment by giving equal consideration to the 'stepchild' soil.

Water and land quality regulations must be integrated and approaches, methodologies and management harmonized without delay. The approaches of the European Water Framework Directive (WFD, 2000) and the USA Clean Water Act (CWA, 1977) may be good models for long-term management of contaminated land/soil. The main advantage of the WFD is that it is a dynamic system which links the hierarchy of water quality with the hierarchy of usage-based classes (drinking, bathing, irrigation or industrially used waters, navigation as a water-use, etc.) with differentiated quality standards for surface water. The parameters to be regulated are specified by a combination of factors, which makes the regulation flexible and dynamic by allowing its scope to change. Successive steps of multi-stage and iterative planning and management ensure the achievement of long-term goals: continuous improvement of water quality.

An iterative management cycle consists of the following steps:

–   Identification of the water body;
–   Identification of (desired) water-uses;

- Setting a quality target;
- Assessment of water quality;
- Comparison of the current water quality to the desired water quality;
- Identification of measures needed to maintain or achieve the quality.

The purpose of WFD is to achieve good ecological and chemical status of surface water bodies. The ecological status is determined by a combination of biological quality elements (aquatic flora, benthic invertebrate fauna and fish fauna) and physicochemical quality elements (oxygenation and nutrient conditions, salinity, and specific pollutants). Good chemical status means that environmental quality standards for 33 priority substances and certain other pollutants (including pesticides, heavy metals, polycyclic aromatic hydrocarbons, and others) are fulfilled (Water quality standards, 2008).

A surface water body can be classified as having good status when the criteria for both good ecological and good chemical status are met. The overall objective of 'good' status represents surface water conditions that are appropriate for all types of water-uses: habitat of the aquatic ecosystem, drinking water, bathing water, fisheries, irrigation, and industrial water. WFD generally requires that land be managed and remedied if contaminated in order to achieve good status of surface waters and the whole watershed.

The WFD dynamic water management system is based on continuous water quality monitoring. Laboratory capacity and financial resources are essential to monitor all parameters at a specified frequency.

The Clean Water Act (CWA, 1977) requires that water quality criteria accurately reflect the latest scientific knowledge and surface waters be classified according to designated uses. In the US the following water uses (classes) are distinguished: potable water supplies, shellfish propagation/harvesting, habitat for fish and human fish consumption (recreational or propagation), agricultural water supplies, navigation, utility and industrial use. Water quality criteria have been established for each class. While some criteria are intended to protect aquatic life, others are designed to protect human health (US NRWQC, 2012). Legal rules point out that site-specific criteria should replace the generic default criteria in cases where site-specific information supports the change.

The CWA includes several topics of the overlapping areas of water and land management such as storm water discharges, use and disposal of biosolids (wastewater sludge), non-point source pollution, oil spills prevention and control, wetland management, agricultural activities such as animal feeding, and diffuse pollution from agriculture.

Great deficiency of current regulations is that soil and land are not regulated similarly to air and water. A uniform European 'soil framework directive' or 'superfund' regulations do not exist yet, although developed national regulations are available. Principles of and experiences gained in risk management of water could be applied to land and soil risk management, but this is still pending in Europe. Another way of making a step forward would be gathering knowledge related to different fields of environmental management (chemicals, natural ecosystems, human epidemiology and targeted environmental monitoring), and applying it in a uniform way.

2.4   Sustainable land and soil management

Land use is characterized by the arrangements, activities and inputs people undertake in a certain land cover type to produce, change or maintain it (FAO/UNEP, 1999). In environmental risk management, typical land uses for natural, residential, agricultural and industrial/commercial purposes are differentiated according to the sensitivity and tolerance of the relevant users.

Contaminated land management should be an integral part of sustainable land management, i.e. managing the land without impairing its ecological services or reducing its biological diversity. The targets of sustainable land management are:

–   Maintaining biodiversity (variety of species, populations, habitats and ecosystems);
–   Preserving ecological integrity (healthy environment with self-recovery potential to compensate for deteriorating impacts);
–   Maintaining the quality of soil, water and air.

2.4.1   Land management in general

Land management must be different according to the use of the land, typically as:

–   A water base (providing drinking water or water used in other ways);
–   Natural land (biodiversity conservation);
–   Agricultural production (arable land, grazing and forest);
–   Residential area (with typical residential uses, homes, apartments, kindergartens, shops, etc.);
–   Less sensitive industrial, mining or commercial uses.

Soil contamination is only one type of deterioration in addition to a number of problems causing an imbalance and decreasing the quantity and quality of soil and terrestrial ecosystem services.

Some of the *degradation processes* may be the consequences of global climate changes, however human activities are responsible for most of the adverse impacts. Climate changes affecting temperature and precipitation influence soil formation (weathering, humification), soil erosion, and diversity of the soil ecosystem including vegetation, element cycles and dynamic balances. Typical anthropogenic impacts on soils are compaction, sealing, increased runoff formation and contamination.

Soil degradation types are physical degradation (erosion, compaction, sealing), chemical degradation (humus degradation, acidification, salinization, sodification, nutrient depletion, microelement depletion, contamination) and biological, ecological degradation (reduced biodiversity and ecological function). The deteriorating impacts lead to adverse textural, structural alterations (instability, improper air and water household, weak secondary structure, desertification, etc.) and functional changes (degraded ecosystem services such as habitat, provisioning, cultural and aesthetic functions, etc.). Physical, chemical and biological/ecological soil deterioration can lead to large-scale disasters, causing mudflows, floods, landslides, or deep subsurface changes.

When discussing risk assessment and risk management concepts and tools, it is also worth differentiating between natural and anthropogenic soils. In the case of agricultural, urban and technosols, human activity is a significant soil-forming factor (FAO, 2006a and 2006b).

Intensive agriculture has a greatly deteriorating effect on soil by extracting nutrients (macro-, meso- and microelements) and only replacing macronutrients; by disturbing the vegetation (crop removal, row cropping, tilling or plowing, planting of monocultures, overgrazing by animals) and leaving the fields naked and exposed to erosion. Mechanized agriculture is responsible for soil compaction caused by the repeated cultivation of the soil surface by heavy machinery. Compaction causes densification in the soil by displacing air from the pores between soil solid particles. Heavy traffic on the soil surface destroys the soil's secondary structure, limits air diffusion, and rainfall infiltration, thus reduces microbiological mineralization, plant root growth and crop yield. Increasing runoff and erosion adversely impact the surface water system by causing floods and abnormal inland water levels. Contaminated runoff containing nutrients (eutrophication) and biocides (emerging pollutants in water with toxic, endocrine and immune disrupting effects) impairs the quality of surface waters (SoCo, 2009).

Urban land use, with an extremely high rate of heterogeneity, sealing and land-take, has a much stronger impact than the urban area itself. Large amounts of contaminated runoff have to be treated and drained. Roads, transport, parks, gardens, building foundations, canalization, sport fields and disposal sites, etc. have a great impact on soil. Adaptive and resistant species exhibit high abundance and distribution within the soil ecosystem. The ecology of urban environments is an emerging scientific topic which looks into urban ecosystems and their interactions including suburban and rural environments (Guntenspergen, 2012 and Meuser, 2010).

Some 1,000 km^2 in the European Union (EU) were subjected to land-take for housing, industry, roads or recreational purposes every year between 1990 and 2006. This area exceeds the size of Berlin, Germany or a quarter of Greater Boston, MA area. About half of this surface is actually sealed by buildings, roads and parking lots (Prokop *et al.*, 2011).

A new science has emerged that studies special aspects of urban soil, e.g. extensive deterioration and the presence of historical and technical layers and patches. The latter contain foreign materials from external sources and contaminated debris and waste material (fly ash, industrial waste, construction rubble, etc.). Large amounts of foreign material and waste may be incorporated into soil as a consequence of earthquakes or bombing (wartime). A special feature of urban soil is that its level is elevated above the natural (original) surface levels due to fillings and waste materials accumulated over centuries. Risk to human (in particular child) health typically characterizes the risk posed by urban land. Socioeconomic, cultural and esthetical considerations play an extremely important role in urban land management.

The green proportion of the surface in towns and cities has decreased significantly in the last 50 years, but some improvement can be seen in the last decade in countries with a high living standard. The proportion of green surface in Europe has been recently assessed and published (Soil Sealing, 2012). Increasing land-take and associated emissions cause higher infrastructure costs for the municipalities.

Soil sealing has several adverse consequences such as:

- Loss of fertile soil;
- Loss of water retention areas;

- Increase in surface water runoff and flood risk;
- Landscape fragmentation;
- Loss of biodiversity;
- Changes of microclimate, e.g. higher surface temperatures compared to natural green surfaces.

Reducing land-take can be recommended as the best practice in European brownfield redevelopment. Brownfields are industrial or commercial properties with high risk of chemical pollution which need redevelopment or reuse; this may be complicated by the presence or potential presence of hazardous substances or contaminants (CERCLA, 2011). Remediation of such sites and reducing the risk to an acceptable level by a new commercial or industrial use might be beneficial. Demanding land uses are questionable and require detailed socioeconomic assessments.

Knowledge and principles for reducing soil sealing have been developed and public acceptance is also improving. Innovative technologies and materials are currently available which ensure infiltration of precipitation or efficient treatment of the runoff. Managing the risk of soil sealing uses the same three tiers as environmental risk management in general: 1. prevention, 2. restriction and 3. remediation (compensation of soil loss in this context) (Prokop, 2011).

Prevention can be achieved by:

- Policy, monitoring, realistic land-take targets;
- Streamlining existing funding policies;
- Steering new developments to land already developed;
- Providing financial incentives for inner urban development;
- Improving the quality of life in large urban centers;
- Shaping inner city centers more attractive;
- Protecting agricultural soils and valuable landscapes.

Limitation can be achieved by:

- Respecting soil quality in planning processes;
- Applying technical mitigation measures to conserve at least a few soil functions (i.e. permeable surfaces on parking areas).

Compensation can be achieved by:

- Establishing qualified compensation measures;
- Facilitating new alternative land uses (Prokop, 2011).

Figure 1.5 shows the percentages of sealed areas in the European countries. The current percentages do not reflect higher consciousness, rather population density, urbanization, traditional land management, and standard of living. The type of trend, i.e. decreasing, stagnating or growing (signified by the colors) says more about the attitude of the countries.

Technosol is a term created for soils of technological origin which lack natural structure, instead, contain a certain amount of artefacts, a constructed geomembrane or technical hard rock (consolidated material resulting from an industrial process).

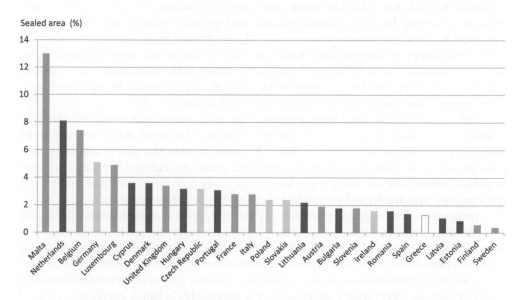

Sealed area (%)

Figure 1.5 Percentage of sealed area in Europe. The trends of changes, shown by the colors, are independent from percentages: green: decreasing, amber: stagnating, red: growing trend, colorless: trend not known (EEA, 2010).

Technosols are poor habitats; air, water and nutrient supply is not suitable for plant growth. Technosols typically occur in industrial and mining areas, their dimension is generally small. Technosols have no standard profile; however, special profiles may develop in old waste dumps, landfills and backfilled areas. Waste dumps and landfills are often covered with good quality soil or other materials: covered and revegetated technosols still remain in the category of technosol. Some experts rank urban soils as technosols (FAO, 2006a and 2006b).

2.4.2 Sustainability means well-balanced environmental, social and economic components

Sustainable land management has developed from risk-based land management. It is the first step from the former unplanned and benefit-based land use toward sustainability.

Risk-based land management specifies and targets the acceptable risk by enhancing land quality or restricting its use, e.g. by excluding sensitive users. Several types of risks, i.e. the probability of the occurrence of a damage due to meteorological, climatic, static, chemical or biological hazards is investigated. Risk is usually considered a negative phenomenon; its positive counterpart is safety. It is recognized that nothing can be 100% safe, so some degree of risk must be accepted. The concept of risk assessment – quantifying and adding up the risks of identified causes – has several shortcomings, for example only identified risk sources and their identified hazards are taken into account. This approach is sometimes satisfactory, e.g. for a single-point source with moderate hazard. But it fails in many cases, especially when, climatic and static hazards, and chemical hazards of identified and unidentified contaminants are combined. This may be the case of human corrosive and reprotoxic effects, aquatic toxicity and bioaccumulation, or when no information is available about long-term impacts. Under these circumstances, risk or safety is associated with a high uncertainty. Nevertheless, risk-based thinking and the accompanying conceptual risk model of a problem or a contaminated site can greatly contribute to a clear picture.

Risk–benefit-based land management relies on a comparison of acceptable environmental risk with the social and economic benefits. One must define a wider scope and apply a uniform methodology that takes into account the short- and long-term risks as well as the human, ecological, economic, social and cultural benefits. This type of management does not work if the site owners only consider short-term and partial risks or only benefits regardless of the risk to others or the environment.

Sustainable land management should place the problem of contaminated soil into a much wider context. All long-term deterioration processes should be integrated, considering the interaction of soil with other environmental factors, including the different land covers and land uses. The original definition from the 1987 Brundtland Report – that was the first to combine social, economic, cultural, and environmental issues and global solutions – says that a development is sustainable when it meets the needs of the present without compromising the ability of future generations to meet their own needs (WCE, 1987). Global activities, such as the Rio Earth Summit in 1992, the Millennium Development Goals (UN, 2000), the World Summit on Sustainable Development held in Johannesburg in 2002, and the Kyoto Protocol of 2005 all enforce a globally harmonized approach. Sustainable land management and sustainable management of contaminated sites should comply with the principles and requirements of these agreements, namely:

– A holistic view;
– The integration and optimization of environmental, social and economic interests (Figure 1.6);
– Avoiding unnecessary work through greater knowledge and a realistic concept;
– High-level quality management, including:
   o Science-based information;
   o Consistent, transparent and ethical decision making, considering all relevant (ecological, economic, social and interdisciplinary) aspects.

Sustainable land management includes sustainable land uses, sustainable assessment and remediation/rehabilitation of already deteriorated or contaminated land. The main target of sustainable land management should be the prevention of soil from contamination and other deteriorating effects as well as maintaining the quality and services of the land while using it. The focus should be on managing potential problems instead of waiting for advanced deterioration. Remediation should be efficient both ecologically and economically, avoiding intensive, destructive, or environmentally unfriendly technologies. Strong, radical interventions are allowed for point source management; otherwise, priority should be given to *in situ*, long-term, ecologically efficient, 'green' technologies. Efficiency is to be measured and proved by verifying the

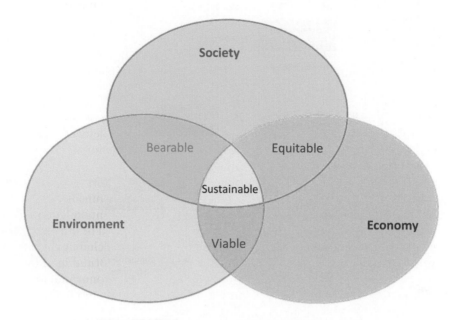

Figure 1.6 Sustainability scheme.

management activities (assessing the efficiency of remediation) and adapting the management plan to the results of continuous monitoring and verification. This topic is discussed in detail in Volume 1 of this book series (Gruiz *et al.*, 2014).

Sustainability can be evaluated in advance to support decision making by comparing the various management options. Sustainability of the subsequently implemented management should be validated by an integrated technological, environmental, social and economic monitoring.

Several methods and indicators are available for sustainability assessment in general and for sustainable management and remediation of contaminated land in particular. SuRF-UK (2015) recommends 15 environmental, social and ecological indicators and provides a downloadable Excel spreadsheet (SuRF-UK, SMP, 2015) to support sustainable management. The open-source decision support system SMARTe (Sustainable Management Approaches and Revitalization Tools – electronic) for developing and evaluating future reuse scenarios for potentially contaminated land contains resources and analysis tools for all aspects of the revitalization process including planning, environmental, economic, and social concerns (SMARTe, 2015). Conventional life cycle assessment has been complemented with social (S-LCA) and cost–benefit assessment with social and environmental aspects (SCBA). Social return on investment (SROI) and the adjusted form of SCBA result in a better evidence framework showing how to achieve good lives and human well-being (Vardakoulias, 2013).

Sustainable land use concerns:

–   The use of land without impairing its ecological balance and services;
–   Maintaining its environmental quality and avoiding deterioration;

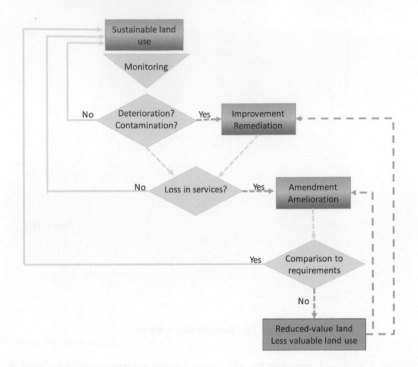

Figure 1.7 Sustainable land use means continuous monitoring and corrective actions.

–   Finding the land use that fits best to its present status and quality;
–   Preventing loss in quality, e.g. not using green fields instead of rehabilitated brownfields for industrial or commercial purposes (see Figure 1.7).

Even if *sustainable land use* (i.e. fulfilling current regulations) is being practiced, the user may be unable to ensure long-term sustainability, e.g. to maintain all element cycles and soil services at the optimal level due to a lack of knowledge and missing tools. In these cases, a slow latent quality decrease may endanger long-term, high-quality use of soils, for example, due to microelement depletion in agricultural soil. Unfortunately, the adverse impact of long-term global changes is not really understood, so the targets and tools of sustainability are accompanied by high uncertainty.

The sustainability of soil remediation projects is a special concept within sustainable land management. Remediation is sustainable when the benefit gained is greater than its impact as indicated by environmental, economic and social indicators. Unfortunately, indicators often provide different results. Typically, human health and safety aspects are only considered, while the whole ecosystem, other social indicators, and long-term aspects are ignored. Spatial sustainability assessment is often limited to the local and neighboring area, while the wider watershed or the global environment is ignored.

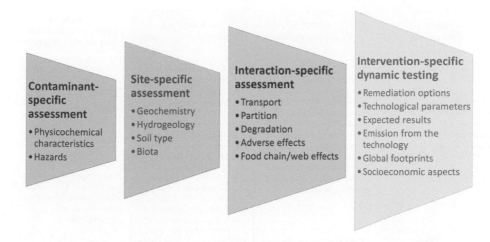

Figure 1.8 Collecting technology-related information in detailed site investigation.

3   BEST PRACTICES IN CONTAMINATED SITE INVESTIGATION

Assessment of contaminated land is a multidimensional task that includes a tiered site investigation and the evaluation of the collected data and information. The risk-based approach calculates the risk value from site investigation data, and determines the subsequent risk management steps.

In site assessment, tiering means that evaluation and interpretation of measured data is the first step, and making decisions about the next task is the second step. Measured data are evaluated and interpreted in comparison with quality standards or other screening criteria. Theoretically, these screening criteria can be chemical model-based concentrations (recognized as harmless) or biological or ecological model-based, measured or extrapolated no-effect values. The ratio of the measured concentration or adverse effect value to the screening value (a certain concentration or an adverse effect value, representing no risk) gives the risk characterizing ratio (RCR). RCR is a multiplier showing how many times the measured value is greater than the screening value.

The assessment phases, also called stages, tiers, or steps, are executed one by one. A decision-making point is inserted between phases of the iterative management cycle to achieve a time- and cost-efficient site investigation (Figures 1.8, 1.9 and 1.10).

The key steps of every site assessment are site and contaminant characterization including their interaction. The risk of a contaminated site can be determined qualitatively or quantitatively from these results. A part of the data (e.g. fate properties of the contaminants) can be used directly for intervention-specific assessment, which helps choose the remediation technology to be used.

This section summarizes theoretical and practical information about site investigation. Section 4 describes sampling and Section 5 the measurements and the evaluation of acquired data. All sections put great emphasis on innovative approaches such as direct toxicity testing, the soil testing triad (STT, i.e. the combined physicochemical,

PHASE 1 Preliminary phase	PHASE 2 Exploratory phase	PHASE 3 Detailed assessment	PHASE 4 Aggregating information long-term management

• Generic site info
• Historical data
• Maps and photos
• Identified risks
• Descriptive RA
• **Conceptual model 1**
• Desktop study report
• Decision on next steps
  • No intervention
  • Phase 2
	• Sampling strategy
	• Tiered sampling
	• Analyses and tests
	• Evaluation of the results
	• Comparison to EQC
	• **Conceptual model 2**
	• Site assessment report
	• Decision on next steps
	  • No intervention
	  • Detailed site assessment
	  • Targeted site assessment
	• Risk communication
		• Assessment of variability
		  • Contaminant occurrence
		  • Spatial distributions
		  • Temporal changes
		  • Long-term processes
		  • Food chain effects, etc.
		• Delineation, boundaries
		• **Conceptual model 3**
		• Risk reduction options
		• Final report including
		  final management action
		• Establishing site-specific
		  target quality criteria
		• Risk reduction, remediation
		  • Technology selection
		  • Outline planning
			• Long-term site monitoring
			  and evaluation of the results
			• Comparative evaluation of
			  risk reduction options
			• Stakeholder consultation
			• Implementation plan
			• Implementing remediation
			• Monitoring
			  • Technology monitoring
			  • Environmental monitoring
			• Evaluation of the final results
			• Verification of the technology
			• Sustainability assessment
			• Final site report
			• Communication

Figure 1.9 Contaminated site investigation and its long-term management – lists of tasks.

biological/ecological and (eco)toxicological assessment methods), *in situ* site assessment methods, and the related dynamic decision making (Triad approach).

To clarify the confusing use of the term 'triad', some explanation is given below. The umbrella term 'triad' for the integrated use of the three types of information from physicochemical, biological/ecological and (eco)toxicological methods, is often used in the literature. Alternative equivalent terms include 'integrated evaluation', 'integrated assessment', 'integrated monitoring' or 'soil testing triad' and 'sediment triad'. It has to be noted that the term 'triad' is used by US EPA as a management approach for *in situ* assessment of and decision making about contaminated land. For this reason, the authors prefer to use the term 'Triad approach' in relation to the concept of '*in situ* site assessment and dynamic decision making' and the integrated application of physicochemical, biological/ecological and toxicological assessment tools and the integrated evaluation of their results called 'integrated site assessment' or the soil testing triad (STT).

3.1   Aims and focus of contaminated site investigation

Contaminated site investigation may have several aspects, depending on the aim of the whole management process and the interest of the stakeholders.

3.1.1   Aims of site investigation

– Identification of contaminants that are (potentially) present on the site;
– Determining their physico-chemical, environmental fate and transport characteristics, hazards and effectuated adverse impacts;
– Characterization of the environment, type of waters, soils, climatic, geochemical, hydrogeological, topographical, etc. conditions, including environmental sensitivity and buffering capacity;

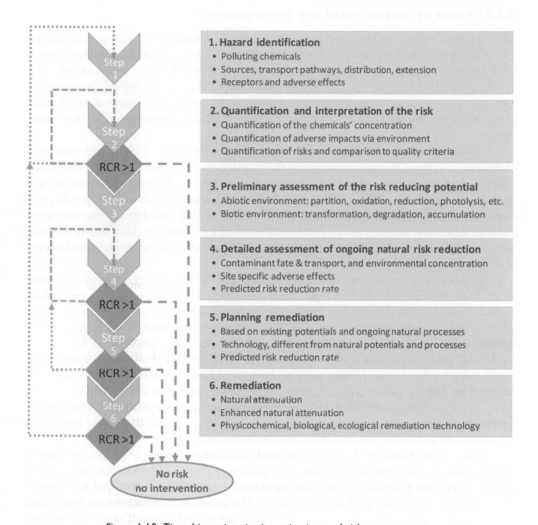

Figure 1.10 Tiered iterative site investigation and risk management.

– Identification and delineation of primary and secondary sources, pathways, potentially or factually contaminated environmental compartments and phases;
– Identification of present and future land uses;
– Identification of potential, and already exposed, receptors;
– Identification of land-use-specific exposure pathways.

Other possible goals relevant for a broader context of land management can be the registration of contaminant land into land tenure, to assess the land for property sale and/or ownership change. Special cases are emergency or natural disaster, when immediate action must be taken after a rapid and partial site investigation.

3.1.2 Focus of contaminated site investigation

The assessment of contaminated land may focus on contaminant(s), site characteristics and on interactions between contaminants and the environmental compartments and phases. In addition to obtaining information about the type and scale of environmental risk, data on risk reduction potential of the site is essential in order to organically connect risk assessment and reduction activities and acquire data for planning the intervention. Figure 1.8 shows the pragmatic classification of the assessment types.

Contaminant-specific assessment: this type of assessment may be relevant to a chemical substance in general to specify the physicochemical and biological characteristics and impacts of the substance. The majority of data used for contaminant-specific assessment can be found in databases, but some information can be site-specific, e.g. concentration in the soil phases that can be measured or calculated using transport models when the site-specific characteristics are already known. One has to identify the contaminants and exclude the presence of other unknown or unidentified agents with similar scale adverse impacts.

Site-specific geochemical, hydrogeological, and biological data provide information about the site's abiotic and biotic characteristics, which determine if a hazardous pollutant will cause any adverse effect and of which scale. The compartments of the abiotic environment responsible for dilution, partition or diffusion of contaminants greatly influence the environmental risk posed by hazardous chemicals. A high water table and sandy soil cause a higher risk for the groundwater than a deep water table and loamy or humic soil for the same hazardous chemical substance. A river of high water flux dilutes a discharge more effectively than a low-flux creek. The biota may have an even stronger influence, the presence of an active biodegrading soil microflora or rhizosphere consortia may eliminate the hazardous component quickly compared to an arid or otherwise inactive degrading microbiota.

Interactions between the environmental stakeholders involve the environmental compartments, their physical phases, the pollutants, and the biota – including food chains. Dynamic thinking and environment-linked modeling are required to examine this multi-parameter network of interactions. The interactions between soil matrix and contaminants, those of contaminants with each other, and between the contaminants and the soil biota have a major influence on the character, type, and extent of risks, and on potential risk reduction measures. The site-specific concentration of the pollutants depends not only on the contaminant itself, but also on the quality of the environment, the environment's sorption, exchange and buffer capacity, and on the mutual effect of the species within the ecosystem. Conventional, equilibrium-based models and testing methods cannot characterize dynamic interactions. Therefore, interactive tests (e.g. direct contact) must be performed, microcosms or mesocosms have to be monitored and evaluated in an integrated way, which means that chemical, biological, and toxicological results need to be combined.

Conventional site assessment methodologies do not include *intervention-specific* assessment. However, intervention-specific assessment is essential in the preparation for risk reduction. Potential risk-reducing transport and fate processes are: dilution, partition, degradation (hydrolysis, photolysis, other chemical degradations, primarily oxidation and reduction, and biodegradation), sorption–desorption, dissolution–precipitation, oxidation–reduction, etc.

In addition to the identified fate processes, technological experiments can be carried out on real soil in the laboratory as well as in pilot experiments at different scales, simulating real circumstances. Other innovative methods can be used to test the effects of *in situ* physical, chemical or biological interventions on a small volume of soil. The response of the soil is measured very closely to, and very soon after, the local intervention (push–pull). The results acquired from this kind of dynamic test can be used directly for the selection and planning of the final remediation technology or emission control, and also for assessing the risks (increased emissions or other adverse effects) posed by the technology itself. After the data are acquired from dynamic tests, the costs of remediation can also be calculated more precisely.

3.2   Phases of site investigation and characterization

The standards and protocols significantly differ from country to country, but the applied site characterization and risk assessment methods agree in using tiering, though the number and content of tiers may be different. Generally speaking, three phases are distinguished and applied: preliminary, exploratory and detailed site assessment.

Figure 1.9 demonstrates the sequence of the three site investigation phases (1 through 3) and the evaluation of the information for use in the long-term management.

3.2.1   *Preliminary site investigation phase*

The aim of preliminary assessment is to decide whether a site is, or may be, contaminated. It requires the following:

– Managerial and legal documentation;
– Information about past, present, planned future land use;
– Statistical and historical data collection – documentation of industrial or mining facilities, production volumes, chemicals used, and technologies applied. The data should include the type and amount of discharges, as well as protective and waste treatment technologies and storage tanks if such were applied. Maps and aerial photos are available in most cases from geographical or military sources. Interviews can be useful for collecting information from former employees and neighboring residents;
– Information whether the pollution stems from one or several point sources or if it is a diffuse one;
– Geochemical and hydrogeological data collection;
– Properties (physical, chemical and adverse impacts) of the probable contaminants;
– A site visit collecting local information, documents and other evidence as well as performing interviews;
– Creation of the first conceptual risk model of the site;
– No sampling is required in this phase.

Sampling cannot be designed or planned before drawing up the conceptual model: sampling does not make any sense without qualitative and possibly quantitative characterization of the source and transport pathways because one cannot see how the sampling point is related in space and time to the source and the transport pathways. A concentration in the source represents a much lower risk than the same concentration at 200 yard distance. Historical data on production and emission volumes from a former chemical plant or gasworks can provide better quality information than analysis results of random samples taken from the site. The physicochemical character, fate and behavior of the contaminant(s) enable transport modeling and calculation of the actual contaminant concentration. Soil sampling is loaded with huge uncertainties due to natural heterogeneities of the soil, illegal uses and decades of waste disposal at inherited contaminated sites, even if the site's conceptual risk model is correct and the sources (point and diffuse) as well as the transport pathways have been identified.

Preliminary assessment can be based on qualitative or quantitative data. When several sites are assessed to prepare the inventory and the priority lists, a qualitative or semiquantitative assessment may be suitable to establish a scoring system for site risk characterization.

The preliminary phase is the most important step in a site investigation: existing data/information generally represents most part of all available information. Well-structured, evaluated and correctly interpreted information from the preliminary phase can validate and support the conceptual model, the sampling plan, and the right selection of the modeling and assessment tools.

Preliminary investigation – also called desktop study – relies on data gathered in the preliminary site investigation phase. The site is visited and the documented hot spots or other points of interest (technological and storage facilities, waste disposal sites, etc.) are identified. Interviews may provide important (perhaps subjective) information. A preliminary conceptual model should be established based on first tier information. The potential output of the preliminary investigation is:

–   Certain no risk: no risk reduction activity is needed; however, some tests and appropriate risk communication might be necessary.
–   Probable no risk: proof of the lack of unacceptable pollution is needed; a reduced exploratory assessment should be carried out.
–   High risk due to unacceptable pollution: the scale and dimension of pollution have to be measured in the exploratory and detailed assessment steps, and the risk of the contaminants should be reduced based on the results.
–   Observable deterioration/damage: the scale and dimension of damage as well as the risk outside the delineated area must be measured in the exploratory phase.
–   The preliminary assessment can deliver new information on unexpected contaminants or unknown environmental conditions, etc.
–   If the suspicion of contamination is confirmed, the next phase, exploratory assessment, is applied.

Figure 1.10 (on page 25) shows the complete management scheme.

3.2.2   *Exploratory phase of site investigation*

The aim of the exploratory phase is further data collection combined with limited sampling to confirm or reject the preliminary assumption(s). The conceptual model is validated and in part quantified in this phase. The sampling plan is based on the conceptual model. If one can find the site hot spots and sample them, one will obtain high contaminant concentration or toxicity (depending on the selected method) and prove the presence of contaminants at the site. This is the aim of the exploratory phase.

Several methods are available for data acquisition on contaminants, ranging from non-intrusive physical methods and chemical analyses of soil and groundwater to biological or toxicological techniques. *In situ*, on-site or laboratory methods can be used to confirm whether or not the site is contaminated. Sampling design should be based on the history and the conceptual risk model of the site and should consider the aim and the accessible assessment tools.

The Dutch standard for exploratory investigation identifies the sampling strategies for the following types of sites (Swartjes, 2011):

- Unsuspected sites;
- Suspected site: identified local soil contamination;
- Suspected site: homogeneously/heterogeneously distributed diffuse soil contamination;
- Suspected site with unknown soil contamination;
- A baseline investigation: for future potential contaminating activities.

Soil sampling protocols should be prepared for each case: sampling strategy, number of samples, sampling depths, analyses and tests, as well as evaluation methods. When the contaminant is an unknown mixture, it is worth screening the site with a cheap ecotoxicological test to identify and exclude negative subsites. If soil is excavated, sampling, qualifying, and possible reuse should also be specified.

This phase yields the confirmation about the suspected presence or absence of contaminants or environmental deterioration and human health impacts.

If a specific contaminant is assumed, chemical identification and concentration measurement are appropriate.

If the contaminant is unknown, assessment of toxicity or other adverse effects is the appropriate method.

If a chemical analytical method proves the absence of the assumed chemical, this must be confirmed by negative toxicity test results. It may happen that the expected contaminants cannot be detected although the toxicity test is positive and a different contaminant or metabolite is present. If both chemical and toxicological results are negative, the site can be declared as 'uncontaminated' or low risk.

Chemical screening, e.g. by an *in situ* rapid method may provide the following outcome:

- Yes, contaminated → sampling, chemical analysis and toxicity assessment to characterize risk of the presence of hazardous components.
- Not contaminated → toxicology to prove no risk.

Biological screening (e.g. ecological assessment, remote sensing, or rapid toxicity testing such as the *in situ* mobile apparatus for *Vibrio fischeri* bioluminescence test).

- Yes, deteriorated/toxic → more detailed integrated (chemical and biological) assessment.
- No → further proof of negative results by other tests, different test organisms.

The application of this simple decision scheme may ensure that inherited, abandoned, and uncontrolled sites with many unidentified contaminants cannot be missed

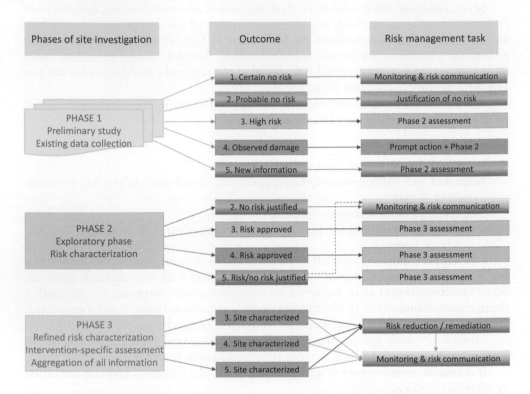

Figure 1.11 The complete decision tree for tiered site investigation. Form outcomes 1–5 of the first phase, outcomes 2–5 continue in phase 2 and 3–5 in phase 3.

and wrongly declared as non-risky, thus endangering the users of the land and enabling a further spread of pollution. This is extremely important when industrial sites are re-developed as residential areas. Besides high environmental and human health damage, it may result in major financial losses for the new owners and investors (see decision tree in Figure 1.11).

The origin of a greatly contaminated or toxic sample, whether it really derives from the hot spot or from other contaminated points of the site, is determined during the detailed assessment step. These types of uncertainties can be minimized by collecting detailed historical data during the preliminary assessment phase to check potential emission sources.

Exploratory phase of site investigation and characterization is based on a simplified assessment with limited sampling to support the preliminary assumption and the fine-tuning of the conceptual model, and finding the boundaries and delineation of the area. The following outcomes may be possible:

–   Probable no risk is approved. No further activity out of risk (no risk) communication is needed.
–   Probable high risk is approved and refined. A detailed investigation should be planned and executed, which should also involve risk reduction-related characterization.

- New suspicion or new risk components are identified: they must be documented and assessed. Risk reduction planning and risk communication should
  be expanded.

3.2.3 Detailed site investigation

A *detailed investigation phase* follows the exploratory phase in positive cases, i.e.
when contamination or deterioration has been identified. This assessment phase itself
generally is a tiered and iterative process with decision points after each step, leading to
the remediation or application of other management actions if unacceptably high risk
has been found. A risk characterization ratio or similar quantitative risk value can be
created after each step. The contaminants themselves, site characteristics, interactions
between contaminants and the abiotic and biotic compartment of soil, as well as the
remediation-specific assessments, are all important.

Detailed site investigation is the final, aggregating phase of the tiered site characterization and risk assessment. In this phase, site investigators must collect all information
from the current and previous phases such as data on contaminants, their site-specific
environmental fate and behavior, concentrations, mobility, bioavailability and effects.
Detailed information about the characteristics of the environment is needed from existing databases or measured data to estimate the trend of changes over space and time.
The receptor-specific risks according to the present and future land uses can also be
estimated.

Detailed characterization is an iterative procedure in itself and the conceptual
model of the contaminated site is refined stepwise. The targeted risk and the corresponding environmental concentrations should be determined (calculated) based on the
conceptual model. Risk communication and decision making about the selection of the
best possible risk management option and the best available risk reduction technology
are the final results of the detailed assessment.

The tiers of site characterization and the exploratory and detailed phases should
be optimized. The optimal number of assessment tiers guarantees the highest time and
cost efficiencies. Fewer tiers cause higher cost (acquiring superfluous data); more steps
are more time-consuming. In some cases, when the risk is high or a beneficial future
land use is expected, the urgency of the assessment justifies one or two steps only. If
there is no urgency, the number of steps can be increased, but too many steps may
offset the savings achieved by eliminating unnecessary analysis.

Pollutant distribution can be mapped using chemical concentration and/or toxicity
data and compared with the conceptual model.

The sampling plan must be prepared prior to the detailed assessment. Sampling,
similarly to other risk management tasks, should be 'risk-based' and targeted sampling, namely based on the site's conceptual risk model. The sampling plan should be
supported by a statistical approach (see later in Section 4).

3.3 *In situ* site assessment combined with dynamic decision
making

The *Triad* approach is a scientific and technical initiative, not a regulatory approach,
supported by US EPA to foster the modernization of technical practices for characterizing and remediating chemically contaminated sites. It resulted from technological

advances in field analytical methods combined with experiences gained from historical contaminated site cleanup work. The core concept is to restore the land and water as efficiently and effectively as possible based on best management practices. The goal of the Triad approach is the identification and management of those uncertainties that may cause excessive or intolerable errors in decision making.

The Triad approach minimizes the likelihood of errors by cost-effectively supporting the development of an accurate conceptual risk model of a contaminated site using a tiered site investigation (Triad approach, 2015).

3.3.1 Technical components of in situ site assessment and dynamic decision making

Triad is not an acronym. The word is intended to convey the notion that there are three elements which are incorporated into a decision support matrix:

−   Systematic project planning includes identification of key decisions to be made, the development of a conceptual risk model of the site to support decision making, and an evaluation of decision uncertainty in the context of the conceptual risk model (US EPA, 2003).
−   Dynamic work strategies are work strategies for contaminated site characterization, remediation, and monitoring that incorporate the flexibility to change or adapt to information generated by real-time measurement technologies (Triad overview website, 2015).
−   Real-time measurement systems provide data quickly enough to affect the progress of field work. Real-time measurement systems represent the third leg of the Triad approach. They are essential for implementing dynamic work strategies because they feed timely data to the decision-making process (Real-Time Measurement Systems, 2016). Real-time measurements include field screening, geophysical techniques, direct sensing technologies, on-site analytical methods and/or rapid turnaround of conventional analytical methods that can be used in conjunction with each other to provide collaborative data sets (Beard et al., 2010) as well as ecological or ecotoxicological methods, which give the most adequate end points for decision making. Real-time measurement technologies return results quickly enough to influence data collection and field activities, given that the results can prompt changes in the previously developed plan, including the next measurement point, the frequency of in-situ measurements or the sampling location for more detailed assessment.

An ideal Triad project would rely heavily on all three elements (Crumbling, 2004).

3.3.2   Benefits of in situ *site assessment and dynamic decision making*

The advantages of the Triad approach override the disadvantages and the efficiency and cost balance are usually positive. Advantages and disadvantages can be summarized based on the online training course of Clue-in as follows (Clue-in, 2005):

Advantages:

– Lower operation costs;
– Better quality of the investigation;
– Faster investigation, remediation and redevelopment of a site;
– Smaller uncertainty, greater confidence in data;
– Better established decisions;
– More efficient risk reduction, clean-up, remediation.

Disadvantages:

– Higher initial costs;
– New tools need to be developed;
– New approach, new thinking, great need of training;
– Negative bias toward *in situ* measured data.

The Triad approach emphasizes systematic planning combined with dynamic work strategies. Unlike a traditional work plan where stakeholders know precisely what will be done before work is initiated, in a dynamic work strategy at least some of the key decisions are deferred until the actual field work takes place. Consequently, *stakeholders concur with a process* rather than a product. For a Triad approach to be successful, it requires stakeholder involvement at some level in the decisions that are being made in the field as work progresses. This guarantees their participation in the characterization and/or remediation process that is beyond what has traditionally been the case. This, in turn, generally results in enhanced stakeholder concurrence with the final decisions derived from data produced by dynamic work strategies.

The ultimate goal of the Triad approach is *improved decision quality*. By focusing on reducing decision uncertainty, rather than simply analytical uncertainty, and making use of collaborative data sets combined with a weight of evidence approach to data evaluation, a Triad approach will typically result in much better decisions being made with the same resource investment. With its emphasis on real-time measurement systems, Triad-based data collection programs can pursue data collection activities for an area until decision quality objectives have been attained. In practical terms, this means a reduced likelihood that contamination is left undiscovered or that resources are spent unnecessarily on portions of a site where contamination concerns in fact do not exist. Improved site decision making also can reduce overall remediation costs through waste stream minimization and adaptive remedial strategies, an important outcome from a site manager's perspective (Triad benefits, 2015).

3.4   Standardized investigation of contaminated sites

Contaminated site investigation has been standardized in the last 20 years worldwide as a result of the US EPA Superfund Program (2015) and European Research Projects such as Common Forum, CARACAS, NICOLE, CLARINET and EUGRIS which laid down the basic theories for contaminated site assessment and management (see Volume 2, Chapter 2 of this book series, Gruiz *et al.*, 2015b).

US EPA pioneered the standardization of contaminated site assessment and established the following ASTM standards for contaminated site management:

–   ASTM E1528-00 (2000) Transaction Screen Process, consisting of a questionnaire and a corresponding guide. The questionnaire comprises three areas of inquiry: (i) interviews with the owner and/or operator of the property, (ii) site visit, and (iii) review of government records and historical sources.
–   ASTM E1527-05 (2005) Phase I Environmental Site Assessment, which is a more comprehensive assessment than the aforementioned questionnaire and must be performed by an environmental professional. It consists of four parts: (i) a thorough review of previous records, (ii) site visit, (iii) interviews with the owner and/or operator of the property, and (iv) records.
–   ASTM E1903-97 (2002) Phase II Environmental Site Assessment is a detailed investigation requiring sampling and analysis. The purpose of Phase II investigations is to estimate the nature and extent of contamination and to provide the basis for a preliminary assessment of the technological alternatives and cost for corrective or preventive action.
–   ASTM E1739-95 (2015) standard guide for 'Risk-Based Corrective Action applied at Petroleum Release Sites' is a framework for consistent management and decision making for a variety of sites. It categorizes the sites according to their risk in order to use resources for maximum protection of human health and the environment by:
    o   Identifying exposure pathways and receptors;
    o   Determining the level and urgency of response required;
    o   Determining the level of appropriate overview;
    o   Incorporating risk analysis into all phases;
    o   Applying the three tiers of: (i) qualitative risk assessment based on general site assessment information, (ii) site-specific data to determine the appropriate risk-based actions and (iii) detailed site characterization;
    o   Selecting the appropriate and cost-effective corrective action measures.

The US superfund practice applies a more detailed procedure:

–   Screening, i.e. compiling existing information: site visit, toxicity assessment, exposure and risk estimation;
–   Problem formulation, i.e. what to assess: contaminant sources, transport and exposure pathways, conceptual model;
–   Study design: workplan and sampling plan;
–   Verification of field sampling design;
–   Detailed site investigation: data collection, evaluation and analysis;
–   Risk characterization: application of the weight-of-evidence approach, qualitative/quantitative risk assessment, risk characterization of all exposed receptors;
–   Uncertainty analysis;

– Decision on the next risk management step: selection of the most appropriate method of risk reduction (if necessary) by evaluating the risk reduction options and planning of the measure.

Australia (NSW, 2000) uses a very simple and clear guideline for consultants on reporting on contaminated sites, differentiating four assessment stages and closely related remediation and site monitoring:

– Stage 1: Preliminary site investigation;
– Stage 2: Detailed site investigation;
– Stage 3: Site remedial action plan (RAP);
– Stage 4: Validation and site monitoring reports.

In the European practice, three phases are generally distinguished and applied in contaminated site assessment: preliminary site assessment, exploratory and detailed site assessment (see also Figure 1.10).

In the United Kingdom, contaminated land regulation is outlined in the Environment Act 1995 (UK, 1995) which created new agencies and standards for environmental management. The Environment Agency of England and Wales has produced a set of guidelines for Contaminated Land Exposure Assessment (CLEA), a standardized approach to the assessment of contaminated land (see also the CLEA model: Jeffries & Martin, 2009; the CLEA software, 2014 and the report of Hosford, 2009 on the human toxicity of contaminants in soil).

In the UK, the site investigation reports must be assembled by an environmental professional.

– Phase 1 desktop study contains all generic information, photos, maps and a preliminary conceptual model explaining how the site may interact with the environment.
– Phase 2 detailed study involves sampling, risk assessment and reporting. The quantitative risk assessment involves comparing the concentration of contaminants found on site with the soil guidance values that are provided by the Environment Agency. A soil guidance value varies depending on the intended use of the site: e.g. commercial land use requires the lowest land quality and residential land use with gardens the highest (Cole & Jeffries, 2009).

In Hungary contaminated site and soil assessment was standardized after the issuance of the 1995 General Environmental Regulation (HU LIII, 1995) and of the Soil and Groundwater Protection Regulation in 2000 (HU 10/2000 and HU 33/2000). These regulations set guidance values for soil and groundwater and the obligation for subsurface water quality protection. The Hungarian contaminated site assessment protocol includes only two steps: preliminary and detailed site investigation. The standard protocol requires to create a site-specific target value, which should be compared to the assessed risk value and fulfilled by remediation or other type of risk reduction.

The standardized methods indicate that the assessment of possibly contaminated land generally has three distinct phases. From the practical point of view, each main phase may have further steps or tiers, depending on the aim of the assessment and the extent of contamination. But, still many open questions remain: How can the contaminating components be aggregated? With simple addition, or should synergism or antagonism also be taken into account? How to include food chain effects and biomagnification? How to deal with uncertainties? How should conservatism be interpreted, which scale of overestimation is acceptable?

Urgency and other management/organizational aspects may influence the tiering of the iterative assessment method. All in all, the available time and the endeavor to avoid superfluous sampling, analysis and testing are in competitive conflict with one another. A stepwise iterative risk assessment and an accompanying cost–benefit assessment can ensure optimal tiering and optimal risk assessment.

4   SAMPLING

Sampling in a general sense means the application of statistical methods for obtaining representative data or observations from a population. Population is a term used in sampling and statistics to describe the total set of the targeted observations to be represented by samples. For example, when targeting the size of clams in Lake Erie, OH, the population covers all the clams in the lake. Then a subset is selected from which one can extrapolate to the whole population. The correct selection of the representative sub-group (called the *sample*) allows the determination of the characteristics of the population without examining the entire population. Sampling errors cause differences between sample and environmental distribution, meaning that the conclusion based on sample analyses may not be valid for the real environment. In addition, time is also an important influencing factor: the environment may be properly characterized at the moment of sampling, but if the trends and the rates of change are unknown, real environmental processes cannot be characterized correctly based on a one-off sample.

Several definitions, standards, and protocols support proper planning and execution of sampling, as well as the selection and compilation of analytical and testing tools. There is a great deal of literature describing terms and definitions, a full overview on sampling strategies, sampling methods and their advantages and disadvantages, for example Carter & Gregorich, 2007, BC Field Sampling, 2003, Schoenenberg *et al.*, 2012, Webster & Lark, 2013, NRCS, 2014, Irish EPA, 2015).

4.1   Aims and strategies of environmental sampling

Sampling is a complex issue in general, and it has special importance in contaminated site assessment and management as the basic information source.

A *sample* in environmental practice is often understood in a narrow sense, as a portion of material that is extracted from its original location for preparation and testing. However, environmental science uses 'sample' not exclusively as a representative piece of the environment in a container, but in a wider sense, it also includes the location of observations of remote sensing, or *in situ* measurements and tests with different scale of disturbance caused by the measuring device. Field observation, field measurement, or sample-taking from environmental compartments and phases or from living organisms applies a wide variety of methodologies, which can be grouped according to statistics (targeted, systematic, random or stratified samplings) or evaluation and analysis (reference sample, control sample, replicate, spiked, point samples, composite sample, incremental sample, etc.).

The *aim of sampling* may be to make predictions on the whole of the environment or to prepare a case study. Probabilistic sampling is applied for the purpose of prediction (e.g. a sampling campaign for mapping the distribution of a contaminant), and mostly non-probabilistic for a case study. The aim of sampling and the region or the site and its variabilities determine the sampling strategy and the protocols for sampling and statistical evaluation. The variables to observe or measure specify sampling methods, samplers and sample handling.

The most common aims of sampling are:

– Soil survey and mapping: geological, pedological, biological studies and subsurface water exploration;
– Agricultural: nutrient content and fertilizer requirements, pesticide residues, effects of seasonal changes, crop type, tillage technology, etc.;
– Environmental:
  o Soil deterioration: long-term studies: geological, pedological, soil structure deterioration, damage of the biota;
  o Soil contamination: short-term studies on soil water contamination and biological responses;
  o Soil remediation or other technology application on soil: targeted parameters sampling;
– Nature conservation: element cycling and species sampling.

Some examples of soil sampling strategies:

– Comparison studies evaluate the changes in time or the effect of land use or engineering interventions. Sampling at different time points is comparable, if using the same sampling patterns and sample size. Comparison to a reference site needs thorough selection of the reference and the sampling pattern.
– Studying the soil profile provides the basic information in a soil survey. Soil evolution, soil type assessment and mapping are the main tools of geological and pedological studies. Soil profile development (chronosequence) is typically used to describe landscape development due to major changes such as deglaciation, volcanic activity, wind deposits, or sedimentation.
– Periodic sampling and measurement is used in order to study chronic effects of climate changes, land uses, natural attenuation or ongoing engineering interventions on soil organic matter content, salinity, pH, desertification and soil deterioration.
– Short-term monitoring of the effect of natural events or engineering interventions on soil properties, e.g., moisture content, groundwater levels, redox potential, nutrient content, respiration rate of the soil microbiota.
– Composite random sampling is the collection of representative soil samples from large, not too variable, land to get information on nutrients' or contaminants' presence and contents. This strategy is chosen to avoid large sample numbers and analysis costs, and to fit sampling to interventions executed on the whole investigated area uniformly, such as application of fertilizers, soil amendments or additives/reagents for contaminated soils. (See also composite sample under definitions, in Section 4.3.).
– Directed sampling is used when the sites or subsites can be delineated as being different in landscape, land use or any soil properties known from maps or

remotely-sensed images. These delineated subareas are then sampled separately using the best fitting sampling patterns.

– Grid sampling is a systematic technique used both for agricultural or environmental screening purposes. The size of grids depends on the variability of the soil and the required resolution of the distribution of the measured parameters. Uniform grids may be the source of systematic errors. Point and composite samples can both be collected from the grids.

– Statistical evaluation of sampling means establishing the relationship between sample statistics (e.g. distribution) and population parameters (distribution). If sample error were zero, sample statistic would be equal to the population parameter. (*Statistics* is the plural of statistic, which is the same quantity for a sample as a *parameter* for a population.)

4.2   Sampling patterns and statistics

Sampling plays a significant role in the exploratory and detailed assessment phases of contaminated sites when highly structured sampling is used. Planning the sampling and selecting the best fitting sampling pattern will determine the quality of the assessment.

In those rare cases when sources and transport routes are completely unknown and, as a consequence, the conceptual risk model of the site has not yet been created, systematic sampling is the most convenient. In a more advanced stage of site investigation, the sampling plan should be based on the conceptual risk model including the transport model from the sources to the impacted environmental compartments and the exposure model based on land uses and receptors. In this case, the conceptual model designates the type of sampling and the best fitting sampling pattern (Figure 1.12):

– *Probability sampling* is the selection of a random sample. In a random sample, each member of the population, meaning any point of the contaminated site, has an equal chance of being selected. Probability sampling may be random sampling, stratified random, systematic random or cluster random sampling.

  o *Simple random sampling* means that all items of a population (the totality of sampled and investigated items or units) have an equal probability of being sampled. In terms of the distribution of the environmental characteristic, random means that the characteristic has an equal probability of occurring in any and all items of the population. After delineating the site and determining the number of samples (depending on the expected variation), sampling locations are chosen randomly at the site. Simple random sampling is rarely applied for contaminant distribution, but rather for basic site characteristics such as geochemical characteristics, soil type, nutrient supply, etc.

  o *Stratified random sampling* is a method which divides the items to be sampled (e.g. the soil) into subgroups, also called strata (with and without vegetation, or flooded and not flooded soil, etc.) and then simple random sampling is applied for each subgroup. The sampling plan should determine how to stratify the site. Optimal stratification may result in efficient site characterization. The size of strata may be similar or different from each other, and the sampling frequency may be different too. If the soil is highly stratified and too many subgroups need to be created, the intensity of sampling/number of samples

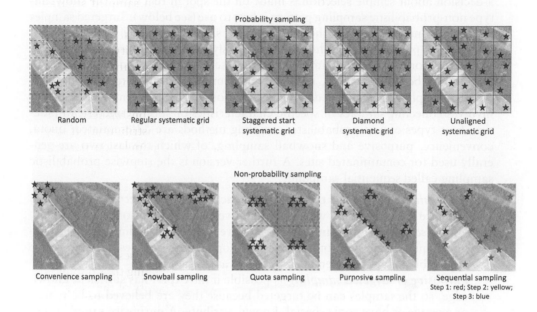

Figure 1.12 Graphical illustration of probability and non-probability sampling patterns.

should be determined according to the expected risk (e.g. larger number of samples for flooded soils).

○ *Systematic random sampling* uses a pattern or grid for selecting the samples from the population, e.g. one soil sample from a sampling grid or every tenth plant from a crop field. This type of sampling is used for the characterization of large areas (large populations) by a small sample size. The application of a 'pattern' increases the efficiency of sampling compared to random sampling. However, this pattern often fails to represent the real situation at a site where hot spots and key transport routes are significantly different and clearly separated from other parts of the site. In such cases, targeted sampling is more adequate, all the more because many of the samples may be negative. To analyze hundreds of negative samples is wasting time and money.

○ *Cluster (area) random sampling* selects a representative cluster from a smaller area instead of the entire population whose sampling would be unfeasible. Wrong selection may bias representativeness. Mesocosm- and microcosm-based predictions may be burdened with such uncertainties.

- *Non-probability sampling* is applied for a case study of a representative group or for such cases where random selection is not appropriate. A case study means the testing of a hypothesis (e.g. the conceptual risk model of the contaminated site) or a law (e.g. transport, partition and fate of the contaminants), targeted points of a contaminated site (already identified hot spots, transport routes, border lines) or just a representative small part of the environment (e.g. a microcosm). Using a portable device for *in situ* contaminant assessment and mapping, whereby a decision about sample selection is made on the spot in real time, the snowball-type non-probabilistic sampling pattern is best to use (see below). Targeted samples (see below) never represent the whole of the site but only the selected special points. Non-probability sampling uses previously collected information about the population for planning the sampling. In the case of contaminated sites, this information can be the hot spots, contaminant sources, transport routes, etc. This type of sampling is also called targeted, purposive, or judgmental sampling; it involves collecting samples of specific characteristics or specific interest. In practice, four types of non-probabilistic sampling methods are differentiated: quota, convenience, purposive and snowball sampling, of which the last two are generally used for contaminated sites. A further version is the stepwise probabilistic sampling called sequential sampling.

o *Convenience sampling*: the easiest-to-reach-sample is selected. There is no evidence that it is in any way representative of the population.
o *Quota sampling*: non-random samples are selected according to a predetermined quota from subgroups. It was developed for market research but is also applicable for environmental surveys and assessments.
o *Purposive or targeted sampling* is possible if the expert has sufficient knowledge, so the samples can be targeted because they are believed to be typical or average or have some special, known attribute. A purposive sample is not representative of the whole of the population (e.g. the site to be assessed) but can be useful in exploratory studies, case studies or contaminated site characterization for taking samples from points where the contaminant is expected.

The results of targeted samples cannot be used for the characterization of the whole of the site, but they give evidence for the presence of the contaminant and for proving the correctness of the preliminary concept.

Using targeted sampling, the assessor can identify hot spots and delineate the boundary of the contaminated site. The concept of the targeted site sampling and the predefined sampling pattern should be in line with the conceptual risk model of the site, allowing the localization of sources, hot spots, transport routes and receptors as well as the assumed border line.

– *Snowball sampling* is a method whereby only the first samples are identified – contaminant sources, hot spots or damaged receptors – and the subsequent members of the sample come by 'identification' or 'nomination' by the first sampling point. It can be applied parallel to *in situ* sample analysis, whereby the decision about the next sampling point depends on the result of the previous one. Snowball sampling does not guarantee a representative sample of the whole site but can map the distribution and transport of the contaminant or other deterioration at the site efficiently.
– *Sequential* sampling means a stepwise probabilistic sampling: taking a single or a group of samples in a given time interval, and after measuring and evaluating the results, taking another group of samples and so on.

Sampling itself should be a stepwise procedure similar to site characterization and risk assessment. The analysis and evaluation of a sample set will show the uncertainties and errors of the taken samples and involves a second sampling round to refine results and lower uncertainties.

Sampling uncertainties can be reduced by additional sampling at the most uncertain sampling points or between two points showing illogical discrepancies, e.g. contradicting prior estimates, geological/hydrogeological characteristics, naturally occurring trends, etc. Geostatistical techniques such as correlograms, covariance functions, variograms or point and block kriging can also be applied to lower uncertainties.

In situ sample analysis significantly amends originally uncertain situations, having immediate result after sampling: one can save a significant amount of time and cost compared to conventional laboratory analyses. More sampling rounds can be integrated into one step, and the handling of uncertainties is also clear: conflicting results can be clarified immediately, and in some cases, the number of sampling points can be adequately increased.

In situ sample analysis is especially valuable when soil is collected for further studies, e.g. for microcosm or mesocosm tests or when the selection of the most representative sample for a special target is essential (high contaminant concentrations). Sampling soil during excavation and remediation may have serious consequences, both from a technological and economic point of view because the selection of the proper remediation technology will be based on the results (depths of the injection into groundwater, or bioremediation influencing the redox potential, etc.), or the precise delineation of the soil to be excavated (smaller excavated quantity results in residual contamination at the site, larger amount causes additional workload, costs and unnecessary loss in quality soil). When the decision is urgent because sampling and resampling otherwise cannot be scheduled, one-step sampling and assessment may be essential from a management perspective (see Chapter 6 for more information).

4.3   Sample types and related terms

Representative samples should accurately and precisely reflect the frequency distribution of the variable of interest (e.g. contaminant concentration). What is required to take representative samples?

– Fitting sampling pattern and sampling scheme in line with the site's conceptual risk model;
– Suitable sample size and variation to enable the assessor to generalize from the results of the samples to the entire site, or the targeted items;
– Generalization from highly representative sample results in a realistic assumption;
– Preliminary information is needed: its quantity and quality are proportional to the success of the sampling and characterization of the site.

The number of necessary samples depends on the aim of sampling, the size of the area to be assessed, the scale of spatial and temporal variations, heterogeneity of the site and the required statistical quality.

The sampling plan should reflect the site's conceptual risk model, primarily including spatial and temporal characteristics of the site (topological, geological, hydrogeological, geochemical, soil type, vegetation), the statistical distribution of the contaminant or related characteristics, transport routes and receptors.

Field samples have significant fate, no matter whether tested *in situ*, on site or packed and stored before laboratory analyses. The history of the sample should be carefully recorded, including the aim, the date and location (coordinates) and type (remote, non-contact, contact non-disturbed, disturbed) of sampling, type of sample container, chemical preservation, storage conditions, holding time, and sample pre-treatment for extracted/removed samples.

The term *reference sample* has many different interpretations in environmental assessment and monitoring practice. A noncontaminated reference sample for a contaminated site should represent the same or be very similar to the site to be assessed in terms of soil type, land use, hydrogeology, geochemistry, etc.

Control site samples or *background samples* are those where the analytes may exist at background levels and are located near to the assessed site. The aim of their application is to prove that the assessed site is really contaminated or different from the background, the baseline representing the norm. Control site samples may represent the local environment or a larger area.

Reference in the laboratory where the samples are analyzed is the clean matrix spiked with known analyte concentration, also known as reference sample. Reference material in the same lab is the known quantity of the pure chemical to be analyzed for the calibration of the analytical method.

In-house, certified reference samples are useful to determine both contaminant and analyte loss that might arise during handling, transport or storage. They also yield estimates of analytical errors.

Field reference samples are sent by laboratories or are parts of *in situ* applicable analytical or test kits: water or soil samples with a known quantity of the targeted contaminant, which are handled and measured identically to a field sample. By comparing field results to this standard or certified field reference, accuracy (bias) and precision can be determined, and the results are reported along with the normal field results.

Field blank, similar to laboratory blank, is an analyte-free matrix (water or soil), treated and exposed to the environment in an identical manner as real samples from the site and analyzed together with real site samples to trace its possible contamination with the analyte during sample handling and analysis.

Replicate samples are taken from the same environmental matrix and handled identically. The aim of replicate sampling and sample analysis/testing is to assess total variance: the variances due to the heterogeneity of the sampled population and the preparatory and measuring activities together.

Grab sample is a single sample taken from the environment, from an effluent or from solid waste over a short period, generally less than 15 minutes. Grab samples are taken from homogeneous materials such as water or other homogeneous liquids, in a single vessel. Grab samples provide a snapshot view of the quality of the sampled environment at the point and time of sampling. Without additional monitoring, the results cannot be extrapolated to other times or to other parts of the sampled soil, river, lake, or groundwater.

Grab samples from soil lack the component of correctness; therefore, they are biased. The so-called grab sample in the case of soil is not really a sample but rather a specimen of the material that may or may not be representative of the population to be sampled.

Bulk sample is taken from a larger quantity of material for analysis. A bulk sample is often taken for microcosm, mesocosm, or treatability studies. Bulk sample can be taken from air, water, soil, sediment or wastes. In mining bulk sample means a huge sized representative block of rock.

Composite sample is a combination of multiple individual samples taken at pre-selected times to represent the integrated composition of the environmental matrices being sampled. Usually all samples added to the composites are equal in size, but flow-rate-proportional composite samples are also collected for waters when the amounts are proportional to flow rate. Composite samples are highly cost-efficient, mainly in the case of uncontaminated samples: only one analysis instead of five or ten individual analyses, all with negative results. This logic has the consequence that both the negative and positive results are considered and accepted based on the assumption that the sampled site is homogeneous from every point of view (soil/water properties and contaminant composition), which may be correct for low variability sites, but not correct in many other cases. There are several practices to refine positive results, e.g. by handling samples individually and preparing the composite just for the purpose of analysis, allowing further analyses of sub-composites or individual samples if necessary.

Incremental sampling means the collection of increments from a sampling unit (e.g. sections of a soil core along the depth), which are combined, processed, and analyzed to get the best picture on the sampled environment. One increment is a portion that is collected with a single operation (see also Incremental Sampling, 2012).

Regular sampling can be intermittent or continuous. The *intermittent sampling* frequency depends on the prevalence and duration of the expected events and the rate of changes in an environmental or technological process. Short peaks of contaminant occurrence in air, water, or groundwater are typical and cannot be detected when using intermittent sampling; real-time continuous online measuring devices can detect such sudden and short-term occurrences of contaminants or other fluctuating parameters. Another option is the use of aggregating type samplers, which selectively collect environmental samples (e.g. groundwater, soil moisture, leachate, and air) or contaminants. Collection of contaminants during a longer period also covers contaminant transients. This type of sampling enables detecting the presence of contaminants which occur in the short term, but without identifying the moment and duration of their emergence.

Continuous sampling is frequently used to monitor the environment. Remote or proximal optical sensors and video cameras can detect all kinds of changes, movements, or behaviors. Air and water flows can be sampled by online-emplaced sensors combined with data loggers and telecommunication (see also Chapter 3). Continuous sampling combined with *in situ* real-time measuring devices is the most advanced and versatile methodology in environmental monitoring. It is equally applicable for physical, chemical, and biological processes as well as for observing long-term changes and identifying environmental trends.

On site, real-time and online sampling and measuring is extremely important for efficient environmental management in order to get real-time information over the long term. Real-time signals closely relate the actual environmental risks, enabling decision making and intervention without delay. Another advantage is the avoidance of sample handling and transportation into labs, as well as the separation of the sample from its original context. It means that the sample can be analyzed without changing its pH, redox potential, the living conditions of the biota and other sources of uncertainties.

Sampling in several phases enables cost-efficient sampling and site investigation. After the estimation of the mean and variance of samples taken in the first phase, a proper sampling plan and an efficient second phase sampling can be implemented. It is worth applying cheap, rapid *in situ* analysis and test methods in the first phase to provide a sufficient amount of information for an efficient sampling and assessment plan for the second phase using more expensive sampling, preparation and analysis methods if necessary. Sample bias can be further reduced when the time delay between the phases is as short as possible. The Triad approach is based on *in situ* site assessment combined with immediate and dynamic decision making, depending on the identified contaminants or adverse effects, and their distribution pattern. The outcome of the decision may be further sampling, changing sampling strategy, or modifying the original sampling and site investigation plan.

Sampling for chemical analysis accentuates the discrepancy between environmental variability and heterogeneity and the extensive improvement in sensitivity and accuracy of chemical analyses, as well as the continuously decreasing sample amount needed for analysis. References and blanks (see above) can only solve a part of the problems originating from sampling.

A general view is that if the error of the analytical method is one-third or less of the sampling error (environmental variability), further reduction of the analytical error has no importance/relevance (the resulting information will not be better). Of course, the purpose of the analysis is also crucial: requirements differ for the regulatory assessment/compliance monitoring of a one-off sample and the long-term observation of certain environmental trends. Proper statistical design enables harmonization between sample taking and analyses to mine as much information from the assessment as possible. Recently, the optimization of the economic aspect has come to the fore to avoid collecting data of little use, unnecessarily high security and the related costs.

Sampling for biological studies is even more demanding than for chemical analysis, as the sample is supposed to represent biological entities, bioindicators, living organisms, members, populations or the community of the biota. *In situ* sampling and measurement allow the elimination of the difficulties of keeping the biota alive and unchanged during sample collection, package, transport and the laboratory process. These problems always arise when biological samples are analyzed and tested in remote laboratories and may have a significant impact on the collected environmental material for micro- and mesocosm studies or laboratory treatability tests.

Biological and environmental samples for bioassaying have special requirements: they should keep their biological properties, activities, diversity and the biota parameters should change as little as possible. As biological activities, species diversity, and the distribution of activities in the community are greatly influenced by environmental parameters such as pH, redox potential, humidity, nutrient supply, light intensity, etc., a small proportion of the whole removed from its original ambient – even if undisturbed – can never produce the same environment to its inhabitants as in the original place. Consider the soil microorganisms or the sediment-dwelling micro- or macro-invertebrates: how can it be ensured that their environment remains unchanged during sample collection, packaging, and transportation? Undoubtedly, the most unbiased measurement result is expected when the sample is not poked, not extracted or removed, but instead a non-invasive measuring tool is gently deployed or remotely applied *in situ*, in real-time in a real situation.

When biological properties and responses cannot be measured by *in situ* emplaced or remote sensors without sample collection, there are two possibilities for an assessment: (i) bioassay in a remote or on-site laboratory, or (ii) a field test with portable test kits. These test kits include all necessary reagents and the live or viable (e.g. lyophylized) test organisms. When using live organisms, viability and activity should be retained during transportation, handling, and field deployment using incubators or other equipment or tools.

Historical information may substitute sampling in many cases, so it is extremely important to collect previous biomonitoring data and study results of similar cases. If historical data do not fulfil the requirement of the current task, complementary assessment and monitoring may be necessary with sampling. Historical information and similar cases generally give efficient support for planning sampling or monitoring.

Proper sampling of biological indicators requires a good risk model of the site which reflects the endangered biological receptors. The space and time domain of sampling means assigning the conceptual risk model to spatial and temporal dimensions, as well as to the context. The following requirements should be adhered to:

- The sampling resolution should fit to the ecological/biological dimensions of the bioindicator.
- The duration of intermittent sampling should fit to the seasonal cycles or other temporal changes and the life cycle of the bioindicator. Continuous sampling can eliminate this difficulty.
- Controls/references and replicates are defined by the statistical regularities valid for the area and the bioindicator.

Sampling for *toxicity and treatability studies* is easier when using a test organism under control. In such a case, only the environmental sample needs to fulfil the requirement of remaining unchanged and retain all characteristics of the original environment in its original place. Otherwise, all the aspects of sampling resolution, frequency, duration, and of lowering uncertainties by references and controls are the same as for chemical or biological sampling. The problem of sampling strategy and replicates may require special attention, as the replicates of such studies are generally determined not based on sampling uncertainties, but on the chemistry and the biology of the test system.

Sampling without direct contact is increasingly applied for physical, chemical, biological, and toxicological data acquisition alike. Remote and proximal sensors can detect emitted or reflected radiation specific for indicator characteristics, elements or molecules of the biotic and abiotic parts of the environment. The surface and the shallow subsurface can be monitored in this way. Remotely measurable indicators can be found in surface geometry, geochemistry, element content, presence of organic molecules, biomolecules, movement and behavior of living organisms, etc.

Quality assurance and quality control are crucial for sampling, as sampling may have the highest error among environmental investigation activities. The smaller the sample or population number, the greater the error. Key features of sampling are: record-keeping and documentation of the sampling method, sample history, specification of the sampler or other used equipment (containers, auxiliaries), sample pretreatment and preparation, the reagents and preservatives, mode of transportation, preventive measures against cross-contamination and sample loss, *in situ* and on-site activities, methods of analysis and tests, the laboratory where the sample has been delivered, and the type of long-term sample storage. The type of analysis and other studies carried out on the sample should also be documented in detail: measurement type, equipment, reagents, calibration, standards and reference samples, sample splitting and spiking, data handling and evaluation, and, finally, overall quality management.

Several standards are available for sampling, prescribing how to plan and perform sampling, dealing with quality assurance, specifying data quality objectives and the requirements on sampling and analysis plans. The prerequisite of the standards application is the existence of a conceptual risk model of the site and the problem because this is the foundation of defining the environmental compartments to be sampled and the transport routes along which the hazardous agents can reach the receptors. Environmental sampling standards cover air, surface water with sediment, soils including groundwater and wastes. Depending on the target environmental compartments and phases, significant differences can be found both in sampling strategy and sampling technologies.

When sampling soil, in addition to the spatial and temporal heterogeneities (which are landscape attributes), one must take into consideration the following issues:

–   Composition or inference error, leading to a fundamental error;
–   Distribution heterogeneities resulting in grouping and segregation error;
–   Short-range errors occurring within the sampling support such as tools (samplers, holders. storage tanks) and sample preparation methods;
–   Long-range heterogeneity, e.g. due to local trends causing fluctuation error (geostatistics deals mainly with this kind of error);
–   Periodic heterogeneity such as rainy season, resulting in fluctuation error (US EPA, 1992).

Sampling errors are especially critical for soil and other granular matters where a complex relationship may exist between the variability of the material, the particle sizes, the distribution of the contaminant according to space and particle size on the one hand, and the size of sample taken on the other.

Another major source of sampling bias is the increment delimitation error. This results from incorrectly defining the shape of the material's volume (i.e. sample) to be extracted. In the case of a core sample, an increment is a section of the core between two horizontal parallel planes passing through the core (i.e. a cylinder of certain height). This mainly applies to waste and soil deposits.

Several other common errors will be added to the sampling errors mentioned such as those linked to sample preparation:

–   Variation produced during grinding, screening, sifting, storage;
–   Cross-contamination and alteration of the sample;
–   Human errors;
–   Loss errors.

The well-known analytical errors are also present, but using standard techniques, these may contribute to the overall error to a lesser extent.

Since soil is particularly heterogeneous, conventional statistics would require too large sample numbers, which would lead to unacceptable cost. Thus, economy is the driving factor in developing new sampling and measuring strategies such as tiered sampling, *in situ* measurements, selective samplers, cost-efficient real-time and online sampling methods ensuring large sample numbers, and rational compromises.

Sampling bias, confidence level and sampling precision are defined below.

Confidence level is the probability that the selected confidence interval will include all the samples with the specific parameters to be analyzed. Confidence levels usually range from 90 to 99 percent. The statistically adequate size of the bulk sample can be defined when variance is known. It is worth also taking the cost of analysis into consideration.

Sampling bias is always introduced to the sampling process, but it may be significant if correct sampling, fitting to the site and the problem, is not applied. The bias that is introduced may be quite small in materials that are relatively uniform in composition and particle size. Mixed materials such as a cobbly clay loam may show a large bias if the cobbles and gravel particles are excluded from the sample. These coarse materials may need to be addressed by double sampling techniques. (Double sampling is applied when the results of the first sample are not conclusive.) The complete sample bias is the sum of sampling, preparation, and analysis bias.

Precision is a measure of the reproducibility of measurements for a particular soil condition or constituent. The statistical techniques used for soil sampling determine the extent of precision. *Accuracy* cannot be interpreted for environmental samples, given that the real contaminant content is not known, and therefore the measured value cannot be compared to any value.

Standards for sampling are widespread such as ISO and ASTM. National standards ensure uniform sampling and comparable results. Laboratories and companies are required to be accredited in most countries especially when they perform sampling (Schreiber, *et al.*, 2006). Some standards are listed in Table 1.1.

Samplers can be classified according to the environmental compartment or phase, the depth of sample collection, the mode of sampling: one-off, intermittent or continuous. Sampling with extraction of the sample significantly differs from sampling for *in situ* non-destructive measurement or remote sensing. In the latter cases the sampler and the detector is the same device contacting or not contacting the soil. Samplers can be conventionally classified as hand tools and power probes, including well-based probes. According to physical phases, the main groups of samplers are: samplers for soil gas, surface and subsurface water, sludge and sediment, solid soil and waste. Sampling for geotechnical purposes is mostly performed using *in situ* tools combined with a measurement tool (penetrometer, moisture sensor, test hammer). The power probes include direct push tools and core probes. Chapters 3 and 4 introduce and classify several innovative sampling and measuring devices. One fairly new type, the *passive sampler*, may reduce the need for and costs of sampling infrastructure and assistance at the sampling location. Passive samplers rely on the unassisted transport (diffusion is typically involved) of mobile phases and/or molecular species through a conductive/diffusive surface to the collector, which is mostly a sorbent. Passive samplers work without active media transport induced by pumping, nevertheless natural flow and

Table 1.1 ISO and ASTM standards for soil sampling.

Identification number	Sampling standards and guidance
ISO 10381-1:2002	Guidance on the design of sampling programs
ISO 10381-2:2002	Guidance on sampling techniques
ISO 10381-4:2003	Guidance on the procedure for investigation of natural, near-natural and cultivated sites
ISO 10381-5:2005	Procedure for the investigation of urban and industrial sites with regard to soil contamination
ISO/NP 10381-6	Collection, handling and storage of soil under aerobic conditions for the assessment of microbiological processes, biomass and diversity in the laboratory
ISO 10381-6:2009	Guidance on the collection, handling and storage of soil under aerobic conditions for the assessment of microbiological processes, biomass and diversity in the laboratory
ISO 10381-7:2005	Guidance on sampling of soil gas
ISO 10381-8:2006	Guidance on sampling of stockpiles
ISO/FDIS 18400-102	Selection and application of sampling techniques
ISO/FDIS 18400-101	Framework for the preparation and application of a sampling plan
ISO/FDIS 18400-103	Safety
ISO/CD 18400-104	Strategies and statistical evaluations
ISO/FDIS 18400-105	Packaging, transport, storage and preservation of samples
ISO/FDIS 18400-106	Quality control and quality assurance
ISO/FDIS 18400-107	Recording and reporting
ISO/FDIS 18400-201	Physical pretreatment in the field
ISO/CD 18400-202	Preliminary investigations
ISO/DIS 18400-203	Investigation of potentially contaminated sites
ISO/DIS 18400-204	Guidance on sampling of soil gas
ISO 18512:2007	Guidance on long and short-term storage of soil samples
ISO 23611-1:2006	Sampling of soil invertebrates – Hand-sorting and formalin extraction of earthworms
ISO 23611-2:2006	Sampling of soil invertebrates – Sampling and extraction of micro-arthropods
ISO 23611-3:2007	Sampling of soil invertebrates – Sampling and soil extraction of enchytraeids
ISO 23611-4:2007	Sampling of soil invertebrates – Sampling, extraction, identification of nematodes
ISO 23611-5:2011	Sampling of soil invertebrates – Sampling and extraction of soil macro-invertebrates
ISO 23611-6:2012	Sampling of soil invertebrates – Design of sampling programs with soil invertebrates
ISO 23909:2008	Preparation of laboratory samples from large sample
ASTM D1452 – 09	Standard Practice for Soil Exploration and Sampling by Auger Borings
ASTM D4700 – 15	Standard Guide for Soil Sampling from the Vadose Zone
ASTM D6519 – 15	Standard Practice for Sampling of Soil Using the Hydraulically Operated Stationary Piston Sampler
ASTM D1587 – 08	Standard Practice for Thin-Walled Tube Sampling of Soils for Geotechnical Purposes
ASTM D422-63 (2007)	Standard Test Method for Particle Size Analysis of Soils

capillary suction are typically applied. They are exposed to media under ambient conditions during the sampler deployment period. Passive samplers collect mobile phases or contaminant species for a certain duration, so the collected amount and concentration is a time-integral from equilibrium, steady-state, or non-equilibrium systems. This should be taken into consideration when the concentration in the sampler is evaluated and the data interpreted. Amounts in the sampler may typically be the result of

- A cumulative procedure (e.g. collecting pore water from unsaturated soil) resulting in the total amount collected by the sampler during the sampling period, from which a time-related average can be calculated;
- A time-related equilibrium concentration in static systems (e.g. in a lake);
- A steady-state equilibrium in a flow-through system if the sampling time is long enough to reach equilibrium in the sampler (without saturation);
- Combination of the above cases.

Passive samplers cannot differentiate between measurands after they have been saturated with the substance sampled. A sampler can be a simple passive collector or a device with slight negative pressure, capillary suction, selective sorption, or chemical interaction. A large number of low-cost passive samplers can be employed, thus allowing high-resolution mapping. Passive samplers can be combined with sensors detecting levels, moisture content, ion content, or the presence of concrete chemical species. Due to their small size, the passive sampler can also be hidden and thereby lower the risk of vandalism. Examples of passive sampling devices are presented in Chapters 3 and 4 together with *in situ* assessment tools and methods.

4.4 Sustainable and efficient sampling

Random, stratified random and statistical sampling are the most frequent sampling strategies in environmental assessments. In random sampling, each sample point within the site must have an equal probability of being selected. Stratified sampling can be used to reduce the variability of the sample. Strata are identified as regions of the site that are expected to be uniform in character. The variance within the strata should be smaller than the variance between strata. In the soil environment, strata are often associated with soil types or interpreted as areas of known pollution versus areas where pollutants are not expected. Stratifying the site according to levels of contamination, for example, to orientate the sampling grid along the axis of the plume, provides a means of identifying the trend of the data and controlling it in the analysis of the developed variograms and in kriging the data (US EPA, 1992).

Systematic sampling is the best strategy if sources and transport pathways cannot be identified, or if the pollution is airborne or diffuse. For this purpose, a default grid based on a pattern of the site map, is best to use. A large number of samples are taken, many of which will be negative. In addition, hot spots and linear sources may be missed. In the case of point sources, detailed assessment of specific areas within the site, for example, the vicinity of the sources and the surroundings of hot spots, is more appropriate.

Geostatistics and kriging are essential tools in site assessment and data evaluation and interpretation. The technique was transferred from mining and exploration to contaminated land management as well as soil and soil contaminant mapping. Kriging or Gaussian regression is an interpolation method which provides the best interpolation function (curve) that runs along the means of the normally distributed confidence intervals. Thus a raster map with best estimates can be generated from a point map.

An error map can also be created. Based on the information from existing sample data, geostatistics can predict likely values in points and subsites which have not been sampled. This can be done within the zone of influence, i.e. in the vicinity of sampled points, but not outside this zone, where all values will be independent from measured sample data. Kriging can be applied both for punctual and block results. Kriging and other geostatistical techniques are used for soil mapping, isopleth development, and for the evaluation of the spatial distribution of soil and waste characteristics and contamination. A comprehensive overview on this topic can be found in Practical Geostatistics by Clark and Harper (2000), Clark (2010) presents a review about sampling errors, and an easy-to-understand explanation is available on the internet (Nederlof, 2015).

Tiered sampling is the most efficient sampling methodology. In the first tier, the areas or points of interest are selected by rapid and cheap *in situ* measurement techniques (mobile XRF for metals or mobile IR or specific detectors for volatile or semi-volatile organic pollutants). After identifying hot spots or other locations of interest, samples are taken by using traditional sampling techniques and laboratory analysis in the second tier.

Another important statistical issue is the **number of samples** to be taken from a site, their size and replicate number. Soil heterogeneity determines the number of samples statistically required or even an oversampling. However, smaller sample numbers are normally only feasible on grounds of cost efficiency.

The standard deviation around a mean estimate obtained from a series of samples taken from a block of soil material is often quite large. A well-homogenized sample made up of a number of increments (see under definitions in Section 4.3) of material or from several samples collected from a block of soil will normally exhibit a smaller variance. This sample is the *composite sample* (see Section 4.3). **Composite sampling**, similar to *discrete sampling*, allows the calculation of a reliable estimate of the arithmetic mean of contaminant concentration in surface soils. Composite sampling involves the physical mixing of soils from multiple locations and then collecting one or more sub-samples from the mixture. The use of composite samples allows the reduction of the costs of sampling significantly.

Pitard (1989a and 1989b) recommends developing a sample by taking a large number of small increments and combining them into a single sample submitted to the laboratory. This sample is then reduced to an analytical subsample by splitting or some other method of volume reduction. This approach provides the benefits of the composite sample and yet avoids the problems of homogenization etc. that one encounters with large volumes of soil. Pitard also advises the homogenization of soil before analysis to reduce distribution and segregation errors.

As soils' variability is very high, several hundred-percentage differences can be measured in soil properties within one yard. In addition, many soil properties are transient, and not randomly, but rather systematically, distributed, changing both spatially (horizontally and vertically) and in time. It is important to know soil variability before choosing a sampling strategy and preparing the sampling plan, also when evaluating soil parameters and contaminant concentrations. The proper tool battery – the analytical tools and test methods, their precision and the adequate statistical evaluation methods – can only be selected when the information regarding the variation-based uncertainty is known and taken into account. When the variations, e.g., in subsurface porosity or in layering of a waste disposal site are known, the assessor can control these variations and minimize their misleading influence on the result and the decisions based on them.

The aim of *subsurface soil sampling* is the assessment of the exposure that occurs when contaminating chemicals migrate up to the soil surface or down to an underlying aquifer. Sampling and subsurface exploration drilling should be performed where the contaminants originate from or are transported to by the groundwater from remote sources. Investigation of pollution from surface sources needs shallow boreholes near the identified source and separate samples must be analyzed depending on the depths of contaminant infiltration. If subsurface pollution is transported by subsurface waters, the boreholes should go down to the impervious stratum beneath the subsurface water.

Multiple contaminants need a sampling and assessment strategy which takes the differences in fate and behavior of these substances into account. Volatile, water-soluble, and sorbable contaminants may all be present at a contaminated site at the same time, requiring different sampling methods and analyses. Depending on the density of a liquid-phase contaminant, the sample may be taken from the surface or from the bottom of the groundwater layer. Interaction with other contaminants can significantly modify contaminant transport, fate, and behavior. For example, biodegradable petroleum hydrocarbons of low risk can solubilize and mobilize otherwise stable contaminants such as DDT or dioxins. When assessing the toxic effects of multiple contaminants, the potential interactions between toxicants should be measured. Some contaminant risks are additive, but they may also show synergism, i.e. risks higher than the sum of the individual ones.

Sampling of water or soil contaminated with volatile chemicals or living organisms, and sampling of soil for biological study purposes, where the microbiological activity of the soil should be retained, pose a special challenge to sampling. Therefore, special care or special devices must be applied to find the proper sampling location and not to disturb, warn of, select or kill soil organisms by the very sampling. During the transport to the place of study, *in situ* conditions must be ensured to keep the activity and viability of the organisms.

4.5 Sampling, *in situ* analysis, testing and immediate decision making

Traditional sampling and analysis programs of contaminated site characterization, remediation and closure rely on work plans specifying the number of samples to be taken, the sample location and the type of laboratory analysis. Many of the national regulations and connected guidelines and standards use this approach to ensure an acceptable quality in contaminated site management. Long turnaround times for laboratory analyses and high costs of analysis lead to decisions being based on a limited number of samples, in turn resulting in high project decision uncertainty.

During the past decade, significant progress has been made in data collection technologies and measurement systems leading to optimal tiering, pragmatic assessment tool batteries and screening methods. For many contaminants of concern, it is now possible to obtain information about their presence and concentration in "real time", or quickly enough to influence or change conventional sampling approaches. Advances in Global Positioning Systems (GPS) enable rapid determination of spatial locations. Traditional laboratory analytical methods have been developed to speed up screening tests and handy mobile devices to measure pH, conductivity, concentration of gases in the atmosphere and soil, dissolved oxygen, metals, volatile organic contaminants in air, water or soil, etc. Special samplers for collecting aggregated or selectively collected samples make the rapid screening of air, surface waters and soil surfaces possible. *Direct push technologies* provide a quicker and cheaper method for retrieving sub-surface samples and enable pushing sensors into the ground for *in-situ* measurements, both for static and dynamic assessments. The latter, so-called 'push and pull' technologies, can follow the effect of (im)pulse-like subsurface interventions. 'Push' means the injection of water or a reagent into a subsurface layer, and 'pull' means the collection of the sample from the same location or from a certain distance downstream from the injection point (Figure 1.13).

The technological advances together with professional knowledge on historical cleanup work opened the way for a new approach aimed at addressing decision making about contaminated sites and the design and implementation of sampling programs to support decisions.

5   MEASUREMENT AND TEST METHODS FOR CONTAMINATED SITE INVESTIGATION

The typical scopes of site investigations are: (i) global ecological risk assessment; (ii) natural conservation; (iii) agriculture's effects on soil structure and nutrient content; (iv) environmental survey and monitoring; (v) the health of workers and residents, generally characterized by chemical exposure assessment and toxicology; (vi) contaminated land and brownfields.

Management of the risk posed by contaminants and the determination of the target quality of a contaminated, or otherwise deteriorated land aim at achieving sustainability. This means that the site should preserve the quantity and quality of long-term ecological services (including provisioning, regulating, supporting and cultural services) (see more in Volume 1 of this book series; Gruiz *et al.*, 2014).

The chemical approach is the traditional method to evaluate contaminated sites: the characterization of a contaminant is considered an analytical task. This is based on the fact that the size of the contaminant's effect is generally proportional to its concentration or doses. This assertion is true only for pure chemicals, in a specific range of concentration and when measured *in vitro*. However, it does not apply to the real environment due to the interactions (i) between different chemicals (synergism, antagonism); (ii) the chemical substances and the environmental parameters (light, temperature, humidity, redox potential); (iii) the chemical substances and the environmental medium (e.g., the solid organic and the inorganic matrices of the soil); (iv) the ecosystem members and contaminants (biotransformation, degradation, accumulation).

The main problems of contaminated site investigation, analysis and testing are summarized below:

–   Most contaminated sites are polluted with mixtures of contaminants.
–   Interactions between contaminants, matrix and biota can significantly modify the risk by terminating or compensating risk, or the opposite; invoking hitherto latent toxicity e.g. by biological mobilization or other kind of transformation.

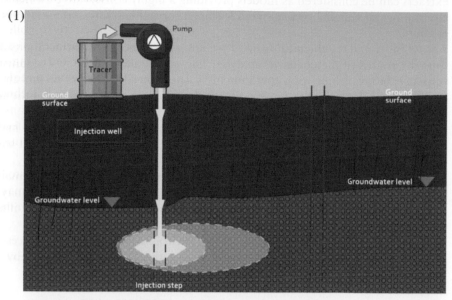

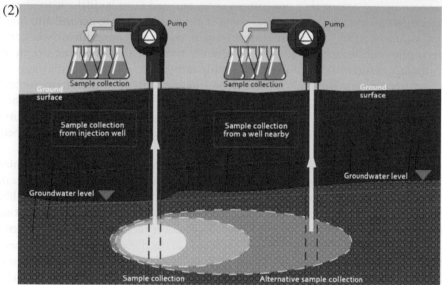

Figure 1.13 Push and pull technology for *in situ* site characterization: 1. Injection to the injection well, 2. Sampling from the injection well and another well nearby.

– Biotransformation of pollutants may lead to products with higher toxicity than the original.

– The medium tested may be a soil extract, a leachate or the whole soil. Accordingly, (i) whole soil represents all phases and interactions, as well as the terrestrial ecosystem as its habitat; (ii) soil leachate represents groundwater contamination originating from the soil above, transported by infiltrated precipitation; (iii) soil extracts can be considered as models providing a highly conservative estimate for environmental runoffs, seepages and leachates, or soil entering surface waters by erosion.

– The mobility and the chemical form of pollutants influence their extractability and thus the results of the chemical analysis. The same holds for the test organism's response: bioavailability of a contaminant is determined by its physical and chemical form and the strength of its binding to the environmental matrix and the type of interaction with the organism.

– The physicochemical and biological availability of soil contaminants are highly variable and differ from each other, therefore the chemical and biological study results may also differ from each other.

– The analytical program generally includes only some of the contaminants, mainly those which are listed in regulations. The number of these contaminants may be small compared to all possible contaminants present at an abandoned or illegal disposal site.

– The biotic and abiotic composition of the environmental sample influences the effect and the risk posed by the contaminants; this is measurable only by the integrated application of physicochemical methods and effect testing.

The complex system that constitutes contaminated land is difficult to characterize by chemical models, even when using a refined (therefore complicated) model that takes into consideration the interactions between contaminants; their partition between physical phases; binding into the solid matrices; their possible degradation, etc. Simple microcosms in combination with integrated monitoring may have high capability to simulate the real environment. Integrated monitoring of the simulation model (microcosm) or a representative part of the real environment (field mesocosm) means the application of biological, ecological and physicochemical methods in parallel. It is the best option in contaminated land investigation in those cases when the aim is to restore biological and ecological functions.

Another problem with most chemical approaches is that the substance must be extracted from the environmental sample before analysis. Extraction allows standardization and comparison to environmental quality criteria, but it is not suitable for modeling the site-specific risk, which is proportional to the actual adverse effect of the bioavailable and non-degradable fraction. Extraction methods can poorly simulate (i) the biological availability of the contaminant; (ii) the dynamic effect of the chemical substance on the ecosystem's adaptation; (iii) the effects of changes in environmental condition, such as the pH and the redox potential, which determine the chemical form of the contaminating substance and influence the diversity of soil microbiota, as well as the interactions between solid matrices, contaminants and the microflora – to mention only the most obvious.

In addition to the above-mentioned problems of extraction, the extractant dilutes the contaminant to be extracted, thus lowering the likelihood of contaminant detectability at low concentrations. Physicochemical analysis without extraction of the contaminants raises the problem of chemical binding, and matrix effects. The solvent-available fraction of the contaminant does not agree with the bioavailable fraction. There are trials using biomimetic solvent systems which extract bio-comparable contaminant proportions from solid environmental matrices to characterize the probable biological response. Such results can be used for screening and properly interpreted only under limited environmental conditions and concentration ranges.

In summary, biological interactions in environmental compartments or phases are hard to model or simulate by chemical interactions.

5.1 Soil characterization using STT

Soil is a complex environmental compartment, and its assessment and characterization needs an integrated approach and innovative methodology. Soil comprises three phases: gases, liquids and solids, but soil microbiota and its habitat with special micro-surfaces and biofilms can be considered as two additional phases with special characteristics and roles.

The contaminants within the soils are generally complex mixtures of chemicals; the interactions between the soil phases, the biota and the components of the contaminants may lead to endless combinations. In addition, there is no equilibrium state in the soil after a contamination event, which means that a continuously changing environment will be responsible for the impacts of the pollutants.

The results of sampling and chemical analysis reflect just one point in space and time; even when samples are properly selected, they represent the entire site only with certain probability. The results of the chemical analyses often do not correlate with the measured adverse effects and actual risks of the contaminant mixture. An integrated approach is needed to obtain a realistic view of the risks: complementary biological testing for ecotoxicity, mutagenicity, reproductive toxicity, food chain effects, etc. The physicochemical, biological and toxicity results should be evaluated together. The (lack of) consistency between chemical analytical (CA), biological/ecological (BE) and toxicological (DTA) results provides information on mobility and biological availability of the contaminant and the interaction between alien and indigenous molecules present. The STT approach draws attention to the chemically unmeasured or non-measurable but existing hazardous components through their (sometimes delayed) adverse effects (Horváth *et al.*, 1996; Gruiz *et al.*, 2000). STT enables a deeper insight into the chemical time bomb situation when contaminants are in an immobile chemical form, unavailable to biological systems, even non-soluble in water, and seemingly harmless under the current conditions. However, the conditions may change, the soil can become wet or change from anaerobic to aerobic or vice versa, a wetland can dry out, sediments can be transported to a different place, and all this may trigger the explosion of the time bomb. Dynamic model systems such as micro- or mesocosms can simulate pessimistic scenarios, e.g. enhanced mobilization, increased bioavailability and toxicity of soil contaminants. These test conditions can be controlled to simulate realistic worst-case scenarios with significant likelihood (see also Chapter 7 in Volume 2 of this book series; Gruiz & Hajdu, 2015).

Following is a summary of the advantages of STT application:

- Integrating interactions between toxic chemicals;
- Integrating interactions between a toxic agent and the matrix;
- Measuring the bioavailable fraction of the pollutants which potentially affects the receptors;
- Measuring the effects of the chemically non-measurable toxins;

– Measuring the effects of those chemicals that have not been included in the analytical program;
– Direct toxicity testing of the soil may give a realistic view on the soil.

The complex organo-mineral colloidal system of the three-phase soil may contain thousands of living organisms and hundreds of contaminant components, and these may interact with each other in different ways. Normally, such a system is treated as a black box, but for soil, one must understand its properties, complex structure, the interactions between its components, and its dynamic response to external and internal impacts.

The soil testing triad (STT) comprises:

– Physico-chemical analytical methods;
– Biological and ecological methods;
– Toxicity tests, including bioassays, simulation tests and micro- and macrocosms (Figure 1.14).

The targets of the physicochemical analyses can be the soil, the contaminant and the physicochemical components of the ecosystem. The biological and ecological methods characterize the living domain: its status or response. The characteristics of an ecosystem are qualitative and quantitative, and the response to a natural or anthropogenic impact provides information about the adaptive nature of the biological system and the ability of the soil to neutralize or compensate for the pollution. The assessment methods can be static (confirming an ongoing situation) or dynamic (reflecting a change).

Toxicity testing is an important part of STT: representative test organisms can provide information on the hazards posed by polluted. Environmental toxicology characterizes the actual effects – an effect directly associated with environmental risk. It is basically measured in the ecosystem members present or by standard test organisms added to the soil to be characterized.

In STT, the physicochemical, biological and toxicity measuring methods are of equal importance, i.e. they are complementary. They provide information on the quality and quantity of contaminants, the characteristics of the soil, the biological status of the soil, and the activity, vitality and adaptive behavior of the soil microbiota, the billions of living cells and microorganisms making up the soil. Data about the effects, mobility, bioavailability and the biodegradability of the contaminant, and about the response of the soil to external factors are also results of STT.

The three STT elements – physicochemical measurements and ecotoxicity tests (Gruiz, 2005) – can be characterized by:

– The spatial definition of the measurement/tests are:
   o *In situ*: the sample is not removed and typically measured without sample preparation. Semi-quantitative rapid tests (e.g. XRF, IR, GC) and mobile instruments are used for site assessment or technology monitoring;
   o On-site: the sample is removed and prepared in place – on or close to the site – and mobile measuring methods or mobile laboratories are used;

o   Laboratory analysis: precise sample preparation, transportation to the laboratory and a long delay from sampling to results, prohibiting direct and immediate decisions or interventions;
− Temporal definitions: real-time or delayed;
− Sampling: heterogeneity, statistics, sampling strategy – see sampling in Section 5.

5.1.1 Physicochemical methods

The chemical concept is most frequently used for environmental monitoring, early warning, emission control and technology monitoring. It is also used for the detection of contaminants and determination of their concentrations in food, human tissue, and in environmental samples. As a method, physicochemical analysis is given priority for the identification of contaminants in generic risk assessment of chemical substances, in the control of pollution sources, for quality control and other regulatory purposes. These fully justified cases served as model/examples for applying the same concept to contaminated sites, despite anticipated difficulties and high uncertainties.

Physicochemical methods can be characterized and grouped according to what is measured (the analyte), how it is measured (analysis method), how sampling and sample preparation is done, and how it can be applied for screening, refined or special assessments:

− Environmental parameter measurement: pH, temperature, redox potential, dissolved oxygen, turbidity, conductivity, CO_2 content, and many others;
− Characterization of soil solid, liquid and gas phases:
   o   Pedological characterization;
   o   Geophysical characterization: e.g. particle size distribution, stability;

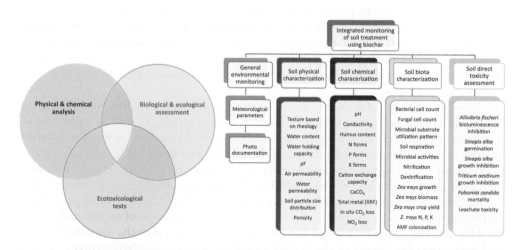

Figure 1.14 Soil testing triad (STT) scheme and a sample STT design for the monitoring of a field demonstration of deteriorated soil treatment using biochar additive.

- o   Geochemical characterization: mineral composition, humus content;
- o   Water content and water balance /water regime characteristics;
- o   Nutrient content: mobile, ionic, exchangeable ions, concrete elements;
– Contaminant analysis: *in situ*, on-site or in the laboratory:
  - o   With sample preparation, extraction and separation;
  - o   Without extraction and separation.
– Destructive and non-destructive analytical methods:
  - o   Destructive methods have poor environmental reality; they do not model the effect and fate of the contaminant, but have better reproducibility.
  - o   Non-destructive methods, due to heterogeneities in chemical forms and the difference between the surface and the inner matrix of the solid phase of the soil, have poor statistics and are burdened by matrix effects. The influence of the matrix on chemical properties of the contaminant is not always correlated with toxicity or other adverse effects on living organisms. It means that direct chemical analysis of whole soil does not increase the reality of the analytical results, like in the case of direct toxicity testing.

The physical and chemical form of the contaminant decides about the environmental risk. Complete sample digestion destroys chemical species that are simultaneously present in the soil. Low-selectivity solvents are *aqua regia* (the 1:3 mixture of nitric acid and hydrochloric acid) or other strong acids such as nitric acid, perchloric acid or their mixtures for 'total' metal extraction, and a 3:1 hexane:acetone mixture for organic chemicals. 'Soft' extraction methods provide partial extracts and may be selective to some extent. These are applied for the fractionation of different chemical forms of contaminants, e.g. sequential extraction using acidic and alkali solutions. Another aim could be modelling natural leaching or washing processes, e.g., with the acetate extractant ethylenediaminetetraacetic acid (EDTA) or diluted nitric acid. Salt solutions are used for the simulation of precipitation and natural leaching. Organic acids apply to model root acidity and plant uptake. Selectivity of solvents does not agree with the selectivity of living organisms, due to their different interactions with contaminated soil. The problem is further exacerbated by the fact that the selectivity and uptake differ from organism to organism. Biomimetic solvents may simulate a quasi-average ecosystem in a reproducible way for certain contaminants in certain concentration ranges and soil types.

Binding of the contaminant into or onto the soil matrix involves sorption, ionic bonds, molecular or atomic lattice, covalent chemical bonds, as well as metal speciation. The fate of the contaminant in soil can be forecasted based on chemical properties, but the chemical properties-based model should be validated by simulation tests or microcosms. Mobility and bioavailability of contaminants in soil greatly influence their biological effects and risk. Mobility of the contaminant is often characterized by the partition between physical phases. Partition is calculated either from K_{ow} and soil sorption capacity or by measuring the equilibrium concentration of sorbed and solubilized forms. This simplified image can be refined by making the study system dynamic and measuring the impact of the change in fluxes, temperature, pH, redox potential and the biota. Mobility of contaminants in soil is influenced by (i) the groundwater flux and the diffusion of soil-air, (ii) its partition between phases and (ii) transport by or with the mobile soil phases. Bioavailability can be characterized through contaminant properties and the dynamic interactions with soil and soil organisms. Plant roots' acidic exudates compete for the metal ions with the negatively charged soil particles (e.g. clay minerals). Bacterial rhamnolipids and other surfactants increase contaminant desorption and biological availability by a shift in partitions toward the aquatic phases. This kind of partition and ion exchange, responsible for plant and microbiological uptake, is influenced by the moisture content and redox potential of the soil.

Biodegradation and bioaccumulation of the contaminant (occurring in the environment) depend on the biodegradability and bioaccumulative potential of the chemical substance. The soil type, which is responsible for the partition of the contaminant between the soil solid phase and soil moisture/pore water, also plays a role. Finally, it is influenced by the transport mechanism into the cells or tissues of living organisms.

Physicochemical methods can be refined by non-destructive, rapid procedures and *in situ* or on-site measurements (Sarkadi *et al.*, 2009). Sampling and sample preparation methods have been developed in order to separate and concentrate physical phases, providing great benefits for cases of diluted pollutants: their selective and concentrated collection increases the sensitivity of the analysis method. Another way to refine the chemical model is to increase the environmental relevance of physicochemical analytical tools. Assessing the risk to the ecosystem or humans based on total metal concentrations (complete digestion using *aqua regia*) results in an overestimate, i.e. this type of risk model is too pessimistic. Nevertheless, a chemical model should be pessimistic, albeit within a biologically relevant range (Hajdu *et al.*, 2009). If the toxic metals are incorporated into the lattice of clay minerals, they will become much less available both to the ecosystem biota and to the digestive processes of humans (Feigl *et al.*, 2009).

The measured physicochemical characteristics are compared to quality criteria expressed in terms of physical and chemical quantities, typically mass or concentration. Based on the comparison of the measured values to a reference or a criterion, the soil is classified (sandy loam, forest soil, etc.), soil status is characterized (nutrient supply or deficiency), and the pollution is assessed, comparing contaminant concentrations to soil screening concentrations (screening concentrations ensure certain safety; when they are exceeded, detailed risk assessment is necessary) or other quality criteria (see Section 7).

5.1.2   Biological and ecological methods for soil characterization

A biological model is used to measure the adverse effects of chemical substances or other types of interactions of chemical substances with biological systems such as biodegradation, biotransformation (changing molecular size, chemical form, redox state or solubility of chemicals), or bioaccumulation. Biological models are the most important tools to estimate the risks posed by adverse biological effects and to make decisions. Biological models measure the risk directly due to contaminant toxicity and not through a chemical model, based on concentrations.

The proper biological method requires scientific knowledge and some experience. The problem of biotesting is that – similar to chemical analysis – it is a model in contaminated site assessment, only certain organisms (selected as representatives) are

Table 1.2 Test types according to the origin of the sample, the contaminant and the organisms.

Test type	Environmental sample	Ecosystem	Contaminant	Environmental conditions
Bioassay for testing chemicals	Artificial, standardized test medium	Standardized test organism	Pure chemical conditions	Standardized
Simulation test	Natural water, soil, sediment	Indigenous biota	Pure chemical	Standardized/ specific conditions
DTA	Natural water, soil, sediment	Standardized test organism	Original contaminants included in the sample	Standardized/ specific conditions
Micro-/mesocosms	Standardized, artificial water, soil, sediment	Standardized test organisms	Pure chemicals	Standardized conditions
Micro-/mesocosms	Natural water, soil, sediment	Standardized test organisms	Pure chemicals	Standardized/ specific conditions
Micro-/mesocosms	Natural water, soil, sediment	Indigenous biota	Original contaminants included in the sample	Standardized/ specific conditions
Micro-/mesocosms	Natural water, soil, sediment	Indigenous biota	Original contaminants included in the sample	Natural conditions

tested, and the effect on the whole ecosystem or humans is an extrapolated value with uncertainties. The selection of the test organism(s) and the exposure scenario is a crucial part of the assessment and predetermines the validity of the test results. It has to be considered whether the higher deviation of (most of) the biological tests would compensate for the low environmental relevance of the chemical or mathematical models. In many cases, biological models are preferable, in spite of their generally poor reproducibility, which can be improved by standardized test organisms and test assemblage. Some models, e.g. the multispecies models such as micro- and mesocosms, provide environmentally relevant, reliable results, but their statistical power is even lower. The most common test types are summarized in Table 1.2.

Ecological methods may use similar bioindicators as toxicity tests, but the interpretation of the results applies a different approach: the assessed characteristics – e.g. species diversity, or key species with known sensitivity/tolerance to contaminants – are compared to the known or hypothesized indices of the healthy environment. Therefore, the biological and ecological methods need their own quality criteria and screening values for risk management. They can be applied alone or parallel to physicochemical and ecotoxicological methods. Biological and ecological models can measure the deterioration of the terrestrial ecosystem, or of the selected true representative members of the ecosystem. Consequently, ecosystem assessment should (also) be considered as a model because one has to extrapolate from partial information to the whole ecosystem. The simplification makes the derived risk values uncertain, similar to the risks based on chemical and ecotoxicological results.

The metagenome – instead of bioindicators – may represent all ecosystem members in a true proportion, but the necessary knowledge is lacking to distinguish between static genetic information and data characterizing interactions and dynamic changes. Depending on the assessment method, biological and ecological results are evaluated by statistical means. An index is generally created at the end of the process, showing the scale of deviation from the healthy environment. Specific characteristics can also be used in comparison with screening levels created specifically for the site. Such parameters may be the presence and frequency (number) of certain organisms such as collembolans, nematodes, and soil microorganisms, and soil activities such as respiration, nitrification. The absolute values of activities or species characteristics can be interpreted only if compared to the proper reference representing the healthy state of the same soil, in the same location, at the same time – which is almost impossible. Long-term records may improve the situation. Some biological/ecological soil characterization methods are:

– Biological/ecological assessment at the site or in the laboratory: species generally cannot visually be assessed or identified *in situ* because microscope and databases may be necessary;
– Qualitative and quantitative assessment of soil-living organisms such as microorganisms, plants and animals (typically insects, spiders, and rodents);
– Determination of the number of species and their distribution, i.e. biodiversity by the application of adequate bioindicators, selective biomarkers:
  o genetic, biochemical, chemical, morphological and physiological markers;
  o biomarkers for early warning and quality control;
  o monitoring the adaptation of the ecosystem and the soil microflora;
– Evaluation of the results using statistical methods;
– Interpretation of the results by creating qualitative indices.

Human and wildlife epidemiology is an important tool for assessing endangered and contaminated land. Unfortunately, the use of epidemiological data is generally limited by data availability, low accuracy, uncertainties related to individual variations. and subjective interpretation, which together generate very high uncertainties in most cases.

5.1.3 Direct environmental toxicity assessment

Direct toxicity assessment of environmental samples can integrate well-controlled toxicity measuring tests and natural circumstances. Simulation tests, direct toxicity assessment of environmental samples, microcosms and mesocosms are available for the practice. Simulation tests use pure chemicals under conditions very similar to those in nature. Direct toxicity assessment uses well-controlled test organisms exposed to real environmental samples in the lab or to the real environment *in situ*. Microcosms and mesocosms are small-scale models of the real environment in terms of their biota and conditions.

The results of direct toxicity assessment and micro- and mesocosms measured under real conditions can be used directly for decision making. The interpretation and the mode of integration of the test results into the assessment procedure should be planned on a case-by-case basis. For example, if standardized surface-water mesocosms (containing all of the relevant types of organism) give negative results for a pesticide, this pesticide can be authorized. If the formerly contaminated soil is not toxic to the soil microbes or other soil-living animals and plants, this soil can be reused after remediation without any restrictions. Biological and environmental toxicological tests use various end points, ranging in scale from DNA-specific changes to the behavior of a whole population or community (see Volume 2 of this book series, Gruiz *et al.*, 2015). The toxicity of soil or other environmental samples can be used as an indicator in early warning systems, for contaminated site assessment, integrated environmental monitoring, environmental technology monitoring and for direct decision making.

Bioassays are widely used to measure the toxicity, mutagenicity and reproductive toxicity of polluted environmental samples, waters, leachates, sediments, or whole soils (Horváth *et al.*, 1996). They can be applied to environmental monitoring, contaminated site assessment, solely to remediation technology monitoring, or combined with physicochemical analyses. The decrease of toxicity in the treated soil or an increase of the toxicity in the surrounding groundwater can serve as the main indicators when chemical analyses are not feasible. Toxicity as early warning is ideal in the case of contaminants showing toxicity at non-measurable low concentrations. The no-effect results of toxicity tests can be used as effect-based (or risk-based) environmental quality criteria or as screening values. The useful end points are the largest sample amount (volume/mass) resulting in no effect, or the smallest sample amount causing an effect. Other toxicological end points, e.g. benchmark values can be read from the sample amount–response curve. The effect-based decision making compares actual adverse effects to the no-effect target.

One of the site-specific, ecosystem-linked indicators and possible end points is species diversity, i.e., the number of species and their relative occurrence. Traditional diversity assessment is based on the identification and counting of collected or assessed species (macrozoobenthos in the sediments, plants or soil microbes in the soil, etc.). Investigation of the metagenome of the environmental compartment using DNA techniques is one of the most promising innovative methods. Both traditional and novel diversity assessments can only be evaluated versus the norm, i.e. the uncontaminated state of the same or a very similar (reference) environment. The result of this comparison can be considered as an index showing the size of risk, but its inclusion into the conventional RCR-based risk management approach has not yet been solved.

Soil toxicity tests measure the actual adverse effects directly on the organism instead of extrapolating from chemical results to ecosystem members. Depending on the test method, biological and toxicological results can either be utilized directly for decision making in risk management or after conversion into a more useful and generic parameter such as an RCR. Direct toxicity measuring methods are classified according to the following:

- Location of the toxicity test may be *in situ*, on-site or in the laboratory.
- *In situ* toxicity testing may be based on active or passive biomonitoring and on mobile and rapid bioassays which produce an immediate result.
- On-site sample preparation and application of mobile test methods/portable devices.
- Laboratory tests: sampling and delivery of samples into the laboratory may cause changes in sample redox potential, pH and microorganism diversity, but the application of standardized and sound test organisms, test assemblage and testing parameters may compensate for it.

- Sampling: heterogeneity influences toxicity measurements in a way similar to physicochemical analysis. Sampling strategy is as important as the physicochemical analysis, but the testing strategy may be different. When applying the STT approach, subsamples of the same or an already homogenized sample should be tested. When planning sampling (location, type of sampling, sample volume, preparation, delivery, storage, etc.), the physicochemical-analytical, biological, ecological, toxicological and microbiological aspects should be considered. Sample delivery and storage should prevent both chemical and biological changes which may influence the toxicity, i.e. the change of redox potential/pH and biodegradation.
- Soil phases: soil phases are separated from each other or the whole soil is sampled and measured. Alternatively, different extracts are prepared, simulating environmental processes by washing, extraction or leaching.
- Contaminant-specific responses of soil are possible to detect through the use of specific omics such as genes responsible for adaptation to substrate utilization or resistance, metabolites (e.g. methallothioneins to mobilize, or large-size proteins to immobilize metals in the cell), genes of the enzymes playing a role in the transformation or degradation of certain contaminants.

The physical and chemical form of the contaminant is the key determinant of toxicity and of other adverse effects: one chemical form may be highly toxic (Cr(VI), methylated Hg), but another form is much less harmful or completely harmless within a certain concentration range (Cr(III) oxide, mercury sulfide).

The binding of the contaminant into the soil matrix and the stability of this binding is another key determinant in the realization of adverse effects of contaminated soil. Sorption–desorption and the partition of a chemical substance between soil phases results in the partition of the toxicity (cf. Section 5.2). Partitioned toxicants endanger soil organisms, depending on whether their habitat is the free water phase or the protected capillaries. The ability of the organisms to mobilize contaminants by biosurfactants or acidic exudates also increases the risk posed by less mobile contaminants. Mobility can change in the samples during delivery and storage because while removing the sample from its natural surroundings, an increased sample surface and new external conditions (temperature, atmospheric air, etc.) can lead to a new equilibrium. This will change the contaminants' conditions and the physicochemical analysis may detect a much higher toxicity. Therefore, *in situ* measurements should be given priority.

Biodegradation and bioaccumulation are typical fate properties of the contaminants which exert their influence during the interaction with living soil organisms. The soil's microbiota starts to adapt to the new substrate immediately after the contamination event. As a result, the diversity of soil microbiota changes, and the genes of enzymes needed for biodegradation concentrate in the metagenome (the whole of the DNA, no matter which individuals or species they compose). Test methods and test organisms applicable for measuring adverse effects, degradation and accumulation are summarized below:

- Bioassays for acute and chronic toxicity, mutagenicity, and reprotoxicity;
- Field assessment, active and passive biomonitoring;
- Biodegradation, bioaccumulation, and biomagnification;

–   Microcosms and mesocosms;
–   Dynamic testing and technological experiments;
–   Test organisms:

   o   Bacterial, fungal, plant and animal test organisms and a selection of three trophic levels for representing a full ecosystem;
   o   Hypersensitive organisms for early warning and for 'conservative testing';
   o   Medium-sensitivity test organisms for representing average of the ecosystem;
   o   Low-sensitivity organisms for screening and selecting hot spots.

Statistical evaluation of the results: ANOVA (ANalysis of VAriance) or multivariate statistical methods (see Chapter 9 in Gruiz *et al.*, 2015a).

Interpretation of the results includes toxicity end points, such as EsM_{20}, EsM_{50} and NOEsM, as well as a toxicity equivalent, or the RCR or similar quantitative risk values. The tested environmental sample represents the highest toxicity, and this adverse effect is compared to the no-effect sample proportion or the proportion causing an acceptable level effect, e.g. 20% or 25% lethality or inhibition.

5.2   Chemical analysis and direct toxicity assessment of soil

The advantages of biological, genetic, biochemical and ecological models and tools are obvious. Appropriate concepts, tests and test organisms may provide a true picture of the environment. However, the quality of the biological test results should be improved which should become visible in their statistics. Proper test organisms, end points and their adequate sensitivity may ensure high quality. Very sensitive methods can be used for early warning, moderately sensitive ones for the representation of the whole ecosystem, and less sensitive methods to distinguish between average and extreme adverse effects, e.g. in the course of screening contaminated land. In addition to sensitivity, selectivity is also an important characteristic of the test organisms. Selective species can be used for monitoring single chemical substances or specific effects. Non-selective species will respond to many of the pollutants and can measure most of the adverse effects and characterize the average sensitivity of an ecosystem. *In situ* toxicity tests may significantly increase the quality of the results.

Once a contaminant has been identified and its concentration measured or calculated using transport and fate models, the risk assessor can apply a fully chemical model and compare the calculated environmental concentration to generic limit values or other quality criteria. The maximum acceptable or maximum allowed concentrations can be the targets. This simplified solution works only when a well-known contaminant with well-known effects must be managed in a relatively simple environment, e.g. in the aqueous phase of surface waters. When the environmental compartment is more complex, for example the contaminant is distributed among more environmental compartments and phases in the soil or sediment, the chemical concentration in itself may be misleading so that good practice involves the assessment of the mobile or bioavailable fraction of the contaminants, and/or the actual response of living organisms (see also direct toxicity assessment = DTA in Chapters 5.4 and 7.2). If there is a consensus between the chemical concentrations and the toxicity test results of the environmental samples, the field ecosystem response may still remain a question due to the adaptability of the ecological community which has been exposed to the contaminant(s). The field response may agree with the exposure and toxicity profile, but may be adapted to the contaminant, in spite of the high concentration and high toxicity measured by standard (unadapted) test organisms. In such cases, the causal relationship should be identified by chemical and toxicological assessment. Risk analyses may detect and aggregate all the relations between physicochemical, biological/ecological and toxicological information.

The problem of the scale of conservatism differs from that discussed in connection with chemical models because, by definition, biological models work within the biological range. This means that if a more conservative approach is needed, sensitivity of the biological method should be increased, e.g., by choosing more sensitive indicators.

5.3 The use of STT in contaminated site investigation

The soil testing triad (STT) represents a versatile tool for:

–   Toxicity screening and mapping;
–   Estimating the proven no-risk level at minimal cost;
–   The identification of sites and environmental compartments which are not toxic or otherwise risky;
–   Refined assessment where bioavailability, biodegradation, bioaccumulation and the partition of toxicity between physical phases play a role.

In addition to risk characterization, STT also points towards remediation, by:

–   Providing data for the remediation plan;
–   Monitoring remedial technologies and
–   Post-monitoring.

STT is essential in the risk management of a potentially contaminated environment. The STT results are evaluated and interpreted in an integrated way, comparing them to each other. Integrated monitoring provides a true and more detailed picture of the environment as opposed to only chemical or only ecological assessment methods. Direct toxicity assessment is particularly important in the STT because it:

–   Provides information about the effects of non-analyzed components resulting in greater safety;
–   Refers to the bioavailable/bioaccessible fraction of the contaminants, thus providing realistic information on actual effects and acute risks;
–   Increases safety by measuring objective effects using standard organisms, unlike the highly adaptable indigenous ecosystem which may get used to the pollution over the long term and appear healthy at first sight even in contaminated land.

When the *chemical approach* is adopted, chemical contaminant analysis is the main tool. DTA and biological assessment provide refined information or verify the negative chemical results (first two cases in Figure 1.15: sites with known or suspected contaminants). The decision about the risk management measure is based on measured and soil screening concentrations of the contaminants. Addition is the simplest way to aggregate the risks.

DTA and biological assessment focus on the health and activity of the soil microbiota. Ecological assessment should check the health status and sustainability of the whole ecosystem over the long term. In addition to a one-off assessment, the same 'collaboration' between STT compartments may be beneficial during the long-term monitoring of all kinds of sites: those contaminated, potentially contaminated, already remedied, or in the process of remediation, e.g. under natural attenuation. Ecological health may become the dominant target in STT, when long-term sustainability is monitored.

The *DTA-based approach* places DTA in first place, and the chemical analysis enters the management process only in the case of positive DTA results (unknown contaminants' cases in Figure 1.15). Screening of suspected or assumed pollution of unknown or multicomponent chemicals or sporadic, incidental contaminants cannot be safely monitored by chemical analytical tools, so the certainty is increased by effect monitoring based on DTA. After toxicity has become certain (confirmed by measured data), the causes and the sources must be identified by physicochemical analytical tools and the consequences by models. The decisions based on DTA results mean a comparison with the 'no-effect' or the 'acceptable scale of effect' levels. The predetermination of the acceptable toxicity is part of the management process.

The *ecological/biological health-based approach* applies for surface waters, natural conservation areas and global ecosystem studies (the last case in Figure 1.15). Long-term sustainability of soil is also in this category, where biological and ecological assessment should have priority, while chemistry may provide evidence on chemically characterizable environmental deteriorations and toxicology on adverse effects of contaminants or biological toxins. The results of the ecological assessments are expressed as indices, which can be interpreted compared to the healthy version of the same or very similar environment. Diversity results need their own 'screening values', indicated in a tolerable percentage of deviation from the diversity of a healthy environment.

There is a large variability in the size and topography of the sites, the types of sources and the distribution of contaminants. Screening and mapping is generally a task of the exploratory site assessment phase and may help to prepare, refine and validate the conceptual site risk model. The scale may vary widely, from (a few) yards to (several) miles. Existing maps and sketches, satellite images, aerial photos, normal photos, land registries and all available spatial documentation of the site can be utilized. The stepwise investigation results more and more measured physicochemical, biological, ecological and toxicological data, which can be put on the maps. Natural areas, potentially contaminated and abandoned industrial sites differ in the design of screening and mapping, as explained below.

–   Contaminated natural sites require ecological assessment of terrestrial ecosystem members (both plants and animals). They have priority in the first assessment step. When pollution is suspected, a detailed site investigation is necessary using chemical and ecotoxicological assessment methods as well as soil microbiota.
–   Potentially contaminated sites must be assessed and monitored to prove they are not contaminated when spatial planning, redevelopment and land use change are in progress. The adequacy of soil quality for a new land use can be checked by chemical methods if the contaminant is known and the presence of others can be excluded with high probability. If the contaminants are not fully identified,

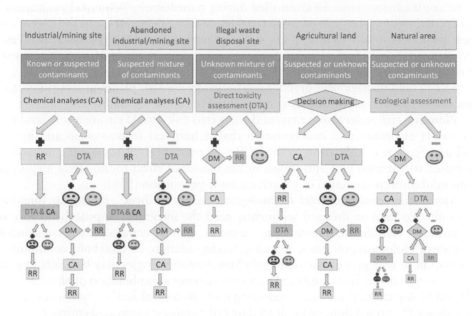

Figure 1.15 Efficient combinations of the chemical analytical and toxicity measuring methods for site investigation.

ecological, ecotoxicological or epidemiological screening must be applied. Every site and every case needs an adequate test battery, similar to the chemical analytical plan for known contaminants. If the toxicity test battery shows negative results, the soil can be considered healthy; if not, the identification of the contaminant is necessary using chemical analysis (see Figure 1.15).

– Inherited abandoned, contaminated sites must be intensively tested using chemical and ecotoxicological methods: chemical assessment helps comply with regulatory requirements, but ecotoxicity is equally important since those contaminants not included in the analytical program can be detected. The quality and activity of soil microbiota may be important information.

5.4 The use of STT data in site remediation

Site assessment to provide data for planning the remediation technology should be an integral unit with exploratory and detailed site assessment. Remediation-related assessment must identify the natural risk-reducing processes and potentials for the technology to reduce the main risk types. Soil ecology and ecotoxicology have a priority in exploring, characterizing and verifying natural risk-reducing processes because biological processes play the main role in natural attenuation based on degradation, transformation or detoxification. Physical assessment may provide data for risk reduction based on transport (dilution, partition, sorption, filtration, etc.), while chemical assessment can complement biological investigation by measuring environmental conditions (redox indicators, pH, nutrient balance, etc.). Chemical analysis can also be used for the validation of contaminant elimination. The risk posed by and the possible transport of the contaminants must be controlled during remediation. When soil is excavated and treated *ex situ*, the volume to be removed must be delineated and the urgency of excavation determined. When *in situ* soil treatment is applied, physico-chemical and environmental toxicological monitoring as well as fate and impact assessments may be necessary during remediation.

STT in land management practice can be widely used for planning and monitoring remediation as well as for quality control of the remedied site and soil.

Planning risk reduction measures the targeted risk should ensure a safe land use. The target environmental concentration that is harmless to ecosystem and human land users can be calculated from the RCR ≤ 1 criterion. RCR represents the ratio of measured to harmless concentrations, or the measured effect to 'no-effect' expressed as an inhibition (percentage) or effective sample proportion (ESP).

Monitoring environmental technologies, especially those applied for inherited contaminated land or dredged sediments, need the integrated approach because of the partially identified contaminating mixtures. STT is certainly the best solution to monitor biological processes, including biodegradation, co-metabolic contaminant elimination or emergence of toxic metabolites. Biotechnologies may be limited by the lack of oxygen or nutrients, and can be monitored using a combination of physicochemical and biological methods, e.g. measuring microbiological activity, nutrient content, microbial CO_2 production, or biodegrading cell numbers. Integrated emission control increases the safety of *in situ* soil remediation, chiefly for those technologies based on contaminant mobilization.

Evaluation of the treated soil, assessment of its quality and possible reutilization provide the basis for effect-based decision making. DTA should take priority when there is a mixture of contaminants in the soil of abandoned sites or bed sediments containing diffuse pollution. On the basis of DTA, a direct decision-making system can be established for the reuse of the soil: if the soil is not toxic to sensitive test organisms from three different trophic levels, its use has not to be restricted.

Evaluation and verification of remedial technologies rely on the results of STT and DTA and the final proof of technological efficiency is the health of the soil and the associated ecosystem, and the lack of adverse impacts on treated soil, groundwater, surface waters or sediments. Besides the toxicological verification of the technological efficiency of environmental remediation, its environmental or eco-efficiency should also be characterized by biological and toxicological methods. The eco-efficiency of a remediation technology is characterized jointly by (i) the improved quality of water and soil as habitat and (ii) the lowest possible impact of the technology application on the environment.

Planning and implementation of *natural attenuation-based* remedial technologies need information about soil ecology, the activities of the microbiota and plants, as well as about geochemical, geotechnical, topographical, pedological, and meteorological characteristics and processes. The stages of technology development are laboratory experiments, field experiments, and field demonstrations.

5.4.1 Technology monitoring

Monitoring of soil remediation is normally based on the concentrations of contaminants and/or their adverse effects. Technological parameters such as temperature, pH and redox conditions, concentration of the additives and reagents should be controlled during remediation to achieve optimum performance. Soil bioremediation can primarily be monitored by microbial activities and by measuring the decrease of the substrate and the electron donor, or the increase of the products (typically gaseous ones: CO_2, NO_x, NH_3, H_2S or NH_4), as well as the technological parameters (soil nutrient content, moisture content, pH and redox potential). Environmental monitoring is extremely important when applying *in situ* remediation to prevent uncontrolled mobilization and transport toward air and water. Technology monitoring goals are summarized below:

– Optimum performance which, depending on the type of technology, can be achieved by:
  o ecological methods measuring key ecological indicators;
  o biological methods for soil microbiota, microfauna and macroplants activities and physico-chemical methods for optimal biological activity;
  o physico-chemical assessment for physico-chemical remediation such as soil venting, pump and treat, soil washing, leaching, chemical oxidation and reduction, stabilization, thermal methods, etc. Environmental monitoring and eco-efficiency assessment of such complex technologies requires biological methods to measure residual biological activity in the soil after treatment to decide on the need for revitalization. Toxicological methods qualify the treated soil to prove its detoxification and to decide on possible utilization.
– Controlling and regulating technological parameters, which can be physicochemical (pressure, flux, temperature, etc.), biological (e.g. respiration) or ecological features (species diversity) as main parameters of the technology, or supplementary information to these main parameters, such as chemically measured nutrient content or redox indicators of the biological activity.
– Controlling emissions from the technology to keep the technological risk at a low level. Both *ex situ* and *in situ* technologies may have emission-related risks such as volatilization of volatile contaminants (higher in the case of *ex situ* than *in situ*), liquid phase releases such as pore water, leachates and groundwater (it must be controlled in both *ex situ* and *in situ* technologies). Dusting afflicts both *ex situ* and *in situ*, whereas erosion mainly affects *in situ* treated soils.
– Proving the elimination or modification of the contaminant and reaching the no-risk target: a few known contaminants are measured by chemical analysis; final verification can be accomplished by ecotoxicity testing. In the case of a large number of unknown components, ecotoxicity may have priority over chemical analysis in quality control and final verification.

5.4.2 Qualifying remedied soil

Qualifying remedied soil is the final verification phase of a successful remediation: negative ecotoxicity test results can confirmed the reuse of the soil. This type of verification is extremely important when dealing with inherited contaminated sites. Large soil treatment plants (soil recycling facilities) where the charges cannot be treated separately are exposed to high uncertainty regarding the origin and contaminant content of the soil, and thus soil quality and the allowable reuse must be verified by accurate DTA. Direct decision making based on measured adverse effects can be applied for soil reuse if the soil is not toxic for bacteria, plant and soil-living animals (three tests by test organisms of three trophic levels, or prescreening with a highly sensitive test organism). Mutagenicity should also be excluded. Uncontaminated soils can be utilized for multifunctional use without restriction or any regulatory concerns. Soils with low-level contamination or toxicity can be utilized as roadway sub-base material or landfill cover. Soils with significant contaminant content and/or toxicity should be considered as hazardous waste and treated, recycled or disposed of.

5.4.3 Verification of the technology

Verification of the product of remediation – the non-hazardous soil – is only one segment of remedial technology verification. The good quality of the product is a partial proof of technological efficiency. In addition to the narrowly defined technological efficiency (i.e. that the technology is working and producing the required product), eco-efficiency and socioeconomic efficiency are essential components of technology verification. A classification of remediation technologies to be verified is therefore much more complex, and, in addition to STT, it needs economic and social metrics and the aggregation of all these in the final step. Eco-efficiency (ecological efficiency) covers the local, regional, and global risks due to emissions, energy, and material use. Economic efficiency measures time and energy requirement, as well as socioeconomic costs and benefits (including human health risk), and tries to find the common denominator (e.g. monetization) for the two. Physicochemical, ecological and toxicological monitoring of the technology fulfils the data requirement of technological and eco-efficiency. Socioeconomic issues require other tools than STT.

6 EVALUATION AND INTERPRETATION OF SITE INVESTIGATION DATA

The collected and measured data must be evaluated and interpreted to extract the information they hold.

6.1 Evaluation tools

Evaluation and interpretation depend on the aim and methodology of the assessment and can be based on the approaches discussed below.

Visual evaluation and verbal description of a site assumed to be contaminated can be a perfect solution in the preliminary assessment phase. A professional who knows the site and the area must inspect the site and interpret his/her observations. The geological, hydrological, hydrogeological, biological, and ecological site characteristics and their association with toxicity indicators and other signs/symptoms of deterioration may provide sufficient information for the initial conceptual model of the site and for a qualitative or semiquantitative risk characterization.

Mapping the observed or measured characteristics may clear up further connections between site characteristics and contaminants with regard to topography, land cover, land uses, neighboring facilities, etc. Spatial trends can be discovered and quantitatively or semi-quantitatively characterized with the help of mapped data.

Applying the *chemical approach*, identification of the contaminant(s) is necessary to properly interpret measured or observed data. The concentration-based *transport and fate model* describes the distribution of a contaminant at the site. Modelling is based on i) the chemicals' intrinsic properties, environmental fate and behavior, (ii) the characteristics of the environment, as well as (iii) the interaction between the two. More sophisticated models include the impact of soil biota, primarily as biodegradation estimates. The source and distribution pathways of point sources are identified by 2- and 3-dimensional transport and fate models, while the transport of diffuse pollution is supported by GIS-based information and digital maps. Transport models are applied to individual contaminants. The predicted concentration is then compared to the predicted no-effect value or to the screening values.

The most complex interpretation of environmental data can be accomplished by a stepwise quantitative risk assessment that includes (i) the transport model to calculate the predictable environmental concentrations and (ii) the exposure models to calculate the exposure of or the intake by the receptors.

When the *DTA-based approach* is applied, mapping of the measured toxicity helps to identify sources, transport pathways and the trends of change based on spatial and time series. The interpretation of the results includes the comparison of the measured toxicity to the no-effect sample proportion.

Bioassay results expressed as a toxicity equivalent – compared to a 'calibrating' or 'equivalencing' chemical substance and termed as equivalent concentration – enables chemical and toxicological models to converge for better interpretation of environmental risk (see Chapter 9, Volume 2 of this book series, Gruiz *et al.*, 2015a).

Biological and ecological assessment may provide useful data for identifying the type and extent of damage. If the observable signals occur early enough, the risk can be assessed before irreversible deterioration. The chemical and the direct toxicity approach can corroborate and refine the image based on biological indicators, and their combination may enable an optimal site investigation.

When assessing contaminated sites, two cases should be distinguished:

– If the contaminant has already been identified or its occurrence verified, the assessment of substance-specific data, site characteristics and data on site-specific interactions will be similarly weighted.
– In the case of inherited contaminated site, or illegal waste disposal sites with unidentified chemical substances, but observable environmental responses, site-specific interaction- and intervention-related characteristics will dominate the assessment (see Figure 1.9).

If the adverse effect has already been proved, the number of tiers can be reduced, and *in situ* applicable assessment tools (remote sensing, *in situ* emplaced sensors, and direct push technologies) can be applied. These types of tools enable the execution of a complete site assessment in a few hours or days using physicochemical characterization, chemical analyses and toxicity tests. *In situ* dynamic tests can simulate the changes under conditions other than the original, including the impacts of interventions (e.g. push and pull or substrate-induced soil respiration studies). The cost of a one-off complete assessment may be higher than that of the tiered one, however additional costs can be compensated by the benefits listed below:

– Complete information retrieval in one or two steps;
– *In situ* corrections in sampling/assessment plan (providing a better fit to the site characteristics);

– The possibility of implementation of immediate risk-reduction measures;
– Urgent utilization of the site.

Inherited abandoned sites with unknown history can be directly tested for toxicity to draw a toxicity map and delineate the site in the exploratory phase. According to the scale of risk, further obligations/actions can be identified such as no intervention, detailed assessment and contaminant(s) identification, or immediate risk mitigation or risk reduction.

6.2   Comparison of chemical analysis and DTA results

The comparison of chemical analysis and DTA results may clarify the details of the 'black box' of the soil, help refine risk assessment and support risk management. Chemical and DTA results can be compared where there is a common denominator such as the risk characterization ratio (see Section 7) that can be interpreted for both. Another option is to express the results of DTA as a toxicity equivalence value.

The relationship between chemical analysis (**CA**) and toxicity (**DTA**) results can be corresponding or contradictory. The following cases may occur:

– **CA ≈ DTA**: chemical and toxicological results agree.
  o   CA+ and DTA+: both sets of results are positive: there is a high contaminant concentration and a strong negative effect, thus a *high risk*.
  o   CA– and DTA–: both sets of results are negative: there is either no contaminant, or it is present at a low concentration. There is no measurable effect, thus *no* or *low risk*.
– **CA+ and DTA–**: high chemical concentration measured by analysis, but no adverse effect on the test organisms and/or the whole ecosystem.
  o   The contaminant is present but is not toxic to the test organisms, i.e. there is *no toxic risk* due to adverse effects.
  o   The contaminant is present, but it is neither mobile nor bioavailable: there is *no short-term* risk, neither due to air nor water transport nor biological uptake. However, this situation does not exclude the presence of a *chemical time bomb* type risk, i.e. a latent risk which manifests itself under certain conditions, e.g. mobilization when redox potential or pH increases.
– **CA– and DTA+**: the contaminant is chemically not measurable or its measurement has not been included into the analytical program, but the soil shows strong toxicity.
  o   The contaminant is very toxic even at low, chemically non-measurable concentrations, i.e. there is a *high risk*.
  o   The toxic substance is present but was not included in the analytical program, i.e. there is a *high risk*.
  o   No analytical method is available to identify and determine the nature of the contaminant(s), i.e. there is a *high risk* posed by unknown compounds.

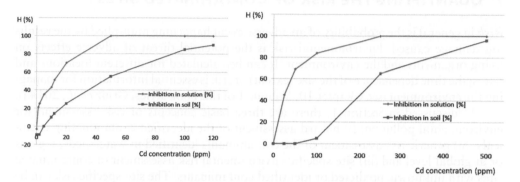

Figure 1.16 Toxicity of the same concentrations in the pore water and in the solid phase soil, measured by luminobacterium *Aliivibrio fischeri* (left) and white mustard *Sinapis alba* (right).

6.3 Integrated assessment of environmental phases

A comparison of the toxicity of dissolved and adsorbed contaminants, combined with parallel testing of the whole soil/sediment and pore water (alternatively leachate or water extract) may provide further details about the nature of the risk.

The *toxicity buffering capacity* of soils/sediments can be characterized by comparing the toxicity of contaminants in dissolved and sorbed form. The toxic effect–concentration curves are different for dissolved and adsorbed contaminants; the area between the two curves is proportional to the adsorption capacity of the soil, as shown in Figure 1.16. The toxicity of toxic metals adsorbed onto a soil with high clay content, for example, was only 1–20% of their toxicity in dissolved form.

The *partition of toxicity* between physical phases can be characterized as toxicities in the pore water (**P**) compared to the toxicity in saturated soil or whole sediment (**S**). Based on the measured data for the Danube river sediment, the nature and fate of the pollutants and their environmental risk can be characterized as follows (Gruiz *et al.*, 1998):

– P+ and S+: the pollutant is toxic and mobile/bioavailable; the partition between the solid and the pore water results in toxicity in the pore water, thus K_p ($= c_{solid}/c_{water}$) is low.
– P+ and S−: the pollutant is toxic and highly mobile/bioavailable; it is characterized by a very low K_p, high water solubility and, as a result, it is mainly present in the pore water. Extreme pH and intensive ion exchange may lead to such a situation.
– P− and S+: the pollutant is toxic, not water-soluble, but bioavailable. Bioavailability may be due to biotensides (microbes), cell or tissue exudates (fungi, plants), or sediment digestion (sediment-dwelling animals).
– P− and S−: the pollutant is either non-toxic or non-bioavailable; it is necessary to compare the data with the chemical analytical results. If both are negative, the sediment is not toxic.

7 QUANTIFYING THE RISK OF CONTAMINATED SITES

Risk in general is the probability of an adverse event happening multiplied by the resulting damage caused. Environmental risk is the potential threat of adverse effects on living organisms and the environment. It can be calculated for different locations and dates; the time dependence of the size (extent) of risk is essential information for managing the environment (see Chapter 10, Volume 1 of this book series, Gruiz *et al.*, 2014a).

As discussed in Section 1, there are three basic concepts of risk assessment for environmental pollution: (i) hazard assessment of the identified contaminating chemicals, (ii) generic risk assessment of the contaminants identified in watersheds, regions or at global level and (iii) site-specific environmental risk assessment of contaminated sites with unknown, predicted or identified contaminants. The site-specific risk can be estimated by the (a) aggregated risk of individual contaminants, (b) diversity, function or service indicators that characterize the health status or extent of deterioration of the ecosystem, (c) toxicity of environmental samples measured directly by representative biological or ecological indicators, and (d) mixture of the three approaches in one or several assessment steps.

7.1 Risk characterization using the chemical model

There is a clear difference between generic and site-specific risk of contaminants: generic risk refers to a single contaminant or a mixture of known contaminants. However, the situation is more complex in the real environment: most of the pollutants is a mixture whose components have different environmental fate and effect properties. Every chemical component has its own evolution in the environment which is specific to the site and would be different on another site. In those rare cases when one single contaminant is present in the environment, 'only' the environmental conditions, the natural air/water/soil components and the living compartment should be considered that can interact with the contaminant, resulting in several physical and chemical forms with variable environmental fate properties (mobility, degradability, toxicity etc.).

A sufficiently reliable value of exposure, e.g. *predicted environmental concentration* (PEC) can be compared to any of the default values such as the maximal concentrations acceptable or tolerable by the receptors, or the guaranteed no-effect concentrations or levels for humans, the ecosystem, or certain members of the ecosystem. The acceptable or the 'no-effect' value should also be estimated or a default value used, – such as the predicted no-effect concentration (PNEC), or the acceptable daily intake (ADI) – in order to quantify the risk. It is 'predicted' because the entire human population or the whole ecosystem comprising countless members with different sensitivities cannot be assessed one by one. Indicator doses and levels from PEC can be derived depending on exposure pathways and exposure parameters; for example, an *ingested dose* (mass) of contaminant can be calculated/estimated from the PEC of water or food crops, taking into account the amount consumed or taken up and metabolized.

The *daily or weekly intake* of humans is the most common parameter, and it can also be used for some protected members of wildlife. The inhaled air, ingested water and food as well as the dermal uptake can be determined and used as a specific value for the groups of the population exposed differently (children, pregnant women, workers, etc.). The risk characterization ratio is the ratio of the dose inhaled, ingested, or taken up dermally (D: mg/kg bw/day) to the acceptable daily intake (ADI: mg/kg bw/day) or to the tolerable weekly intake (TWI: mg/kg bw/week) (see also Chapter 7, Volume 1 of this book series, Gruiz et al., 2014):

$$RCR_{human\ intake} = D/ADI \text{ or } RCR_{human\ intake} = D/TWI$$

The dose taken in is calculated from the environmental concentrations and the generic exposure parameters (inhaled air volume/day, consumed water and food mass/kg body weight/day, contact time with and adsorption from contaminated water and soil, etc.). As an example, the RCR of mercury in fish is calculated from mercury concentration in fish and the tolerable weekly intake:

$c_{Hg} = 0.150$ mg total Hg/kg fish (tuna, mean) (FDA, 2014)
Fish consumption: 17 kg/year (world average) and 58 kg/year (Portugal or South Korea)
Yearly Hg intake via fish: $D_{Hg,17} = 2.55$ mg/year $D_{Hg,58} = 8.7$ mg/year
$TWI_{Hg} = 4\,\mu g/kg$ bw/week (JECFA, 2010);
Body weight $= 70$ kg $TWI_{Hg} = 14.6$ mg/70 kg/year
$RCR_{17\,kg/year} = 2.55/14.6 = 0.17$ $RCR_{58\,kg/year} = 8.7/14.6 = 0.6$

The RCR value calculated for mercury intake via fish consumption should be summed up with other mercury contaminated intakes, i.e. drinking water, food and inhaled air. The aggregated RCR should remain under 1, so people with high fish consumtion have a very narrow safety margin for mercury intake other than fish.

Non-threshold chemicals, such as carcinogens and endocrine disruptors may pose high risk in any small concentration, so their 'no-effect threshold' cannot be defined. Several approximations are used to find a reference value for comparison. Such value is the CSF, the *cancer slope factor*, $(mg/kg/day)^{-1}$, which gives the lifetime increase of cancer risk (percentage) due to a certain dose of a chemical substance (mg/kg body weight/day). It is determined by linear extrapolation from the point of departure to 0% response. To characterize the risk, the lifetime average daily dose (LADD: mg/kg/day) is multiplied by the cancer slope factor and one obtains the excess lifetime cancer risk (ELCR):

$$ELCR = LADD \times CSF$$

The acceptable excess cancer risk of 10^{-5} is generally specified for non-threshold chemicals.

As an example, the ingestion exposure to nitrosamines in chlorinated drinking water of the Korean Chuncheon is shown quoting the paper of Kim & Han (2011).

$LADD_{oral,\ NDMA} = C_w \times IR \times EF \times ED/bw \times LT$, where
$LADD_{oral}$ is the lifetime average daily dose from ingestion exposure (mg/kg/day);
C_{NDMAW}: concentration of NDMA in chlorinated tap water (52.9 ng/L in average);
IR: average ingestion rate (1.5 L/day);
EF: exposure frequency (365 days/year);
ED: exposure duration (80.1 years);
bw: body weight of a reference Korean adult (62.8 kg);

LT: expected lifetime of a Korean person (80.1 years × 365 days/year).
$LADD_{oral, NDMA} = 1.42 \times 10^{-6}$ mg/kg/day for an average NDMA concentration.
$ELCR = LADD \times CSF$, where
CSF: oral slope factor $= 51$ (mg/kg/day)$^{-1}$ (IRIS, 2010)
$ELCR = 1.42 \times 10^{-6} \times 51 = 7.2 \times 10^{-5}$. This estimate exceeded the baseline risk
of 10^{-5} which is usually considered acceptable.

The *risk to the ecosystem posed by chemicals* can be given by the ratio of the
contaminant concentration to the no-effect concentration:

$RCR = PEC/PNEC.$

Normally, a multipathway exposure and contact between the living organisms and
the environment is assumed for the ecosystems. An RCR in this case specifies how
many times the no-effect (environmental) concentration the predicted environmental
concentration is, without measuring or calculating the amounts inhaled, ingested, or
taken up dermally.

An example is the Toka Valley (Northern Hungary): both agricultural and natural
land are contaminated with mine waste. The average Cd concentration was 3.2 mg/kg
soil of a hobby-garden and 2.2 mg/kg soil of the natural area. The predicted no effect
concentration (PNEC) of Cd for agricultural soil varies between 0.4 and 1.4 mg/kg
referring to various regulatory and scientific resources. The Hungarian threshold
value is 1 mg/kg. Four different no effect values are recommended by the OSWER
Directive (Eco-SSL, 2005) to protect terrestrial plants: 32 mg Cd/kg, invertebrates:
140 mg Cd/kg, avian: 0.77 mg Cd/kg and terrestrial mammals: 0.36 mg Cd/kg soil.

Using the different PNEC values to calculate RCR for agricultural (hobby-garden)
soil, the RCR was above 1 in all cases, representing high risk:

– Based on the range of 0.4–1.4 mg/kg: $RCR_{Cd\,agric} = 3.2$ mg/kg/1.4–0.4 mg/kg $=$
  2.3–8.0 $>> 1$
– Based on the HU threshold of 1 mg/kg: $RCR_{Cd\,agric} = 3.2$ mg/kg/1 mg/kg $= 3.2 >> 1$

Terrestrial ecosystem is also highly endangered as demonstrated by the $RCR = 6.1$
value. RCR estimation was based on the most sensitive taxonomic group of terrestrial
mammals:

$RCR_{Cd, eco, mammals} = 2.2$ mg/kg/0.36 mg/kg $= 6.1 >> 1$

ADI and PNEC are generic, effect-based reference values, and constitute conservative estimates. ADI is a human reference value based on toxicological and/or
epidemiological studies, while PNEC represents a safe concentration for the aquatic
or terrestrial ecosystem as a whole. These are highly conservative, non-site-specific
values, but serve as default environmental quality criteria (EQC) used as screening values and are integrated into national or regional legislation. These generic EQC values
can be applied to every area and site in spite of the fact that they are not completely
valid for any site. In the detailed or refined site assessment and risk assessment phases,
site-specific quality criteria can be created based on specific land uses and unusual
habits as well as specific receptors and exposure parameters. For example, the quality

criteria for mercury and biomagnifiable organic contaminant content in fish must be lower in Portugal than in Hungary because an average Portuguese consumes 58 kg fish annually, while a Hungarian person only 5–6 kg/year. Criteria for metal content in a hobby garden soil (meeting the total consumption of the family) must be lower than in average agricultural soil.

7.1.1 Soil quality criteria (SQC)

Risk-based soil quality criteria (RBSQC) may be derived from ADI (for humans) and PNEC values (for the ecosystem) which are the results of a chemical approach: the estimation of one single chemical concentration in the environment or a dose intake by humans, which does not pose an unacceptable risk. If there is an exposure to several chemicals and other adverse effects of non-chemical nature, the different risks must be aggregated, and the target quality criteria for one single chemical substance depend on the other agents present, something that is not considered when assessing the risk of a contaminated site.

A number of soil quality criteria are used in practice such as soil quality standards, guideline values, contaminant thresholds, screening values, target levels, intervention values, clean-up standards, cut-off values, limit values, etc., enabling a distinction between acceptable and non-acceptable soils or assigning certain risk levels to the soils (Carlon, 2007).

Concentration-based quality criteria are highly variable all over the world, resulting in significantly different generic quality criteria. Carlon & Swartjes (2007b) give a comprehensive overview of the topic. A collection of national guidelines and standards can be found on the website of the ESdat environmental database system (ESdat.com, 2015).

Some old type SQCs are based on expert judgements, others are taken over from other countries. Many European countries tried to adapt the 'old' Dutch List in the 90s with little success because it was based on multifunctionality, which could not be adhered to in practice. Even the Dutch developed the concept further, and differentiated screening values, remediation target values and site-specific guideline values for contaminated site remediation. Today most European countries apply these quality criteria in contaminated land management. A *screening value* is defined as a chemical concentration in environmental media below which no additional regulatory attention is warranted. If chemical concentrations at a site exceed the screening values, additional investigation and evaluation of that chemical and/or remediation may be warranted (ITRC, 2005). The soil remediation *intervention values* indicate a level of contamination above which there is a serious case of risk due to soil contamination (VROM, 2000), thus further assessment and risk reduction is required. *Site-specific quality criteria* are based on the site-specific risk, mirroring the local sources, transport, and fate of the chemical compounds. *Site-specific remediation target* is a value that should guarantee the safety of local land uses by local human and ecological receptors.

Soil quality criteria are generic or site-specific risk-based values enabling the assignment of different risk levels (multifunctionality, human risks, and natural protection) or just a distinction between acceptable and non-acceptable soils for certain purposes. Human health has priority when creating and using SQC for soil, and they are exclusively applied to groundwater used as drinking water. The terrestrial ecosystem is equally considered in more demanding regulations.

In addition to the above threshold values, background concentrations are important data for contaminated site management. Background values are sometimes applied for screening, but since they are not risk-based, their use may lead to costly overprotection. In management practice, screening values based on analytical detection limits are also used, but they do not represent the scale of risk.

Carlon and Swartjes (2007a) distinguished three functional classes based on the management activity that the criterion belongs to: (i) prevention, i.e. ensuring long-term sustainable soil quality (most stringent criteria), (ii) warning, i.e. performing more detailed assessment, specific studies, (iii) rehabilitation, i.e. triggering intervention, e.g. risk assessment or risk reduction (highest threshold values).

A general problem with existing soil and groundwater criteria is that they are independently created and not harmonized, in spite of the fact that these two values are always dependent on each other.

SQCs are differentiated in some refined systems, e.g. according to soil types based on the differences in their toxicity buffering capacity. Bos *et al.* (2005) determined maximum permissible concentrations (MPC) for Zn in the three main Dutch soil types. Differences in background concentrations between soil types were up to a factor of 7.5, whereas the 'maximum permissible addition' (compared to the bioavailable fraction of the background value) varied by up to a factor of 3 between the soil types. This resulted MPCs that vary from 44 mg Zn/kg soil to 208 mg Zn/kg soil for the three predominant soil types.

Another important parameter providing differences in risk-based SQCs is land use: there are less demanding land uses (industrial, commercial), and highly sensitive ones (hobby gardens, kindergartens, natural protection areas, etc.) requiring stricter SQCs. The Dutch regulation for example differentiates between the maximum concentrations for residential and for industrial land uses. In addition to these, an intervention value is also fixed. These quality standards for Cd are the following: residential: 1.2 mg Cd/kg, industrial: 4.3 mg Cd/kg and intervention: 13 mg Cd/kg. The average background concentration is 0.6 mg Cd/kg.

Some further practical 'rules' in the use of EQC are:

–   Concentration- or level-based environmental quality criteria are only valid for one chemical substance; the number of contaminants determines how many assessments are necessary.
–   When calculating risk and target concentrations (assigned to 'no risk'), all contaminants, all exposure pathways and all types of adverse effects should be aggregated.
–   The risk reduction measure should address all contaminants with concentrations above the screening/target value.
–   The intervention values triggering detailed site assessment or risk reduction are much above the target values and indicate that regulations tolerate a certain level of risk and the socio-economic implications.
–   The individual target quality criteria can be above or below the generic EQC.
–   Risk reduction (RR) other than remediation, is acceptable and recommendable. For example, a brownfield should not be remedied to multifunctionality, but

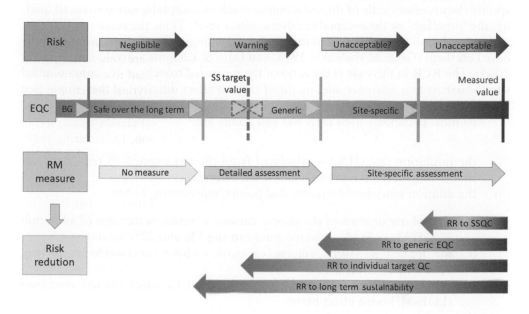

Figure 1.17 Environmental quality criteria (EQC) vs. risk management (RM), risk reduction (RR), and risk levels.

designed to become an industrial park, where no sensitive land uses are permitted. This represents the risk reduction option of restriction in land use.

Figure 1.17 shows the environmental quality criteria on the horizontal axis, representing (i) background levels (BG), (ii) relative low risk levels fulfilling long-term sustainability of the soil, (iii) the risk-based generic quality criteria (typically based on human health risk) and (iv) the site-specific quality criteria (SSQC). SSQC are in most cases less stringent compared to the generic criteria because the generic EQC is a rather conservative estimate. Nevertheless, the SSQC of a very sensitive site may be lower than the generic QC. The target EQC may arbitrarily be defined considering arguments other than human risk: it is symbolized by the moving dotted-line and arrows.

A sophisticated management scheme requires a large number of SQCs (special values for clayey and sandy soils, acidic and non-acidic soils etc.). This approaches a scenario of fully site-specific criteria, and is necessitated by just one single chemical substance.

7.2   Risk characterization applying the direct toxicity model

Direct toxicity assessment (DTA) enables the characterization of the aggregated effects of unknown contaminants on organisms typical in aquatic and terrestrial ecosystems. It measures the effect of complex mixtures of chemically and toxicologically different chemical forms and species, showing variable bioavailability and interactions typical in the environment. It characterizes the health and environmental services (supporting and regulating services) of the habitat. The adverse effects measured by DTA directly specify the necessary scale of the reduction to reach an acceptable environmental quality, the 'no-effect' or the acceptable adverse effect level. Thus, the results of DTA are in direct relation to risk, and therefore decision making and risk management can be based on them (Gruiz & Vodicska, 1993, van Dam & Chapman, 2001, Tinsley *et al.*, 2004). The RCR in this case is the ratio of the measured toxicity at the contaminated site to that (i) at a reference site, or (ii) of the 'no-effect dilution' of the sample (see Chapter 9, Volume 2 of this book series, Gruiz *et al.*, 2015a).

The most commonly used DTA test end points are:

(i)   the inhibition rate (H%) – calculated from the test organism's response scale in the sample, compared to the non-toxic reference;
(ii)   the dilution series-based toxicity end points, representing either:

    a.   the volume or mass of the sample causing a certain percentage of inhibition (e.g. EsM_5 or EsM_{20}: sample mass causing 5% and 20% inhibition);
    b.   the highest measured volume or mass without measurable effect (e.g. NOEsM: no-effect mass);
    c.   the lowest measured volume or mass found to affect the test organism (LOEsM: lowest effect mass).

The *inhibition rate* is recommended to be used for samples with low toxicity or to monitor processes where toxicity is expected to decrease. If toxicity is high, dilution series should be tested, the dose–response curve plotted and a certain maximum inhibition rate (%) should be specified as a screening level. Samples which do not fulfil the 0% inhibition criterion can be characterized by the risk characterization ratio based on DTA results (RCR_{DTA}):

$$RCR_{DTA} = \text{sample toxicity/reference toxicity}$$

The necessary risk reduction rate (RRR) may be the same or may differ from RCR_{DTA} because it is calculated as RRR = sample toxicity/target toxicity. The targeted toxicity can be measured as inhibition rate: H%. It can be 0% but it can be determined optionally as 5, 20, 25 or even higher percentages depending on land use, receptor sensitivity and the management concept.

As the effective amount of the sample (e.g. EsM_{20}) is test-dependent, it cannot be used for the characterization of toxicity in this form. Instead, the *rate of dilution* (e.g. 10-fold) or the *effective sample proportion* (e.g. 60%) of the water or soil which causes the maximum acceptable inhibition (e.g. 0% or 20%) is used. These values are equivalent to the necessary attenuation/reduction rate of sample toxicity to reach the targeted effect level. Thus, the same quantitative end point characterizes the toxicity of the sample and the scale of the risk reduction.

The DTA based risk characterization ratio (RCR_{DTA}) gives the relative risk of the sample, compared to the (non-toxic) reference.

$$RCR_{DTA} = ESP_{sample}/ESP_{reference}$$

where ESP is the sample proportion in percentage, resulting in no inhibition. RCR_{DTA} equals the necessary risk reduction rate to reach reference i.e. no inhibition:

$$RRR_0 = RCR_{DTA}$$

If the acceptable inhibition rate is 20%, the risk reduction rate equals the necessary toxicity attenuation rate to reach < 20% inhibition:

$$RRR_{20} \leq RCR_{DTA}$$

Examples for the DTA-based risk assessment will be introduced in Section 7.2.1.

Another option for the interpretation of DTA results is the *equivalence method*. It enables toxicity originating from unknown mixtures to be converted into the concentration of an equivalently toxic reference substance chosen for comparison. This procedure called *equivalencing* is more of a practical support tool than a precise analytical method because the user cannot decide whether or not the components are additive, whether their dose–response curves run parallel (they typically do not), or if they have the same or different effect mechanisms – since the contaminants have not yet been identified. In spite of all this, our experience is that equivalencing can be very useful in tiered risk assessments for screening large contaminated areas with similar contaminants and for classifying a great number of environmental samples tested using different tests. An additional benefit is that the equivalence method enables the translation of environmental toxicity results into the language of chemistry – by converting toxicity results into concentrations – to make them understandable for non-ecotoxicologists and other professionals whose knowledge is based on the chemical model. The standardized calibration tools used for equivalencing may also act as a toxicity reference enabling the control of the test organisms, their biological status and sensitivity, and the repeatability of tests. Different test organisms can also be compared to each other (see more in Chapter 9, Volume 2 of this book series (Gruiz *et al.*, 2015a)).

In summary, DTA in contaminated land assessment can be used for:

– Non-targeted screening of uncertain contaminants and site history. It is the assessment of general toxicity for the water and soil ecosystem;
– Non-targeted screening applied to mixtures with known contaminants, but with unknown or non-existing screening concentrations;
– Targeted screening of known contaminants or specific effects, e.g. using contaminant-specific or effect-specific biosensors;
– Integrated evaluation by parallel chemical analysis and biological/ecological and toxicological characterization.
– Equivalencing of an unknown contaminant or mixture of contaminants to a representative species of known quantity.

The integration of DTA results into environmental management is the prerequisite of the widespread application of *in situ*, rapid effect-measuring methods which enable efficient contaminated site investigation, immediate decision making and intervention. This leads to a higher-level understanding of environmental risks and currently used models.

The advantages of DTA could be utilized widely by using the above introduced toxicity-based evaluation methodology, i.e. comparing sample toxicity to DTA-based environmental quality criteria. Quantification of the risk based on DTA results may help to overcome the currently existing obstacles i.e. the measured toxicity cannot be expressed in concentration, thus it does not fit into the chemical model-based risk assessment (RA) procedure and regulatory screening concentrations cannot be applied either.

7.2.1 Examples for DTA application in risk assessment

Direct Toxicity Assessment (DTA) is the most adequate method for measuring the environmental risk (toxicity) posed by discharged or disposed wastes (contaminated water or soil) by assessing their toxicity using standardized toxicity tests or test batteries. Aquatic toxicity tests (algae, daphnids, oyster embryos, fish, etc.) are used for wastewaters discharged into surface waters, and terrestrial test organisms (bacteria, collembolans, spiders, ants, plants, etc.) are applied to contaminated soil or solid waste material received by soil. To plan the acceptable load, the highest concentration not yet causing measurable adverse acute effects in the receiving water (effluents) or soil (waste disposal) is referred to as the *toxicity threshold*, expressed as a percentage (volume effluent/volume receiving water or kg waste/kg receiving soil). Contaminated soil is often used by the ecosystem and humans, thus the risk-based threshold is not based on a load forecast (such as for effluents and wastes to be disposed), but should be read directly from the soil amount–response curve. This curve is identical to the concentration–response or dose–response curves, but instead of contaminant concentration or dose, the sample volume or mass is represented on the horizontal axis (see also Chapter 9 in Volume 2 of this book series, Gruiz *et al.*, 2015a).

The inclusion of DTA into policies has started in the 1990s aiming to increase surface water quality by controlling wastewater discharge in the US (US EPA, 1995 and 2002), in Australia and New Zealand (van Dam & Chapman, 2001; ANZECC/ARMCANZ, 2000) and in the UK (Hunt *et al.*, 1989). New Zealand laid down standard methods for whole effluent toxicity testing as far back as in 1998 (Hall and Golding, 1998), the UK in 2000 (UK DTA, 2000a and 2000b) and the Environment Agency UK (EA) harmonized its DTA protocol with the European IPPC Regulation (Integrated Pollution Prevention and Control) in 2006 (DTA in PPC, 2006). Three aquatic test methods were generally applied: oyster embryo development (Oyster test for DTA, 2007) and growth inhibition of two algal strains (Algal test for DTA, 2007 & Algal test for DTA 2008). The Monitoring Certification Scheme for Direct Toxicity Assessment was issued by EA in 2010 (MCERTS, 2010). DTA is recommended for (i) effluent screening and characterization, (ii) monitoring effluent toxicity against a toxicity limit, (iii) assessing the impact of point source discharges on receiving waters, (iv) providing a general quality assessment of receiving waters (for example, within monitoring programs). Canada (2012) recommends specified toxicity-based screening levels (see more in Gruiz *et al.*, 2016).

After US EPA had published methods for direct effluent testing and the use of the toxicity results in environmental risk management (Weber, 1991), intensive research and application began which provided several new ideas and test methods for rapid *in situ* DTA of waters and wastewaters. These new methods include the modification of conventional laboratory bioassays such as the rapid Microtox with lyophilized test organisms (Weltens *et al.*, 2014), or the rapid algal bioassays and other microbiological responses (Baran and Tarnawski, 2013). Contaminant-specific detection methods appeared based on omics and genetically manipulated organisms carrying substance-specific promoters (regulator genes) (Köhler *et al.*, 2000). DTA is especially important in the risk management of contaminated soils, given that the actual effects and risks in soils can hardly be extrapolated from the toxicity of pure chemicals (Fernández *et al.*, 2005). Direct toxicity tests for soils are also known as contact tests, solid-phase tests, direct contact biotests, interactive bioassays, etc. (Kwan, 1993; Campbell *et al.*, 1997; Chapman, 2000; Gruiz, 2005).

In spite of significant developments, DTA could not make a breakthrough, has not become widespread in risk-based environmental management of contaminated sites, and remains second to the chemical methods. Some of the reasons are that professionals were accustomed to the chemical approach; concentration-based quality criteria can be used almost everywhere, while toxicity-based end points are not uniform and difficult to interpret for non-ecotoxicologists. This is why decision makers do not trust DTA-based information.

7.2.2 DTA for effluents/wastewaters

For industrial effluents endangering surface waters, the European IPPC requires direct toxicity assessments. Among others, the UK Environment Agency prepared guidance for the DTA in PPC Impact Assessments (Leverett, 2006), where the advantages, the methods and the use of DTA for environmental management are summarized. The UK guidance describes a three-tiered risk-based approach in which whole effluent ecotoxicity data are used as a 'trigger' to investigate and reduce effluent toxicity. The three tiers are (Leverett, 2006):

– Toxicity screening and estimation of risk;
– Toxicity characterization and refined risk assessment by further toxicity testing or chemical analyses;
– Toxicity reduction, if necessary.

The target value in the effluent cannot be higher than the acceptable risk in the recipient. Where unacceptable risk in the receiving water is predicted, it is important to specify the level of toxicity that would be acceptable for a particular discharge.

ECORIVER (2002–2005) – a European Life project on ecotoxicological evaluation of municipal and industrial wastewaters – proposed a selected test battery with bacteria, algae, and crustaceans for wastewater DTA. The project demonstrated that an integrated approach including DTA is an effective strategy for minimizing ecotoxicological pressure on the environment. It plays a role in the risk assessment and control of discharges, protecting aquatic life and retaining the good ecological status of the receiving waters. The project proposed toxicity criteria to be included in national level legislation.

The Environment Agency UK recommends freshwater and marine algal, crustacean and oyster test organisms for direct toxicity assessment. The concept of whole effluent toxicity testing was further developed by the Energy Institute (2010), which utilized DTA for petroleum refinery effluent toxicity assessment.

7.2.3 DTA for sediments

The US EPA contaminated sediment management strategy (US EPA CSMS, 1998) is aimed at developing scientifically sound sediment management tools for use in pollution prevention, source control, remediation, and dredged material management. Standardized bioassays and field-based testing are required to evaluate the bioaccumulative and toxic potential of sediments. Standard test protocols are used in a hierarchical, tiered manner from simple acute toxicity assessments to chronic and sublethal test end points.

Three sediment testing scenarios can be considered:

– Prevention of the water phase from adverse effects due to contaminated sediments. – The applied technique includes the separation of pore water or the preparation of an aqueous leachate from the sediment for modeling contaminant partition between water and sediment. The pore water is tested using aquatic organisms such as algae, crustaceans, and fish to measure aquatic toxicity.

– Assessing actual sediment toxicity on sediment-dwelling organisms. – The ostracod *Heterocypris incongruens* is a suitable test organism because it feeds by ingesting particulate matter. Sediment-dwelling mussels and fish are also in close contact with sediments and their interaction with the sediments makes them suitable for testing.

– Disposal of dredged sediment on soil. – In this case, soil and groundwater are tested as target objects. Leachate toxicity can be used to predict groundwater toxicity, and toxicity of the dredged material on soil-dwelling organisms predicts the potential adverse or beneficial effect when applying it on soil. For characterizing the risk of chemicals in soil, adverse effects, and bioaccumulation can be tested by bacteria, earthworms, collembolans, and plants. The appropriate groundwater model depends on its transport, its contact with surface waters or drinking water bases, as well as on its use for irrigation, drinking water, and other purposes.

7.2.4 DTA for solid waste

DTA proved to be efficient for industrial wastewaters and other toxic discharges to protect receiving water bodies and their ecosystems. The same concept can be applied to solid wastes in order to protect terrestrial or aquatic ecosystems and ensure the safe use of soil and groundwater.

The European Waste Framework Directive (Waste Directive, 2008) applies DTA to differentiate between hazardous and non-hazardous wastes of the same origin, with the same European Waste Code (EWC).

Standardized test methods of aquatic and terrestrial ecosystem representatives are used for testing ecotoxicity of wastes. Based on the results, further testing or direct decisions on classification, risk assessment or risk reduction are possible. The most frequent management measures relying on environmental and human toxicity assessments are: classification as hazardous, restriction in use, establishment of environmental monitoring or application of a risk-reducing technology.

The Hungarian waste regulation (Waste HU, 2001), based on the EU waste directive, prescribes toxicological (animal toxicity and mutagenicity), hygienic (potential pathogens) and ecotoxicological (algae, daphnia, fish, plant germination and soil test) assessment of wastes and applies threshold criteria to the classification of wastes into H6, H9, H11 and H14 hazard categories.

– Toxic waste (H6): if inhaled or ingested or if penetrating the skin, may involve serious, acute or chronic health risks and even death. Threshold: lethal for cell cultures in 40-fold dilution.
– Infectious waste (H9): containing viable microorganisms or their toxins known or reliably believed to cause disease in man or other living organisms. Threshold for *Salmonella*, fecal coliform, fecal streptococci and/or parasitic worms of the digestive system is 200/g waste.
– Mutagenic (H11): by inhalation or digestion. Threshold is 500 mutant/g waste.
– Ecotoxic wastes (H14): cause acute or chronic toxicity for one or more sectors of the environment. Threshold: measurable adverse effect in 100-fold dilution (plant germination and daphnia tests) or 50-fold dilution (bacterial, algae, and fish tests) of water extract.

German researchers initiated a European interlaboratory test to develop standardized tests for the H14 category and recommended a tiered test procedure (Moser & Römbke, 2009 and UBA, 2014). The 'basic test battery' contains three aquatic and two terrestrial methods, duplicate algal and plant test organisms, with a total of seven test organisms.

7.2.5 DTA for soil and groundwater

Direct toxicity testing of soil is even more important than of liquid compartments due to the strong interactions between contaminants and the soil matrix. Soil is an effective sorbent, able to bind organic and inorganic chemical substances, including ionic ones. The quality and composition of a soil matrix influences the partition of chemicals (both nutrients and contaminants) between soil solid, soil liquid, and the biota. The soil organic fraction (humus) is mainly responsible for the sorption of organic contaminants, adsorbed on solid surfaces or in the biofilms of the soil micropores or even bound covalently to giant humus molecules. Inorganic chemicals, plants and microbial nutrients and contaminating metals and metal ions are often bound to the soil's mineral component, primarily to the clayey and loamy fractions. The mobility of chemicals cannot be properly described by soil chemical models as a function of physical or chemical parameters. In addition, the impact of soil living organisms, the effect of root exudates, root and microbial exoenzymes, complexing agents and surfactants significantly influences contaminant mobility and bioavailability. Direct toxicity testing gives a result proportional to mobility and bioavailability of toxic components.

Biodegradation in soil depends on soil characteristics and the status of soil microbiota. Chemical models based on the dominant influence of microbial adaptation have failed in predicting site-specific biodegradation in soil, therefore direct biodegradation assessment is the only way to acquire realistic information.

Hunt *et al.* (2009) applied DTA to groundwater contaminated by volatile chlorinated hydrocarbons. They prepared a site-specific guideline in Australia for assessing potential risk to aquatic organisms of surface waters contaminated by groundwater. The test organisms included microalgae (*Nitzschia closterium*), amphipods (*Allorchestes compressa*), polychaete worms (*Diopatra dentata*), sea urchin (*Heliocidaris tuberculata*), and oyster larvae (*Saccostrea commercialis*). A trigger value was calculated for a mixture of 14 volatile chlorinated hydrocarbons (VCHs) from toxicity data. The authors applied DTA to several contaminated sites with success to identify sources and transport pathways, to map toxicity and prove non-toxicity for

– abandoned mining sites with dispersed tailings, mine waste, and ore, both on the surface and as sediments containing a variety of chemical species of 8–10 identified metals;
– brownfields with hundreds of unidentified contaminants;
– floodplains, former disposal sites before spatial planning;
– artificial cover layers and illegal disposal sites;
– treated soil monitoring to check remediation results.

Direct toxicity results were integrated into the site/soil management scheme in themselves, together with physicochemical data or utilized in tiered assessment schemes.

7.2.6 Screening values based on DTA

Direct toxicity-based screening values can be created and used for decision making. The simplest way of obtaining an easy-to-understand DTA result is the comparison of the effect to a non-contaminated, i.e. non-toxic reference sample. The screening value (a protective level of toxicity, below which no additional regulatory attention is warranted) suggests that the acceptable rate of inhibition is typically 20% for acute toxicity and 25% for chronic toxicity. In statistical terms, the null hypothesis is 0.8 and 0.75, respectively. When the deviation from the null hypothesis is greater than 20% or 25% (more information on statistical evaluation of toxicity data is available in Chapter 9, Volume 2 of this book series, Gruiz *et al.*, 2015a), further assessment (e.g. identification of the hazards and their sources) and risk reduction must be performed.

If the null hypothesis is rejected by the response (e.g. the inhibition rate < 0.8 or < 0.75), the 'no-effect' sample amount can be read from the sample amount–response curve (Figure 1.18, cf. Figure 9.4, p. 455, Chapter 9 in Volume 2 of this book series, Gruiz *et al.*, 2015a) and compared to the tested sample amount. Thus the sample proportion showing no effect can be determined, which is identical to the scale of dilution of a liquid sample necessary to make it non-effective. Not only does this information provide the degree of toxicity, but also the necessary scale of toxicity reduction. The toxicity result to be used for decision making can be the average of the toxicities of several test organisms (e.g. three organisms from three different trophic levels) or the most sensitive one of them.

Table 1.3 shows an example from the authors' work on the soil of a contaminated cover in a mine waste dump.

Based on the inhibition rate results, one can see that none of the soils meet the maximum 20% inhibition threshold. However, this inhibition rate allows a judgment as to how far the quality of the sample is from the acceptable level. Thus, smaller proportions from the soil were also tested (mixed into inert artificial soil), and the soil proportion causing 20% inhibition (EsM$_{20}$) was read from the statistically fitted decreasing soil mass–inhibition curve (such a curve is shown in Figure 1.18). The necessary toxicity reduction rate was calculated as the reciprocal of the soil proportion.

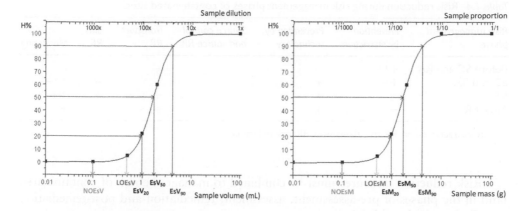

Figure 1.18 Test end points on the sample volume–response and the sample mass–response curves of environmental water and soil samples containing unidentified contaminants (Gruiz *et al.*, 2015c).

Table 1.3 Contaminated soil DTA: results of the most sensitive test organism.

Sample number	1	2	3	4
Inhibition rate of the original sample (%)	70	68	40	28
Soil proportion causing 20% inhibition (%)	6	32	75	85
Necessary rate of toxicity reduction (times)	17	3	1.3	1.2

In practice, it must be decided if 1.2x, 1.3x, 3x or 17x toxicity reduction is feasible or not. This can be decided using information about the contaminants, for example whether they are metals or biodegradable organic contaminants, etc.

One can directly compare RCR_{DTA} (based on toxicity) and RCR (based on chemical concentration), and draw the conclusions from the difference (cf. Section 6.2). Differences in origin, purport and the unit of measurement do not preclude the comparison any longer, and one can decide on the next management step.

7.2.7 Selection of the best risk reduction option

The site-specific risk value and the intervention-specific site investigation results support the decision about the best possible risk reduction option. If RCR > 1, risk should be reduced by risk reduction measures, most frequently prevention, restriction and remediation or their combination. A combination of the three measures is always necessary with one possibly dominating the procedure. For example, when remediating a site contaminated from a point source, contaminant transport and the pollution of further areas should be prevented and the use of the contaminated site should also be restricted until the risk has been reduced.

The pollution of a site is a dynamic process because transport and fate processes produce changes and gradients. Management of the pollution process needs a dynamic approach and model.

Table 1.4 Risk reduction during risk management phases of contaminated sites.

Risk management phase	Prevention by restriction	Prevention by technology	Source control and source RR	Transport RR	Site RR	Monitoring
Before SC and RA	+/−	−	−	−	−	−
SC and RA	++	+	−/+	−	−	+/−
RR	+	+	+	+	+	++
After RR	−	+	−	+/−	−	+

SC: site characterization, RA: risk assessment, RR: risk reduction

Risk reduction measures must be combined to manage the risk of contaminated land in the phases of pre-assessment, assessment, remediation and post-remediation (see Table 1.4). In addition to site management, socioeconomic assessment and risk communication should be included.

How can risk be reduced to reach $RCR \leq 1$? As RCR is the ratio PEC/PNEC or D/ADI, there are two clear alternatives: reducing the numerator (PEC or D) or increasing the denominator (PNEC, ADI).

$$\downarrow RCR = \frac{\downarrow D}{\uparrow ADI} \qquad \downarrow RCR = \frac{\downarrow PEC}{\uparrow PNEC} \qquad \downarrow RCR_{DTA} = \frac{\downarrow Sample\ toxicity}{\uparrow Target\ toxicity}$$

The numerator can be reduced by:

–   Reducing emissions by prohibiting the use of the chemical substance (and using a substitute chemical solely);
–   Reducing emissions by using less of the chemical substance (partially moving to the substitute chemical or substitute technology);
–   Reducing emissions and wastes from manufacturing technologies and uses of the products (recycling, innovative technologies for production or use of the products);
–   Controlling the sources (emission control, control measures and equipment);
–   Restricting environmental transport of the chemical substance (controlling transport pathways by e.g. landfill liners for solid waste and permeable reactive barriers for groundwater);
–   Removing the contaminant from the environment (remediation, clean-up);
–   Modifying mobility and/or activity (adverse effect) of the contaminant in the environment (remediation by stabilization and deactivation);
–   Containing uptake by using protective equipment. This solution is the transition to the next point, i.e. reducing exposure (e.g. of workers) by reducing concentration in the inhaled, or dermally contacted environment and changing the use of the environment at the same time.

The denominator can be increased by

–   Changing land use to less demanding uses (from natural or residential to industrial or commercial uses);
–   Restricting land use (no entry, no swimming, non-drinking water, etc.);

- Decreasing the daily dose by reducing the intake of inhaled contaminated air (mask), contaminated drinking water (bottled water), contaminated food (safe foodstuffs, organic food, etc.) and by avoiding contaminated land;
- Excluding sensitive receptors (children, pregnant women);
- Changing risky habits (e.g. eating less redfish, not growing vegetables in one's hobby garden that is likely to be contaminated).

The interface between a contaminated environment and the living organisms is a very important factor influencing interactions and uptake of contaminants by the receptors. Ecosystem members in water can hardly exclude contact with contaminants dissolved in water, but soil-living organisms may have a special sense of avoiding contaminants, e.g. small particles of the contaminant or sorbed chemical substances on soil particles (e.g. avoidance of plant roots or soil-dwelling animals). Human beings have a wide range of tools for avoiding a contaminated environment and protecting themselves. The most apparent tools are protective wear and equipment used by workers (ventilation, indoor air treatment, gloves, masks, glasses, overalls, etc.). Anyone can use similar wear when working with hazardous chemicals in households, gardens, or urban traffic. Humans are able, in most cases, to control their contact with the indoor and outdoor environment, but knowledge is necessary to decide what is risky and what is not, and how to apply restrictions or protective devices to priority cases. This knowledge is very often poor due to insufficient education and consumerism-related manipulation. Traditions and short-sightedness are also responsible for many environmental risks that could have been avoided. For example, villagers living near Chernobyl refused to leave their homes after the nuclear power plant disaster, and in so doing exposed themselves to enormous risks.

7.3   Summary of contaminated site risk management

The environment, including its biotic and abiotic segments plus their interactions, cannot be characterized truly because of its complexity and the shortcomings of the measurement techniques. A contaminated environment is even more complex, data collection tools and their interpretation must therefore be optimized to acquire the best possible information.

Site-specific assessment of the physicochemical, biological, and ecological consequences of pollution and the acquisition of high-quality information from the collected historical and measured data need innovative concepts and monitoring methods. Tiered site assessment allows efficient resource management; *in situ* assessment and decision making accelerate site assessment and the application of STT provides refined, risk-based information. Real-time information makes decision making and intervention as well as precise regulation possible without delay. Remote, proximate, and contact sensors based on physicochemical and biological signal detection open new horizons in environmental monitoring by unifying chemical and biological methods and by detecting biospecific chemical signals and contaminant-specific biological responses in the form of electromagnetic signals. These innovations and available tools are introduced in Chapters 2, 3 and 4.

Quantification of risk raises the issue priority receptors, humans (children) are typically given priority. Nature (ecosystem or certain ecosystem members) may have priority in natural conservation areas. Long-term monitoring shows the trends in risk changes and helps to select the most appropriate risk management tool battery.

The risk characterization ratio being the key value of quantitative risk assessment, characterizes one chemical substance–receptor relationship when chemical models are applied, thus several chemistry-based RCR values must be calculated and aggregated. The resulting aggregated uncertainty (due to uncertain RCR estimates and inappropriate aggregation (adding the RCR values of non-additive, i.e. synergistic or antagonistic chemicals) can be handled using conservative estimates. Distribution of contaminants between soil phases may cause major changes in the size of risk, e.g. when dissolved contaminants are sorbed on the contacting solid phase: it may reduce the risk in water, however, if the sorbent phase remains in the environment, and the sorption is not irreversible, a long-term pollutant source or a chemical time bomb may have been created.

An RCR_{DTA} value, based on measured toxicity, eliminates a significant part of the problems of the chemical models because it includes all contaminants and their interactions with the soil components, aggregates all effects, and responds to partition, mobility, and bioavailability of the contaminant. One shortcoming of the DTA-based risk management is that one, or even three, test organisms cannot accurately represent the whole of the ecosystem; but innovative test systems (e.g. simulations in microcosms, mesocosms), and end points (e.g. *in situ* measurable community indicators) may increase environmental realism. Environmental heterogeneities and other natural uncertainties cannot be excluded, they should be managed using appropriate statistical methods. RCR is the key point in risk management, but is only the first orienteering point. The size of the contaminated site, its type (environmental, social, ore economic), the risks and costs compared to the benefits of risk management should be evaluated in a complex socio-economic and sustainability assessment procedure (see more in Volume 1 of this book series, Gruiz *et al.*, 2014).

REFERENCES

Algal test for DTA (2007) *The direct toxicity assessment of aqueous environmental samples* using the marine algal growth inhibition test with *Skeletonema costatum*. EA, UK. [Online] Available from: www.gov.uk/government/uploads/system/uploads/attachment_data/file/316803/oyster209jan30_1388168.pdf. [Accessed 24th Aug 2015].

Algal test for DTA (2008) *The direct toxicity assessment of aqueous environmental samples* using the freshwater algal growth inhibition test with *Pseudokirchneriella subcapitata*. EA, UK. [Online] Available from: www.gov.uk/government/uploads/system/uploads/attachment_data/file/316789/bluebook219_2060295.pdf. [Accessed 24th Aug 2015].

ANZECC/ARMCANZ (2000) *Australian and New Zealand guidelines for fresh and marine water quality*. Volume 1 – The Guidelines. ANZ Environment and Conservation Council, Agriculture and Resource Management Council of ANZ. [Online] Available from: https://www.environment.gov.au/system/files/resources/53cda9ea-7ec2-49d4-af29-d1dde09e96ef/files/nwqms-guidelines-4-vol1.pdf. [Accessed 24th September 2015].

ASTM E1528-00 (2000) *Transaction Screen Process*, ICS Number Code 13.020.30, ASTM International, West Conshohocken, PA. DOI: 10.1520/E1528-00. [Online] Available from: www.astm.org/DATABASE.CART/HISTORICAL/E1528-00.htm. [Accessed 24th June 2015].

ASTM E1527-05 (2005) *Standard Practice for Environmental Site Assessments*: *Phase I Environmental Site Assessment Process*, ICS Number Code 13.020.30, ASTM International, West Conshohocken, PA. DOI: 10.1520/E1527-05. [Online] Available from: www.astm.org/Standards/E1527.htm. [Accessed 24th June 2015].

ASTM E1903-97 (2002) *Phase II Environmental Site Assessment*, ICS Number Code 13.020.30, ASTM International, West Conshohocken, PA. DOI: 10.1520/E1903-97. [Online] Available from: www.astm.org/DATABASE.CART/HISTORICAL/E1903-97.htm. [Accessed 24th September 2015].

ASTM E1739-95 (2015) Standard Guide for Risk-Based Corrective Action Applied at Petroleum Release Sites. [Online] Available from: www.astm.org/Standards/E1739.htm. [Accessed 24th December 2015].

Baran, A. & Tarnawski, M. (2013) Phytotoxkit/Phytotestkit and Microtox® as tools for toxicity assessment of sediments. *Ecotoxicology and Environmental Safety*, 98, 19–27.

Beard, M., Gray, A.L. & Ruf, J.C. (2010) Implementing the Triad Approach. Using multiple programs to optimize the management and visualization of environmental data sets. *Pollution Engineering*. [Online] Available from: www.pollutionengineering.com/Articles/Feature_Article/BNP_GUID_9-5-2006_A_10000000000000747222. [Accessed 14th December 2014].

BC Field Sampling (2003) *British Columbia field sampling manual*. [Online] Available from: www.env.gov.bc.ca/wsd/data_searches/field_sampling_manual/field_man_pdfs/fld_man_03.pdf. [Accessed 24th September 2015].

Bos, R., Huijbregts, M. & Peijnenburg, W. (2005) Soil type-specific environmental quality standards for zinc in Dutch soil. *Integrated Environmental Assessment and Management*, 1(3), 252–258.

Campbell, C.D., Warren, A., Cameron, C.M. & Hope, S.J. (1997) Direct toxicity assessment of two soils amended with sewage sludge contaminated with heavy metals using a protozoan (*Colpoda steinii*) bioassay. *Chemosphere*, 34(3), 501–514.

CARACAS (1996–1998) *Concerted Action on Risk Assessment for Contaminated Sites in the European Union*. [Online] Available from: www.commonforum.eu/Documents/DOC/Caracas/caracas_publ1.pdf. [Accessed 24th September 2015].

Carlon, C. (ed.) (2007) *Derivation methods of soil screening values in Europe*. A review and evaluation of national procedures towards harmonization. Ispra, European Commission, Joint Research Centre, EUR 22805-EN, pp. 306. [Online] Available from: http://eusoils.jrc.ec.europa.eu/esdb_archive/eusoils_docs/other/EUR22805.pdf. [Accessed 10 July 2011].

Carlon, C. & Swartjes, F. (2007a) Analysis of variability and reasons of differences. In: Carlon, C. (ed.) *Derivation methods of soil screening values in Europe. A review of national procedures towards harmonisation opportunities*. JRC PUBSY 7123, HERACLES. Ispra, European Commission Joint Research Centre.

Carlon, C. & Swartjes, F. (2007b) Rationale and methods of the review. In: Carlon, C. (ed.) *Derivation methods of soil screening values in Europe. A review of national procedures towards harmonisation opportunities*. JRC PUBSY 7123, HERACLES. Ispra, European Commission Joint Research Centre.

Carter, M.R. & Gregorich E.G. (2007) *Soil Sampling and Methods of Analysis*, 2nd Ed. Boca Raton, CRC Press.

CERCLA (2011) *Comprehensive Environmental Response, Compensation, and Liability Act*, 1980, updated in 2011. US EPA. [Online] Available from: www.epw.senate.gov/cercla.pdf and www2.epa.gov/superfund. [Accessed 24th September 2015].

Chapman, P.M. (2000) Whole effluent toxicity (WET) testing – usefulness, level of protection and risk assessment. *Environmental Toxicology and Chemistry*, 19, 3–13.

CLARINET (1998–2001) *Contaminated Land Rehabilitation Network for Environmental Technologies*. [Online] Available from: www.commonforum.eu/Documents/DOC/Clarinet/rblm_report.pdf. [Accessed 4th June 2015].

Clark, I. (2010) Statistics or geostatistics? Sampling error or nugget effect? *The Journal of the Southern African Institute of Mining and Metallurgy*, 110(6), 307–312.

Clark, I. & Harper, W.V. (2000) *Practical Geostatistics*. Columbus, Ohio, Ecosse North American Llc.

CLEA software (2014) Contaminated land exposure assessment (CLEA) tool. [Online] Available from: www.environment-agency.gov.uk/research/planning/33732.aspx. [Accessed 24th September 2015].

Clue-in (2005) *Technical and Regulatory Guidance for the Triad Approach: A New Paradigm for Environmental Project Management*. ITRC's Internet-based Training Program, co-sponsored by the EPA Office of Superfund Remediation and Technology Innovation. [Online] Available from: www.clu-in.org/conf/itrc/triad_021005. [Accessed 14th March 2015].

Cole, S. & Jeffries, J. (2009) *Using Soil Guideline Values*. Environment Agency, ISBN: 978-1-84911-037-2. [Online] Available from: www.environment-agency.gov.uk/static/documents/Research/SCHO0309BPQM-e-e.pdf. [Accessed 24th September 2015].

Crumbling, D.M. (2004) *Summary of the Triad approach*. [Online] Available from: www.triadcentral.org/ref/doc/triadsummary.pdf. [Accessed 14th December 2014].

CWA (1977) *Clean Water Act*, Federal Water Pollution Control Act, as amended by the Clean Water Act, Section 304(a)(1). [Online] Available from: www.epa.gov/npdes/pubs/cwatxt.txt. [Accessed 4th June 2015].

Dictionary.com (2015) *Dictionary – land*. [Online] Available from: http://dictionary.reference.com/browse/land. [Accessed 24th September 2015].

DTA in PPC. (2006) *Integrated Pollution Prevention & Control. Guidance on the use of DTA in PPC Impact Assessments*. Environment Agency, UK.

ECORIVER (2002–2005) *Ecotoxicological evaluation of municipal and industrial waste waters*. Europen Life project, LIFE02 ENV/P/000416. [Online] Available from: http://ec.europa.eu/environment/life/project/ Projects/index.cfm?fuseaction=search.dspPage&n_proj_id=2146. [Accessed 24th April 2016].

Eco-SSL (2005) *Ecological Soil Screening Levels for Cadmium*. Interim Final OSWER Directive 9285.7-65, U.S. EPA, Office of Solid Waste and Emergency Response. [Online] Available from: https://www.epa.gov/ sites/production/files/2015-09/documents/eco-ssl_cadmium.pdf. [Accessed 24th September 2015].

EEA (2010) *Overview of best practices for limiting soil sealing or mitigating its effects in EU-27*. [Online] Available from: http://ec.europa.eu/environment/soil/sealing.htm. [Accessed 24th September 2015].

Energy Institute (2010) *Viability of using direct toxicity assessment for petroleum refinery effluent toxicity assessment*. Research report. [Online] Available from: http://publishing.energyinst.org/__data/assets/file/0005/6737/Pages-from-Viability-of-using-direct-toxicity-assessment-for-petroleum-refinery-effluent-toxicity-assessment-2.pdf. [Accessed 24th Aug 2015].

ESdat.com (2015) *Environmental Guidelines and Standards*. ESdat, [Online] Available from: http://www.esdat.com.au/Environmental_Standards.aspx. [Accessed 24th September 2015].

FAO (2006a) *Guidelines for soil description*. Food and Agriculture Organization of the United Nations, Rome. [Online] Available from: ftp://ftp.fao.org/agl/agll/docs/guidel_soil_descr.pdf. [Accessed 24th September 2015].

FAO (2006b) *World Reference Base for Soil Resources*. [Online] Available from: www.fao.org/nr/land/soils/soil/en. [Accessed 24th September 2015].

FAO/UNEP (1999) *Land Use*. [Online] Available from: http://www.fao.org/nr/land/use/fr/. [Accessed 24th September 2015].

FDA (2014) *Mercury Levels in Commercial Fish and Shellfish (1990–2010)*. FDA. [Online] Available from: www.fda.gov/Food/FoodborneIllnessContaminants/Metals/ucm115644.htm. [Accessed 24th September 2015].

Feigl, V., Uzinger, N. & Gruiz, K. (2009) Chemical stabilization of toxic metals in soil microcosms. *Land Contamination & Reclamation*, 17(3), 483–495.

Gruiz, K. (2005) Biological tools for the soil ecotoxicity evaluation: Soil Testing Triad and the interactive ecotoxicity tests for contaminated soil. In: Fava, F. & Canepa, P. (eds.) *Innovative Approaches to the Bioremediation of Contaminated Sites, Soil Remediation Series No. 6.* Venice, Italy, INCA. pp. 45–70.

Gruiz, K. (2014) *Abandoned and contaminated land.* In: Gruiz, K., Meggyes, T. & Fenyvesi, E. (eds.) (2014) *Engineering tools for environmental risk management. Volume 1. Environmental Deterioration and Contamination – Problems and their Management.* Boca Raton, Fl., CRC Press, pp. 77–92.

Gruiz, K. (2015) Environmental toxicology – A general overview. In: Gruiz, K., Meggyes, T. & Fenyvesi, E. (eds.) *Engineering tools for environmental risk management. Volume 2. Environmental Toxicology.* Boca Raton, Fl., CRC Press, pp. 1–63.

Gruiz, K. & Hajdu, Cs. (2015) Bioaccessibility and bioavailability in risk assessment. In: Gruiz, K., Meggyes, T. & Fenyvesi, E. (eds.) *Engineering tools for environmental risk management. Volume 2. Environmental Toxicology.* Boca Raton, Fl., CRC Press. pp. 337–400.

Gruiz, K. & Meggyes, T. (2009) (eds.) Innovative decision-support tools for risk-based environmental management. *Land Contamination & Reclamation.* 17 (3–4), 315–736.

Gruiz, K., Fekete-Kertész, I., Kunglné-Nagy, Zs., Hajdu, Cs., Feigl, V., Vaszita, E. & Molnár, M. (2016) Direct toxicity assessment – Methods, evaluation, interpretation. *Science of the Total Environment.* [Online] Available from: DOI: 10.1016/j.scitotenv.2016.01.007. [Accessed 04th May 2016].

Gruiz, K. & Vodicska, M. (1993) Assessing Heavy-metal Contamination in Soil Applying a Bacterial Biotest and Xray Fluorescent Spectroscopy – In: Arendt, F., Annokkée, G.J., Bosman, R. and van den Brink, W.J. (eds.) *Contaminated Soil '93,* Kluwer Academic Publ., The Netherlands, pp. 931–932.

Gruiz, K., Hajdu, Cs. & Meggyes, T. (2015a) Data evaluation and interpretation in environmental toxicology. In: Gruiz, K., Meggyes, T. & Fenyvesi, E. (eds.) *Engineering tools for environmental risk management. Volume 2. Environmental Toxicology.* Boca Raton, Fl., CRC Press. pp. 445–544.

Gruiz, K., Horváth, B., Molnár, M. & Sipter, E. (2000) When the chemical time bomb explodes. Chronic risk of toxic metals at a former mining site. In: *ConSoil 2000.* Leipzig, Thomas Telford. pp. 662–670.

Gruiz, K., Meggyes, T. & Fenyvesi, E. (eds.) (2014) *Engineering tools for environmental risk management. Volume 1. Environmental Deterioration and Contamination – Problems and their Management.* Boca Raton, Fl., CRC Press.

Gruiz, K., Meggyes, T. & Fenyvesi, E. (eds.) (2015a) *Engineering tools for environmental risk management. Volume 2. Environmental Toxicology.* Boca Raton, Fl., CRC Press.

Gruiz, K., Molnár, M., Nagy, Zs.M. & Hajdu, Cs. (2015b) Fate and behavior of chemical substances in the environment. In: Gruiz, K., Meggyes, T. & Fenyvesi, E. (eds.) *Engineering tools for environmental risk management. Volume 2. Environmental Toxicology.* Boca Raton, Fl., CRC Press. pp. 71–124.

Gruiz, K., Meggyes, T. & Fenyvesi, E. (eds.) (2015c) Data evaluation and interpretation in environmental toxicology. In: *Engineering tools for environmental risk management. Volume 2. Environmental Toxicology.* Boca Raton, Fl., CRC Press, pp. 455 and 465.

Gruiz, K., Murányi, A., Molnár, M. & Horváth, B. (1998) Risk assessment of heavy metal contamination in the Danube sediments from Hungary. *Water Science and Technology,* 37 (6–7), 273–281.

Gruiz, K., Vaszita, E. & Clement, O. (2014a) Site-specific risk assessment and management of point and diffuse sources. In: Gruiz, K., Meggyes, T. & Fenyvesi, E. (eds.) (2014) *Engineering tools for environmental risk management. Volume 1. Environmental*

Deterioration and Contamination – Problems and their Management. Boca Raton, Fl., CRC Press.

Guntenspergen, G.R. (2012) (ed.) *Urban Ecosystems.* Journal No. 11252. ISSN: 1083-8155 (print version) ISSN: 1573-1642 (electronic version). [Online] Available from: www.springerlink.com/content/1083-8155. [Accessed 24th September 2015].

Hajdu, Cs., Gruiz, K. & Fenyvesi, É. (2009) Bioavailability and bioaccessibility dependent mutagenicity of pentachlorophenol (PCP). *Land Contamination & Reclamation,* 17 (3–4), 473–482.

Hall, J.A. & Golding, L. (1998) *Standard Methods for Whole Effluent Toxicity Testing: Development and Application.* Prepared for the Ministry for the Environment by National Institute of Water & Atmospheric Research Ltd., NZ. [Online] Available from: https: //www.mfe.govt.nz/sites/default/files/media/Marine/Standard%20methods%20for%20whole%20effluent%20toxicity%20testing%20Development%20and%20application%20full%20report.pdf. [Accessed 24th September 2015].

Horváth, B., Gruiz, K. & Sára, B. (1996) Ecotoxicological testing of soil by four bacterial biotests. *Toxicological and Environmental Chemistry,* 58, 223–235.

Hosford, M. (2009) *Human health toxicological assessment of contaminants in soil.* Science Report Final SC050021/SR2. Bristol: Environment Agency. [Online] Available from: http://publications.environment-agency.gov.uk/pdf/SCHO0508BNQY-e-e.pdf. [Accessed 24th September 2015].

HU 10/2000 (VI. 2.) *Hungarian Soil and Groundwater Quality Criteria,* KöM-EüM-FVM-KHVM. [Online] Available from: www.kvvm.hu/index.php?pid=10&sid=53&hid=353. [Accessed 24th September 2015].

HU 33/2000 (III.17.) *Certain tasks relating to the activities affecting the quality of groundwater.* Govt. Order. [Online] Available from: www.kvvm.hu/index.php?pid=10&sid=53&hid=352. [Accessed 4th June 2015].

HU LIII (1995) *General Rules of Environmental Protection,* Law 1995. LIII. [Online] Available from: http://net.jogtar.hu/jr/gen/hjegy_doc.cgi?docid=99500053.TV. [Accessed 24th September 2015].

Hunt, D.T.E., Cooper, V.A., Johnson, D. & Grandy, N.J. (1989) *Discharge Control by Direct Toxicity Assessment (DTA).* [Online] Available from: http://ea-lit.freshwaterlife. org/archive/ealit:2243/OBJ/19000244.pdf. [Accessed 24th Aug 2015].

Incremental Sampling (2012) *Incremental Sampling Strategy.* Online training material, ITRC. [Online] Available from: http://www.itrcweb.org/ism-1/2_1_Introduction_Nature_of_Soil_ Sampling_and_ Incremental_Samping_Principles.html. [Accessed 24th December 2015].

IPPC (2008) *Integrated pollution prevention and control.* Directive 2008/1/EC of the European Parliament and of the Council. [Online] Available from: http://eur-lex.europa.eu/legal-content/EN/TXT/HTML/?uri=CELEX:32008L0001&from=EN. [Accessed 24th September 2015].

IRIS (2010) *Integrated Risk Information System* (IRIS). United States Environmental Protection Agency. [Online] Available from: http://www.epa.gov/IRIS. [Accessed 24th September 2015].

Irish EPA (2015) *Ensuring high quality in sampling.* [Online] Available from: www.epa.ie/ enforcement/ensuringhighqualityaqueousemissionsmonitoringdata/sampling/#.Vf8j4pfLlKU. [Accessed 24th September 2015].

ITRC (2005) *Examination of Risk-Based Screening Values and Approaches of Selected States.* Interstate Technology & Regulatory Council, Risk Assessment Resources Team. [Accessed 24th September 2015].

JECFA (2010) *Joint FAO/WHO Expert Committee on Food Additives.* Seventy-second meeting Rome, 16–25 February 2010. JECFA/72/SC. [Online] Available from: http://www.who.int/foodsafety/chem/summary72_rev.pdf. [Accessed 24th September 2015].

Jeffries, J. & Martin, I. (2009) *Updated technical background to the CLEA model.* Science Report SC050021/SR3. Bristol: Environment Agency, ISBN 978-1-84432-856-7. [Online] Available from: http://publications.environment-agency.gov.uk/pdf/SCHO0508BNQW-e-e.pdf. [Accessed 24th September 2015].

Kim, H. & Han, K. (2011) Ingestion exposure to nitrosamines in chlorinated drinking water. *Environmental Health and Toxicology,* 2011, 26: e2011003. [Online] Available: DOI: 10.5620/eht.2011.26.e2011003.

Köhler, S., Belkin, S.C. & Schmid, R.D. (2000) Reporter gene bioassays in environmental analysis. *Fresenius' Journal of Anaytical Chemistry,* 366(6), 769–779.

Kwan, K.K. (1993) Direct toxicity assessment of solid phase samples using the toxi-chromotest kit. *Environmental Toxicology and Water Quality,* 8(2), 223–230.

Leverett, D. (2006) *Direct Toxicity Assessment Proficiency Scheme (DTAPS).* Environmental Agency, UK. [Online] Available from: https://www.gov.uk/government/uploads/system/uploads/attachment_data/file/296742/geho1105bkac-e-e.pdf. [Accessed 24th Aug 2015].

MCERTS for DTA (2010) *The MCERTS scheme for Direct Toxicity Assessment (DTA) of effluents.* Environment Agency, UK, 2010. [Online] Available from: www.environment-agency.gov.uk/business/regulation/38783.aspx. [Accessed 24th Aug 2015].

Meuser, H. (2010) *Contaminated urban soils.* Environmental Pollution 18. Dordrecht, Heidelberg, London, New York, Springer.

MOKKA Project (2004–2008) *Innovative decision support tools for risk based environmental management.* [Online] Available from: http://enfo.hu/mokka/index.php?lang=eng&body=mokka. [Accessed 24th April 2016].

Moser, H. & Römbke, J. (2009) *Ecotoxicological characterization of waste.* Results and experiences of an international ring test. New York, Springer. [Online] Available from: DOI: 10.1007/978-0-387-88959-7. [Accessed 14th Aug 2015].

Nederlof, M.H. (2015) *Geostatistics and kriging.* [Online] Available from: www.mhnederlof.nl/geostatistics.html. [Accessed 24th September 2015].

NICOLE (1996–1999) *Network for Industry Contaminated in Europe (1996–99).* [Online] Available from: www.nicole.org. [Accessed 24th September 2015].

NRCS (2014) *Soil Survey Field and Laboratory Methods Manual.* [Online] Available from: www.nrcs.usda.gov/Internet/FSE_DOCUMENTS/stelprdb1244466.pdf. [Accessed 24th September 2015].

NSW (2000) *Guidelines for Consultants Reporting on Contaminated Sites.* NSW Environment Protection Authority Australia, ISBN 0 7310 3892 4, EPA 97/104. [Online] Available from: www.environment.nsw.gov.au/resources/clm/97104consultantsglines.pdf. [Accessed 24th September 2015].

Oil Pollution Act (1990) 33 U.S.C. § 2701–2761, Amended in 2000. [Online] Available from: www.epw.senate.gov/opa90.pdf. [Accessed 24th September 2015].

Oyster test for DTA (2007) *The direct toxicity assessment of aqueous environmental samples* using the oyster *Crassostrea gigas* embryo-larval development test. Environment Agency, UK, 2007. [Online] Available from: www.gov.uk/government/uploads/system/uploads/attachment_data/file/316785/DTAmarinewateralgae-225.pdf. [Accessed 24th Aug 2015].

Pitard, F.F. (1989a) *Pierre Gy's Sampling Theory and Sampling Practice.* Inc., Boca Raton, CRC Press.

Pitard, F.F. (1989b) *Sampling Methodologies for Monitoring the Environment: Theory and Practice.* Class Handouts. Pierre Gy & Francis Pitard Sampling Consultants. Broomfield, CO 80020.

Prokop, G. (2011) *Reducing Soil Sealing and Land Take – Best practices in the EU.* Green Week, Brussels, 24–27 May 2011. [Online] Available from: http://ec.europa.

eu/environment/greenweek2011/sites/default/files/3.2_prokop.pdf. [Accessed 24th September 2015].

Prokop, G.; Jobstmann, H. & Schönbauer, A. (2011) *Report on best practices for limiting soil sealing and mitigating its effects in EU-27*. Final Report. European Communities. [Online] Available from: http://ec.europa.eu/environment/archives/soil/pdf/sealing/Soil%20 sealing%20-%20Final%20Report.pdf. [Accessed 24th September 2015].

RCRA (1976) *Resource Conservation and Recovery Act*. [Online] Available from: www.epa.gov/lawsregs/laws/rcra.html. [Accessed 24th September 2015].

RCRA Corrective Action Program (2003) *Final Guidance on Completion of Corrective Action Activities at RCRA Facilities*. [Online] Available from: www.epa.gov/epawaste/hazard/ correctiveaction/resources/guidance/gen_ca/compfedr.pdf. [Accessed 24th September 2015].

Real-Time Measurement Systems (2016) *Triad resource center* [Online] Available from: https://triadcentral.clu-in.org/mgmt/meas/index.cfm. [Accessed 13th April 2016].

SARA (1986) *Superfund Amendments and Reauthorization Act*. [Online] Available from: www.epw.senate.gov/sara.pdf. [Accessed 24th September 2015].

Sarkadi, A., Vaszita, E., Tolner, M. & Gruiz, K. (2009) *In situ* site assessment: short overview and description of the field portable XRF and its application. *Land Contamination and Reclamation*, 17(3–4), 431–442.

Schoeneberger, P.J., Wysocki, D.A., Benham, E.C. & Soil Survey Staff (2012) *Field book for describing and sampling soils*, 3.0. Natural Resources Conservation Service, National Soil Survey Center, Lincoln, NE. [Online] Available from: www.nrcs.usda.gov/ Internet/FSE_DOCUMENTS/nrcs142p2_052523.pdf. [Accessed 24th September 2015].

Schreiber, M.P., Komppa, V., Wahlström, M. & Laine-Ylijoki, J. (2006) Chemical and environmental sampling: quality through accreditation, certification and industrial standards. *Accreditation and Quality Assurance*, 10(9), 510–514.

Small Business Liability Relief and Brownfields Revitalization Act (2002) [Online] Available from: www.gpo.gov/fdsys/pkg/PLAW-107publ118/html/PLAW-107publ118.htm. [Accessed 24th September 2015].

SMARTe (2015) *Sustainable Management Approaches and Revitalization Tools*. [Online] Available from: www.smarte.org/smarte/structure/getsmart.xml?layout=no-sidebar-no-mainnav. [Accessed 4th June 2015].

SoCo (2009) Louwagie, G., Hubertus, S. & Alison Burrell, G. (eds.) *Final report on the project 'Sustainable Agriculture and Soil Conservation'*. European Commission Joint Research Centre. [Online] Available from: http://eusoils.jrc.ec.europa.eu/esdb_archive/eusoils_docs/other/ EUR23820.pdf. [Accessed 4th September 2015].

Soil Sealing (2012) *Science for Environment Policy* – In-depth Reports. DG Environment, EC. [Online] Available from: http://ec.europa.eu/environment/integration/research/ newsalert/pdf/IR2.pdf. [Accessed 4th June 2015].

Spira, Y., Gruiz, K., Uzinger, N. & Anton, A. (2014) Management of abandoned and contaminated land. In: Gruiz, K., Meggyes, T. & Fenyvesi, É. (eds.) *Engineering tools for environmental risk management. Volume 1. Environmental Deterioration and Contamination – Problems and their Management*. Boca Raton, CRC Press.

SuRF-UK (2015) *The SuRF-UK Indicator Set For Sustainable Remediation Assessment*. [Online] Available from: www.claire.co.uk/index.php?option=com_phocadownload& view=file&id=262:initiatives&Itemid=230. [Accessed 24th September 2015].

SuRF-UK, SMP (2015) *SuRF-UK, sustainable management practices*. [Online] Available from: www.claire.co.uk/index.php?option=com_phocadownload&view=file&id=403& Itemid=230. [Accessed 24th September 2015].

Swartjes, F.A. (2011) (Ed.) *Dealing with Contaminated Sites: From Theory towards Practical Application*. Springer Dordrecht Heidelberg London New York, Springer. DOI: 10.1007/978-90-481-9757-6.

Triad approach (2015) *A new paradigm for environmental project management.* [Online] Available from: http://www.clu-in.org/conf/itrc/triad_021005/ [Accessed 14th March 2015].

Triad benefits (2015) *Triad resource center.* [Online] Available from: http://www.triadcentral.org/over/key/#1 [Accessed 14th March 2015].

Triad overview (2015) Triad resource center. [Online] Available from: http://www.triadcentral.org/over/index.cfm. [Accessed 14th March 2015].

Triad requirements (2015) Triad resource center [Online] Available from: http://www.triadcentral.org/over/req/index.cfm [Accessed 14th March 2015].

UBA (2014) *Weiterentwicklung der UBA-Handlungsempfehlung zur ökotoxikologischen Charakterisierung von Abfällen.* [Online] Available from: www.umweltbundesamt.de/sites/default/files/medien/378/publikationen/texte_19_2014_weiterentwicklung_der_uba_handlungsempfehlung_1.pdf. [Accessed 14th Aug 2015].

UK (1995) *Environmental Act 1995.* [Online] Available from: www.legislation.gov.uk/ukpga/1995/25/contents. [Accessed 24th September 2015].

UK DTA (2000a) *Direct Toxicity Assessment (DTA)* – Demonstration Programme. Technical Guidance. Report 00/TX/02/07. Water Industry Research, UK.

UK DTA (2000b) *Direct Toxicity Assessment (DTA)* – Demonstration Programme. Recommendations from the Steering Group to the Environmental Regulators. Report: 00/TX/02/06. Water Industry Research, UK.

UN (2000) *Millennium Development Goals.* An agreement to a set of time-bound and measurable goals for combating poverty, hunger, disease, illiteracy, environmental degradation and discrimination against women, to be achieved by 2015. [Online] Available from: wwf.panda.org/what_we_do/how_we_work/people_and_conservation/our_work/post_2015_development_agenda/mdg. [Accessed 24th September 2015].

US EPA (1992) *Preparation of Soil Sampling Protocols. Sampling Techniques and Strategies.* [Online] Available from: http://nepis.epa.gov/Exe/ZyPDF.cgi/20008O5T.PDF?Dockey=20008O5T.PDF. [Accessed 4th June 2015].

US EPA (1995) *National policy regarding whole effluent toxicity enforcement.* Washington DC: US EPA, Offices of Regulatory Enforcement and Wastewater Management.

US EPA (2002) *Methods for Measuring the Acute Toxicity of Effluents and Receiving Waters to Freshwater and Marine Organisms.* Fifth Edition. [Online] Available from: http://water.epa.gov/scitech/methods/ cwa/wet/upload/2007_07_10_methods_wet_disk2_atx.pdf. [Accessed 14th December 2014].

US EPA (2003) *Using the triad approach to streamline brownfields site assessment and cleanup – Brownfields Technology Primer Series.* Brownfields Technology Support Center [Online] Available from: http://www.brownfieldstsc.org/pdfs/Triadprimer.pdf. [Accessed 14th December 2014].

US EPA (2015) *Superfund program.* [Online] Available from: www.epa.gov/superfund. [Accessed 24th May 2015].

US EPA CSMS (1998) *EPA's Contaminated Sediment Management Strategy.* Office of Water, 4305. [Online] Available from: https://clu-in.org/download/contaminantfocus/sediments/EPA-contaminated-sediment-strategy1999.pdf. [Accessed 24th September 2015].

US NRWQC (2012) *US National Recommended Water Quality Criteria.* [Online] Available from: http://water.epa.gov/scitech/swguidance/standards/criteria/current/index.cfm. [Accessed 24th September 2015].

UST (2005) *Underground Storage Tanks* (UST) provisions of the Energy Policy Act. [Online] Available from: www2.epa.gov/ust. [Accessed 24th September 2015].

Van Dam, R.A. & Chapman, J.C. (2001) Direct toxicity assessment (DTA) for water quality guidelines in Australia and New Zealand. *Australasian Journal of Ecotoxicology – Direct toxicity assessment,* 7, 175–198.

Vardakoulias, O. (2013) *Social CBA and SROI.* MSEP Project to build the socio-economic capacity of marine NGOs. nef (the new economics foundation). [Online] Available from: http://b.3cdn.net/nefoundation/ff182a6ba487095ac6_yrm6bx9o6.pdf. [Accessed 24th September 2015].

Waste HU (2001) *Hungarian decree on the rules of certain activities related to hazardous waste* (modified as of 225/2015. (VIII. 7.) governmental decree. [Online] Available from: http://net.jogtar.hu/jr/gen/hjegy_doc.cgi?docid=A1500225.KOR. [Accessed 14th Aug 2015].

Waste Directive (2008) *Waste Framework Directive 2008/98/EC of the European Parliament and of the Council of 19 November 2008 on waste, and repealing European Waste Directive and Hazardous Waste Directive.* [Online] Available from: http://eur-lex.europa.eu/legal-content/EN/TXT/?uri=URISERV%3Aev0010. [Accessed 24th Aug 2015].

VROM (2000) *Dutch Target and Intervention Values – the New Dutch List.* Dutch Ministry of Housing, Spatial Planning and the Environment. [Online] Available from: www.esdat.net/Environmental%20Standards/Dutch/annexS_I2000Dutch%20 Environmental%20Standards.pdf. [Accessed 24th November 2015].

Water quality standards (2008) *Directive 2008/105/EC of the European Parliament and of the Council on environmental quality standards in the field of water policy.* [Online] Available from: http://eur-lex.europa.eu/legal-content/EN/TXT/HTML/?uri=CELEX:32008L0105& from=EN. [Accessed 24th September 2015].

WCE (1987) *Our Common Future – the Brundtland Report.* World Commission on Environment and Development. [Online] Available from: www.un-documents.net/wced-ocf.htm. [Accessed 24th September 2015].

Weber, C.I. (1991) *Methods for Measuring the Acute Toxicity of Effluents and Receiving Waters to Freshwater and Marine Organisms.* 4th edition. EPA/600/4-90/027. Environmental Monitoring Systems Laboratory, Office of Research and Development, U.S. EPA.

Webster, R. & Lark, M.R. (2013) *Field Sampling for Environmental Science and Management.* Abingdon, Oxon, New York, Earthscan from Routledge.

Weltens, R., Deprez, K. & Michiels, L. (2014) Validation of Microtox as a first screening tool for waste classification. *Waste Management,* 34(12), 2427–2433.

WFD (2000) *Water Framework Directive 2000/60/EC of the European Parliament and of the Council of 23 October 2000 establishing a framework for Community action in the field of water policy.* [Online] Available from: http://eur-lex.europa.eu/ LexUriServ/LexUriServ.do?uri=CELEX:32000L0060:en:HTML. [Accessed 4th June 2015].

WICE (2012) *World Institute for Conservation and Environment.* [Online] Available from: www.monitoring-nature.info. [Accessed 24th September 2015].

Chapter 2

Monitoring and early warning in environmental management

K. Gruiz

Department of Applied Biotechnology and Food Science, Budapest University of Technology and Economics, Budapest, Hungary

ABSTRACT

Measurement and continuous monitoring of representative indicators and early-warning signals belong to the most effective engineering tools of modern environmental management. Emissions caused by the use and production of chemical substances may seriously affect the environment, the ecosystem and humans. Environmental managers must undertake preventive measures at the earliest possible stage to avoid irreversible damage. Efficient early warning requires special bioindicators and detecting tools to enable immediate interventions. Time series of data obtained from long-term monitoring support sustainable environmental management and provide a basis for long-term planning and legislation.

It is extremely important to define early-warning indicators for those chronic environmental risks which are not yet defined. Chemical substances that do not show short-term adverse effects but are extremely risky in the long run, and whose chronic effects are not proportional to their concentrations, are rather difficult to manage. Such obscure chemicals and elusive chronic effects require efficient biochemical indicators.

In situ and possibly online monitoring methods which enable immediate control and intervention may increase early warning efficiency.

This chapter introduces geological, hydrogeological, geochemical, physicochemical, biochemical, ecotoxicological, biological and ecological monitoring methods. It also discusses planning and implementation of the monitoring and early warning systems as well as their role in environmental management. The tools, instruments and equipment developed, as well as their commercial availability, are introduced in the next two chapters (3 and 4).

I MONITORING AND EARLY WARNING IN ENVIRONMENTAL MANAGEMENT

Monitoring is the key activity in environmental management. It provides continuous and updated information on the health status of the environment as well as on the trends of changes. An efficient monitoring system maps the environment truly and reflects every impact that may cause significant change or deterioration in the ecosystem and humans. The planning of a properly working monitoring system requires that the environment and the possible problems are known in detail and the baseline be set.

If we are not prepared to accept an imperfect monitoring system, we must iteratively approach the perfect design by continuously incorporating new information from long-term monitoring.

1.1 Definitions and the basics of monitoring, biomonitoring and early warning

Environmental monitoring is the systematically designed sampling of air, water, soil and biota in order to collect data from the environment for studying its state and observing the changes, as well as to generate new knowledge from the information and utilize this information in environmental management (Artiola *et al.*, 2004; Wiersma, 2004). The two main groups of information are: meteorological or other natural changes of global origin, and anthropogenic impacts. The latter may be unintentional or caused by technologies in direct contact with the environment.

The *purpose of monitoring* can be nature conservation, resource management, fulfillment of quality criteria, measurement of discharges from point or diffuse sources, discovery of spatial and temporal trends and many other problem- or site-specific requirements. Some typical terms for differentiation monitoring according to purpose are:

– *Surveillance monitoring* is a continual surveillance to determine if the monitored environmental compartment or phase (e.g. water or air) contains harmful levels of contaminants or fulfills the 'no effect' requirement;
– *Compliance monitoring* confirms that industrial, mining, agricultural, etc. activities comply with regulations.

Compartments to monitor can be air, water, sediment, soil and biota alone, or the system of all of them combined. A problem-specific compilation of selected indicators can also be monitored. Many scientists classify the earth's ecosystem as atmosphere, hydrosphere, biosphere, lithosphere and cryosphere, and select the scope of monitoring based on these spheres and their interactions. It is worth separately mentioning *global surveillance*, which is nowadays supported by global observation systems of the atmosphere or the oceans and managed amongst others by:

– United Nations Environment Programme (UNEP, 2011) – the World Conservation Monitoring Centre (WCMC, 2015);
– WHO and World Weather Watch – Global Environment Monitoring System (GEMS, 2015);
– World Meteorological Organization (WMO, 2015);
– Global Atmosphere Watch (GAW, 2015).

The *spatial and temporal scale* of monitoring should fit to the scale of the problem, based on the impacted area's risk model from the source through the transport pathways to the receptor organisms. The basic rule which applies is that the location of the environmental monitoring should be as close as possible to the source of the problem and the system based on adequately sensitive indicators. The monitoring system may include geological, geochemical, physicochemical and biological indicators from atomic (even subatomic) and molecular size to subcellular (virus, cell organelles) and cellular levels, microorganisms, tissues, mesoscale organisms, macro organisms (animals, plants, fungi), populations and communities, all at the relevant spatial scales. Monitoring may cover point sources, contaminated sites, waste disposal sites, agricultural fields, watersheds, states, regions, continents or Earth itself. The temporal scale may range from seconds, years and decades to hundreds or tens of thousands of years.

The environmental situation can be clarified by differentiating between:

- healthy conditions and significantly differing variables;
- stress conditions with compensated effects: response to the stressor + compensatory response of the organism to the primary response;
- curable deterioration: deterioration as outcome of environmental stress + curing response of the organism, population or community;
- irreversible deterioration.

Response is any answer (biochemical, physiological or behavioral) of a living organism that results from an internal or external stimulus, unusual impacts, stress, environmental contamination or disease. Most of the response signals are not measurable because the changes do not differ significantly from natural variables. Other responses may reach a meaningful cause-related, measurable value: they can be considered as biomarkers.

Biomarker – this term is generally used for a response signal upon exposure of organisms to stressors (e.g. contaminants) at lower levels of biological organization (genes, enzymes, other biomolecules, biochemical and immunological reactions, physiological and metabolic changes, etc.), while *bioindicators* are ecologically relevant, observable responses at higher levels of organization (population, community). Some extremely or selectively sensitive or other key species are also called bioindicators, meaning that these species are able to indicate the deterioration of the community, the food web or other community-level function.

Biomarkers are key molecular or cellular components or processes that link a specific environmental exposure to an ecosystem or human health outcome. The change in biological response may range from molecular through cellular and physiological responses to behavioral changes, which can be related to environmental chemicals. The best known biomarkers are the oxidative stress biomarkers, triggered by a wide variety of pollutants. In addition to oxidative stress biomarkers, a number of other molecules proved to be applicable for pollution-linked stress indication. In addition to biotransformation enzymes and products, several stress proteins, metallothioneins and the multixenobiotic resistance protein (MXR) may serve as indicator molecules. Parameters of hematological, immunological, reproductive, endocrine, genotoxic, neuromuscular, physiological, histological and morphological conditions can serve as early-warning end points, simply in fish (van der Oost *et al.*, 2003). Biomarkers of DNA damage can be evaluated by the comet assay or the micronucleus test – both are standardized methods and applied to pollution indication in zebra mussels. Biomarkers for long-term exposure can be the accumulating tissues or species – for example, hair for some toxic metals such as mercury, or bone for lead and cadmium. New knowledge in 'omics' such as genomics, metabolomics and proteomics is exploring new frontiers in biomarker research. The Integrated Biomarker Approach combines genomic, proteomic, metabolomic and lipid data from coordinated experiments under controlled and stressed conditions to measure and identify biosignatures. Using this approach, novel biomarkers of exposure, early indicators and new exposure–response pathway relationships can be identified (PNNL, 2015).

Bioindicators are biological processes, organism- or community-level symptoms, or measurable responses that reveal the presence of contaminants, for example:

–   body burden: the content of metals or persistent organic compounds in the cells or tissues of organisms;
–   changes in morphological or cellular structures;
–   metabolic process parameters;
–   behavior;
–   population structure, species diversity, etc.

Not every process or organism is suitable for bioindication. A bioindicator species generally has moderate tolerance to environmental stress. Overly sensitive species with narrow tolerance response to too many environmental effects, and overly tolerant species or communities cannot represent the average or the whole of a community. A good bioindicator provides measurable and sublethal responses, represents the whole population or the community response, and its response is proportional to the exposure. One species can never adequately represent the whole ecosystem. Selection of the proper bioindicators depends on the type of stress, environment, indigenous species, etc.

Genetic bioindicators, or bioreporter genes, are genes that can create easily detectable products; thus, they can enhance selectivity and amplify the signal because they only work under certain conditions.

Bioreporter organisms are genetically manipulated organisms which contain bioreporter genes.

Early-warning indicators amongst environmental indicators are those that are more sensitive or provide an earlier response than the main part of the ecosystem. The early-warning indicator gives a measurable response before significant adverse effects occur. Early-warning signals may be physicochemical, meteorological or climatic parameters such as water quality, water level, erosion and salinity. Biological early-warning indicators only differ from bioindicators in their scale of response.

Early warning systems are complete management systems for climate change detection, natural resource management, environmental change detection, food security monitoring (drought, famine and pest), water resource assessment and hazard identification/mitigation. An early warning system includes: the identification of risk to manage, plan and maintain an adequate monitoring system; an efficient communication system; preparedness for damage mitigation (coordination, good governance, appropriate action plans, public awareness and education). UNEP (2012) differentiates between early warning for rapid- and slow-onset environmental threats:

–   *Rapid*: oil spills, chemical and nuclear accidents, geological hazards, earthquakes, landslides, tsunami, volcanic eruptions, hydrometeorological hazards, floods, epidemics, wildfires;
–   *Slow*: air and water quality, droughts, desertification, food security, impact of climate variability, location-specific environmental changes.

The application of early warning systems is only reasonable if prompt action follows early detection.

End points of bioindication should be measurable values which are directly linked with the biomarker molecule, the key process or the response of the indicator organisms, population or community. Such directly fitting end points are fluorescence for the chlorophyll content, color change of indicators for a redox reaction, heat or electron flux for an energy-producing reaction, growth or growth inhibition for microbial cells or higher organisms, respiration rate or mobility, avoidance or other abnormal behavior for animals, etc.

Biomonitoring, or biological monitoring, is the use of biological responses to environmental impacts. The sensing element of biomonitoring may be either a biological/biochemical molecule or a living organism (or parts of it).

The target of biomonitoring can be:

– The direct evidence of the hazardous agent's presence (e.g. a toxicant detected by a very selective biomolecule or organism).
– The presence of a biomolecule as the specific product of the biological organism's response (enzyme response, immune response, genetic response, resistance, etc.) to the hazardous agent.
– The presence of the living cell or organism which indicates the environmental effects or changes. Biomonitoring in this case may work with: (i) sampled indigenous organisms (passive biomonitoring); (ii) controlled populations of organisms relocated into the environment (active biomonitoring); (iii) controlled organisms placed into the removed aliquot of the environmental compartment or phases on-site or in the lab (side-flow, microcosm, bioassay); (iv) parts of the organism – tissues, isolated cells or artificial models mimicking tissues.

Tools ranging from fully manual methods, automated readers, image analyzers and *in situ* microsensors to remote sensing are available for measuring an end point. A comprehensive overview can be found in Chapters 3 and 4. Evaluation of the results requires statistical methods and proper interpretation, as detailed in Volume 2 of this book series (Gruiz *et al.*, 2015).

The many advantages of biomonitoring and early warning have already been mentioned and several more are worth emphasizing; for example, they are environmentally realistic, give a rapid response and make dynamic decision making and immediate intervention possible. Their weaknesses may originate from improper planning or imperfect knowledge and, as a consequence, oversimplification of a complex system, using nonrepresentative indicator species, using only one or too few indicator species, or not acquiring time series.

1.2 Orientation of monitoring from sources to the receptors

The efficiency of monitoring depends on the quality of data, the soundness of the concept and the monitoring plan. Lovett *et al.* (2007) emphasize the importance of focused, relevant, and adaptive questions for guiding the development of a monitoring plan. Data collection, processing and handling should fulfill the requirements of quality management. The creation of information from data and utilization of the information by different professionals should fulfill the requirements of modern knowledge management.

The earliest warnings are forecasts which apply mathematical models based on the results of geochemical, hydrological, physicochemical, biological and/or ecological assessments. One can expect an improvement in the quality of the assessment result through application of an integrated methodology (Gruiz, 2009). Compared with traditional environmental monitoring, early warning applies more specific and selective indicators and/or super-sensitive measurement methods. Most of the early indicators in ecosystem monitoring tend to be 'omics'. These are molecular markers produced as a response to environmental stress, but which do not yet cause an observable change in growth, mobility, reproduction, etc. of the traditionally used toxicity end points. Early warning systems maximize and bring together knowledge of the environment, the ecological system and human health, and then apply the latest techniques for the characterization of the environment (physicochemical analysis, biological and ecotoxicological testing) and forecasting.

The best-fitting risk model and reliable prediction form the basis of monitoring and early warning. The conceptual risk model integrating the transport and exposure models can be used locally, regionally and globally, and it serves as the conceptual basis of environmental monitoring.

Two types of concepts are worth differentiating:

–   **Stressor-oriented** monitoring focuses on hazardous agents and the sources of contaminants. It is based on the risk model established for a single identified/known hazardous agent, or of several individual models of a number of hazards if faced by a multicausal case, typical for chemical contaminations. This type of targeted monitoring follows the source–transport pathway–receptor track and concentrates on the hazardous agent identified, its sources and its specific adverse effects. In the case of unacceptable risk, it can be managed by terminating or decreasing emissions from the source, preventing its transport and/or eliminating it from the environment of the targeted receptor.
–   **Impact-oriented** monitoring derives from the biological status of organisms, populations or communities. The aim of this type of monitoring is the continuous observation of the biological status and changes of the receptors, i.e. ecosystem and man. Recording significant differences relative to reference values provides the warning signal which triggers the search for the causes. The biomonitoring of workers in occupational risk management may be highly targeted when the hazardous agents likely to occur are known. It is rather difficult to find an unambiguous correlation between biochemical indicators and environmental pollution for an average population because of multicausal chronic effects, differences in individual sensitivity, compensatory responses and many other reasons. Improving statistical methods and tools gives hope for better correlation of the presence of hazards and the detected impacts, and identifying and managing the causes. Correlating spatial distributions may help in this task. The problems of ecosystem observation are similar to those of the human population but there are more possibilities for ecosystem data collection: as a great number of species can be observed, including sentinel species, accumulator species help to increase the sensitivity of the assessments.

The integrated risk model describes the fate of a hazardous agent in the environment, either in a virtual or real setting. The stressor-oriented risk model can be applied to one chemical substance or one infectious organism, invasive species or other source of harm. The impact-oriented model derives from the deterioration measured or observed and is generally *receptor-oriented*, using damage or adverse effects on humans or ecosystems.

1.2.1 Stressor-oriented risk model and monitoring

Stressor-oriented monitoring aims to follow the transport, fate of, and exposure to certain hazardous agents such as environmental contaminants or pathogens.

This kind of model focuses on the source of the hazardous agent, typically a chemical or biological contaminant which can be a diffuse or point source. The contaminant spreads from the source along pathways which are influenced by the characteristics of the contaminant and the environment. Transport models can describe the transport of the contaminant from the source to the targeted environmental compartment.

The integrated risk model also incorporates the exposure model and shows how the ecosystem or humans are exposed to the hazardous (chemical) agent within the endangered (contaminated) environmental compartment, e.g. by inhalation, digestion or dermal contact. A quantitative model can characterize the content/concentration of the contaminant in the source or contaminated environment by quantity (piece, mg, kg, ton, liter, etc.), concentration (piece/L, mg/kg or μg/L), or body burden (mg/kg body mass), and the transport itself is usually characterized by mass transfer/flux (e.g. kg/hour). Exposure type and rate is determined by land uses such as natural, residential, recreational, agricultural or industrial.

The transport model maps the potential transport pathways and determines the contaminant's occurrence and concentration at any time and point. The actual prevalence or concentration calculated by the model can be compared to the not yet harmful (allowable, acceptable) no-effect levels, yielding a quantitative parameter, the risk characterization ratio (RCR). Long-term monitoring can validate the model's results and refine it.

After drawing up an integrated risk model, the most expedient and 'earliest' points in space (relative to the source) and time (relative to emission and transport) of the monitoring systems in the environment can be identified. The red stars in Figure 2.1 mark these locations in relation to the source, on transport pathways or in the vicinity of the receptors.

The early warning system must be placed as close as possible to the location and time at which the emissions are likely to occur. Another requirement is that the selected indicator or the monitoring method must be highly sensitive. If the detection limit is not adequate, enrichment or other preparative methods must be applied to increase sensitivity. The selectivity requirement often goes hand in hand with sensitivity, especially in targeted monitoring. Selective early-warning signals, such as chemical analytical results or the immunochemical evidence of a pathogen's presence, enable targeted and source-specific rapid action or, in an ideal case, immediate intervention.

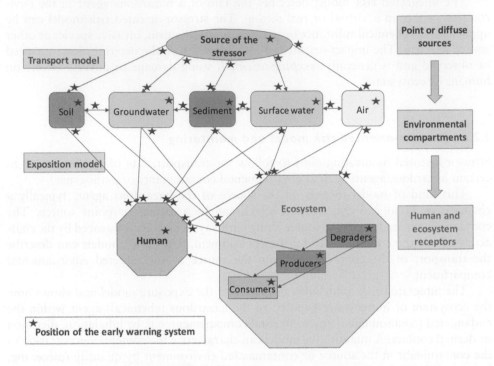

Figure 2.1 Locations for early warning systems on the conceptual risk model of the area in scope; direction of extrapolation: source–transport pathway–receptor.

Targeted monitoring, including early warning, can serve general surveillance or specific contaminant-oriented, activity-oriented and land-use compliance monitoring purposes. The first step is the listing of hazardous contaminants and other stress factors, whereupon priorities have to be set:

– If the source(s) of a hazard is/are precisely known and the hazardous agent has been identified, the particular agents close to the source can and should be monitored. Chemical analytical methods or hazard-specific effects are most appropriate for monitoring in this case. Measuring the actual effects is of high importance for contaminants with limited bioavailability. A contaminant that is immobile, non-water-transportable and unavailable for the biological entities has either no, or no high, acute risk and thus has no priority.
– If the source is diffuse and cannot be precisely identified and localized, key points such as influxes, confluences or junctions of transport pathways should be identified in order for the monitoring systems to be placed. Combining targeted (pollutant-oriented) and impact/receptor models may provide an appropriate concept for risk assessment and monitoring of diffuse pollutions.
– Both chemical monitoring and biomonitoring of the contaminant-specific biological responses are feasible for known contaminants, agents and stress factors. A pesticide, for example, can be detected by electrochemical or enzyme sensors

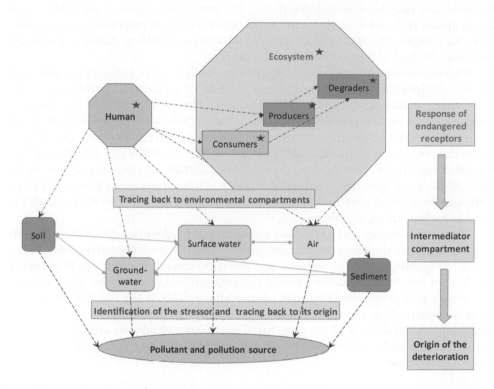

Figure 2.2 Receptor-oriented monitoring and tracing from the result back to contaminated environmental compartments, contaminants and contaminant sources.

placed *in situ*, immunological methods or sensitive test organisms, e.g. algae or luminobacteria. Sensors may be more specific to the targeted analyte, while test organisms have a broader sensitivity spectrum, and can integrate the responses of several adverse effects.

1.2.2 Receptor-oriented monitoring

Receptor-oriented monitoring aims to measure the health status and its changes for the exposed target receptor, which can be an organism, a population, a community or the whole ecosystem. Receptor-oriented monitoring applies direct-effect assessment by measuring the responses – possibly early responses – of biomarkers/bioindicators at the level of molecules, cells, organisms, populations or communities, in order to acquire information on adverse effects proportional to actual risk. This concept applies both to human and ecosystem health.

Receptor-oriented methods (e.g. human clinical-chemical or bioindicators, ecosystem bioindicators, extinction of certain sensitive species, contaminant content of accumulator organisms, diversity) usually indicate that there is a problem, but they do not provide unambiguous information about the causes (neither the type and amount of contaminant or other stress factor, nor the location of the sources: see also Figure 2.2).

Nonetheless, receptor-oriented monitoring has great environmental importance as it can pinpoint real problems via quantitative responses. These responses are integrative and can combine the effects of all causes and sources: weather, climate, season and unknown biological (e.g. virus epidemic) causes.

Receptor-oriented monitoring can perform both general surveillance and compliance monitoring. The receptors are at the end of a transport pathway, and the monitoring points are placed in that location. The concept should be based on measuring the earliest possible emerging, unacceptable level of an adverse effect. Receptor-oriented monitoring can only be aimed at selectivity in certain special cases when the response is unambiguously linked to a stress factor, e.g. a chemical or biological contaminant. This may be the case when certain enzymes appear due to the effect of their substrate (the contaminant), or an immune response is given to a certain pathogen by a target organism. These stressor-specific selective responses can also be used as early-warning signals.

If the stress factors and hazardous agents cannot be identified by selective bioindicators, the targeted environmental compartments and their ecosystem (protected area, protected species, species abundance and diversity) should be observed. Biomonitoring human inhabitants/residents within the area affected can measure the incidence, prevalence and geographic range of adverse effects. This is the domain of environmental epidemiology. Lower sensitivity, more complex biochemistry and physiology, individual vulnerability and the aggregated appearance of adverse effects of environmental, social and psychological origin make orientation more difficult than in the case of ecosystems.

1.2.3 Efficient monitoring

Selection of the proper measurement and test methods and their combination is the key to efficient monitoring. The list of recommendations below is representative but by no means comprehensive:

–   It is advisable to use direct physical measurements, chemical analyses or specific and selective biotests for identified sources, contaminants or agents. Cheap, *in situ*, online methods and devices are recommended.
–   If the problem is located in defined environmental compartments (e.g. surface water or soil), the solution is integrated physico-chemical, ecotoxicological and biological/ecological assessment of the environmental compartment and its ecosystem.
–   If the harmful effect only appears at the receptor level, the target organism, or if there is no specific target organism, key species and populations or the ecosystem's community-level indicators should be observed, e.g. the diversity of the ecosystem or the presence or lack of certain indicator species and their metabolic status/changes. Modern gene technologies are increasingly becoming part of ecosystem monitoring. The combination of gene technologies with traditional biological, microbiological, and ecological knowledge, as well as the translation of traditional knowledge to genetic information, are all of great importance for ecosystem monitoring.

People handling hazardous agents or working in a polluted environment, e.g. during environmental remedial works, may be an appropriate target population for biomonitoring. The advanced state of practice in occupational safety and health and human toxicology enables contaminant-specific indicators to be investigated through regular clinical-chemical control tests.

A combination of source-oriented and receptor-oriented monitoring is the most efficient solution in many cases. Lack of harmonization and poor aggregability may hamper success.

The collection, storage and use of environmental data, as well as uniform, harmonized and standardized methods for data acquisition, play a key role in monitoring. Unfortunately, epidemiological studies and biomonitoring as well as environmental and ecosystem monitoring activities are not harmonized and the acquired information has not yet been aggregated. Data handling based on geographical information systems (GIS) and statistical systems may improve the situation in the future.

Environmental monitoring is overseen throughout most of the world by governmental agencies, authorities or special environmental departments. However, several measurements and observations are made by companies and civilians and their results often fail to be included in environmental databases – a practice which has not been remedied yet. An encouraging exception is the national monitoring program of the Swedish Environmental Protection Agency (SEPA), which is a continuous monitoring program compiling valuable observation series (for the longest timescale of any existing observation series in the world), and ensuring maximum efficiency of monitoring programs across the country. All possible monitoring data are collected by national and municipal government agencies, private industry consultants and non-governmental organizations. Detailed monitoring guidance criteria and regulation are provided by SEPA to ensure consistency, quality assurance and quality control of data collected by different agencies and organizations. Data records are readily available through the Agency's website (SEPA, 2010).

The Community Based Environmental Monitoring Network (CBEMN) represents a new approach. CBEMN encourages individuals and groups to initiate monitoring activities, lends out equipment through the Environmental Stewardship Equipment Bank (ESEB, 2015) and provides long-term support for monitoring (CBEMN, 2015).

2 EARLY WARNING, FORECAST AND ENVIRONMENTAL RISK ASSESSMENT OF CHEMICALS

Early warning covers two basic areas:

1. Immediate or short-term concerns – weather events (hurricanes, cyclones, tornadoes), climatic variations (El Niño, droughts), geophysical events (earthquake, tsunami), accidental pollution or illegal discharges. These are often unpredictable events whose management requires specific monitoring and preparedness. Specific national and international organizations focus on the early warning of these events (DEWA, 2015).
2. Recognizing 'emerging' environmental threats that can now be observed due to new scientific knowledge or higher level of awareness. They also may derive

from new phenomena, newly increased scale, accelerated rate of environmental processes, or a new way of managing the problem. Some typical emerging issues:

- environmental degradation;
- increased ecosystem vulnerability;
- cumulative environmental stress, e.g. mutagenic, reprotoxic effects and bioaccumulation;
- stress/contamination formerly not perceived as risky, such as of many the endocrine and immune disruptors.

Many of the adverse impacts cannot be immediately observed due to the ecosystem's high adaptive potential. These kind of initially hidden impacts insidiously degrade the environment and the ecosystem's functioning and may lead to the 'explosion of a time bomb', e.g. famine and plague.

Forecasting environmental harm/damage can be considered a generic form of early warning. In the case of chemical pollution, forecasting is based on the amounts of chemical substances produced and/or used, their behavior in the environment, and their harmful effects. The environmental parameters of temperature, wind strength, precipitation, drought, flood, etc. play an important role in damage forecast. The results of forecasts, the predicted exposure levels or risk values, provide a warning signal to take precautions or make interventions. Precautions can be implemented when the contaminating substance or other stress factor is known. Assuming, for example, that chemical substances are produced at a chemical plant's known locations, the prediction of exposure is based on the amount produced and the proportion of the chemical released (release factor). The predicted exposure level provides some direction, too, for the monitoring design.

Unfortunately, chemical products, e.g. pesticides or paints, may be used at highly uncertain locations where only the region of distribution is usually known and the diffuse sources and transport pathways are impossible to trace. Quite often the amounts of the substance produced, sold, exported or imported, together with its physicochemical and ecotoxicological properties, constitute the only information available. This information only fulfils the requirements of a hazard assessment and not those of risk assessment. If the spatial scope is a country, a watershed or a region, generic risk assessment can be carried out based on hazard information for the chemical substance and the generic characteristics of the targeted area, which is considered as a homogeneous and average entity. The generic risk value obtained enables early warning even during the planning period of production and use, and countermeasures, restrictions, or risk-reduction interventions can be implemented to prevent an unacceptable level of risk. This is the aim of many regulations such as the European REACH (2006), CLP (2008) and most environmental licensing procedures.

The globally harmonized warning system (GHS, 2011) for classifying and labeling hazardous chemicals established a uniform signal system for warning and advising manufacturers, suppliers, dealers and buyers of chemical substances. The EU REACH regulation applies a generic risk assessment in its chemical-substance-linked risk management, introducing more stringent regulations, limitations, risk reduction and more widespread risk management for hazardous substances. Some substances, such as pesticides, which are released into the environment in large quantities and in an almost uncontrollably diffuse distribution pattern, are also controlled by this type of

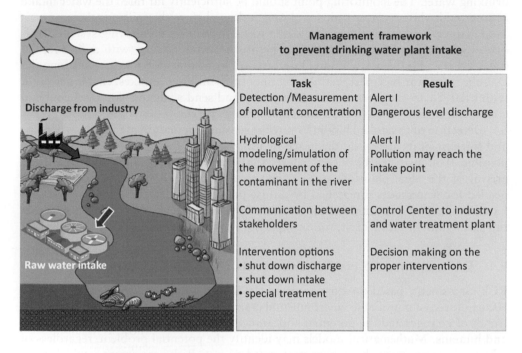

Figure 2.3 Monitoring, early warning and preventive measures.

forecast-based, generic warning system. In such cases, additional environmental monitoring should be implemented for the most exposed environmental compartments and ecosystems as required by the US Clean Water Act (CWA, 1972) and the European Water Framework Directive (WFD, 2000) for waters.

Thus, forecasting plays an important role in early warning. Forecasts concerning production, transportation, distribution and use of chemical substances, and their role as waste in the environment, can be based on databases, statistics and actual measurement results.

Quantitative environmental risk assessment of chemical substances is one of the most effective forecasting methods, which can issue a warning about potential problems prior to the production and utilization of a substance.

Risk assessment is a forecast based on static data and modeling (transport and exposure models), and the models are based in part on monitoring data, database data and mathematical models. Monitoring data can be used for the validation of the models, whereby modeling and monitoring are in close interaction. Modeling helps to find the proper monitoring approach and the best location for the monitoring points. Information on preventive actions (e.g. time requirement from the alert to the intervention) and the transport of the contaminant (e.g. transit time from source to the compliance point) determines the optimum location of the monitoring point. In return, monitoring data can fine-tune and validate the model.

Figure 2.3 illustrates a raw water intake at a drinking water treatment plant. The discharge from an industrial facility located upstream may impair the quality of drinking water. The monitoring point should be sufficiently far from the water intake as to provide the necessary time for closing it down before the spill gets there.

Chapter 3 describes the example of a metal-measuring monitoring system developed by Mezei and Cserfalvi (2007). The *in situ* metal sensors (with a Zn 0.1, Cd 0.05, Cu 0.2, Pb 0.2 and Cr 1 mg/L sensitivity) located at different points along the Danube river near Budapest, mainly at the hot spots of sewer influxes, enable detection with a 10- to 15-minute measuring frequency and send a distress signal when metal concentrations are close to environmental quality criteria.

Detection of toxic algal bloom by satellite sensing, simulation of the spread of algae and forecast of its arrival on the beach by using weather forecast models has become possible. If intensity reaches a certain level, the beach will be closed before the predicted arrival of the algal plume to prevent human exposure. Longer-term monitoring is needed for watersheds threatened by diffuse pollution. Thus, the creation of a GIS-based model and its calibration as necessary by real data, e.g. water balance data, for the prediction and control of, for example, nutrient and pesticide use in agriculture.

2.1 Environmental risk of chemicals as early warning

Risk assessment, based on existing or newly acquired data and mathematical models, can forecast a probable environmental concentration of a chemical substance, the trends of changes and the potential appearance of an adverse effect on the ecosystem and humans. Mathematical models may identify the potential problem, regardless of whether measurements have been performed or not. Risk assessment – which is an approximate calculation yielding an estimate – is an efficient and inexpensive method that can work without any measurements; consequently, no traceability limit (compared to chemical analysis) and no minimum effective concentration or dose (compared to toxicity testing) are needed. Today, a large part of the workload and cost is made up of the establishment of databases (needed for risk assessment models) which have to be compiled for most of the survey cases because the currently available databases are incomplete, inaccessible or contain poor-quality data.

The earliest alert can be achieved using the forecasts based on risk estimation. At the same time, however, the environment's heterogeneity and uncertain usage of chemical substances limit the quality of the risk values. Thus, there is an inaccuracy in the spatial and temporal forecast at a local level, and the uncertainty increases with an increase in the area or duration. This problem is similar to that of individuals' exposure and vulnerability. If there are a few people whose lifestyle, living space or personal metabolism indicates that their risk is above average, it does not help if we say that the pollution caused by a chemical substance is generally negligible. The personal differences can only be considered when using individual assessment, e.g. by considering individual response characteristics, but specific safety factors are applied for extrapolation due to the lack of this information. When using safety factors, the average is considered as the baseline, and depending on the sensitivity of a certain population or individual, smaller or greater factors are applied. These are multiplication factors giving a result proportional to the average, which simplifies the real situation. The size of the safety factor depends on the spatial (geological, hydrogeological, land cover, land use), time-related (acute, chronic exposures, lifespan) and personal (age, sex, pregnancy) differences. As the name suggests, these factors make our decisions safer. At the same time, it may result in significant overestimation of risk and lower efficiency due to the increased costs of environmental management.

2.2   Environmental risk management of chemicals as an early warning system

Environmental risk assessment (ERA) is used to forecast the damage caused by a chemical substance at a general (global, regional) or local level. At *general level*, a generic risk value is calculated that shows unequivocally whether the risk due to the production, use or environmental presence of a chemical substance exceeds the acceptable risk level; thus, it indicates that the problem requires attention and a complex management strategy.

The forecast at general level concerns a regional (Europe) or global (planet) level: it is not location-specific and it does not provide concrete information for local levels. This is because it includes neither the geographical distribution of the chemical substance nor the characteristics and uses of the exposed environment.

A *local-level* forecast can be made based on data about local production, usage and environmental presence of chemical substances. Their whole life cycle has to be taken into account, and it is worthwhile to separate the processes of production, use, and their life as wastes. The usage and waste phases have to be subdivided into further categories – for example, professional (industrial, agricultural), household, point, diffuse, etc. The characteristics of the local environment are equally important factors in this kind of risk assessment. Locally occurring point-source emissions represent the simplest case. A typical example of this type is the production of the chemical itself, and the local pollution and risks associated with the chemical.

2.3   Risk of production and use of chemicals

The risks associated with the *production of chemicals* can be forecast very precisely. The basis is a good conceptual model which needs information on the production technology, the emissions of the chemical substance and the local environment. The forecast can be continuously validated with the help of a monitoring system placed in or close to the source. The dominant transport pathways originating in the source are determined by the characteristics of the chemical substance (volatility, water solubility, biodegradability, etc.), the production technology and the characteristics of the environment (temperature, wind, proximity of surface water, soil type, etc.). If the substance is volatile, the air in the factory and the amount of vapor coming from the production equipment have to be monitored; while in the case of water-soluble substances, the wastewater needs to be monitored at the point where it is produced and/or before it is discharged into the surface water from the wastewater treatment plant. If there is a risk of underground leakage, the groundwater has to be monitored close to the storage tank or pipe(s). If the substance produced is sorbable to solids, the soil phases at the hot spot need to be monitored.

The production process of a chlorinated hydrocarbon is, for example, well known: the technological steps, production processes, reactors, intermediates, substance forms, transport processes, the possibilities of exposure to the environment and workers, etc. The physicochemical and environmental characteristics of the substance are also known (its volatility, water solubility, sorbability, its partition between the physical phases, its Henry and K_{ow} constants, its propensity to photodegradation and hydrolysis, its biodegradability, toxicity and other effects, etc.). Thus, the 'hot spots' in the production environment can be identified and the early-warning monitoring points can be arranged in positions that take these hot spots into account.

The hot spots have to be ranked according to the level of risk and calculated from the maximum amount of substance released into the environment at the potential emission points. Measurement and monitoring points have to be set up in these places, and measurements have to be arranged that determine the physicochemical or biological end points with adequate levels of sensitivity. If the measurable concentration in the air or water derives from the discharge, then a direct measurement is carried out by setting up selective sensors for continuous online signal production, or by sampling and analyzing the environment periodically using *in situ*, on-site or laboratory methods. If the emissions do not cause a directly measurable concentration, then the air and/or water have to be led (sucked) through selective sorbents to enrich/concentrate the contaminant to a detectable level. By placing selective and sensitive sensors into the environment or into the sorbents of the samplers, continuous signals can even be obtained. Knowing the partition of the contaminant between the mobile (gas, liquid) and solid environmental phases (soil) that are in contact with each other, we can extrapolate to the contaminant concentration in the solid phases from the results obtained in the mobile phases.

The problem becomes more difficult if the issue is the *use* of a chlorinated substance. If it is a professional, e.g. industrial, utilization then the hot spots (identified in the site- and use-specific risk assessment) have to be monitored in order to supervise their emissions, analogously to production monitoring. If a chlorinated hydrocarbon, e.g. tetrachloroethene (PCE), is used for washing clothes in a laundry, the concentration of the vapor from the washing machines and the concentration of the PCE discharged with the wash and rinse water need to be measured at the production and possible emission points, in addition to the emissions during the residue's treatment, storage and transport. In fact, these near-source control measurements are only needed to validate and prove that there is no emission greater than the values forecast (which were precisely determined according to the characteristics of the chemical substance and the environment, and whose values were accepted as permissible when issuing the permit). If the emission is bigger than that forecast, then countermeasures must be taken.

If the chemical substance is used in households or in agriculture without documented locations, its use is considered to be diffuse.

Figure 2.4 shows the flow chart of site-specific risk assessment and risk-based monitoring using early-warning indicators. Both procedures are based on the integrated risk model shown in Figure 2.1.

In our example, a generic risk assessment methodology is applied with respect to sensitive land uses and receptors (children, kindergartens, sensitive ecosystems) to assess the risk of the use of a chemical substance. A highly site- and use-specific assessment could also be developed, for example, dealing with the special needs and living conditions of families, or the work methods and traditions of farmers, but this is not common yet. However, it is very likely that the improvement and sophistication of risk assessment methods and the prioritization of individual needs will result in such individualized forecast systems.

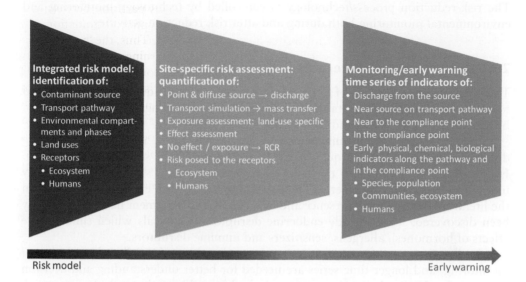

Figure 2.4 From the concept of risk assessment to monitoring and early warning.

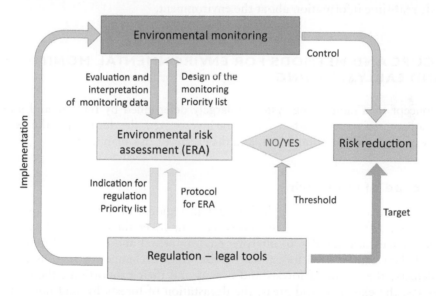

Figure 2.5 The role of monitoring in environmental risk management.

Environmental monitoring plays a key role in every phase and every task of environmental management. Figure 2.5 illustrates the interactions between monitoring, risk assessment, risk reduction and legislation. Monitoring data provide information for risk assessment and monitoring targets are selected and the system designed based on risk values. Monitoring specifies priority hazards via ERA and hazardous contaminants are monitored in compliance with statutory obligations.

The risk-reduction process/technology is controlled by technology monitoring and environmental monitoring both during and after risk reduction activities.

2.4   Data for exact forecasts

The precision and adequacy of forecasts mainly depend on the quality of the data used for the calculations. If no suitable data are available in existing databases, they must be provided by assessment and monitoring.

Currently, there are a number of gaps in the databases, and even the existing data are not always of high quality, i.e. they do not provide sufficient information. For example, in addition to the known effects of common substances, previously unknown effects may also emerge. These substances are called 'emerging pollutants' in the literature because their presence and effects in the environment have only recently been discovered. They include endocrine disruptors (chemicals which simulate the effects of hormones), allergens, sensitizers and immune disruptors.

As there is a requirement to increase the quality, level of detail and resolution of monitoring, and longer time series are needed for better understanding and decision making, the data and data acquisition methods should continuously be improved. A number of new measuring methods and IT-supported statistical evaluation tools can provide real-time information about the environment.

3   SCOPE AND METHODS FOR ENVIRONMENTAL MONITORING AND EARLY WARNING

The concept of a monitoring system is largely determined by the size and intensity of the source, the territory affected or to be investigated, the duration of the response, the intervention's time requirement, and the extent of risk management.

3.1   Scope of monitoring

The indicators and methods applied at global, regional or local levels are different (Figure 2.6). At a global scale, methods that can cover the earth's surface, such as satellite images, are suitable to interpret environmental and ecological trends. Time series of data can be used to estimate global trends, changes in land use, population density, the extent of forests and oceans and their proportions, the expansion of deserts, the extent of arid areas, the devastation of forests by acid rain, soil erosion, atmospheric risks, the health of vegetation, etc. The primary goal is to identify endangered global functions and the shifts in the borderlines. The indicators applied are mainly the dimensions, e.g. the size of an area with a certain function (forests, cultivated land, arid areas, etc.), and the function-bound quality, which can be observed from great distances (soil moisture content, vegetation density, chlorophyll content, etc.).

Regional-level monitoring systems will mainly use those indicators that provide information about the catchment area. Remote-sensing methods, catchment-scale models and methods identifying and tracing the transport pathways have priority

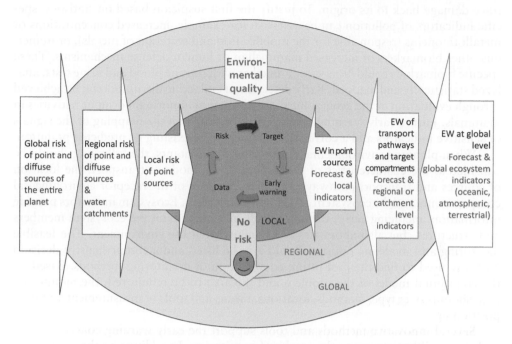

Figure 2.6 Local, regional and global level emission and early warning (EW).

importance in regional-level monitoring, although observation of the target environment of the surface-water network and sensitive local indicators of the water quality of the watershed are of similar importance. Physicochemical indicators, e.g. indicators that show the presence of a chemical substance, have the same significance as measurements of their harmful effects on the ecosystem, on parts of the ecosystem, or on humans.

Local-level early warning systems (EWS) are based on the observation of pollution sources: mainly point sources and their emissions. The priority indicators used here are the selective physicochemical measurement end points. If there is no identified source or substance responsible for the local risk, an assessment of the effect may provide the solution.

An adequate risk model is needed to configure the concept of a good environmental monitoring or early warning system. The subsequent steps are: selection of the relevant indicators; selection of the respective techniques for measuring the selected indicators; data analysis and interpretation in order to use the measured data for decision-making.

Modelling is a basic element of monitoring; timeline data can be interpolated and extrapolated both in space and time, which makes mapping and forecasting possible.

Human health indicators can also be used as early-warning signals. Based on morbidity and mortality statistics, and on those for congenital diseases, the presence of unacceptable levels of toxic, mutagenic and reprotoxic chemicals can be predicted in the environment. If high-quality, reliable, residence-based epidemiological and clinical chemistry lab data were systematically collected, it would be possible in many cases to trace damage back to its origin. To justify the first suspicion based on statistics, specific indicators of pollution can be assessed; for example, increased concentrations of metallothioneins (responsible for the mobilization and secretion of metals), or numerous other biomarkers of increased mammalian or human defense mechanisms. These specific biomarkers could be assessed, together with routine blood and urine tests, analyzed statistically and mapped. Early detection of endocrine disruptors can be achieved through evaluation of the geographical distribution of abnormal hormone activities in mammals. Future early warning systems may be based on the mapping of the signals of sensitive indicators and a correlation to the map of existing or predicted pollution.

When planning environmental monitoring of pollutants, the following should be considered: the characteristics of the pollutant; the mode of transport; the location of sources and transport pathways; the characteristics of the receptor environmental compartments and the ecosystem and people using them. Ecosystem indicators provide information, including early-warning signals, not only about the ecosystem members and structures, but also about risks to humans from the environment. The feasible measuring and modeling options should also be listed and a comprehensive decision pathway used to find the best-fitting combination of monitoring methods. Based on the conceptual model of the problem and the area to be monitored, the monitoring plan should cover type, methods, locations, means and tools of measurement and their positioning.

Several innovative methods and tools support the early warning concept and an earliest possible response to the results of monitoring. In addition to the monitoring of targeted hazardous agents, e.g. detection of contaminants by chemical analysis, developments are increasingly focusing on biomonitoring and ecotoxicology. *In situ* measuring/testing of toxicity is going to become a priority tool in environmental management. Mobile laboratories, toxicity test kits and online applicable toxicity sensors based on molecular or whole-cell responses provide the technical basis. The theoretical bases of these methods are discussed in Section 5 and their application to environmental practice in Chapters 3 and 4.

Monitoring practice approaches environmental risk from two sides: from the perspective of the hazardous chemical substances discharged into the environment, and from the perspective of living organisms, i.e. the receptors impacted by the pollutants' adverse effects. These are discussed below through the example of chemical pollution:

1.   The chemical model is stressor-oriented and applies physicochemical monitoring. It measures the concentration of the priority contaminants and calculates their risk one by one. This model is based on the assumption that the measured chemical concentrations are proportional to the adverse effects and the probable damage, i.e. the risk. To calculate the risk, in addition to the actual concentrations, more information is needed about the chemical substances, such as their physico-chemical characteristics, biological (adverse) effects, fate and behavior in the environment, degradation characteristics, reactivity, metabolites and behavior in the food chain. These data are available for industrially produced chemicals, but information is imperfect for many existing chemicals used in large quantities and so risk assessment will also be imperfect. Aggregation of the risks of individual pollutants is not solved in this model either. Moreover, only the risks of those chemicals which were included in the monitoring program are taken into account.

2. Ecosystem monitoring considers the ecosystem as a whole, theoretically, but in practice it looks at representative or key species or populations and trophic communities. Species abundance and diversity, or other bioindicators such as accumulated contaminants, can be determined by the indigenous ecosystem members. The monitoring result, in this case, is an end point aggregating all manifested environmental conditions and adverse effects without assignments to contaminants present or differentiation between environmental conditions and anthropogenic impacts. The resulting index is proportional to the scale of deterioration compared to a reference. The reference is practically unidentifiable and assessment is generally time- and labor-intensive. The causes of deterioration can be assumed after detailed pollution analysis and mapping.

3. Instead of monitoring the ecosystem as a whole through its indigenous organisms, controlled representative organisms can be monitored under natural conditions (active biomonitoring) or under controlled conditions (microcosm tests, simulation tests, bioassays, biosensors). In these cases, a reduced but statistically still appropriate number of bioindicators (biomarker molecules or bioindicator species) are used and contaminant-specific effects measured. Test conditions can be standardized or they can simulate the real environment. These test options – which fall within the category of ecotoxicology (see Volume 2 in this book series (Gruiz *et al.*, 2015)) – are based on the overall or selective response of given bioindicators to all, or only to one specific, environmental deterioration. The main challenge of this method is in identifying how to extrapolate from the results to the ecosystem or humans and find the gateway to chemical models.

4. The integrated approach applies chemical monitoring, ecosystem assessment and ecotoxicology in one complex system and evaluates the three types of information in an integrated way.

4 MONITORING BASED ON GEOPHYSICAL, GEOCHEMICAL AND CHEMICAL DATA

Geophysical, geochemical and chemical analytical data represent the basic information in environmental monitoring. Most of these data do not vary over time widely, and the existing databases and reference values can be applied for environmental risk assessment. Contamination and anthropogenic stress, however, do vary both spatially and temporally, and monitoring thus often only focuses on these parameters.

4.1 Geophysical and hydrogeological methods

Geophysics and hydrogeology traditionally use *in situ* methods. Physical methods are applied for positioning, for invasive and non-invasive exploration of the surface and subsurface, and for testing the water phase and the sediment of surface waters. The aim is to get an overview of the character and the contamination of these environmental compartments.

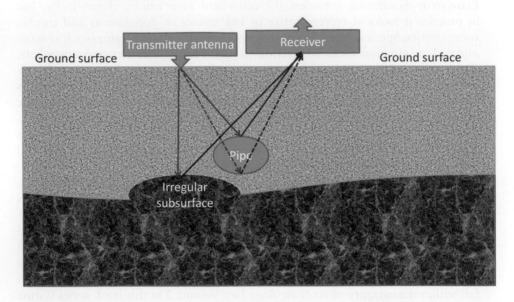

Figure 2.7 Ground-penetrating radar: transmitter antenna sends electromagnetic waves into the ground and the reflected waves are detected by the receiver, sensing subsurface irregularities, interfaces or buried objects such as pipes, canals and cables.

4.1.1 Applications

Positioning is the first step in every environmental assessment: the exact identification of the location, using pocket survey transits, compasses, binoculars and *GPS devices*.

Non-invasive exploration of the subsurface is possible by measuring gravity, magnetic potential, electrical field (e.g. electrical resistivity), the earth's self-potential or induced polarization, as well as electromagnetic field.

Ground-penetrating radar (GPR) is able to identify buried objects or chemical pollution; seismographs can identify movements in the earth by measuring sound waves. Sounds of the earth (earthquakes, volcanic disruption, etc.) detected by geophones, seismic reflection and refraction may also provide useful information about the subsurface (Figures 2.7, 2.8 and 2.9).

Passive magnetic resonance subsurface exploration (PMRSE, 2010) is an innovative technology based on the non-intrusive reception of the earth's natural electromagnetic fields and extraction of a useful signal from electromagnetic noise using the method of stochastic resonance. The objects of investigations are subsurface irregularities, direct exploration of minerals and investigation of contaminants at polluted sites.

Several other engineering tools – including innovative ones – support geophysical and hydrogeological observation and assessment of the environment:

–   water level and water flux meters for surface waters;
–   water level and interface meters, e.g. for oil/water interface;

Figure 2.8 A touchable interface for georadar management and data acquisition.

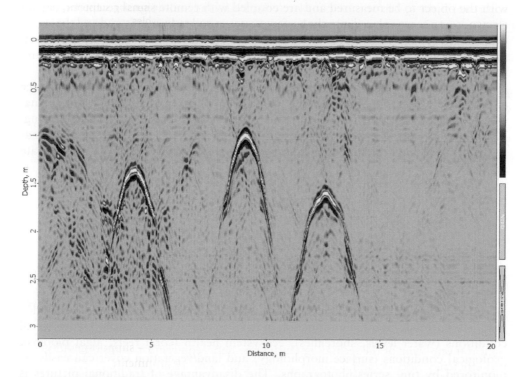

Figure 2.9 Detection of three metal pipes buried in soil at a depth of 1.0 to 1.5 m. Each pipe generates a path signal with a hyperbolic shape whose apex corresponds to the pipe location. Sounding frequency: 900 MHz. Location area: vicinity of Daugavpils, Latvia (Radar Systems, 2015).

- depth and flow meters for subsurface waters;
- turbidity meters;
- temperature, pH, redox potential, conductivity and turbidity meters in mobile design;
- drilling rigs, direct push equipment;
- visual assessment tools, e.g. borehole cameras;
- portable data storage, processing and handling based on information technologies.

4.1.2   Data sensing, storage and processing

The demand for remote data sensing, storage and processing has resulted in several combinations of *in situ* and remote technological solutions.

Remote data storage, processing and handling systems based on information technologies may be coupled to sensors either placed *in situ* or remotely. 'Remote' generally refers to observing, perceiving or sensing objects or events in faraway places. However, a sensor may also be 'remote', in a lab far from the measuring point located in the field, in an airplane or a drone, for example, as well as a satellite connected with a point on Earth. A remotely applied tool can be both the sensor itself and the signal receptor of a field-placed sensor, meaning that either the detection or the data acquisition and processing may be implemented remotely. Sensors placed *in situ* are in direct contact with the object to be measured and are coupled with remote signal receptors, e.g. an electrode or a sensor placed into the surface water or soil to be monitored and the signal transmitted to the data receptor in a water quality laboratory. In this case, traditional physical measuring methods are coupled with remote data collection, processing and handling.

Remote sensing is a method in which the sensor itself is located at a remote place and is not in direct contact with the location to be assessed or monitored. Visual observation of objects without contacting them is also considered remote sensing. The human eye is the sensor in this case, and the brain is the data processor. Alternatively, primary or reflected electromagnetic waves are detected remotely by appropriate sensors.

The *primary signal* in remote sensing (e.g. visible, infrared or fluorescence spectra) is obtained from a remote place (e.g. a satellite, spacecraft, airplane, drone, balloon or high tower). The products of remote sensing are digital images.

4.1.3   Remote sensing from the air and space

Some widespread airborne remote sensing methods are reviewed in this section (see also NRC, 2000).

Aerial photography: historical photographs of industrial, mining and waste disposal sites provide essential information in environmental management. Hydrological conditions (water levels, shorelines), ecosystem health (coral reefs, algal blooms), geological conditions (surface morphology) and land/vegetation cover can easily be monitored by time series photographs. The disadvantage of traditional pictures is that they cannot harmoniously be integrated into digital tool systems; therefore satellite-produced digital images are gradually replacing earlier aerial photographs (Figure 2.10).

Figure 2.10 Aerial photography/video from plane, drone (OPC, 2015) and balloon (KAPshop, 2015).

Drones – small unmanned aircraft – have revolutionized aerial monitoring. The terms Unmanned Air Vehicle (UAV), Unmanned Air System (UAS) or Remote Piloted Aircraft Systems (RPAS) reflect the differences not in technology, but rather more in attitude – emphasizing the role of the control personnel, the pilots on the ground. These originally exclusively military tools have become widespread scientific and engineering tools in the last five years. Drones are extremely versatile and mobile, remotely controllable devices which can hover in midair, and are able to flip and spin and maneuver with precision. The most sophisticated ones may be equipped not only with a stabilized video camera but several other sensors; for example, high-power zoom lenses, infrared, ultraviolet and thermal imaging, LIDAR (Light Detection and Ranging), different radar technologies, video analytics for comparative image evaluation, or biometric recognition to identify living organisms, including human individuals. Drones can efficiently support scientific research by tracking and monitoring environmental characteristics and changes. Compared to satellite imagery, UAV aerial imagery provides more detailed (higher resolution) and continuous information. It may be beneficial in the follow-up of rapidly changing environments, accidents, construction or remediation and rehabilitation activities.

Balloons are traditionally applied to meteorological purposes and atmospheric research. Weather balloons can reach a 40 km altitude and are conventionally used for pressure monitoring in meteorology. Balloons equipped with specialized radiosondes are used for determining ozone concentrations and other airborne contaminants or radioactivity. Equipped with a photo or video camera, tethered small balloons are constructed for environmental applications such as the follow-up of clean-up processes, mining activities, surface transport of contaminants or accidents.

Figure 2.11   Earth observation satellite ERS 2 (Poppy, 2015).

A *high-altitude platform* (HAP) is a specially designed, quasi-stationary aircraft that deploys sensing and measuring devices while staying at an altitude of 17–22 km (higher than powered aircraft) for hours or even days. The new generation of HAPs are planned to stay in their orbit for several years.

Stratospheric airships are in a developmental phase and will operate at an altitude of 10–20 km. These are remotely operated, unmanned aerial vehicles equipped with versatile remote sensing devices and data recording and transmitting tools.

A great number of *satellites* are in operation for the purpose of spaceborne environmental monitoring. Approximately 3600 have been operating recently, about 500 of which are in low-Earth orbit (160–2000 km), 50 in medium-Earth orbit (at 2000–20,000 km), and the rest in geostationary orbit (at 36,000 km). A comprehensive collection of satellite missions is published by the Earth Observation portal (EO, 2015) of the European Space Agency. A few examples of such satellites are described below:

–   Landsat-8 is the oldest satellite for global environmental studies and management, the imagery of which has been used for several land surface monitoring tasks aimed at resource management and global change issues. It has provided information on spectral and spatial characteristics of the earth's surface since 1970, thus enabling the long-term study of land cover and land-use changes (Landsat, 2015).
–   ERS-2, the European Remote Sensing Satellite-2, was launched in 1995 and retired in 2011 (see Figure 2.11). It was an enhanced copy of ESR-1, with the mission of environmental monitoring in the microwave spectrum, in particular, for the observation of oceans, polar ice, land ecology, geology, forestry, wave phenomena,

bathymetry (underwater depth), atmospheric physics and chemistry, as well as meteorology (ERS-2, 2015).

- Copernicus (2015) missions (e.g. Sentinels 1, 2 and 3) represent the EU's contribution to GEOSS (2015), the Global Earth Observation System of Systems. Two existing Copernicus satellites are equipped with a Synthetic Aperture Radar (SAR) imaging constellation for land and ocean services, and are responsible for continuous radar mapping of the Earth, mainly targeting global changes.
- TerraSAR-X (2015) is a German SAR satellite mission started in 2002 with the scientific objective of making multi-mode and high-resolution X-band (a microwave region) data available for scientific applications in hydrology, geology, climatology, oceanography, environmental and disaster monitoring, and cartography.
- Many other satellites serve environmental monitoring and research. In 2015 alone, the following nine new land observation satellites and minisatellites were launched (EO, 2015): BIROS, Copernicus Sentinel-2, Copernicus Sentinel-3, Iridium NEXT, KOMPSAT-3A, NovaSAR, PISat (PESIT Imaging Satellite), SEOSat and SMAP (Soil Moisture Active Passive).

Real-time satellite information is available on the public website of IDEA (2015), a former partnership – between NASA, the US Environmental Protection Agency (EPA) and the National Oceanic and Atmospheric Administration (NOAA) – to improve air quality assessment, management and prediction by integrating NASA satellite measurements into EPA and NOAA analyses. Since 2008, IDEA has been running at NOAA's Center for Satellite Applications and Research (STAR), hosted on its web server.

4.1.4 Airborne and spaceborne sensors

Airborne and space-borne sensors detect reflected radiation that originates from the sun or from artificial electromagnetic radiation sources. Sensors can be selective to certain wavelengths of sunlight.

Multispectral scanning means the simultaneous creation of digital images in different wavelength bands. Reflectance data or thermal infrared data are quite useful in environmental management. They can be processed by image analysis algorithms using statistical models based on laboratory-measured physical data of the objects and empirical correlation between the image and the real character of the environment, e.g. type, density and health indicators of the vegetation cover (Figure 2.12).

Multispectral imaging deals with several images in discrete and somewhat narrow bands. The Landsat-8 satellite is an example of the use of multispectral imaging.

Hyperspectral imaging (HSI) covers narrow spectral bands over a continuous spectral range and produces the spectra of all pixels in the scene. The distinguishing characteristic of HSI is that it is able to detect more than 1,000 bands (human eyes can detect only three bands: red, green, and blue). The fine wavelength resolution enables the identification of plant characteristics, contaminants, minerals and, combined with remote sensing, one can monitor endangered land, urban areas, agricultural lands, harbors, industrial or mining areas, atmospheric dust and contaminating gases, as well as natural land and forest. A hyperspectral sensor/camera can provide immediate and accurate results, which can be handled similarly to other imaging systems. Environmental compartments, geological formations, chemical substances, living organisms,

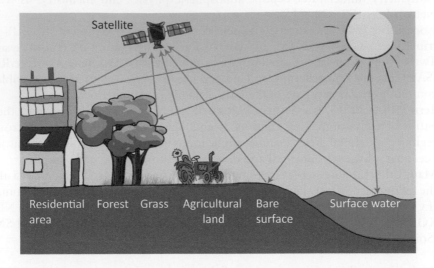

Figure 2.12 Multispectral scanning of reflected solar radiation by a satellite.

plant species and many other objects can be characterized with a specific hyperspectral fingerprint, which enables their identification on any of the hyperspectral images. If one has the fingerprint of, for example, an invasive plant species, its presence can be detected in the airborne hyperspectral image if the data processor has been trained for its previously recorded HSI 'signature'. Given the specific HSI signature of a contaminant or a mineral, an airborne or space-borne HSI image can prove its presence in the environment. The sensor resolution must be aligned to the size of the object.

Sensing *electromagnetic radiation* remotely requires an energy source (e.g. electromagnetic radiation) to illuminate the target. The sun produces the whole spectrum of electromagnetic radiation. Sensors can measure the electromagnetic radiation returned by the earth's natural and anthropogenic features. The source of the transmitted radiation can be the sun (passive sensing) or an artificial source such as a radar transmitter (active sensing). Different objects return different types and amounts of electromagnetic radiation, and the sensors can detect these differences using specific instruments (Figure 2.13).

Imaging spectrometry is useful in mapping surface minerals – for example, acid-generating rock or mine waste. Some other innovative remote-sensing techniques are *passive microwave radiometry*, *radar interferometry* and the application of different *lasers*.

4.2 Geochemical and chemical analytical methods

Geochemical and chemical analytical methods are used to describe the chemical composition of the earth's gas, liquid and solid-phase compartments in the atmosphere, surface waters and sediments, soil and subsurface waters. Physico-chemical identification and measurement of contaminants are among the other goals of these techniques.

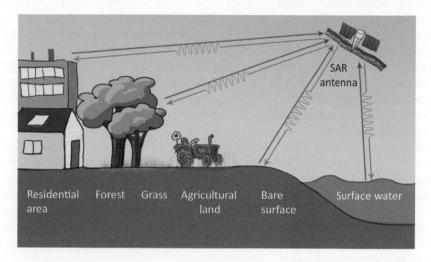

Figure 2.13 Receiving reflected radar signals with a SAR antenna, utilizing the relative motion between the antenna and the target surface to achieve finer spatial resolution.

Both sensors placed *in situ*, with portable or remote reading, and remotely working sensors can be applied.

4.2.1 Gas phase monitoring

Normal air components and airborne particles can be detected by *in situ* sensing – coupled with an *in situ* or remote data logging and processing system – or remote sensing of different layers of the *atmosphere*, including the earth's surface.

Ozone, methane, CO_2, SO_x, NO_x, particulate matter, volatile organic chemicals (VOCs) and polycyclic aromatic hydrocarbons (PAHs) in the atmosphere are regularly measured by fully automated ground-placed monitoring stations, which forward their signals to computer centers via satellite communication for processing and immediate publication through the Internet. The same parameters can be 'sensed' remotely for the whole Earth or for particular areas by satellite-based sensors, and results can be shown on satellite images.

Soil gas can be relatively easily detected by sensors working with diffusion or by applying small ventilators or air pumps to collect soil air and direct a flow through the detector. Methane, ammonia and CO_2 concentrations give information about natural microbiological activity, volatile organic contaminants in soil or groundwater pollution.

4.2.2 Water phase monitoring

Water bodies, catchments, seas and oceans are exposed to high risks and therefore environmental management is focused on them. In addition to hydrological characteristics such as water levels or flow rates, chemical composition can also be monitored by fully automated equipment with remote signal receptors/processors.

The *portable, miniaturized* versions of laboratory water analytical methods are popular for *in situ* exploratory investigations. Complete and ready-to-use kits are provided by several developers and dealers for manual analysis of chemical parameters in water. Many of these tools and devices are described in Chapter 3. As well as dispensing with the need for sample packaging and transportation, the advantages of these analytical tools are their ease of use, stable quality, and lack of reagent handling. The kits contain everything that is needed to run the analysis. The methods are mainly based on colorimetry, but many innovative molecular techniques based on DNA, RNA, enzyme and immune reactions.

Sensors, by detecting electrical signals, can selectively measure several chemical indicators of environmental problems, e.g. charged cations or anions for salinization, sodification, dissolved salts and metal contamination from leachates, seepages, run-off waters and mine drainage. Dissolved oxygen in surface waters, dissolved redox indicators, and contaminants in groundwater can be monitored by sensors appropriately placed into surface and subsurface waters. Groundwater monitoring is generally carried out by sensors in conventional monitoring wells and in shallow or deep holes prepared by direct push tools.

Electrochemical sensors (electrodes) have widespread application and their coupling to remotely working data processors is increasingly used in monitoring of both the aquatic environment and industrial technologies and discharges.

Specific chemical sensors have been developed on a case-by-case basis. They are mainly used for source or near-source control of suspected organic pollutants and employ empirical calibration. Certain sensors, known as biosensors, rely on specific biochemical or biomimetic reactions. The basis of the reaction in this case is the complementary spatial structure of the analyte and the built-in probe molecule. The mass or energy released as the result of chemical binding is transformed into an easily measured signal such as electron or photon flows and detected by electrochemical or optical detectors (see Section 4.3).

Remote systems – local sensors coupled with remote data loggers/processors, or airborne and spaceborne remote sensors – can observe water characteristics of surface waters, run-offs, effluents, water treatment technologies or hard-to-reach remote places of, for example, oceans. They can monitor important indicators of ecological status, of ongoing meteorological processes such as storms, hurricanes or tsunami, and of ongoing chemical processes such as eutrophication and other disadvantageous changes accompanied by algal or cyanobacterial growth as well as by chemical contamination. Remote sensing and the generation and analysis of hyperspectral images are some of the most promising technologies in this field. Depending on the resolution of the images, remote sensing can be used as an early warning system (Szomolányi *et al.*, 2009).

4.2.3   Solid environmental phases

The geochemistry of solid environmental phases, namely the soil and sediments, plays an important role in the development of environmental risk and its management. Remote sensing can provide information on land surface, while *in situ* assessment of natural and contaminating elements and organic chemical compounds may characterize both the surface and subsurface soils and sediments.

Portable instruments applicable for elements (e.g. handheld X-ray fluorescence analysers) and organic soil pollutants (e.g. portable sensors) with various detectors (ultraviolet, infrared, near-infrared and mass spectrophotometry (MS), gas chromatography (GC) and GC/MS) are introduced in Chapter 4.

In addition to a helpful buffering capacity, many of the long-term risks and potential chemical time bombs are also associated with solid environmental phases. Their quality has a significant impact on our waters, as the waters actively interact with solids: surface water with suspended matter and bed sediments as well as with the soils of the watershed, and groundwater with the soil. The partition of mobile substances between liquid and solid phases leads to their relatively large concentration and accumulation in the solid phases.

4.3 Chemical sensors

Chemical sensors play a key role in monitoring chemical pollution. Chemical sensors installed *in situ* are primarily used for source and emission control at known locations or for the detection of air and water pollution.

4.3.1 Air quality detection

Good air quality is an essential requirement for humans and wildlife. Environmental impact via the atmosphere aggregates exposure directly through a large amount of airborne particles (solid deposition) and dissolved contaminants coming with the precipitation.

Localized sensing, coupled with remote data processing, is the most popular solution for solvent vapors. Photoionization detectors (PIDs), ion mobility sensors (IMS) and surface acoustic wave (SAW) sensors are the most widely used portable gas detectors for measuring VOCs. While PIDs are not selective, IMS and SAW sensors give both qualitative and quantitative results.

Oxygen-selective sensors can measure oxygen concentrations not only in ambient air, but also down to 0.01 mg/L in liquids or 0.02% by volume in gases (TNO, 2014).

Mobile laboratories are capable of real-time sampling and analysis in the bottom range of ppb (parts per billion) concentrations of outdoor air or emissions from various environmental sources. Sampling and analysis is close to real-time and detection/measurement is carried out by sensors, transportable kit sets, handheld devices and other portable equipment belonging to complete mobile labs in a bus, truck or trailer. Mobile labs may be equipped with gas chromatographs, spectrophotometers, GC/MS for organic compounds and X-ray fluorescence for inorganic metals.

4.3.2 Water quality detection

Water quality monitoring using optical and electrochemical sensors is mainly applied to determine the presence of disinfectants, nitrates, dissolved oxygen, organic pollutants, toxins and pathogenic microorganisms. Substance-specific chemical sensors are available for pesticides, volatile amines such as trimethylamine, volatile organic acids and heavy metals such as lead, mercury and cadmium in water. Electrochemical sensors with ion-selective PVC polymer membranes are utilized for measuring nitrate, chloride, ammonium, potassium, sodium ions, HCO_3, phosphate, urea, creatinine, Cd, Cu, Pb and Zn in water.

Wastewater/sewage analysis applies non-contact optical techniques to the determination of biochemical oxygen demand (BOD), chemical oxygen demand (COD), total organic carbon (TOC), color and turbidity, and several contact methods and biosensors for monitoring technological parameters and biological activities.

Several commercially available and innovative monitoring tools for drinking waters, surface waters and wastewaters are introduced in Chapter 3.

Miniaturized chemical sensing systems (Micro Total Analysis Systems (µTAS)) are used for field and process monitoring of industrial and domestic waters as portable off-line or online systems. With this lab-on-a-chip technology, entire complex chemical management and analysis systems with sampling and measurement can be created in a microfluidic chip and interfaced with electronics and optical detection systems. Chemical microsensors apply ion-selective polysiloxane-based membranes for K^+, Na^+, Ca^{2+}, Pb^{2+} and Ag^+, and photochemical sensors for organic contaminants. Sensitivity, selectivity and long lifetimes characterize these chemical sensors (van den Berg et al., 1994). In addition, 2D GC has been miniaturized to a chip by using micro-fabrication methods, providing planar arrangements of the injector, capillaries and junctions necessary for comprehensive GC and the modified photoionization detectors (Halliday et al., 2010).

In water and wet environments, early-warning methods use pH and redox sensors, instruments measuring conductivity and chemically selective, hypersensitive sensors, as well as their combinations with in situ contaminant-specific enrichment, e.g. methods based on solubility, sorbability, emulsion formation, micro-encapsulation, chemical reactions or nanotechnology. Many developments aim to purposefully combine different sensors into a set, built on a common sensor platform. The European RADAR project has developed such a robust, sensitive, and versatile label-free biosensor platform for spot measurements and online monitoring of endocrine disruptors and polycyclic aromatics in the aquatic environment. The biosensor developed applies engineered recombinant bioreceptors (estrogen and aryl hydrocarbon receptors) from aquatic organisms. The combination of isotachophoretic preconcentration and surface nanostructuring ensures high sensitivity. The sensor set is applicable online and linked to a wireless communication system (RADAR, 2015).

4.3.3 Soil and groundwater monitoring

Soil and groundwater characteristics and contamination can be detected and monitored by chemical sensors based on the contaminants in the gas and liquid phases of the soil. Built-in sensors for pH, redox potential, conductivity, resistance, ion concentration, salt concentration or substance-specific sensors can be applied to soil water and groundwater (see also biochemical and microbiological sensors). Chapters 3 and 4 contain detailed information on soil and groundwater investigations and the monitoring of soil remediation.

Microelectrodes or microsensors deserve mention; they can measure the changes in the soil microstructure (emergence of ions) by transforming physico-chemical or biological signals into electrical ones. For instance, by placing a microsensor into the soil on the boundary of the pollution (where respiratory quotient (RQ) should

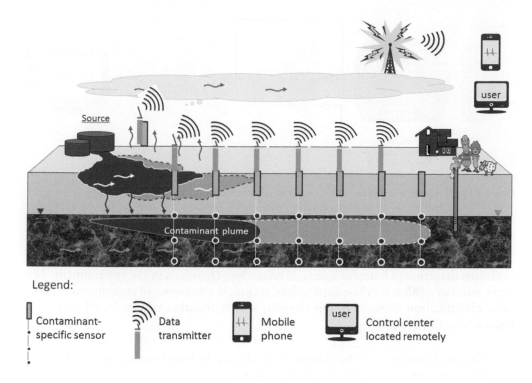

Figure 2.14 Microsensor network in the soil for the early warning of groundwater pollution.

be less than one, e.g. on the perimeter of a factory site), we can measure the presence and concentration of contaminants at a resolution of one millimeter. Chemical microsensors have many advantages: they are small, cost-efficient, robust, reliable and sensitive. A network can be built up from microsensors and the sensor network used for early warning. A large number of contaminants are the target of sensor-network applications; for example, the emergence of toxic metals and other soil contaminants, nitrate or arsenic in groundwater, etc. in connection with salinization and sodification. Figure 2.14 illustrates *in situ* monitoring of an endangered site due to an underground contaminant plume. The Center for Embedded Networked Sensing (CENS) developed and constructed a network from micromachined amperometric nitrate sensors for the early warning of groundwater nitrate contamination (CENS, 2014).

4.4 Biochemical sensors

Biochemical sensors consist of a biochemical element and a signal processor. The biological element can be an enzyme, an immune molecule or any typical receptor molecule of cells or tissues that can selectively bind the analyte. The biochemical element is immobilized (fixed) on a membrane and connected to a transducer (probe). The reaction occurs on the membrane where the substrate of interest is converted to a product that causes an electrical or optical response, which is measured by the transducer, then amplified, processed and displayed (Figure 2.15).

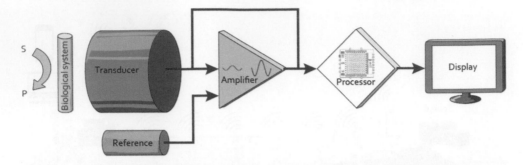

Figure 2.15 Theoretical scheme of an enzyme biochemical sensor: the enzyme converts the substrate (S) to product (P); the signal of this reaction is converted by the transducer to an electrical signal.

Biochemical sensors, also called biosensors, are the most suitable tools for *in situ* real-time detection of harmful agents or hazardous chemicals in the environment. The large number of aims for their application, technical solutions and combinations make their classification difficult. In this chapter, a categorization scheme based on sensor functions is introduced:

– Chemical sensors for the detection of chemicals based on a physico-chemical reaction between the sensor and the analyte.
– Biochemical sensors for the detection of chemicals based on a specific biochemical (enzymatic, immunological or nucleotide) reaction between two biologically active molecules.
– Biochemical sensors for the detection of specific molecular responses (resistance genes, adaptive enzymes) that are selectively triggered by certain chemical substances and are based on the biochemical reaction. This is the indirect detection/indication of a biologically effective chemical substance via the response of the target organism.
– Biochemical sensors for the detection of the presence of certain biological species based on their specific molecular structure elements, e.g. gene, RNA, surface antigens or specific products. The strength of some of these signals can be increased by inducing the gene, RNA, enzyme or specific product production of the organisms, e.g. a substrate-induced enzyme response.
– Selective biological sensors including whole cells or organisms applied for the detection of certain specified stressors. The criterion of selectivity means that the response is generated by the living cell exclusively on the effect of this stressor. It is most efficiently addressed by a genetically engineered built-in reporter gene and promoter which can only be activated by the analyte (see also Figure 2.20).
– Non-selective biological sensors (whole-cell or whole-organism biosensors) for the detection of stimulatory or inhibitory environmental effects. It is typically an aggregating response, showing stimulation in the presence of nutrients or inhibition of one or more activities in the presence of toxicants. They are discussed in detail in Section 5.1.

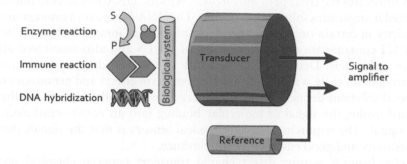

Figure 2.16 Different types of biochemical reactions serve as the basis of biosensing: enzyme reaction, antibody–antigen coupling and DNA hybridization.

The sensors referred to above as biochemical sensors are often termed biosensors. This wider interpretation of biosensors is based on the fact that the molecules fixed in the sensor or the analyte molecules are closely linked to living organisms.

The signal of the sensor may be identical to the adverse effect of the chemical substance on living organisms. In such a case, the binding, selectivity and sensitivity of the sensor to the contaminant is based on the same reaction as the binding of the hazardous agent to the receptor site of a living organism. These signals may be produced by the binding between an enzyme and its substrate (enzyme reaction), an antibody and antigen (immune reaction), or DNA and a complementary probe (DNA hybridization) (Figure 2.16).

Signals produced by interactions can be based on enzyme reactions or other molecular affinities and also on the response of living microorganisms. These types of sensors are named whole-cell microbial sensors. The interaction of the analyte with the enzyme may result in chemical transformation of the substance or activation or inactivation of the enzyme. Bioaffinity-based sensors contain antibodies/antigens or nucleic acids to detect and/or measure contaminant molecules, viruses or microorganisms. Their selectivity makes their use possible in the environment and in wastewaters, where specific agents need to be detected amongst numerous other components. Whole-cell microbial sensors differ from enzymes or immune and nucleic acid probes of high affinity and selectivity because living cells integrate all impacts and give a complex, broad-spectrum response to many of the nutrients and toxic substances, as well as to environmental conditions and technological parameters. Nevertheless, there are some whole-cell biosensors developed for a contaminant-specific response, such as cells containing contaminant-specific resistance genes or those that are gene-manipulated to switch on indicator genes upon the effect of certain contaminants.

Whole-cell biosensors are discussed in Section 5.2.3, and a few biochemical sensor developments are introduced below.

The *functional basis of biochemical sensors* is the well-known selective reaction between molecules able to recognize each other (biorecognition), such as biologically active substances and whole living cells or tissues, enzymes and substrates, antibodies and antigens, biotin and streptavidin, protein A and protein G, DNA and complementer DNA, or DNA and RNA. The example of organic pollutants tending to bind to DNA illustrates the concept of biochemical sensors. DNA-bound contaminants can be detected in organisms following exposure. The DNA-bound contaminant may cause genotoxicity in certain organisms or in their offspring. Gompertz *et al.* (1996) monitored PAH contamination in coastal areas of the US and also monitored aflatoxin exposure in humans. Detecting contaminants via their binding to sensor-integrated DNA provides an early warning of possible long-term effects and genotoxicity.

Several solutions defined by the transducer type are available for fixing the sensor and transforming the signal of molecular binding into an easy-to-read electrical or optical signal. The requirement of biochemical sensors is that the signals must have high sensitivity and good environmental relevance.

Electrochemical sensors detect charge transport between chemical phases or changes of electrical properties due to chemical reactions. These sensor types are the most widespread and have many advantages: rapid and sensitive measurement, low-cost mass production and miniaturization are all possible. Ion-selective PVC polymer membranes are used for measuring ionic contaminants.

Modified electrodes have been applied for the enzymes of xanthine oxidase to detect hypoxanthine, and tyrosinase to detect polyphenols in wine. The detection of beta-galactosidase enabled the indication of the presence of coliforms in water. Carralero *et al.* (2006) developed enzyme sensors for catechol, phenol, 3,4-dimethylphenol, 4-chloro-3-methylphenol, 4-chlorophenol, 4-chloro-2-methylphenol, 3-methylphenol and 4-methylphenol in water samples.

Many sensor developments use *horseradish peroxidase*: the enzyme is immobilized onto the carbon working electrode modified by an aryl diazonium salt. The formation of amide bonds between the amino and carboxylic groups of the enzyme surface, catalyzed by hydroxysuccinimide and carbodiimide, leads to electrode functionalization. This type of biosensor has been used to determine the presence of levetiracetam (an epilepsy medication) in complex matrixes, and of hydrogen peroxide in coastal waters.

Alkaline phosphatase bound to the electrode is suitable for the identification from biological samples of human pathogens such as *Streptococcus pneumoniae* or the SARS (severe acute respiratory syndrome) virus.

A *remote electrochemical biosensor* for field monitoring of organophosphate nerve agents was developed by Wang *et al.* (1999).

A *genosensor* for the detection of nucleic acid sequences specific to *Legionella pneumophila* contains the immobilized thiolated hairpin probe combined with a sandwich-type hybridization assay, using biotin as a tracer in the signaling probe, and streptavidin-alkaline phosphatase as reporter molecule.

Molecular beacons (MBs) are oligonucleotide hybridization probes that can report the presence of specific nucleic acids. Oligonucleotides have a stem-and-loop structure and are labeled with a fluorophore at one end and a quencher on the other end of the stem that become fluorescent upon hybridization.

Optical sensors measure easy-to-use end points that can detect the intensity of photon radiation generated by an interaction of the analyte with a receptor. Fiber optic sensors are the most common type of optical sensors, applied for direct spectroscopy ranging from UV to IR, from absorbance to fluorescence, Raman and surface plasmon resonance. They work by directing light waves to the interface between a metal and a dielectric.

Opto-chemical sensors developed at TNO (2014) consist of a selective coating coupled with an optical read-out system. Measured responses can be a change in color, in fluorescent properties, including fluorescence lifetime, or a change in refractive index at an optical interface. Their selectivity comes from specific biochemical interactions.

Micromechanical cantilevers are used to detect small quantities of biochemical molecules by mechanical deflection of a cantilever upon binding to its gold-coated and chemically functionalized surface. The extent of deflection depends on the cleanliness of the microcantilever surface (Tabard-Cossa *et al.*, 2007).

Piezoelectric sensors are mass-sensitive sensors which transform the mass change at a specially modified surface into a property change in the support material. The mass change is caused by accumulation of the analyte. Piezoelectric devices and surface acoustic wave devices can be grouped in this category. The vibration of piezoelectric crystals produces an oscillating electric field. The frequency depends on the chemical nature, size, shape and mass of the crystal. By placing the crystal in an oscillating circuit the frequency can be measured as a function of mass (Plata *et al.*, 2010).

Thermal microsensors and microprobes have the advantage that they require neither labeling nor immobilization and can sensitively measure heat transport connected to chemical or biochemical (enzymatic) reactions by nanocalorimetry (Todd & Gomez, 2001). Cantilever-based thermal microprobes have been applied successfully for memory storage by Vettigera *et al.* (1999) and flow sensing by de Bree *et al.* (1996). As an innovation, Bruyker *et al.* (2010) integrated cantilevers with vanadium oxide (V_2O_5) microprobes for creating enthalpy arrays.

Several *nanostructures* are currently used in the development of nanosensors such as carbon nanotubes, noble metal nanoparticles, quantum dots (colloidal nanocrystalline semiconductors), magnetic beads, metal nanoclusters, and nanofilms (Riu *et al.*, 2006). Molecular imprinted polymers (MIP), metal complexes, cyclodextrin derivatives (see Chapter 7), sol-gel material and organic ligands are suitable materials for sensing.

Nanoparticle-based biochemical sensors are functionalized with antibodies as markers for proteins. Gold, silver, magnetic and semiconductor materials (quantum dots) can be applied as nanoparticles in optical (absorbance, luminescence, surface-enhanced Raman spectroscopy, surface plasmon resonance), electrochemical and mass-sensitive sensors.

Optical biosensors are introduced in Section 4.6 in greater detail, and whole cell biosensors in Section 5.2.3.

Printed electronics has emerged as an exciting technology and as a complement to silicon-based electronics that now enables the printing of materials and electronic devices such as power sources, biosensors and displays for different types of applications (Turner, 2013). Norberg *et al.* (2015) developed integrated printed biosensor platforms to create disposable sensor systems that are easy and inexpensive to manufacture and suitable for environmental monitoring and use in agriculture. The entire system of the integrated biosensor, including power source, sensor and display, is printed on a sheet of flexible plastic or paper. Circuitries to drive the electronics were later replaced with a chip. The concept can be utilized for any sensor, analyte and mechanism, including enzymatic or affinity with microfluidics, provided there is an electrochemical transduction mechanism. The platform is versatile and adaptable to specific user needs and applications, including the addition of communication technology.

4.5 Immunosensors

Immunoassays are well-established analytical tools and various immunoassay kits are available for environmental contaminants. Immunosensors work on the principle of known immune reactions but the size of the 'reactor' and the transformation of biological response are specific. Some examples of immunosensors are described below.

Estradiol and *ethinyl estradiol* immunosensors for wastewater use magnetic beads as solid support for immobilization of synthetic estrogens (the contaminant) and screen-printed electrodes as sensing platforms. The assay is based on the competition between the free and immobilized estrogen for the binding sites of the primary antibody, with subsequent revelation using alkaline phosphatase-labeled secondary antibody (Kanso *et al.*, 2013).

Penicillin G and other *β-lactam antibiotics* can be detected by a newly developed, highly sensitive amperometric immunosensor. It is suitable for the detection of these drugs in rivers and wastewaters using an amperometric electrode for hydrogen peroxide as transducer and the peroxidase enzyme as marker (Merola *et al.*, 2014).

Bisphenol A (BPA) can be detected at 0.1 nM concentration in wastewater using fluorophore-tagged antibodies on a chip surface (Zhou *et al.*, 2014).

The use of *pathogens* and *bacterial toxins* as biological weapons represents a realistic hazard, so immune analytical methods, including sensors, have been developed for all of the known pathogens and toxins. QTL Biosystems (2015) has developed biosensors for the detection of anthrax, ricin and other potential bioterrorism pathogens and toxins in water.

Pyrethroids and *DDT* were analyzed by an optical immunosensor (AQUA-OPTOSENSOR) in Nairobi river water and sediment. The field results were compared to the conventional enzyme-linked immunosorbent assays (ELISA) in the laboratory (Krämer *et al.*, 2007). Similarly to ELISA, the immunosensor, which employed fluorophore-labeled monoclonal antibodies immobilized on disposable chips, allows determinations at the ppb level.

Further examples are described in several comprehensive books such as those edited by van Emon (2006) and Moretto and Kalcher (2014).

4.6 Optical biosensors

Optical biosensors utilize light absorption, fluorescence, luminescence, reflectance, Raman scattering or refractive index combined with a biological sensing element such as enzymes, antibodies, oligonucleotides, aptamers, subcellular components or whole cells. The biological element is usually immobilized on a solid support or nanoparticles (e.g. gold nanoparticles, quantum dots, graphene or graphene oxide, mesoporous silica nanoparticles), to be built into a test strip or a chip (Long *et al.*, 2013). Miniaturized, rapid, ultrasensitive and inexpensive, nanostructured optical biosensing platforms have been developed for rapid toxicity screening and multianalyte testing. The most popular target molecules are toxic metals, persistent organic pollutants (POPs), endocrine disrupting chemicals (EDCs), toxins and viruses.

The enzyme cholinesterase, for example, can be inhibited by several toxic chemicals, e.g. organophosphate pesticides. However, other pesticides, toxic metals and toxins may also inhibit cholinesterase. Based on this, cholinesterase biosensors can be developed for generic toxicity monitoring and early warning.

Some existing sensor technologies make possible the combination of several different enzymes, e.g. cholinesterase for pesticides and urease for toxic metals, in a single device (Ligler, 2009; Dorst *et al.*, 2010). Some examples of *enzyme-based optical biosensors* follow:

–   *Toluene* oxidation is catalyzed by *toluene ortho-monooxygenase*, which is used in the reaction where the consumption of oxygen is measured by an oxygen-sensitive ruthenium-based phosphorescent dye. The enzymatic biosensor can detect toluene in wastewater with a limit of detection (LOD) of $3\,\mu M$, although with a long response time (approximately 1 h) (Zhong *et al.*, 2011).
–   *Adrenaline determination* by a fiber optic biosensor is possible with a shorter response time (30 s) in addition to very good sensitivity (10 nM). It was developed using *laccase* and immobilized on nanoparticles (Huang *et al.*, 2008).
–   *Cholinesterase biosensors* are useful for generic toxicity monitoring because they can be inhibited by several toxic chemicals such as organophosphates and pesticides, heavy metals and toxins (Borisov & Wolfbeis, 2008; Ispas *et al.*, 2012).

Optical immunosensors provide a repeatable and highly specific analytical option. The proper application of antibodies enables recognition of all environmental contaminants with immunogenic properties, i.e. triggering an immune response in the animal used for antibody production. Non-immunogenic toxicants can be changed to become immunogenic, e.g. by conjugation to larger-size carrier proteins. Most of the water contaminants, such as pesticides, POPs and EDCs, are small molecules with low immunogenic activity, so their direct binding to the surface of the sensor proceeds with low efficiency. In such cases, the immobilization of the conjugate of the small contaminant molecule and carrier-protein onto the surface of the immunosensor obtains a stable and reusable surface for binding the antibody surplus instead of the contaminant itself (Long *et al.*, 2013).

Immunosensors such as the fluorimetric chips used for the determination of BPA – discussed above in Section 2.6 – are also optical biosensors, i.e. optical immunosensors.

Fibre-optic evanescent wave immunosensors (EWIs) apply a planar waveguide as a transducer. Immunoreactions proceed on the outer surface of this waveguide. An evanescent (waning) wave is a near-field wave with an intensity that exhibits exponential decay without absorption as a function of the distance from the boundary at which the wave was formed. One type is the surface plasmon resonance (SPR) immunosensor, which can sense changes in the refractive index within the evanescent field caused by mass adsorption due to immune complex formation. Based on this phenomenon, *in situ* and label-free sensors have been developed.

Specifically sensitized films on the waveguide surface – evanescent wave excitation systems (EWES) – may achieve greater sensitivity, as shown by the new RIANA (River

Analyzer) and AWACSS (Automated Water Analyzer Computer Supported System) devices (Rodriguez-Mozaz et al., 2009).

Sensitivity could be increased significantly by luminescence-based sensors, which detect signals independent of molecular size. The detection limit of the EWES for BPA was reported to be 0.014 µg/L, which was one to two orders of magnitude smaller than measured by the label-free SPR system (Rodriguez-Mozaz et al., 2005).

Zhou et al. (2014) developed a portable, highly sensitive, reusable EWI for BPA detection. The planar optical waveguide chip can be regenerated and this allows the performance of more than 300 assay cycles with an analysis time of 20 min per cycle. By application of an effective pre-treatment procedure, BPA recovery in real water samples varied from 88.3% ± 8.5% to 103.7% ± 3.5%, confirming its application potential for BPA measurement in practice.

EWIs were developed for measuring 2,4-D, microcystin-LR, chlorobenzene and nitrobenzene in real waters (Long et al., 2008a, 2008b; Shi & Long, 2015). The evanescent wave decays exponentially after being generated at the surface of the probe. It could excite the fluorophores in the labeled, surface-bound immunocomplex of the analyte and antibody. It can discriminate between unbound and bound fluorescent complexes, so no washing procedure is necessary to eliminate the unbound proportion.

SPR immunosensors utilize refractive index changes to sensitively detect mass changes at precious metal sensor surface interfaces. This is why they can be applied to immunoassays of large molecules. A sandwich immunoassay can provide excellent sensitivity. An SPR sensor and an antigen-antibody interaction have been developed for sensitive and selective detection of explosives, called an 'electronic dog nose' by the developers. Onodera and Toko (2014) developed innovative surface fabrications which include variations of a self-assembled monolayer containing oligo (ethylene glycol), dendrimer and hydrophilic polymers.

Immunosensors have many advantages, as demonstrated by the examples, but the preparation of the antibody is still a costly and time-consuming undertaking. Sensors can only be applied under natural conditions to avoid denaturation or other damage of the sensitive protein molecules. Two innovative options, aptamers and DNAzymes, will be introduced briefly below.

Aptamers are single-stranded DNA or RNA sequences. They are useful alternatives to antibodies for sensing molecules because they are more stable and more resistant to denaturation and degradation than antibodies (Long et al., 2013). Aptamers can be chemically synthesized easily and do not require complicated or expensive purification steps, which are necessary in antibody production. Fluorescence-based DNA aptamer sensors have been developed for the detection of Hg^{2+}, Pb^{2+} and other trace pollutants (Liu et al., 2008; Wu et al., 2012).

DNAzyme-based optical biosensors have been developed for sensitive monitoring of various metals (Long et al., 2013). DNAzymes (catalytic DNAs) are functional nucleic acids which can fold into a well-defined three-dimensional structure to bind to specific targets.

By combining DNAzymes that can perform chemical modifications on nucleic acids with aptamers that can bind with a broad range of molecules, a new class of functional nucleic acids known as allosteric DNAzymes or *aptazymes* (Hollenstein et al., 2008) can be created.

5   MONITORING AND EARLY WARNING BASED ON BIOLOGICAL ACTIVITY AND TOXICITY

Depending on the nature of the activity, monitoring can build upon the source's characteristics, the transport pathways or the targeted environmental compartment's receptors, i.e. the ecosystem and/or humans. Some data are available in databases, such as the amount of the chemical produced and used, the known effects of the substance and the characteristics of the affected environment. Another type of data that can serve as the basis for environmental monitoring is measured data which can be obtained from the source and its vicinity, the transport pathway, or the targeted environment. Measured data may be readings from instruments placed *in situ* in hot spots, laboratory measurement results of environmental samples, or the response of the affected native receptors or relocated test organisms. The monitoring point on the transport pathway can be close to the source, thus characterizing the emission, or close to the receptor, the user of the exposed environmental compartment (Figure 2.1).

The further the monitoring point is from the source, the more sensitive and more selective the methods generally needed, due to extensive dilution. Chemical analyses to detect and measure very low contaminant concentrations with different environmental fates and behaviors after a long transport route, are problematic. Biological monitoring systems placed at the location of the receptors may be effective if the biological signal, as response to the chemical substance, can be amplified, for example, through the effect on extremely sensitive or accumulating species. A selective collecting system integrated into the early warning can increase sensitivity. It can be either a physico-chemical or a biological method, or their combination. Selectivity plays a crucial role due to the high level of noise.

The aim of environmental monitoring is to forecast potential or detect existing damage to the ecosystem and humans. Forecast enables prevention, while existing damage needs remediation. Failed prevention may result in damage. The forecast can be based on statistics (hundreds of tons are produced and used in a watershed), chemical information (the substance is highly persistent), hazardous effect information (the substance is reprotoxic) or on the potential (very sensitive, protective ecosystem) or detectable response (early indicators or already existing damage) of the ecosystem. Ecosystem response measured *in situ* and in real time is in direct relationship with environmental risk, showing existing adverse impacts (e.g. the concentration of algae in the river is 20% smaller than required for ecosystem health). Direct toxicity assessment with controlled test organisms characterizes actual toxicity by a reproducible value. The 'no effect' proportion of the sample is suitable for the characterization of the risk (Figure 2.17). The result of chemical analysis should be 'translated' and extrapolated to a risk value for decision making. The measured concentration can be translated into the risk of this specific substance by comparing it to the ineffective/unharmful concentration. The risk of more contaminants needs the aggregation of the individual risks.

Adverse effects provide more information in many cases, e.g. when the contaminating substance is very powerful in even extremely small quantities, making analytical detection impossible, but its effects can be measured. It may happen that the contaminant does not show any effect acutely (immediately after exposure) but that the adverse effect arises over the long term, after a great delay (Figure 2.18). During the

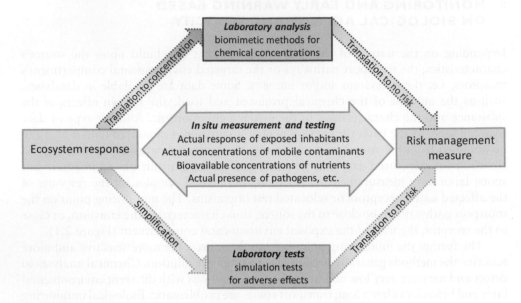

Figure 2.17 *In situ* signals proportional to risk can be used for risk management and decision-making.

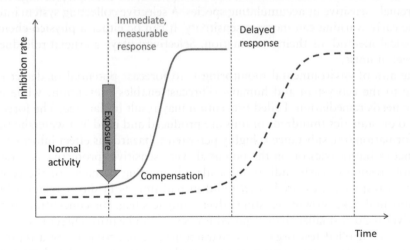

Figure 2.18 Illustration of immediate and delayed measurable response.

time interval between exposure and measurable response, several processes may occur in compensation for the adverse effect. In such cases, the solution is to measure chronic effects or find a measurable end point proportional to the compensatory activity, e.g. heat or neutralizing enzyme production or heartbeat rate. The contaminating substance may be ineffective by itself but demonstrate adverse effects when occurring together with other substances (synergism). The simultaneous effect of several contaminants and their activation in the environment is often a hazardous case, where chemical analysis itself cannot provide a proper solution.

The ideal monitoring solution is the integrated application of physico-chemical, biological and environmental–toxicological methods because simultaneous data facilitate interpretation (Gruiz, 2005, 2009).

5.1   Biological and ecological assessment methods

Similarly to geophysical methods, biological and ecological methods are traditionally applied *in situ* during field work. Of course, they can be accompanied by laboratory or other *ex situ* work (e.g. species identification in the laboratory). The key goal of these methods is the observation and follow-up of the ecosystem's health conditions.

Ecosystem health can be defined as the status of the ecosystem including its natural assets such as biodiversity and geomorphology, its functional quality, and its capacity to maintain its assets and function for the future (i.e. sustainability). Thus ecosystem health can be characterized by measuring its status, activity and sustainability:

- *Status*: quality of the habitat (mapping extent, land use), ecological status of the water and key species diversity;
- *Function*: element cycling, habitat function, critical loads, buffering capacity;
- *Sustainability*: climate change, native and invasive species, restoration/remediation, wastewater treatment, water uses and impacts.

Species abundance and distribution, also called species diversity or biodiversity, is the key indicator. If there is a 95% concordance with healthy diversity, the changes are considered not to cause significant ecosystem deterioration. However, our waters and aquatic ecosystems are significantly endangered due to environmental and anthropogenic impacts. Regulations concentrate on drinking water supply and aquatic ecosystems. Both quantity and quality of the waters are important factors.

The European Water Framework Directive's (WFD, 2000) main objective is the good chemical and ecological status and the conservation of species density and diversity of marine (coastal) and freshwater ecosystems. With respect to species diversity, WFD covers phytoplankton, phytobenthos and macrophytes, benthic invertebrates and fish. The monitoring program includes general surveillance at least once every six years and active monitoring of endangered and deteriorated water bodies.

The Clean Water Act in the US (CWA, 1972), and the relevant regulations all over the world, require new methods and tools for both regulatory purposes (such as the creation of quality criteria) and continuous monitoring and early warning.

New tools and rapid, effective and possibly uniform monitoring methods are necessary to fulfil the requirements of efficient water quality management because traditional diversity assessments are extremely time- and labor-intensive and the results are often questionable.

One of the problems of traditional ecological assessments is the lack of baseline data and detailed knowledge of the healthy ecosystem, such as its natural and seasonal changes and the impacts of climate change. Seemingly obvious information, such as how much water is in rivers and lakes, is generally not available. Another problem is that only some and not all of the species can be assessed. Consequently, the health of the whole ecosystem must be extrapolated from partial information. In addition, the selection of the species or any other indicator organisms as representatives is based on

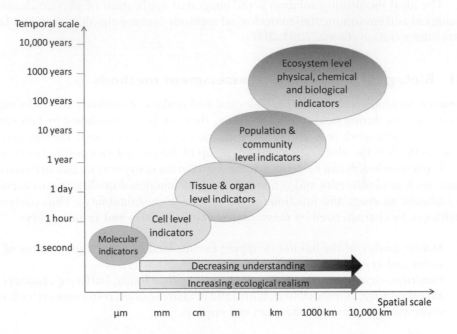

Figure 2.19 Bioindicators' temporal- and spatial-scale relationship, ecological significance and interpretability.

incomplete information about the presence and role of other species. Uncertainties are multiplied by unpredictable emerging contaminants and their mixtures.

Targeted biological or ecological assessments represent an easier task. In this case the biological cause of an environmental problem is known, e.g. eutrophication or pests. For pests, the presence of one or a few species has to be monitored to obtain an indication of the problem, and the number of harmful organisms kept under a certain threshold.

Indicators are properly selected, measurable parameters. It is true that the use of indicators simplifies the picture of a situation, but it also makes environmental management and decision-making possible (Figure 2.19).

Aquatic ecosystem health indicators include physical indicators (e.g. flood frequency), chemical indicators (e.g. water quality), and biological indicators such as fish health, benthic invertebrates (aquatic bugs) and vegetation. Biological indicators can be affected by the environment they live in and can provide information about changing water quality and quantity over longer periods of time. Species at the bottom of the food web, like benthic invertebrates and algae, can provide early warning about contaminants and other environmental stressors. Fish health is linked to human health since humans eat fish and other aquatic wildlife.

Bioindicators may be structural, functional or system-level symptoms of environmental, chemical or biological stress. Stress may be caused by global climate change (temperature), global pollution (acidification) or contaminants at local or watershed scale, such as industrial chemicals, petroleum products, toxic metals, pesticides, personal care products or pharmaceuticals including endocrine and immune disruptors. Indicators include:

- *Individual-level*: condition, growth, reproduction, mortality and behavioral strategies – the principal properties related to fitness (Begon *et al.*, 1996);
- *Population-level*: abundance, biomass, productivity, reproduction and mortality rates, age structure, sex ratio and genetic diversity;
- *Community-level*: in addition to abundance, biomass and productivity, dominance and composition;
- *Ecosystem-level*: physical (flow, flood, turbidity, landscape), chemical (nutrient and element cycling) and biological (biomass production, trophic structure, diversity) components.

5.2   Indicators in biomonitoring

Certain sensitive species may be about to disappear; the distribution of the species can change; the metabolism and the yield in the ecosystem can decrease, etc. These are all signals of environmental deterioration but may not be bioindicators suitable for monitoring. The signals of the best selective bioindicators also provide information on the cause of the problem. Cause-selective bioindicators are needed which can help to find or prove the correlation between an impact and the stress factor causing it. Statistics provide extensive information for reaching a decision on the proper intervention. There are several known selective signals among the biological responses, since in many cases the cells and organisms respond selectively with regard to their metabolism, e.g. resistance against certain metals or biologically active agents, special biodegradation or co-metabolic abilities in order to combat organic pollutants. It is worth distinguishing between biological responses which provide unambiguous evidence of the presence of a certain stressor and those that assume the presence with a certain likelihood. In the latter case, justification by chemical analysis is necessary.

5.2.1   Whole organism inhibition

Mussels are very sensitive sentinel organisms, responding immediately to environmental changes. Mussels can be used as early-warning indicators of contaminants in surface waters. Active biomonitoring has used mussels for many years, so it can be considered as the traditional form of mussel-based biomonitoring. A cage technique developed by Oertel (2000) was used for sediment monitoring in the Danube river (Gruiz *et al.*, 1998). *Dreissena polymorpha* synchronous cultures, raised under controlled circumstances, were placed in flow-through cages and deployed into the sedimentation zones and pollution hot spots in the Danube. After a certain time the mussels were re-collected and their health and accumulated metal content examined. The high sensitivity movement/behavioral responses of mussels made them applicable for the '*Musselmonitor*', an online automated measuring method giving real-time information on water quality. The response of the living mussels is based on the notion that when mussels come into contact with contaminated water, they close their shells (as compared to their normal, open state) in order to protect themselves from the contaminants and to shorten the time of exposure.

Using an inductive electromagnet, Kramer and Foekema (2000) measured the signals generated by the opening and closing of the shell, and they forwarded the signal to a data-processing unit, which could be several kilometers away. This signal does not indicate the cause of what has triggered the protective mechanism in the mussels, but the early-warning signal may initiate a series of risk-reducing actions such as special treatment, recycling or simply not releasing the poor-quality water from a drinking water treatment plant (Musselmonitor, 2015). In Hungary, drinking water managed by the waterworks is controlled by Musselmonitors placed in the great drinking water reservoirs of Budapest, such as the one under the Gellért Hill.

Musselmonitors can be used to control water quality in rivers, outlet systems for wastewater treatment plants and cooling water systems in power stations. The species used as Musselmonitors are *Dreissena polymorpha* or *Unio pictorum* for freshwater, and *Mytilus edulis* for saltwater. Commercially available equipment is introduced in Chapter 3.

Other organisms, such as fish and *Daphnia* or other crustaceans, can be used as sensitive bioindicators. One technical solution is to place the organisms, raised under controlled conditions, into the surface water in flow-through cells or cages. Another solution is to establish a bypass for monitoring and place the flow-through cell with the animals there. As test end points, the number of living animals, their mobility, activities, behavior, proliferation, number and quality of offspring, etc. can be measured. Some of the end points can continuously be observed visually, for example by cameras, and the measurements can be evaluated using an automatic evaluation system. After fitting the data to a statistical analysis, an automatic warning signal will be obtained if the system detects an anomaly that exceeds the standard deviation. The frequency of the gill movement of the fish and the opening and closing of the mussel's shell are monitoring end points that are relatively easy to detect.

The members of the *Tubificidae* family are aquatic organisms that mainly occur in waters contaminated by organic substances. Half of their body burrows into the bottom sediment while the rest floats in the water. In the presence of certain contaminants, these organisms retract a large portion of their bodies and burrow deeper into the bottom sediment. This behavior is proportional to the concentration of the contaminating substance. The retraction can be observed both visually and with the help of a camera using digital image-analyzing systems, quantitative analysis and evaluation. Based on the first measurements, Leynen *et al.* (1999) came to the conclusion that the movement/behavior of *Tubifex* worms can be reproduced, and by tracing and evaluating their movement, an early warning system can be developed.

5.2.2 Accumulator organisms

Bioaccumulators are organisms that accumulate pollutants within their tissues. They may be less sensitive to the accumulated contaminant than average, due to their ability to store neutralized contaminants, e.g. metals packed into large stable protein molecules. This self-protecting mechanism of the accumulators is highly risky for others that are higher up the food chain/food web than the accumulator.

These organisms can increase biomonitoring sensitivity by multiplying the signal to be detected and, what is even more important, they can aggregate and average long-term exposures. The types of organisms suitable for bioaccumulation are those that have high-rate uptake, low-rate elimination and a dynamic equilibrium resulting in long-term storage of the contaminant. The dynamic equilibrium body burden or organ burden of individual contaminants seems to be species-specific, so the average environmental concentration of the contaminant can be estimated from the amount accumulated in the accumulator species.

Accumulation-type bioindicators can significantly increase the sensitivity of monitoring. To survey the toxic metal contamination of rivers and lakes, their indigenous plants are the most suitable. Those that are rooted in the bottom sediment are more likely to bioaccumulate metals and thus signal the sediment's metal content. *Potamogeton pectinatus*, or sago pondweed, is such a plant. The work of several scientists, e.g. Whitton *et al.* (1981) and the Hungarians Kovács and Podani (1986), have proved the applicability of these aquatic plants as early-warning indicators both in the UK and in Hungary. They chiefly accumulate lead, chromium, nickel, silver, cobalt and cadmium. The only disadvantage is that precise chemical analysis is needed to measure the metal content of the accumulator plants.

Typical aquatic and benthic bioaccumulators are fish, bivalves, crabs and shrimp, snails, mussels and macro- and meso-invertebrates, as well as aquatic plants. The terrestrial ecosystem is exposed and can accumulate via soil and air. Plants, mosses, soil invertebrates, ants, collembolans and other insects are typical target organisms of persistent chemicals. Seabirds, predator fish (tuna, salmon, shark) and seals in the aquatic ecosystem, and terrestrial invertebrates, birds (eagles, hawks, songbirds) and top predators in the terrestrial ecosystem, are the most exposed victims of biomagnification.

Metals, pesticides and other persistent organic chemicals are primary accumulating agents, but some natural substances, including vitamins, alkaloids and toxins of biological origin, can also be accumulated due to the specific diet of wild animals. Some organisms accumulate the contaminants in specific organs, e.g. apolar organic contaminants in the liver, metals in bones and hair, etc.

Accumulator organisms can be used both in active and passive biomonitoring.

5.2.3 Whole-cell biosensors and bioreporters

The idea behind the development of whole-cell biosensors and microprobes is the conversion of molecular-level responses into an electrical signal with the help of living microorganisms built into a sensor or electrode. These microbiological methods are suitable for the selective detection of molecular interactions between cellular elements or products of living organisms and the analyte. If specificity must be achieved, instead of whole cells the responsive molecular element of the cell can be separately used and built into the sensor.

Two types of whole-cell biosensors are worth differentiating: (i) contaminant-oriented, selective ones, including a reporter gene within the cell whose signal is triggered only by the analyte; (ii) effect-oriented, non-selective ones which contain a cell with broad range sensitivity, providing signals for most of the contaminants even at low concentrations. By applying stressor-selective whole-cell sensors, the results can be used for decision-making and management, and the additional chemical analysis (if necessary) should only prove the presence of the contaminant. When using non-selective sensors, the cause of toxicity should be identified by additional measurements.

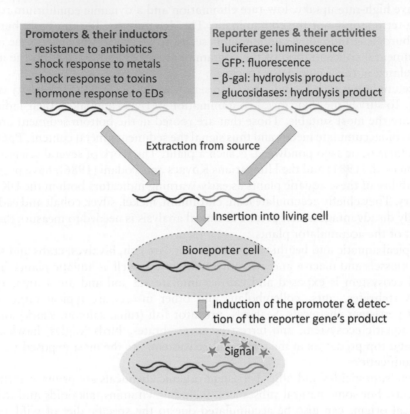

Figure 2.20 In vitro manipulated indicator cells containing inducible promoters and reporter genes responsible for the synthesis of easy-to-measure products such as light or reactive molecules (indirectly detectable).

Optical *whole-cell biosensors* are suitable to be made a part of portable systems for fast detection of toxicity in water or soil. In selective sensors the built-in bacteria may be genetically engineered to respond to the presence of specific chemicals (Hg, Cd, Ni, As, etc.) or physiological stressors by synthesizing a reporter protein such as luciferase, beta-galactosidase or green fluorescent protein (Yagi, 2007), which is the signal to detect. Non-selective whole-cell biosensors analyze the sample through processes in which many enzymes are involved, such as the systems of respiration, fermentation or anaerobic degradation. These whole-cell biosensors are less sensitive and show lower specificity compared to the enzyme-based biosensors.

Bioreporters are genetically engineered microbial cells that can produce a measurable signal in response to a specific chemical or physical agent. This is possible thanks to the new elements built into the reporter organism's genome. These elements are the promoter (a regulator gene) and the reporter gene that is responsible for the synthesis of the gene product. The promoter gene switches on in the presence of the analyte such as antibiotics, metals, toxins or endocrine disruptors (EDs), and the transcription of the reporter gene starts to synthesize the detectable/measurable product. This signal indicates that the analyte-specific promoter has sensed the analyte in its environment (Figure 2.20).

The recipient cell of the new genetic compilation is chosen for its suitability to integration within a sensor and for long-term stable functioning. The promoter region of the gene constructed *in vitro* is selectively sensitive to the analyte, e.g. a metal, hormone or toxin, or even to physical parameters such as temperature. The product of the reporter gene should easily be detected (light or colored products). If the product is an enzyme, for example, it can be detected indirectly: by giving its labeled substrate to the analyzed sample and measuring the labeled product of the enzyme reaction.

Visible or fluorescent light provides easy-to-measure end points. Thus, the most widespread bioreporter genes are those responsible for the emission of light and GFP (green fluorescent protein), which is responsible for green fluorescence. By planting these genes into selectively sensitive microorganisms, they will sense and signal the damage in real time as part of the photosensitive biosensor through a decreased emission of light. The lux gene of *Vibrio fischeri* or *Vibrio harveyi* is only seven kilobases long. It can be built into any selectively sensitive microorganism. The GFP gained from the jellyfish species *Aequorea victoria* can be built into either a prokaryote or a eukaryote, and will result in an easily detectable, strong green light being emitted by the host.

Some examples of whole-cell biosensors:

– The cells of *Chlorella vulgaris* microalgae are immobilized and the fluorescence of the algal chlorophyll is measured, making it possible to monitor herbicides such as atrazine at sub-ppb level (Védrine *et al.*, 2003).
– Genetically engineered *Escherichia coli* bacteria are used in luminescence-based biosensors for detection of toxic compounds in water (Ramiz *et al.*, 2008).
– Genetically engineered bioluminescent magnetotactic bacteria (BL-MTB) are integrated into a microfluidic analytical device to create a portable toxicity detection system (Roda *et al.*, 2013).

5.2.4 Species diversity and other community-level indicators

Aquatic and terrestrial ecosystems try to adapt to environmental circumstances, climatic conditions, seasons and contaminating substances, thus showing great flexibility. Changes in species distribution are caused by the fact that those species that are sensitive to the contaminating substance will decrease and eventually disappear, while those species that can tolerate or utilize the contaminant will gain an advantage, proliferate and their relative numbers will increase within the community. Since certain species have genes that are responsible for their coexistence with the contaminating substance, these genes will naturally proliferate in the community in a contaminated area, not just through an increase in the species population, but also due to other mechanisms such as horizontal gene transfer between members of the community. As a global ecological trend, diversity of the microbiota in waters and soils is growing continuously, even if the contaminants exert a detrimental effect at a local level. However, all in all, they will trigger the evolution of further genes in the metagenome, and thus the quantity and information content in the genes will gradually increase. Healthy diversity and a metagenome dominated by protective and compensatory genes should be differentiated and the difference can be used as a warning signal.

There are two main concepts that can be used to characterize the biota in the environment. The first concept suggests the examination of all of the genes, gene products or gene activities of a community, e.g. a soil microbiota, irrespective of which species' genome they belong to. This examination can take place at the DNA level with the help of DNA chips, real-time polymerase chain reactions (PCRs), or through the measurement of gene products (generally enzymes) and their metabolic activities (e.g. the patterns of the community's substrate utilization). The aim is to characterize the community, and statistical evaluation is needed for the correct interpretation of the result (Myrold et al., 2014; Myrold & Nannipieri, 2014; Metagenomics, 2007; Dobler et al., 2001). It can be used as an early warning system if the harmful effect can be statistically separated from seasonal and climatic anomalies. The second concept is based on the detection of the only gene that selectively appears as a direct consequence of the harmful effect to be expected and observed or, alternatively, detection of the product, metabolic product (of the gene product, i.e. an enzyme) or owner of the gene (the organism itself). In cases of selective genes, selective analytical methods are also used; for example, DNA hybridization, fluorescent in situ hybridization or PCR. If the particular gene can be detected, it can be assumed that the effect, too, is the consequence of the harmful substance/agent.

5.2.4.1 Algal diversity

Algal concentration in surface waters is an important community-level indicator. Density and diversity of green and blue-green algae are important aquatic indicators. Both the response of green algae to toxicity and of blue-green algae to excess nutrients can be measured by fluorometry. Fluorometers (or fluorimeters) with discrete and continuous sampling (flow-through or immersible sensors) make the monitoring simple and sensitive. Differentiation between the typical pigments of cyanobacteria and chlorophylls is possible using suitable optical filters. Characteristic cyanobacterium pigments of phycocyanin (dominant in freshwater) and phycoerythrin (in marine ecosystems) allow for the selective detection of cyanobacteria in vivo by using special optical filters both for excitation and emission.

5.2.4.2 Physiological profile of the microbial community

A microbial community analysis can provide useful information about environmental changes. Microorganisms are present in all environments and are typically the first organisms to react to chemical and physical changes in the environment. Changes in microbial communities are often precursors to changes in the health and viability of the environment as a whole. Community-level physiological profile and the metabolic activity of the water, wastewater and soil microbiota can be assessed using EcoPlates[TM] of the Biolog system (Biolog, 2015).

The Biolog system is a practical tool in the hands of ecologists, specifically for community analysis and microbial ecological studies. Both qualitative and quantitative conclusions can be drawn from the result, and a reliable index can be created for characterizing the changes. The microplate-based assay measures the metabolism of 31 carbon sources per assay in three replicates. It works with a simple colorimetric readout (any microplate reader is suitable) thanks to the colour reagent used for the indication of the substrate's utilization (see also Volume 2 of this book series – Gruiz et al., 2015).

Community-level physiological Biolog profiling has been demonstrated to be effective both in ecological research and engineering practice because both the stability of a normal community and changes following the onset of an environmental change or a technological intervention can be followed. Several publications (Echavarri-Bravoa *et al.*, 2015; Gryta *et al.*, 2014; Nagy *et al.*, 2013; Christian & Lind, 2006; Dobler *et al.*, 2000, 2001) reported its useful application in wastewater treatment, activated sludge characterization, assessment of contaminated waters and soil and remedial activities.

5.2.5 Molecular methods: General overview

Molecular methods are selective and sensitive in themselves, and their traceability can be further boosted using sensitivity- and detection-enhancing signals. One of the most widespread techniques, PCR, is based on multiplication. New developments can further enhance sensitivity and serve as a basis of quantitative methods. However, conventional molecular methods are expensive, require special laboratories, long periods of time and high workload.

Since early warning of the receptors takes place far from the source in many cases, the application of these methods is justified only when the monitoring cannot be arranged close(r) to the source, or when the hazard or the source is either unidentifiable or unstable.

Many of the molecular methods are used for the detection and investigation of harmful species, and very good examples can be found for every type of DNA technique, thus facilitating the integration of genetic engineering and early warning systems:

- fluorescent *in situ* hybridization (FISH);
- sandwich hybridization, which is a variant of FISH;
- sequential hybridization for the amplification of the fluorescent signal in FISH;
- antibodies and lectins for the immune-fluorescent detection of individual proteins;
- PCR for the multiplication of genes for identification;
- quantitative PCR (qPCR) to measure the relative frequency of genes;
- quantitative competitive PCR for the quantification of genes by means of an inner standard;
- quantitative reverse transcription PCR (qRT–PCR) – reverse transcriptase applying PCR to the detection of messenger RNA (mRNA);
- restriction fragment length polymorphism (RFLP) for the detection of the existence or absence of specific DNA sequences;
- random amplified polymorphic DNA (RAPD) expression – this is a type of PCR that does not require previous knowledge of the nucleic acid sequence of the DNA;
- microsatellite variable number tandem repeat (VNTR) marker-based bioassays for examining population dynamics;
- amplified fragment length polymorphism (AFLP) bioassay involving a combination of RFLP and PCR;
- denaturing gradient gel electrophoresis (DGGE)-based assays for investigating the diversity of complex ecosystems;
- heteroduplex method – a type of DGGE evaluation based on the detection of heteroduplex DNA generation.

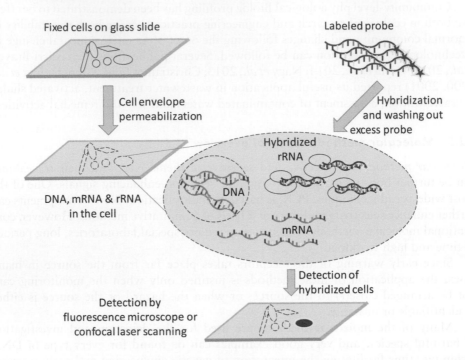

Figure 2.21 Fluorescent *in situ* hybridization (FISH) for the early detection of specific gene activities or the appearance of hazardous organisms.

5.2.5.1   Identification of species and activities by molecular methods

Typical molecular-level biological signals are enzyme responses and metabolites of energy production and respiration, biosynthesis, reproduction, resistance and any other biomarker of adaptation to contaminants. The application of DNA techniques such as PCR or DNA hybridization (e.g. FISH) are worth mentioning as new techniques in environmental monitoring. These methods still have significant drawbacks: for example, DNA has to be obtained from all kinds of complicated matrices; they are time-consuming; it takes a number of hours to obtain results; they are more expensive than traditional analytical methods. Many innovative methods appear in this field from year to year, so the environmental application of such methods is becoming widespread.

In situ hybridization is a powerful technique for identifying specific DNA, mRNA or ribosomal RNA (rRNA) sequences within individual cells and tissues and provides insights into genetic and physiological processes (Figure 2.21). FISH has many advantages:

– It can be applied directly to the biota, without cultivation.
– It may be carried out on a microscope slide or in a suspension and, accordingly, microscopic or flow cytometry is applied for the evaluation.
– It enables the visualization of whole cells so that, in addition to fluorescence, the morphology of cells and the arrangement of the community's components can be observed.

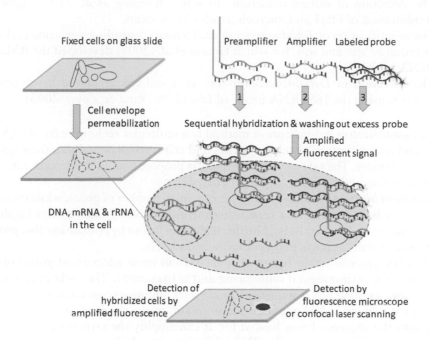

Figure 2.22 Sequential hybridization.

– Several genes can be detected at the same time by using probes labeled with different dyes. The combination of genes may indicate certain potentials and activities, e.g. the degradation of persistent contaminants or the diversity of genes or activities of a community.

However, FISH also has some shortcomings: precisely optimized protocols and suitable crosslinking fixatives are required, and the fluorescent signal is sometimes not strong enough for microscopic or flow cytometric detection. The latter is eliminated by *sequential hybridization,* a relatively new technique for the amplification of the fluorescent signal. Multiple hybridization steps provide the amplification. In the first round, the pre-amplifier molecule hybridizes to the target DNA or the mRNA transcript, and then amplifier molecules hybridize to each pre-amplifier. The labelled probes hybridize to each amplifier molecule. As many as 400 binding sites for labelled probes can be linked to one nucleotide at the end. The amplified signal is visualized using a fluorescence or bright-field microscope (see Figure 2.22).

Probes suitable for environmental studies have been developed extensively. Some good examples of the environmental application of FISH are:

– The summary study of Takashi & Sekiguchi (2011) on methanogenic archaea.
– Marine sediments, which are popular targets of molecular techniques due to the diagnostic value of the sediment microbiota (MacGregor *et al.*, 2003).

- The detection of sulfate reduction, to which Ramsing *et al.* (1993) applied a combination of FISH and microelectrodes for biofilms.
- The identification of *Rhodococcus wratislaviensis* degrading s-triazine herbicides in groundwater and soil, for which Grenni *et al.* (2009) developed the RhLu 16S rRNA probe.
- The detection of *Dehalococcoides* spp. in a soil contaminated by chlorinated solvents using the 16S rRNA probe of Dhe1259t (Yang & Zeyer, 2003).

The *quantitative hybridization* method is a molecular technique for the identification and quantification of selected genes and cells/species containing these genes in complex mixtures. The rate of RNA synthesis is proportional to the growth rate of the species (see also under community-level analysis).

Northern blotting usually looks at one or a small number of genes, while thousands of genes can be visualized at a time using the alternative of microarrays (again, see under community-level analysis). Northern blotting before hybridization also provides quantitative results at particular gene expression levels.

PCR (polymerase chain reaction) is the second most widespread group of techniques used in environmental monitoring and management. The technique amplifies one or more existing DNA sequences *in vitro* by adding excess nucleotides, a DNA polymerase enzyme and oligonucleotide primer(s) to the single-stranded form of the DNA with the sequence being looked for. It can amplify the targeted DNA sequence as many as one billion times. The DNA of interest may derive from organisms, populations, communities or from environmental water or soil samples. The 16S rRNA, or other target sequences of FISH, can also be used for PCR; as primer for the detection and identification of genes, as well as for exploring their diversity in complex environmental matrices. PCR enables the rapid identification of microorganisms, both beneficial and pathogenic.

Gene transcription can be quantified by measuring the amount of mRNA, an end point proportional to growth and/or activities. Primers and primer sets serve to detect specialized strains or groups of microbial species or to characterize communities.

In addition to the methods mentioned, gene expression can be analyzed by real-time PCR or quantitative PCR to detect and quantify the amplified PCR product by incorporation of a fluorescent reporter dye. The fluorescent signal appears and increases proportionally with the amount of the PCR product, i.e. the abundance of the particular PCR product which can be assigned to a gene or to a species.

RNA sequencing (RNA-seq), also called whole transcriptome shotgun sequencing, produces an activity-related end point, a snapshot of RNA presence and quantity from a genome at a given point in time.

5.2.5.2 Molecular techniques for the characterization of communities

FISH can be utilized for community characterization and analysis in several ways by using a set of gene probes responsible for a certain activity, e.g. resistance or biodegradation, and then evaluating occurrence, wholeness and frequency of the gene set in a preparation made from a microbial community, e.g. wastewater sludge, contaminated soil or a specific biofilm. The probable activity or resistance of the community in terms of biodegradation can be determined from the result (compared to a reference).

Quantitative hybridization enables the identification and quantification of selected species within communities. Labelled DNA probes are hybridized to community rRNAs for the quantitative detection of specific groups of microbes within the community. This kind of community analysis reflects the real-time activity of the targeted species: the rate of RNA synthesis is proportional to the growth rate of the species/population. A microbial group- or activity-specific set of probes enables the investigation and control of key actors' activities in the environment or in an environmental technology.

DNA arrays and *DNA chips* can visualize thousands of genes simultaneously. By applying these techniques, as many as one million different DNAs can be synthesized on a membrane surface and can hybridize with the RNA of the environmental sample under test. Preparation of a DNA microarray requires a glass slide (or membrane) which is 'arrayed' with DNA fragments or oligonucleotides that represent specific gene coding regions. 'Arrayed' means that *in vitro* synthesized DNA pieces are attached to the surface of the membrane by covalent bonds or, alternatively, the DNA is synthesized to the surface. Each spot contains a few picomoles (10^{-12} moles) of DNA, a specific sequence functioning as a probe in the following hybridization step. RNA extracted from the (environmental) sample and fluorescently or radioactively labelled is hybridized to the DNA fixed to the glass slide/membrane. After thoroughly washing off the non-hybridized labelled RNA, the hybridization is detected by laser scanning or autoradiography and thus creates an image from the data according to labeling (Figure 2.23).

PCR is a suitable technique for community analysis due to its high selectivity and sensitivity. Its prime use is in the identification of genes and the exploration of their frequency in the environment. The frequency of a set of genes can be translated into the genetic diversity of a community. The metagenome of the community is the analyte and the primer is the probe. Primers and primer sets can be used for detecting specialized genes or groups of genes characteristic of activities or microbial species. PCR can be combined with several other techniques to characterize the metagenome of a community. These methods are also called *community fingerprinting*.

Group-specific primers, such as the ribosomal sequences of alpha-, beta- and deltaproteobacteria, bacilli, actinobacteria, basidiomycota, etc., support the identification and quantification of certain bacterial groups in the community. Primer sets can also be compiled for commonly occurring and important microbial groups in the environment and in environmental technologies.

Terminal restriction fragment length polymorphism (T-RFLP) uses PCR for community fingerprinting. A common gene of the community is amplified by using 5'-fluorescently-labeled primers so that the copied DNAs will also be labelled. The idea behind this method is that the same gene differs from microbe to microbe in a community, demonstrating large variability in the nucleotide sequence. As a consequence, restriction endonuclease enzymes cut the amplified DNA at different places and fragments with different lengths are obtained. The terminal fragments are visualized using fluorescent labeling (see Figure 2.24). Each length is assigned to different types of microorganism. Many different fragment lengths indicate a large number of microorganism types, i.e. the community has a high degree of diversity. The restriction fragments are separated by gel or capillary electrophoresis and detected according to the label type. DNA standards of known size and fluorescence are included in

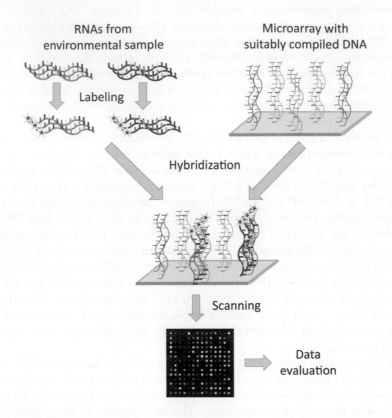

Figure 2.23 DNA microarray: a labeled sample RNA selectively binds to the complementary DNA on the microarray surface.

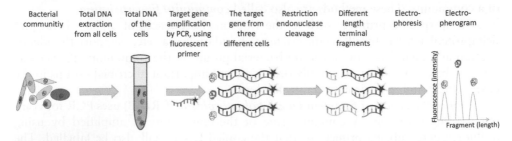

Figure 2.24 Terminal restriction fragment length polymorphism (T-RFLP) analysis: a simplified case with only 3 genotypes.

the analysis as references. The electropherogram shows a series of peaks appearing in time related to fragment length, i.e. microorganism type, and fluorescence intensity (area under the peak) on the vertical axis, proportional to relative abundance of the fragment length, i.e. microorganism type. Minor components often fail to be detected.

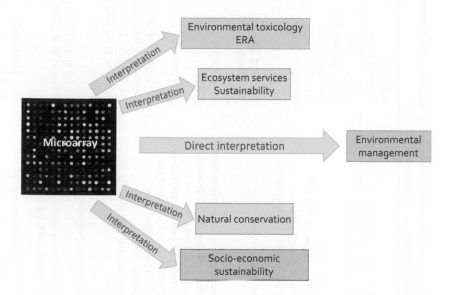

Figure 2.25 DNA techniques and DNA fingerprinting in early warning and environmental management.

T-RFLP is a rapid and sensitive technique for characterizing diversity of complex communities without sequencing the genome. It is a useful tool for examining microbial community structures and dynamics. Changes in the natural bacterial community in the habitats of a marine, freshwater or soil environment due to environmental impact or stress can be monitored. Bacterial communities with key roles in environmental technologies can be studied under different technological systems (activated sludge, remediated soil). Human health conditions depend to a large extent on bacteria, such as the natural microbiota of the digestive tract or the skin, which will exhibit changes when the human body is not in equilibrium.

Community fingerprinting covers different rapid molecular techniques for characterizing and analyzing the diversity of a microbial community based on genomic DNA. It quantifies the variability of the genes present in the environmental community without identifying or counting the individual cells. These methods are of great importance in detecting uncultivable microorganisms which may be dominant in soil and in the tissues and organs of organisms. Community fingerprinting creates an overall picture of a microbial community instead of identifying and individually characterizing community members. The results may provide guidance for more detailed and species-specific assessments as well as new, rapid bioindication and early-warning indicators (Figure 2.25).

5.3 Summary of biomonitoring and bioindication

Tables 2.1–2.4 summarize those environmental health and performance indicators that can be measured by *in situ* rapid methods and other innovative techniques. Indicator types are grouped according to the recommended categories of Odum (1985), Rapport and Whitford (1999) and the OECD (1993).

Table 2.1 Organism-level bioindicators for rapid *in situ*/on-site ecosystem assessment.
Organism level: cell, tissue, organ, organism
Biomarkers: generic and specific biomarkers
Responses: – Stress responses: genes, enzymes, immune response, resistance; – Biochemical or behavioral compensatory response; – The presence of pathogens, opportunists, resistant species; – Bioaccumulation; – Behavior, fitness and reproduction of the organisms.

Indicator type	Stress/toxicity	Available in situ/rapid/innovative method	Reference Chapter/Section
Chemical	Molecules triggered by stress, e.g. ROS on oxidative stress	Reactive oxygen species (ROS) analyzed by sensors	2/4.2 and 2/4.3 4/5.4.1 and 4/5.4.2
Genetic	Presence of stress-specific genes	DNA hybridization, rapid PCR electrochemical/optical biosensors: MB, SPR, quantum dot, colored strip, piezo, etc.	2/4.4 and 2/5.2.5 3/3.1.2.3 4/5.6.2
Genetic	Toxic substance-specific RNA production of the exposed organism	Hybridization, electrochemical/optical sensors	2/5.2.5
Immunological response	Stress protein, stress hormone and neurohormone production	Rapid immunological methods and immunosensors	2/4.5 and 2/4.6 3/2.2.3 and 3/5.2.6 4/6.1.2
Enzymes	Presence of hazardous organisms: detection of their enzyme activity	Fluorogenic substrates hydrolyzed by the enzyme and fluorescence measured (e.g. *E. coli*), presence of beta-galactosidase for coliforms	2/4.4 and 3/2.2.2 4/6.1.2
Enzymes	Generic or toxic substance-specific enzyme production, e.g. oxidative stress biomarkers	Produced enzyme quantity and quality and quality, oxyradical capacity and antioxidant enzymes, e.g. oxidases, esterases, catalase, superoxide dismutase, ROS scavenging enzymes, etc.	2/4.4 and 4/7.1.3 McKenzie *et al.*, 2012; Onodera & Toko, 2014
Enzymes	Toxic substance-specific enzyme inhibition	Activity of the enzyme, e.g. acetyl cholinesterase, inhibition by pesticides	2/4.4 and 3/2.2.2 Galloway *et al.*, 2002 4/6.1.2
Metabolic activity	Decreased activity/inhibition	Special indication, e.g. indicator microspheres	3/3.1
Metabolites	Metabolites of toxic substances, e.g. PAHs	Sampling of aquatic or benthic animals' bile or urine, and analyses of metabolites, e.g. soluble PAH metabolites	4/6.1.3 and 4/7.1.3 Gonzalez *et al.*, 2009
Hormones	Response to estrogenic contaminants	Vitellogenin assessment by rapid immunological methods or immunosensor	2/4.3.2 and 2/4.5 3/2.2.3

Indicator	Characteristic	Method	Reference
Lysosomes	Lysosome destabilization and membrane damage as response to toxicants	Microscopic observation with computer-aided image analyses; lysosomal membrane stability and neutral red retention time	Dondero et al., 2006; Domouhtsidou et al., 2004; Dailianis, 2010
Micronucleus	Formation & frequency as response to genotoxicants	Micronucleus formation and frequency in the gills and other organs of mussels and fish	Cakal Arslan et al., 2010
Resistance to metals	Metallothionein production	FISH for measuring metallothionein mRNA Electrochemical sensor for metallothionein	Sevcikova et al., 2013; Stejskal et al., 2008 3/2.2.3 and 3/3.2
Presence of pathogens	Presence of specific genes of the pathogen	Rapid gene techniques and sensors	2/5.2.5 and 2/5.2.5.1 3/5.2.8 and 4/6.2
	Ecosystem and human health risk indicators	Rapid immune-analytical kits and sensors to detect the presence of the pathogen	2/4.5 and 3/2.2.3 3/3.1.2.2
	Toxins of the pathogen	Rapid ELISA kits and sensors for the detection of bacterial toxins	3/3.1.2.3 and 3/5.2.6 2/4.4 and 2/4.5
Presence of Legionella spp.	Special genes or cell wall antigens of Legionella spp.	Genosensor or immunosensor detection, e.g. rapid lateral flow immunochromatographic assay and combined magnetic immunocapture and enzyme immunoassay	3/2.2.3 and 3/3.1.2.2 3/3.1.2.3 and 3/5.2.6 2/4.4 3/3.1.2.3
Presence of parasites	Ecosystem and human health risk of the parasites	Filtration and microscopic investigation of parasites' cysts and eggs, e.g. analysis of wastewater for agricultural use	WHO, 1996
Movement	Fish, Daphnia, mussel movement	Musselmonitor, fishtox, daphtox with digital image analyses	2/3.1 and 2/5.2.1 3/3.2
Morphology	Fish, Daphnia, mussel	Morphometry	3/3.2
Heart rate	Fish, Daphnia	Heart rate monitoring by computer-aided video analysis	Finn et al., 2012; Brette et al., 2014 3/3.2
Behavior	Avoidance, burrowing behavior of amphipods	Fish, daphnids, amphipods	Boyd et al., 2002

Table 2.2 Population-level bioindicators for rapid in situ/on-site ecosystem assessment.
Biomarkers: mainly generic biomarkers
Responses: – Stress responses; – Genetic response; – Morphological: size and age distribution; – Biochemical response; – Behavioral/compensatory responses, fitness; – Reproduction; – Bioaccumulation.

Indicator type	Stress/toxicity	Available in situ/rapid/innovative method	Reference Chapter/Section
Biomass	Algae g/L	Chlorophyll concentration via chlorophyll fluorescence measured by photometer, by optical biosensors or by remote sensing	2/5.2.3 2/5.2.4.1 3/1.3 and 3/3.1.2.1 3/3.2 and 3/4.1.1
Productivity	Algae g/L/time	Chlorophyll fluorescence in time	3/3
Average size	Fish, *Daphnia*, algae	Morphometry	3/3.2
Age structure and lifespan	Fish, *Daphnia*, algae	Size structure	3/3.2
Abundance	Algae cell counts; plankton, benthos counts	Flow cytometry and filter flow cytometry	Stauber & Adams, 2013
	Cyanobacteria density	Measuring fluorescence for chlorophyll a, phycocyanin and phycoerythrin by fluorometry, optical sensors or by remote sensing	2/4.2.2 and 2/5.1 2/5.2.4.1 3/3
	Plants	Remote and proximal sensing	4/7.2
Photosynthetic activity	Algae chlorophyll	Autofluorescence of chlorophyll a, an indicator of electron transport efficiency	2/5.1 and 2/5.2 3/3
	Plant chlorophyll	Leaf reflectance	4/7.2
Toxic products	Cyanobacterial toxin	Rapid immunoassays, immunosensors	2/3.2 and 2/5.2.4.1 3/2.2.3 and 3/5.2.6
Activities and inhibition	Energy production, respiration, luminescence, growth, enzyme activity inhibition	Respiration tests; Substrate-induced respiration (SIR); Rapid toxicity tests	3.3.1 3/5.2.2 and 3/5.3.3 4/7.1.1 and 4/7.1.2 2/4.4 and 2/4.6
	Enzyme activities of the microorganisms present	Optical biosensor measures the fluorescent products from the substrate cleaved by the enzyme	4/7.1.2 and 4/7.1.3 3/5.2.2 and 3/5.2.5
	Nitrification of nitrifying bacteria	ToxAlarm toximeter quantifying nitrification	3/5.3.1 and 3/5.4 3/5.5 and 4/2.2.3
	Activity of acidophilic bacteria	Bioluminescence in bioleachates	3/3.1 and 3/3.1.2.3 3/5.3.2 and 3/5.3.3
	Bacterial cell numbers, presence of bacteria such as coliforms and *E. coli*	Fluorogenic substrates are hydrolyzed by the enzymes of bacteria and measured by biosensors	2/4.4 3/5 4/7.1.2
Reproductive health	Decrease in reproductivity caused by endocrine disrupting chemicals	Index of intersex assessment through molecular markers	Bizarro et al., 2014
Bioaccumulation	Bioaccumulation in the tissue	Toxicant content in tissue measured by chemical analysis	2/5.2.2 and 4/7.3.1 Yarsan & Yipel, 2013
	Plant bioaccumulation	Toxicant content in plants by remote sensing	4/7.2 and 4/7.3
	Early response signals for bioaccumulation	Bioaccumulation-specific biomarkers in the sentinel organism (e.g. stress/heat shock proteins, oxidative enzymes)	2/1.3.2 2/4 and 2/5.2.2 Yarsan & Yipel, 2013

Table 2.3 Community-level bioindicators for rapid *in situ*/on-site ecosystem assessment.

Biomarkers: – Representative deterioration/pollution indicators: stress-specific effects, food chain deterioration, bioaccumulation, biomagnification; – Type of organisms: sentinels, detectors, exploiters, accumulators, bioassay organisms; – Ecosystem health: compliance, deviations, trends in function and activities.

Responses: – Community-level stress responses; – Energy production; – Community-level genetic (metagenomic), morphological and biochemical responses; – Reproduction; – Adaptation, densities and activities of certain taxa, extinction, size and age distribution; – Bioaccumulation, biomagnification; – Community-level biodiversity and changes in diversity.

Tools: – Early warning, diagnostics; Diversity indices.

Indicator type	Stress/toxicity	Available in situ/rapid/innovative method	Reference Chapter/Section
Activities inhibition	Community energy production, enzyme activities, metabolic activities	Respiration rate; Primary production (algae); Rapid toxicity tests	3/5.2.2 and 3/5.3.1 4/2.2.3 4/7.1.2 and 4/7.1.3
Species diversity	Inducible enzyme synthesis	Substrate-induced respiration (SIR) in wastewater	3/5.2.2 and 3/5.3.1
	Microbial community	Physiological profiling: spatial and temporal changes in microbial communities; Gene technologies.	2/5.2.4 and 2/5.2.5.1 2/5.2.5.2
	Algae taxonomic groups	Chlorophyll fluorescence: excitation spectrum for taxonomic algae classes	4/6.2 and 4/7.1.3 2/5.2.3 and 2/5.2.4.1
	Diatoms	Taxa richness, diversity, taxonomic composition	Li et al., 2010; Torrisi et al., 2010 3/3.1
	Microbial species density & diversity in drinking water	Filter cytometry combined with fluorescence *in situ* hybridization (AquaScope®)	2/1.2.2 and 2/5.1
	One species dominance	Traditional field assessment or metagenome analysis/ variability	2/5.2.4 and 2/5.2.5 2/1.2.2 and 2/5.1
	Short-lived species' dominance	Traditional field assessment or special DNA, immune probe of known short-lived species	2/5.2.4 and 2/5.2.5 2/1.2.2 and 2/5.1
	Exotic species increase	Traditional field assessment, special DNA probe, immune probe, other omics of exotic species	2/5.2.4 and 2/5.2.5 2/1.2.2 and 2/5.1
	Extinction of habitat specialists	Traditional field assessment or special DNA probe, immune probe, omics of known specialist	2/5.2.4 and 2/5.2.5 2/1.2.2 and 2/5.1
	Diversity of selected indicator taxons	Periphyton, benthic community; macroinvertebrates, planktonic community, chironomids, oligochates, fish, etc. using traditional or molecular techniques	2/5.2.4 and 2/5.2.5 Gerhardt et al., 1998; Li et al., 2010
	Bioassessment and creation of indices from diversity, sensitivity and tolerance, etc.	Shannon-Wiener Index & Simpson Index; Trent Biotic Index (TBI); Biotic Integrity Index (IBI); Multivariate indices, e.g. UK RIVPACS; Functional indices, e.g. trophic completeness (ITC); *Chironomidae* and *Oligochaeta* indices; Soil invertebrates density and diversity	Gerhardt et al., 1998; Li et al., 2010 4/7.3.2
	Metagenomic analysis	Species abundance, genetic parameters	Szabó et al., 2007; Sharley et al., 2004
	Nutrient uptake and cycling Food chain and food-web size and efficiency	Biochemical analyses, sensor techniques Integrated methodology	Gerhardt et al., 1998 Wang et al., 2014 4/7.3.2
Biomagnification	Bioaccumulation along food chains/within food web	Early response signals of sentinels; Chemical indicators in top predators	Yarsan & Yipel, 2013

Table 2.4 Ecosystem-level bioindicators for rapid *in situ*/on-site ecosystem assessment.

Indicator types: biological, chemical, physical (social is not included here)

Biomarkers: – Representative ecosystem characteristics; – Water and land uses, anthropogenic loads.

Responses: – Mass & energy balances; – Element and nutrient cycling; – Energy transfer; – Food chains and food webs; – Biomagnification; – Ecosystem-level diversities.

Tools: – Early warning; – Long-term trends; – Aggregated impacts.

Indicator type	Stress/toxicity	Available in situ/rapid/innovative method	Reference Chapter/Section
Biological indicators	Biomass and productivity	Oxygen production/consumption during an incubation period; Radioactive carbon incorporation; Chlorophyll a and other pigments by laser fluorosensor remote sensing	2/1.2.2 and 2/5.1 2/5.2.4 and 2/5.2.5
	Activity; Energetics; Enzyme activities	Energy production – ATP luminometry; Metabolic activity – enzymes/metabolites; Respiration rate – respirometry; Substrate induction – respirometry; Respiration to biomass ratio (R/B); Heterotroph to autotroph biomass ratio; Chlorophyll a to protein ratio; General enzyme level	3/3.1.2 and 3/3.1.2.1 3/5.2.6 and 3/5.3.3 3/5
	Bioaccumulation; Biomagnification Element cycling	Tissue content; key enzymes; other proteins Oxygen, nitrogen, phosphorus, sulfur	2/1.3.2 and 2/5.2.2 4/7.2 and 4/7.3 3/5.1
	Nutrient turnover	Increase, decrease	3/4 and 3/5.4 Tyrell & Law, 1997; Peierls et al., 1991
Chemical indicators: Nutrient & element cycling	Nutrient loss/increase; Nutrient/intermediate accumulation; Organic matter content in water, sediment and soil;	COD, BOD and chemical analysis of the concentrations of nitrogen, phosphorus, dissolved oxygen (DO) in water and sediment;	3/4 and 3/5 Christensen et al., 1989; De Beer & Heuvel, 1988; Jensen et al., 1993; Revsbech & Jorgensen, 1986; Revsbech, 1989
	Dissolved oxygen in water Food web reduction Biomagnification	Nutrient and organic material content of soil Integrated methodology Early response signals of sentinels; Chemical indicators in top predators	4/6.1 Wang et al., 2014 Yarsan & Yipel, 2013
Physical indicators	Increased input of sewage or nutrients; Increased disturbance of the sediments Soil as habitat	Water level and flux Turbidity Soil deteriorations	2/4.1 and 3/4 3/5 4/2.2.3 and 4/3

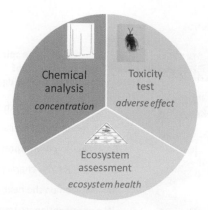

Figure 2.26 Integrated ecosystem assessment: combined chemical, toxicological and biological/ecological assessment and evaluation.

The first step in building up an ecosystem monitoring system is the creation of a conceptual model of the problem/area to be monitored. The conceptual model should include the potential causes occurring along the source–transport pathway–receptor (or reached habitats) track, and the expected responses from the receptors using the endangered habitats. As a second step, as complete a list as possible of the potential physico-chemical, biological and ecological (including toxicological) indicators must be compiled. Finally, the optimum set of informative, easy-to-measure and cost-efficient indicators must be selected on which to then base the monitoring system. The possible intervention methods are an important part of any warning system.

The integrated approach to ecosystem assessment and monitoring, illustrated by Figure 2.26, includes the problem-specific combination of the most relevant and easy-to-acquire physico-chemical, toxicological and ecosystem indicators, environmentally realistic real-time measurement methods and evaluation, and an integrated interpretation of the results (see also Volume 2 in this book series, Gruiz et al., 2015).

6 POSITION OF THE MONITORING SYSTEM

The ideal monitoring system is organically embedded into the complex environmental management and should provide information for management activities and decision making. An additional important task of environmental monitoring is to expand the body of knowledge on the ecosystem and its functioning.

Technical implementation of monitoring is similar to the planning of the statistics-driven sampling in general, discussed in Chapter 1. Planning of the monitoring system covers the determination of the location of the monitoring points, the frequency of sampling, deciding whether the sampling will be intermittent or continuous and, of course, determining the indicators to be measured. Optimal monitoring parameters can be selected based on the conceptual risk model of the area/problem studied. The risk model includes information on the probable contaminants and other stress factors,

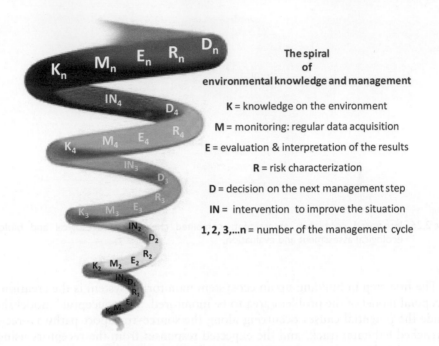

The spiral
of
environmental knowledge and management

K = knowledge on the environment

M = monitoring: regular data acquisition

E = evaluation & interpretation of the results

R = risk characterization

D = decision on the next management step

IN = intervention to improve the situation

1, 2, 3,...n = number of the management cycle

Figure 2.27 The spiral of environmental knowledge and management.

and on the geography, hydrology, hydrogeology, geochemistry, biology and ecology of the area. Planning the monitoring system and placing the measurement points is based on the maximum available preliminary information. After the first monitoring results have been obtained and a statistical evaluation has been made, the monitoring system should be modified and fitted to the situation revealed. The conceptual risk model should also be refined based on the new monitoring results (Figure 2.27). This cycle demonstrates the dynamic and iterative nature of monitoring: an iterative cycle of planning, data acquisition and evaluation makes a gradual refinement of the system possible and enables the application of the best possible methods and tools.

The measuring points of the monitoring system for a point source have to be placed as close as possible to the emission source (see also Figure 2.1). This is extremely important when early warning is being planned. The indicators should be source-oriented. The source can typically be the air space of the workplace, any industrial or mining facility, wastewater discharges, solid-waste yards or contaminant plumes. The only cases when it is worth placing the observation point within the endangered environmental compartment itself is when the discharge point cannot be identified, or if there are too many of them, such as in the case of diffuse pollution.

For the analysis of the atmosphere, air contaminant sources or selected hot spots are monitored, the latter based on meteorological characteristics and forecasts.

For groundwater, the key points are the aquifer and the extraction wells. When planning an early warning system, the characteristics of a sub-surface water flow, e.g. its direction and flux, should be taken into consideration.

The environmental status of surface waters is highly influenced by hydrological and climatic conditions. Surface water monitoring may be concerned with oceans, marine and coastal zones, inland waters and watersheds. Further differentiation is necessary: freshwaters are classified as running and standing waters, springs, run-offs or leachates. Stream size, as well as land and water uses, represent equally important information for planning the monitoring system. Comparison of two points, one before and another after the confluences along the surface water system, may provide useful information for tracking contaminants or adverse effects.

If monitoring sediments, the suspended solids in the water, the solids in the sedimentation zones or the upper soil layer of frequently flooded areas can be regularly assessed.

For diffuse or non-localized sources, the monitoring points need to be placed at the confluence of the mobile environmental compartments and at the junctions of the transport pathways. Pollution of diffuse origin is generally managed and monitored at watershed or regional scales.

Planning the monitoring system can consist of placing the monitoring points on maps, aggregating topography, hydrology, atmosphere, climate, land cover, land use, discharges and their possible transport routes, hot spots of the area to be monitored.

6.1 Remote sensing, GIS-based methods and hyperspectral evaluation

Remote sensing and GIS-based mapping & modelling support the monitoring of large areas, watersheds, and continental or global environmental changes.

Optical methods traditionally play an important role in monitoring, but their evaluation and interpretation require a great deal of experience. Remote sensing is a major step forward in this field, especially if hyperspectral evaluation is used. Any known substance's spectrum can be selected from the spectrum of a hyperspectral image of the right resolution using digital technologies. Thus, the presence and spread of a substance can be traced on a digital map of the area. The basis of the image is usually an interaction between photons and the molecular structure of the observed surface. Reflected and radiated oscillations are sensitive to certain chemical bonds and, like other types of spectroscopy, they can be used to identify the substances on the surface. This method also works with images acquired by remote sensing, in just the same way as it would work in the laboratory or under the microscope. The principle of the measurement is shown in Figure 2.28.

The method enables the differentiation and identification of surfaces with or without vegetation, and with and without contaminants. *Hyperspectral remote sensing* enables observation of the spread of pollution, inspection of emissions, and recognition of different agro-techniques on soil (e.g. to determine the ideal time for irrigation, fertilizer or pesticide application, etc.). Monitoring can keep track of desertification, damage to vegetation by drought or pollution, damage to forests by acid rain, and damage to any kind of surface.

Catchment-scale risk management typically needs transport and fate models and contamination maps based on GIS. Digital maps, together with hydrological and erosion modelling, enable the estimation of chemical concentration in multiple environmental media as well as the mapping of pollution and its transport. Chemical fluxes

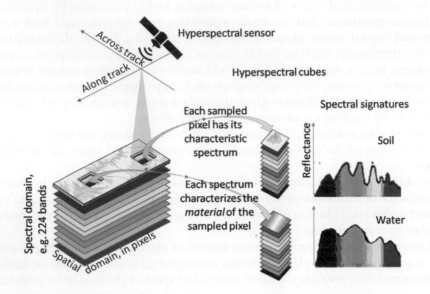

Figure 2.28 Hyperspectral imaging: spatial and spectral sampling of the data cube.

and concentrations are calculated by solving mass balance equations and creating a combination of maps of emissions and removal or transfer rates.

A monitoring system based on the map of the pollution and its transport provides results for the validation and refinement of the GIS-based pollution model and a risk-optimized monitoring system. GIS modeling extends beyond monitoring to the intervention, enabling the calculation of the maximum allowable emission or the necessary reduction rate in order to protect receptors. A GIS-based transport model of an area, e.g. a watershed, can be considered as a primary early warning system for diffuse pollution sources.

The basics of hyperspectral sensing, and the term itself, were first introduced by Goetz *et al.* (1985). As described in the literature, the technique has developed rapidly from its beginnings (Staenz, 1992) until today (Goetz, 2009). Several examples of the application of GIS-based pollution mapping can be identified, such as the overview of Pistocchi (2008) or the case study by Jordan *et al.* (2009) in a watershed contaminated by toxic metals due to diffusely disposed waste of mining origin.

6.2 Near-point source indicators and methods

Pollutant characteristics, the mode of emission and the properties of the environment (temperature, wind, proximity of surface water, soil type, etc.) determine the transport pathways. It is advisable to place the early warning system near the source where emissions are expected and from which the contaminant transport pathway originates.

Volatile pollutants are monitored in the vapor space or the air close to the source using air pollution monitoring equipment. The sampler and the sensor can be placed inside the potentially contaminated vapor/air space. It is advisable to use sorbent-containing collectors when the concentration levels are near the traceability limit and measure the amount of the absorbed pollutant over a longer period of time.

For *water-soluble pollutants*, the inflow and outflow from the sewage treatment plant or the contaminant plume should be monitored and the concentrations in the compliance point calculated with a transport model. Simple transport and fate models can calculate the exposure of the receptors based on the load to the water (or other medium), distribution, degradation or accumulation. GIS-based models are necessary to model watershed-scale or diffuse pollution transport.

Sorbable pollutants are partitioned between the solid and liquid phases and can be detected at locations where solid environmental phases (soil, suspended solid, sediment, solid waste) are in contact with liquid phases (soil waters, surface waters, run-off waters and leachates). In such situations, one can test both the sorbing, i.e. the solid phase, and the desorbing (i.e. leaching, solubilizing, mobilizing) water phase affected by the pollutants. Chemical substances released into the soil are difficult to measure because of the soil's heterogeneity, unless the source and the receiving soil are very exactly known. Continuous direct solid-phase analysis is almost impossible due to high heterogeneity in spatial resolution. However, groundwater and soil air can, as mobile phases, demonstrate a spatial average for the pollutant mobilized from the solid phase (characteristic for the actual distribution of the pollutant between solid and liquid phases in a non-equilibrium situation). The average contaminant concentration of the solid phase can be calculated based on the measured values in soil air, soil moisture, pore water and groundwater. It is also possible to draw conclusions from an adverse effect measured in groundwater or soil moisture as to the toxicity of the solid phase, because toxicity is also distributed between physical phases. Estimation in the opposite direction is also possible, provided that the solid sample analyzed is a realistic representation of the whole.

Sources can be identified from the measured concentrations or adverse effects by using the transport model retrospectively. For *in situ* measurement of the soil's mobile phases, implanted sensors that respond to physico-chemical or biological signals can be used. Alternatively, sensors can be installed on-site in the extracted fraction (groundwater, soil moisture, soil gas, vapor, etc.) of the mobile soil phase. Combining conventional chemical analysis with biosensors and laboratory bioassays provides additional information and enables an integrated evaluation (see Chapters 3 and 4). Pollutant measurement in the water phase (surface waters, runoffs, leachates) using GIS-based transport models enables the management (monitoring and risk assessment) of diffuse or non-identifiable sources.

When sorbable contaminant concentrations and effects are evaluated, it is important to know whether the solid phase is the source and desorption/mobilization is the cause of water contamination, or the opposite applies, i.e. a surface or underground contaminant plume is the actual source and the solid phase is contaminated by sorption. When estimating contaminant concentration in a soil phase based on the measured value of the other phase, the direction of the sorption–desorption process, the flow parameters and the duration of the pollution period should also be considered, in addition to the equilibrium partition of the contaminant and the properties of the soil.

Early warning of chemical pollution needs the enhancement of the assessment by (i) targeted placement of the monitoring point in the source or as near to it as possible, (ii) highly sensitive analytical methods or (iii) an enrichment step before analysis. Enrichment can rely on selective or non-specific sorption, electrokinetic methods (concentrating electrically charged contaminants to electrodes) or selective chemical reaction-based trapping (making reactive compounds soluble or insoluble, oxidized or reduced, condensed, polymerized, etc.). Chemical transformation before analysis not only serves the enrichment and provides a higher concentration, but also helps in developing forms that are easy to detect and enhance sensitivity.

Cyclodextrin is a typical enrichment intermediate with a certain level of specificity. Cyclodextrin-based chemical traps for early warning systems were developed in the MOKKA project. The method patented by CycloLab Kft. employs cyclodextrin polymer-containing traps for binding radioactive iodine to reduce emissions from nuclear power plants (Fenyvesi *et al.*, 1999). Small-scale cyclodextrin traps can be used for early warning to control industrial emissions. Other researchers have described cyclodextrin-based traps to bind polychlorinated dioxins and dibenzofurans from the soil, to bind VOCs (e.g. carbon tetrachloride) from the air (Fourmentin *et al.*, 2006) and to bind PAH compounds, drug residues and pesticides from water (Orprecio & Evans, 2003). Gruiz *et al.* (2011) have developed cyclodextrin polymer-filled filters to collect emerging pollutants such as BPA and waste-phase pharmaceuticals from surface waters. The bound substances can be analyzed using chromatography or, if they are volatile, direct vapor analysis after solvent extraction from the trap (see Chapter 7).

Distribution conveys contaminants not only to the air, water and soil, but also to the ecosystem: thus the ecosystem can be the target of the pollutants but also a part of the transport pathway. Microorganisms and plants grow in the three physical phases of the soil and provide an opportunity for near-source monitoring systems. The soil microbiota responds immediately to every contaminant by changing its metabolism or diversity, e.g. by spreading special genes (responsible for tolerating or degrading a substance) to the metagenome (the sum of the genes in a community). Such genetic markers can be traced, found and identified using DNA techniques. *In situ* whole-cell sensors with natural or gene-manipulated living cells can measure the harmful effects of those contaminants that occur in the mobile soil phases. The task becomes somewhat easier if visual observation of the surface vegetation near the potential contaminant source can be used as an indicator. Optical sensors can transform visual observation into an objective measurement with end points such as vegetation color, size and density. Remote sensing has great potential in this field, especially through hyperspectral evaluation.

The evaluation of aerial photographs or other types of images, preferably as a time series, is the most efficient method to identify and monitor point and diffuse pollution sources. The extent and size of the source, even its material content, can be identified based on these images and hyperspectral evaluation. Transport models can be created based on serial images, the risk can be quantitatively estimated and the urgency of the intervention determined. The erosion and distribution of waste deposits, tailings ponds, industrial and mine wastes and illegal dumps can be traced fairly accurately with the help of timelines taken from aerial or spaceborne images.

6.3 Monitoring methods and indicators applicable to transport pathways

Early warning systems located in transport pathways can be similar to the ones used at sources, but more selective and sensitive methods are needed since they are deployed at a greater distance from the source.

The integrated risk model is of particular importance, since monitoring devices placed on the transport pathway have to focus on the dominant risks, dominant transport pathways and the proper choice of location of the monitoring point.

Near-source monitoring points have to be distinguished from the near-receptor points. The latter chiefly characterize the contaminant and its source, while transport and exposure models are used to determine the risk posed to the receptors. The probability that the monitoring system will work efficiently decreases with increasing dilution, attenuation and retention along the transport pathway. It is important to know where the concentration gradient reaches the limit of physico-chemical or biological traceability along this pathway. Passive biomonitoring should take into account biological adaptation and the time dependency of adverse effects. Many pollutants may have long-term effects without triggering an immediate (acute) response.

Some other phenomena such as physico-chemical concentration (accumulation in the solid phase due to the substance's partitioning or the decrease in redox potential) or bioaccumulation of contaminants may influence the monitoring plan and offer a monitoring option (some organisms are able to concentrate contaminants to 500 times the environmental concentration). Increasing bioaccumulation along food chains, for example, may increase the contaminant concentration several thousand times (biomagnification) and provide natural pre-concentration of the analyte. The toxicokinetic properties of a bioaccumulator species should be known if it is to be applied as a monitoring indicator, i.e. which organs or body parts (fat, liver, spleen, etc.) are likely to accumulate the contaminant and what the uptake, metabolism and excretion rate is. Secondary effects (other than that targeted, e.g. hormone activity of pesticides) and chronic (long-term) effects should also be considered when planning the monitoring system. Very few monitoring tools are available for those chronic effects that do not directly relate to the contaminant concentration. Biochemical assays or rapid simulation tests provide information only about the likelihood of occurrence, not the actual effect. This may lead to great overestimations of pollutant concentration and environmental risk.

The advantage of observation points placed along the transport pathways is that a natural gradient (decreasing and increasing) can be observed; thus, the results confirming the gradients can substantiate the measurement's statistics. Such two-dimensional or matrix-like multidimensional monitoring systems can multiply the chance of identifying the significant and correlating results.

The appearance of a new gene (e.g. responsible for resistance) or diversity changes along the transport pathway may indicate the presence of unacceptable risks and can sometimes identify the cause. The ecosystem obviously changes on long transport pathways, so ecosystem assessments may face high uncertainties. The use of contaminant-selective ecosystem markers may provide a remedy for natural variability. In fairly constant ecosystems, such as most lakes, species diversity can also be a good indicator suitable for early warning. Of course, diversity indices should be interpreted when the aim is to determine the necessary extent of risk reduction.

If a risk model has been established and the relationships inside the model are known, easy-to-measure indicators can be selected for monitoring. A case study of the small Toka Valley watershed in northern Hungary detected serious toxic metal pollution originating from mine waste containing cadmium, copper, zinc and lead sulfides. The waste's weathering caused not only high metal concentrations but also acidic pH because chemolithotrophic anaerobic bacteria oxidized the sulfides. The mobilization of metal cations and the production of sulfuric acid were closely related: the acid dissolved the metal content in the waste. The pH values of the surface runoffs and the creek showed an unequivocal correlation with the toxic metal concentration; thus, pH proved to be a suitable indicator in most cases. The pH decreases downstream for two reasons: acidic water trickling into the runoff water, and the activity of the sulfuric acid-producing microorganisms that spread along the route of surface water flow.

Negative deviation of water ecosystem diversity from normal values can indicate the damage caused to the surface water ecosystem if the characteristics of the non-contaminated, healthy ecosystem is known. Unfortunately, however, this is almost never the case on contaminated sites, where no detailed environmental monitoring took place before contamination, and no proper reference site can be designated. Measurements along the transport pathway of surface waters and the identification of certain gradients can help. Testing the diversity of the macrozoobenthos (sediment-dwelling community) along the Toka Creek, northern Hungary, provided an interesting result: the diversity index of the macrozoobenthos did not show any decrease in quality downstream of the settlements as might have been expected from a normal runnel, but in fact it increased. The diversity data showed that the macrozoobenthos communities were healthier further downstream of the creek's origin. It has been found that the source area of the creek is polluted by a large number of point and diffuse mine waste piles. Thus, the diversity index correlates better with the toxic metal content than with the household waste waters and agricultural run-offs from the village.

Just how important it is to select the proper methods and analytes is also demonstrated by the Toka Creek catchment area. This creek acts both as the pathway and the secondary source of dissolved and solid-bound toxic metals. In addition to the dissolved ionic metals, eroded mine waste is carried by the creek; it spreads via the creek as surface water sediment, settles in the creek's sedimentation zones or on flooded soils in flood areas. Traditional chemical analysis of the sediment (atomic absorption spectroscopy (AAS) or inductively coupled plasma atomic emission spectroscopy (ICP–AES) after complete digestion of the sample) often failed to provide results supported by other evidence. The sediment samples were found to contain the toxic metals in a random distribution, which did not correlate with the damage to the creek's ecosystem and with the toxicity of bed sediments or flooded soils. The picture did not clear up after a sequential extraction either; no gradients along the creek were found. In the end, the simplest and cheapest laboratory toxicity test provided the best information: direct-contact bacterial bioassay unequivocally measured the acute toxicity of the sediment samples. The hot spots could be identified based on the toxicity map of the watershed sediment. Five hidden mine waste disposal sites were discovered and areas were delineated where diffusely dispersed mine waste, sometimes waste ore, had been scattered (Gruiz & Vodicska, 1993). A careful follow-up investigation at a later date indicated that the bioavailability of the toxic metal content of the solid waste transported from the original location changed depending on the environmental parameters, mainly redox potential, pH and grain size. These parameters, and the presence of the sediment mainly at sedimentation zones and floodplains, showed good correlation with toxicity.

Aerial photos, remote sensing and hyperspectral evaluation make it possible to follow the transport of pollutants in watersheds. The most common signals perceived by remote sensing are connected with biological and biophysical signals and the coverage of the surface. Hyperspectral evaluation is capable of identifying anomalies in the vegetation (density, species diversity, and production and content of pigments such as chlorophyll, carotenoids, anthocyanins, etc.), deviations in the mineral composition of the soil and the uncovered surface, and differences in temperature. Geochemical spectral anomalies measured in the vegetation and on the surface draw attention to the pollution and its spread and can be used as an early-warning indicator. The speed and extent of transport can be estimated from the timelines.

6.4 Indicators and methods applicable in the receptors' environment

The receptor environment is the only compartment that can be monitored if the source is diffuse or unknown, such as illegal dumps or abandoned contaminated sites. Even if the pollutant and, to some extent the transport pathway, are known, dilution or very low concentrations can still cause problems when monitoring applies chemical analysis. In such cases, physico-chemical analytical detection of chemical substances plays a lesser role in environmental monitoring, and the role of the biological and ecosystem responses becomes significant. Surface and subsurface waters are suitable as receptor compartments for the monitoring of catchment-level problems when neither the (diffuse) source nor the transport pathways (surface runoff) can be identified.

Solid-phase environmental compartments such as soils and sediments would load monitoring results with high uncertainty due to their heterogeneity, and result in poor statistics. The low transport velocity and high toxicity buffering capacity of the solid environmental phases contraindicate the choice of chemical early-warning indicators in soil or sediments at the end point of transport routes. Indicators of persistency, chemical accumulation and long-term impacts can be linked to stable minerals, persistent organic contaminants, the presence of resistant and accumulator organisms, and chronic toxic effects on population and the community in soils and sediments. Heterogeneities of the solid phases can be compensated for by monitoring soil air or groundwater. These more homogeneous mobile phases (as compared to the solid) may represent the average of a larger area. The result of the back-calculation for soil from soil water is valid for a hypothetical area or volume, but not for any point. Vegetation may be another significant indicator in indirect soil monitoring (analysis of the flora growing on the soil). Tolerance level and degradative capabilities (e.g. in the rhizosphere), as well as bioaccumulation, are useful indicators for soil monitoring. Sediment-dwelling organisms and aquatic plants play the same role. Species distribution mainly provides information about the health of soils (actually, also about the quality of the air and the precipitation) and sediments. Some indicator species, populations and communities (e.g. food webs) give sensitive responses to stresses and can be used as early-warning indicators.

An ecosystem response is highly impact-oriented; it is a benefit on the one hand, but a disadvantage on the other, given that it is an aggregated effect that includes climatic, seasonal, biological and chemical factors. Forecasting is possible, but only based on the time series of actually occurring impacts or obvious damage compared to references. That is why it is so important to find early indicators that can warn the very first time adverse changes occur, before the whole ecosystem deteriorates. Investigation or monitoring based on bioindicators can be conducted either in a passive way, which means that biomonitoring is carried out on autochthonous specimens, or as an active biomonitoring process, using well-controlled, possibly synchronous, cultures of the test organism grown under controlled circumstances.

Species distribution of microorganisms, traditional enzyme activities or molecular and biochemical markers are able to indicate quality changes in the environment, and their response is preferably transformed into an electrical signal, typically with the help of sensors.

As discussed in this section, when the monitoring points are at great distances from the source, sensitivity and efficiency of the monitoring and the warning system will drop: the targeted signal will decrease, the level of noise will increase and it is more difficult to find a sufficiently selective indicator that can signal the concentration and harmful effects of hazardous agents. Contaminants or other chemical pollution indicators may occur irregularly or *ad hoc*: monitoring in these cases can be improved by continuous online or inline sensors and remote signal processing. The heavy-metal-measuring monitor (ELCAD) developed by Aqua Concorde and described in Chapter 3 can facilitate continuous measurement and enhance sensitivity by multiple factors because it can record signals that only last for extremely short periods. Moreover, by integrating the emissions (concentrations), ELCAD is capable of measuring the longer-term loads (mass), which are proportional to the environmental risk.

Sensitive biosensors which are able to detect concentrations in the ppb range can also monitor specific stressors even far away from the sources, in the vicinity of the users of water or soil. The indication of the presence of a stressor or a toxicant only makes sense when the rapid implementation of risk-reducing measures is feasible. If risk management is not prepared for a rapid response or the nature of the hazard does not allow the prevention of the emission or its dispersion, then the only possibility is restriction, which may protect humans adequately but only protect aquatic and terrestrial ecosystems to a very limited extent.

It should be emphasized, as a concluding remark, that risk-based early warning must be the fundamental tool in environmental management. Source-oriented early-warning signals have the best chance of triggering a rapid implementation of damage mitigation and risk-reduction measures. Both chemical and biological indicators can generate warning signals, but the signals must exhibit high sensitivity and specificity and must efficiently convey information about the problem's origin and extent.

REFERENCES

Artiola, J.F., Pepper, I.L. & Brusseau, M. (eds.) (2004) *Environmental Monitoring and Characterization*. Burlington, MA, Elsevier Academic Press.
Begon, M., Harper, J.L. & Townsend, C.R. (1996) *Ecology*, 3rd edn. Oxford, Blackwell.

Biolog (2015) *Microbial Community Analysis with EcoPlates™*. [Online] Available from: http://www.biolog.com/products-static/microbial_community_overview.php. [Accessed 28th May 2015].

Bizarro, C., Ros, O., Vallejo, A., Prieto, A., Etxebarria, N., Cajaraville, M.P. & Ortiz-Zarragoitia, M. (2014) Intersex condition and molecular markers of endocrine disruption in relation with burdens of emerging pollutants in thicklip grey mullets (Chelon labrosus) from Basque estuaries (South-East Bay of Biscay). Pollutant Responses in Marine Organisms (PRIMO17). *Marine Environmental Research*, 96, 19–28.

Borisov, S.M. & Wolfbeis, O.S. (2008) Optical biosensors. *Chemical Reviews*, 108, 423–461.

Boyd, W.A., Brewer, S.K. & Williams, P.L. (2002) Altered behaviour of invertebrates living in polluted environments. In: Dell'Omo, G. (ed.) *Behavioural Ecotoxicology*. Hoboken, NJ, John Wiley & Sons.

Brette, F., Machado, B., Cros, C., Incardona, J.P., Scholz, N.L. & Block, B.A. (2014) Crude oil impairs cardiac excitation-contraction coupling in fish. *Science*, 343(6172), 772–776. DOI: 10.1126/science.1242747

Bruyker, D.D., Recht, M.I., Torres, F.E., Bell, A.G. & Bruce, R.H. (2010) Vanadium Oxide Thermal Microprobes for Nanocalorimetry. In: *Conference book of Sensors, 2010, IEEE, 1–4 November in Kona, HI*. pp. 2358–2362. DOI: 10.1109/ICSENS.2010.5690948

Cakal Arslan O., Parlak, H., Katalay, S., Boyacioglu, M., Karaaslan, M.A. & Guner, H. (2010) Detecting micronuclei frequency in some aquatic organisms for monitoring pollution of Izmir Bay (Western Turkey). *Environmental Monitoring and Assessment*, 165(1–4), 55–66. DOI: 10.1007/s10661–009–0926–5

Carralero, V., Mena, M.L., González-Cortes, A., Yanez-Sedeno, P. & Pingarrón, J.M. (2006) Development of a high analytical performance-tyrosinase biosensor based on a composite graphite-teflon electrode modified with gold nanoparticles. *Biosensors and Bioelectronics*, 22, 730–736.

CBEMN (2015) *Community Based Environmental Monitoring*. [Online] Available from: http://cbemn.ca. [Accessed 29th July 2015]

CENS (2014) *Center for Embedded Networked Sensing*. [Online] Available from: http://www.cens.ucla.edu. [Accessed 3rd December 2014].

Christensen, P.D., Nielsen, L.P., Revsbech, M.P. & Sorensen, J. (1989) Microzonation of denitrification activity in stream sediments as studied with a combined oxygen and nitrous oxide microsensor. *Applied and Environmental Microbiology*, 55, 1234–1241.

Christian, B.W. & Lind, O.T. (2006) Key issues concerning Biolog use for aerobic and anaerobic freshwater bacterial community-level physiological profiling. *International Review of Hydrobiology*, 91(3), 257–268. DOI: 10.1002/iroh.200510838

CLP (2008) *Classification, Labelling and Packaging of Substances and Mixtures. Regulation (EC) 1272/2008 of the European Parliament and of the Council*. [Online] Available from: http://eur-lex.europa.eu/LexUriServ/LexUriServ.do?uri=CELEX:32008R1272: EN:NOT. [Accessed 29th July 2015].

Copernicus (2015) *Copernicus: Sentinel-1*. Earth Observation Portal of the European Space Agency. [Online] Available from: https://eoportal.org/web/eoportal/satellite-missions/content/-/article/sentinel1. [Accessed 14th March 2015].

CWA (1972) *Summary of the Clean Water Act*. [Online] Available from: http://www2.epa.gov/laws-regulations/summary-clean-water-act. [Accessed 28th May 2015].

Dailianis, S. (2010) Environmental impact of anthropogenic activities: The use of mussels as a reliable tool for monitoring marine pollution. In: McGevin, L.E. (ed.) *Mussels: Anatomy, Habitat and Environmental Impact*. New York, Nova Science. pp. 2–30.

De Beer, D. & Van den Heuvel, J.C. (1988) Response of ammonium-selective microelectrodes based on the neutral carrier nonactin. *Talanta*, 35, 728–730.

de Bree, H.E., Leussink, P., Korthorst, T., Jansen, H., Lammerink, T.S.J. & Elwenspoek, M. (1996) The μ-flown: A novel device for measuring acoustic flows. *Sensors and Actuators A: Physical*, 54, 552–557.

DEWA (2015) *Division of Early Warning and Assessment*. United Nations Environment Programme, [Online] Available from: http://www.unep.org/dewa. [Accessed 29th July 2015].

Dobler, R., Saner, M. & Bachofen, R. (2000) Population changes of soil microbial communities induced by hydrocarbon and metal contamination. *Bioremediation Journal*, 4(1), 41–56. DOI: 10.1080/10588330008951131

Dobler, R., Burri, P., Gruiz, K., Brandl, H. & Bachofen, R. (2001) Variability in microbial population in soil highly polluted with heavy metals on the basis of substrate utilization pattern analysis. *Journal of Soils and Sediments*, 1(3), 151–158.

Domouhtsidou, G.P., Dailianis, S., Kaloyianni, M. & Dimitriadis, V.K. (2004) Lysosomal membrane stability and metallothionein content in Mytilus galloprovincialis (L.), as biomarkers. Combination with trace metal concentrations. *Marine Pollution Bulletin*, 48, 572–586.

Dondero, F., Dagnino, A., Jonsson, H., Capri, F., Gastaldi, L. & Viarengo, A. (2006) Assessing the occurrence of a stress syndrome in mussels (*Mytilus edulis*) using a combined biomarker/gene expression approach. *Aquatic Toxicology*, 78S, S13–S24.

Dorst, B.V., Mehta, J., Bekaertb, K., Rouah-Martin, E., Coen, W.D., Dubruelc, P., Blusta, R. & Robbens J. (2010) Recent advances in recognition elements of food and environmental biosensors: A review. *Biosensors and Bioelectronics*, 26, 1178–1194.

Echavarri-Bravoa, L.V., Aspraya, T.J., Portera, J.S., Winsona, M.K., Thorntonc, B. & Hartla, M.G.J. (2015) Shifts in the metabolic function of a benthic estuarine microbial community following a single pulse exposure to silver nanoparticles. *Environmental Pollution*, 201, 91–99.

EO (2015) *Satellite Missions Database*. Earth Observation Portal of the European Space Agency. [Online] Available from: https://eoportal.org/web/eoportal/satellite-missions. [Accessed 14th March 2015].

ERS-2 (2015) *European Remote-Sensing Satellite2*. Earth Observation Portal of the European Space Agency. [Online] Available from:https://eoportal.org/web/eoportal/satellite-missions/e/ers-2. [Accessed 24th March 2015].

ESEB (2015) *Environmental Stewardship Equipment Bank*. [Online] Available from: www.envnetwork.smu.ca/equipment.html. [Accessed 29th July 2015].

Fenyvesi, E., Szente, L. & Szejtli, J. (1999) Entrapment of iodine with cyclodextrins: Potential application of cyclodextrins in nuclear waste management. *Environmental Science and Technology*, 33(24), 4495–4498. DOI: 10.1021/es981287r.

Finn, J., Hui, M., Li, V., Lorenzi, V., de la Paz, N., Cheng, S.H., Lai-Chan, L. & Schlenk, D. (2012) Effects of propranolol on heart rate and development in Japanese medaka (*Oryzias latipes*) and zebrafish (*Danio rerio*). *AquaticToxicology*, 15(122–123), 214–221. DOI: 10.1016/j.aquatox.2012.06.013.

Fourmentin, S., Surpateanu, G.G., Blach, P., Landy, D., Decock, P. & Surpateanu, G. (2006) Experimental and theoretical study on the inclusion capability of a fluorescent indolizine beta-cyclodextrin sensor towards volatile and semi-volatile organic guest. *Journal of Inclusion Phenomena and Macrocyclic Chemistry*, 55(3–4), 263–269.

Galloway, T.S., Millward, N., Browne, M.A. & Depledge, M.H. (2002) Rapid assessment of organophosphorus/carbamate exposure in the bivalve mollusc Mytilus edulis using combined esterase activities as biomarkers. *Aquatic Toxicololology*, 61, 169–180.

GAW (2015) *Global Atmosphere Watch*. World Meteorological Association. [Online] Available from: www.wmo.int/gaw. [Accessed 29th July 2015].

GEMS (2015) *GEMStat*. Global Environment Monitoring System. [Online] Available from: www.gemstat.org. [Accessed 29th July 2015].

GEOSS (2015) *GEOSS Portal*. Global Earth Observation System of Systems. [Online] Available from: http://www.geoportal.org/web/guest/geo_home_stp. [Accessed 14th March 2015].

Gerhardt, A., Carlsson, A., Ressemann, C. & Stich, K.P. (1998) New online biomonitoring system for Gammarus pulex (L.) (Crustacea): In situ test below a copper effluent in South Sweden. *Environmental Science and Technology*, 32, 150–156.

GHS (2011) *Globally Harmonized System of Classification and Labelling of Chemicals (GHS)*. 4th rev. edn. New York, United Nations. [Online] Available from: https://www.unece.org/fileadmin/DAM/trans/danger/publi/ghs/ghs_rev04/English/ST-SG-AC10- 30-Rev4e.pdf. [Accessed 29th July 2015].

Goetz, A.F.H. (2009) Three decades of hyperspectral remote sensing of the Earth: A personal view. *Remote Sensing of Environment*, 113, 5–16.

Goetz, A.F.H., Vane, G., Solomon, J.E. & Rock, B.N. (1985) Imaging spectrometry for earth remote sensing, *Science*, 228(4704), 1147–1153.

Gompertz, D., Shuker, D., Toniolo, P., Shuker, L., Bingham, S., Calow, P., Coggon, D., King, N., McMichael, A. & Walker, C. (1996) *The use of biomarkers in environmental exposure assessment*. Leicester, Institute for Environment and Health. [Online] Available from: http://www.iehconsulting.co.uk/IEH_Consulting/IEHCPubs/HumExpRiskAssess/ r5.pdf. [Accessed 29th July 2015].

Gonzalez, C., Greenwood, R. & Quevauviller, P. (2009) Existing and new methods for chemical and ecological status monitoring under the WFD. In: Roig, B., Allan, I., Mills, G.A., Guigues, N., Greenwood, R. & Gonzalez, C. (eds.) *Rapid Chemical and Biological Techniques for Water Monitoring*. Chichester, Wiley. DOI: 10.1002/9780470745427.ch1c

Grenni, P., Gibello, A., Barra Caracciolo, A., Fajardo, C., Nande, M., Vargas, R., Saccà, M.L., Martinez-Iñigo, M.J., Ciccoli, R. & Martín, M. (2009) A new fluorescent oligonucleotide probe for in situ detection of s-triazine-degrading *Rhodococcus wratislaviensis* in contaminated groundwater and soil samples. *Water Research*, 43(12), 2999–3008.

Gruiz, K. (2005) Biological tools for soil ecotoxicity evaluation: Soil testing triad and the interactive ecotoxicity tests for contaminated soil. In: Fava, F. & Canepa, P. (eds.) *Soil Remediation Series, No. 6*. Venice, INCA. pp. 45–70.

Gruiz, K. (2009) Integrated and efficient assessment of polluted sites. *Land Contamination & Reclamation*, 17(3), 371–384.

Gruiz, K. & Vodicska, M. (1993) Assessing heavy-metal contamination in soil applying a bacterial biotest and X-ray fluorescent spectroscopy. In: Arendt, F., Annokkée, G.J., Bosman, R. & van den Brink, W.J. (eds.) *Contaminated Soil '93*. Dordrecht, Netherlands, Kluwer. pp. 931–932.

Gruiz, K., Murányi, A., Molnár, M. & Horváth, B. (1998) Risk assessment of heavy metal contamination in the Danube sediments from Hungary. *Water Science and Technology*, 37(6–7), 273–281.

Gruiz, K., Molnar, M., Fenyvesi, É., Hajdu, Cs., Atkári, A. & Barkács, K. (2011) Cyclodextrins in innovative engineering tools for risk-based environmental management. *Journal of Inclusion Phenomena and Macrocyclic Chemistry*, 70(3–4), 299–306.

Gruiz, K., Meggyes, T. & Fenyvesi, E. (eds.) (2015) *Engineering tools for environmental risk management: 2. Environmental toxicology*. Boca Raton, FL, CRC Press.

Gryta, A., Frąc, M. & Oszust, K. (2014) The application of the Biolog EcoPlate approach in ecotoxicological evaluation of dairy sewage sludge. *Applied Biochemistry and Biotechnology*, 174(4), 1434–1443. DOI: 10.1007/s12010-014-1131-8

Halliday, J., Lewis, A.C., Hamilton, J.F., Rhodes, C., Bartie, K.D., Homewood, P. *et al.* (2010) Lab-on-a-chip GC for environmental research. *LC-GC Europe*, 23, 514–523.

Hollenstein, M., Hipolito, C., Lam, C., Dietrich, D. & Perrin, D.M. (2008) A highly selective DNAzyme sensor for mercuric ions. *Angewandte Chemie International Edition*, 47, 4346–4350.

Huang, J., Fang, H., Liu, C., Gu, E. & Jiang, D. (2008) A novel fiber optic biosensor for the determination of adrenaline based on immobilized laccase catalysis. *Analytical Letters*, 41, 1430–1439.

IDEA (2015) *Infusing satellite Data into Environmental air quality Applications*. [Online] Available from: http://www.star.nesdis.noaa.gov/smcd/spb/aq/index.php?product_id=1. [Accessed 14th March 2015].

Ispas, C.R., Crivat, G. & Andreescu, S. (2012) Review: Recent developments in enzyme-based biosensors for biomedical analysis. *Analytical Letters*, 45, 168–186.

Jensen, K., Revsbech, N.P. & Nielsen, L.P. (1993) Microscale distribution of nitrification activity in sediment determined with a shielded microsensor for nitrate. *Applied and Environmental Microbiology*, 59, 3287–3296.

Jordan, G., van Rompaey, A., Somody, A., Fügedi, U. & Farsang, A. (2009) Spatial modelling of contamination in a catchment impacted by mining: A case study for the Recsk copper mines, Hungary. *Land Contamination & Reclamation*, 17(3), 413–422.

Kanso, H., Barthelmebs, L., Inguimbert, N. & Noguer, T. (2013) Immunosensors for estradiol and ethinylestradiol based on new synthetic estrogen derivatives: Application to wastewater analysis. *Analytical Chemistry*, 85(4), 2397–2404. DOI: 10.1021/ac303406c.

KAPshop (2015) *Balloon* [photo]. [Online] Available from: http://www.kapshop.com/Lifters-Balloons-&-Blimps/c75_32/index.html. [Accessed 14th May 2015].

Kovács, M. & Podani, J. (1986) Bioindication: A short review on the use of plants as indicators of heavy metals. *Acta Biologica Hungarica*, 37, 19–29.

Kramer, K.J.M. & Foekema, E.M. (2000) The 'Mussel monitor' as biological early warning system: The first decade. In: Butterworth, F., Gunatilaka, A. & Gonsebatt, M. (eds.) *Biomonitors and Biomarkers as Indicators of Environmental Change*. New York, Kluwer. pp. 59–87.

Krämer, P.M., Weber, C.M., Kremmer, E., Räuber, C., Martens, D., Forster, S., Stanker, L.H., Rauch, P., Shiundu, P.M. & Mulaa, F.J. (2007) Optical immunosensor and ELISA for the analysis of pyrethroids and DDT in environmental samples. In: *Rational Environmental Management of Agrochemicals*. ACS Symposium Series, Volume 966, Chapter 12, pp. 186–202. DOI: 10.1021/bk-2007-0966.ch012.

Landsat (2015) *Landsat-8/LDCM – Landsat Data Continuity Mission*. Earth Observation Portal of the European Space Agency. [Online] Available from: https://eoportal.org/web/eoportal/satellite-missions/content/-/article/landsat-8-ldcm. [Accessed 14th March 2015].

Leynen, M., Van den Berckt, T., Aerts, J.M., Castelein, B., Berckmans, D. & Ollevier, F. (1999) The use of Tubificidae in a biological early warning system. *Environmental Pollution*, 105(1), 151–154.

Li, L., Zheng, B. & Liu, L. (2010) *Biomonitoring and Bioindicators Used for River Ecosystems: Definitions, Approaches and Trends*. Amsterdam, Elsevier. DOI: 10.1016/j.proenv.2010.10.164.

Ligler, F.S. (2009) Perspective on optical biosensors and integrated sensor systems. *Analytical Chemistry*, 81, 519–526.

Liu, C.W., Hsieh, Y.T., Huang, C.C., Lin, Z.H. & Chang, H.T. (2008) Detection of mercury(II) based on Hg^{2+} – DNA complexes inducing the aggregation of gold nanoparticles. *Chemical Communications*, 19, 2242–2244.

Long, F., He, M., Shi, H.C. & Zhu, A.N. (2008a) Development of evanescent wave all-fiber immunosensor for environmental water analysis. *Biosensors and Bioelectronics*, 23(7), 952–958. DOI: 10.1016/j.bios.2007.09.013.

Long, F., Shi, H.C., He, M. & Zhu, A.N. (2008b) Sensitive and rapid detection of 2,4-dicholorophenoxyacetic acid in water samples by using evanescent wave all-fiber immunosensor. *Biosensors and Bioelectronics*, 23(9), 1361–1366. DOI: 10.1016/j.bios.2007.12.004

Long, F., Zhu, A. & Shi, H. (2013) Recent advances in optical biosensors for environmental monitoring and early warning. *Sensors*, 13, 13928–13948. DOI: 10.3390/s131013928.

Lovett, G.M., Burns, D.A., Driscoll, C.T., Jenkins, J.C., Mitchell, M.J., Rustad, L., Shanley, J.B., Likens, G.E. & Haeuber, R. (2007) Who Needs Environmental Monitoring? *Frontiers in Ecology and the Environment*, 5(5), 253–260.

MacGregor, B.J., Ravenschlag, K. & Amman, R. (2003) Nucleic acid-based techniques for analyzing the diversity, structure, and function of microbial communities in marine waters and sediments. In: Wefer, G., Billet, D., Hebbeln, D., Jorgensen, B.B., Schlüter, M. & Van Weering, T.C.E. (eds.) *Ocean Margin Systems*. Berlin, Springer.

McKenzie, D.H., Hyatt, D.E. & McDonald, J.V. (eds.) (2012) *Ecological Indicators 1*. London, Elsevier Applied Science.

Merola, G., Martini, E., Tomassetti, M. & Campanella, L. (2014) New immunosensor for β-lactam antibiotics determination in river waste waters. *Sensors and Actuators B: Chemical*, 199, 301–313.

Metagenomics (2007) *The new science of metagenomics – Revealing the secrets of our microbial planet*. Committee on Metagenomics National Research Council of the National Academies, Washington, DC, National Academies Press. [Online] Available from: http://dels.nas.edu/resources/static-assets/materials-based-on-reports/booklets/metagenomics_final.pdf. [Accessed 19th July 2015].

Mezei, P. & Cserfalvi, T. (2007) Electrolyte cathode atmospheric glow discharges for direct solution analysis. *Applied Spectroscopy Reviews*, 42, 573–604.

Moretto, L. & Kalcher, K. (eds.) (2014) *Environmental Analysis by Electrochemical Sensors and Biosensors: Fundamentals*. Berlin, Springer. [Online] Available from: www.springer.com/gp/book/9781489974808. [Accessed 4th May 2015].

Musselmonitor (2015) *Mosselmonitor*. [Online] Available from: http://www.mosselmonitor.nl. [Accessed 19th July 2015].

Myrold, D.D. & Nannipieri, P. (2014) Classical techniques versus omics approaches. In: Nannipieri, P., Pietramellara, G. & Renella, G. (eds.) *Omics in Soil Science*. Hethersett, UK, Caister Academic Press. pp. 179–187.

Myrold, D.D., Zeglin, L.H. & Jansson J.K. (2014) The potential of metagenomic, and other omic, approaches for understanding soil microbial processes. *Soil Science Society of America Journal*, 78, 3–10. DOI: 10.2136/sssaj2013.07.0287dgs.

Nagy Zs.M., Gruiz, K., Molnár, M. & Fenyvesi, É. (2013) Comparative evaluation of microbial and chemical methods for assessing 4-chlorophenol biodegradation in soil. *Periodica Polytechnica Chemical Engineering*, 57(1–2), 25–35. DOI: 10.3311/PPch.2167.

Norberg, P., Bogren, S. & Nilsson, D. (2015) *Integrated printed biosensor platforms*. Acreo Swedish ICT. [Online] Available from: https://www.acreo.se/projects/integrated-printed-biosensor-platforms. [Accessed 14th November 2015]. DOI: 10.1002/polb.23213 & DOI: 10.1016/j.orgel.2013.02.027.

NRC (2000) *Seeing into the Earth: Noninvasive Characterization of the Shallow Subsurface for Environmental and Engineering Applications*. National Research Council; Commission on Geosciences, Environment, and Resources; Board on Earth Sciences and Resources; Water Science and Technology Board. Washington, DC, National Academies Press. [Online] Available from: www.nap.edu/openbook.php?record_id=5786&page=91 [Accessed 14th December 2014].

Odum, E.P. (1985) Trends expected in stressed ecosystems. *BioScience*, 35(7), 419–422.

OECD (1993). OECD core set of indicators for environmental performance reviews: A synthesis report by the group on the state of the environment. *Environment monographs 83*. OECD/GD 179, Paris.

Oertel, N. (2000) Dreissena-Basket – a powerful technique to monitor and control heavy metals in the river Danube. *International Association for Danube Research*, 33, 383–390.

Onodera, T. & Toko, K. (2014) Towards an electronic dog nose: Surface plasmon resonance immunosensor for security and safety. *Sensors*, 14(9), 16586–16616. DOI: 10.3390/s140916586.

OPC (2015) *Drone* [photo]. Office of the Privacy Commissioner of Canada. [Online] Available from: http://www.priv.gc.ca/information/research-recherche/2013/drones_201303_e.asp. [Accessed 14th March 2015].

Orprecio, R. & Evans, C.H. (2003) Polymer-immobilized cyclodextrin trapping of model organic pollutants in flowing water streams. *Journal of Applied Polymer Science*, 90(8), 2103–2110.

Peierls, B.L., Caruco, N.F., Pace, M.L. & Cole, J.J. (1991) Human influence on river nitrogen. *Nature*, 350, 386–87.

Pistocchi, A. (2008) A GIS-based Approach for Modeling the Fate and Transport of Pollutants in Europe. *Environmental Science & Technology*, 42(10), 3640–3647. DOI: 10.1021/es071548+.

Plata, M.R., Contento, A.M. & Ríos, A. (2010) State-of-the-Art of (Bio)Chemical Sensor Developments in Analytical Spanish Groups. *Sensors*, 10, 2511–2576. DOI: 10.3390/s100402511.

PMRSE (2010) *Innovative Subsurface Investigations – Passive Magnetic Resonance Subsurface Exploration Technology (PMRSE)*. [Online] Available from: http://www.pmrse.com/en. [Accessed 14th December 2014].

PNNL (2015) *Biomarkers for Health & Environmental Sustainability. Integrated Biomarker Approach*. [Online] Available from: http://biomarkers.pnnl.gov/. [Accessed 29th July 2015].

Poppy (2015) *Full-size model of ERS-2* [photo]. [Online] Available from: https://en.wikiversity.org/wiki/Lofting_technology#/media/File:ERS_2. jpg. [Accessed 29th July 2015].

QTL Biosystems (2015) *QTL Biosensor*. [Online] Available from: http://meridiansix.com/qtl/QTL/qtl/biodefense.htm. [Accessed 28th May 2015].

RADAR (2015) *Rationally Designed Aquatic Receptors integrated in label-free biosensor platforms for remote surveillance of toxins and pollutants*, CORDIS Project., European Commission, [Online] Available from: http://cordis.europa.eu/project/rcn/97836_en.html. [Accessed 29th July 2015]

Radar Systems (2015) *Zond-12e GPR*. Radar Systems, Inc. [Online] Available from: http://www.radsys.lv/en/products-soft/products/prod/4. [Accessed 14th April 2014].

Ramiz, D., Ronen, A., Amit, R., Shimshon, B. & Yosi, S.D. (2008) Modeling and measurement of a whole-cell bioluminescent biosensor based on a single photon avalanche diode. *Biosensors & Bioelectronics*, 24(4), 888–893. DOI: 10.1016/j.bios.2008.07.026.

Ramsing, N.B., Kuhl, M. & Jorgensen, B.B. (1993) Distribution of sulfate-reducing bacteria, O_2 and H_2S in photosynthetic biofilms determined by oligonucleotide probes and microelectrodes. *Applied and Environmental Microbiology*, 59, 3840–3849.

Rapport, D.J. & Whitford, W.G. (1999) How ecosystems respond to stress. *BioScience*, 49(3), 193–203.

REACH (2006) *Registration, Evaluation, Authorisation and Restriction of Chemicals (REACH). Regulation (EC) 1907/2006 of the European Parliament and of the Council*. [Online] Available from: http://eur-lex.europa.eu/legal-content/EN/ALL/?uri=CELEX: 32006R1907. [Accessed 14th December 2014].

Revsbech, N.P. (1989) An oxygen microsensor with a guard cathode. *Limnology and Oceanography*, 34, 474–478.

Revsbech, N.P. & Jorgensen, B.B. (1986) Microelectrodes: Their use in microbial ecology. *Advances in Microbial Ecology*, 9, 293–352.

Riu, J., Maroto, A. & Rius, F.X. (2006) Nanosensors in environmental analysis. *Talanta*, 69, 288–301.

Roda, A., Cevenini, L., Borg, S., Michelini, E., Calabretta, M.M. & Schüler, D. (2013) Bioengineered bioluminescent magnetotactic bacteria as a powerful tool for chip-based whole-cell biosensors. *Lab Chip*, 13(24), 4881–4889. DOI: 10.1039/c3lc50868d.

Rodriguez-Mozaz, S., de Alda, M.L. & Barceló, D. (2005) Analysis of bisphenol A in natural waters by means of an optical immunosensor. *Water Research*, 39, 5071–5079.

Rodriguez-Mozaz, S., de Alda, M.L. & Barceló, D. (2009) Achievements of the RIANA and AWACSS EU Projects: Immunosensors for the Determination of Pesticides, Endocrine Disrupting Chemicals and Pharmaceuticals. In: *The Handbook of Environmental Chemistry 5J*. pp. 33–46. DOI: 10.1007/978-3-540-36253-1_2.

SEPA (2010) *Environmental Monitoring*. Swedish Environmental Protection Agency. [Online] Available from: http://www.naturvardsverket.se/en/In-English/Menu/State-of-the-environment/Environmental-monitoring. [Accessed 29th July 2015].

Sevcikova, M., Modra, H., Kruzikova, K., Zitka, O., Hynek, D., Adam, V., Celechovska, O., Kizek, R. & Svobodova, Z. (2013) Effect of metals on metallothionein content in fish from Skalka and Želivka reservoirs. *International Journal of Electrochemical Science*, 8, 1650–1663.

Sharley, D.J., Pettigrove, V. & Parsons, Y.M. (2004) Molecular identification of *Chironomus* spp. (Diptera) for biomonitoring of aquatic ecosystems. *Australian Journal of Entomology*, 43(4), 359–365. DOI: 10.1111/j.1440–6055.2004.00417.x.

Shi, H. & Long, F. (2015) *Development of biosensors by using nanoscale sensitive materials for the detection of pollutants in water environment*. Tsinghua University. [Online] Available from: http://www.civil.hku.hk/uk-china-forum/pdf/11-Prof_Pengyi_ZHANG.pdf. [Accessed 28th May 2015].

Staenz, K. (1992) A decade of imaging spectrometry in Canada. *Canadian Journal of Remote Sensing*, 18(4), 187–197.

Stauber, J. & Adams, M. (2013) Flow cytometry applications in aquatic toxicology. In: Férard, J.F. & Blaise, C. (eds.) *Encyclopedia of Aquatic Ecotoxicology*. Berlin, Springer. pp. 521–532.

Stejskal, K., Křížková, S., Sures, B., Adam, V., Trnková, L., Zehnálek, J., Hubálek, J., Beklová, M., Hanuštiak, P., Svobodová, Z., Horna, A. & Kizek, R. (2008) Bio-assessing of environmental pollution via monitoring of metallothionein level using electrochemical detection. *IEEE Sensors Journal*, 9, 1578–1585.

Szabó, K.É., Ács, É., Kiss, K.T., Eiler, A., Makk, J., Plenković-Moraj, A., Tóth, B. & Bertilsson, S. (2007) Periphyton-based water quality analysis of a large river (River Danube, Hungary): Exploring the potential of molecular fingerprinting for biomonitoring. *Archiv für Hydrogiologie, Supplement*, 161(3–4), 365–382.

Szomolányi, R.M., Frombach, G. & Nagy, A. (2009) Remote sensing as a promising tool of the environmental assessment. *Land Contamination & Reclamation*, 17(3), 423–430.

Tabard-Cossa, V., Godin, M., Burgess, I., Monga, T., Lennox, R.B. & Grutter, P. (2007) Microcantilever-based sensors: Effect of morphology, adhesion, and cleanliness of the sensing surface on surface stress. *Analytical Chemistry*, 79, 8136–8143.

Takashi, N. & Sekiguchi, Y. (2011) Oligonucleotide primers, probes and molecular methods for the environmental monitoring of methanogenic archaea. *Microbial Biotechnology*, 4(5), 585–602.

TerraSAR-X (2015) *TSX (TerraSAR-X) Mission*. Earth Observation Portal of the European Space Agency. [Online] Available from: https://eoportal.org/web/eoportal/satellite-missions/content/- /article/terrasar-x. [Accessed 14th March 2015].

TNO (2014) *Biochemical sensors*. TNO Triskelion bv. [Online] Available from: http://www.triskelion.nl/food-feed/services-food-feed/biochemical- sensors. [Accessed 14th December 2014].

Todd, M.J. & Gomez, J. (2001) Enzyme kinetics determined using calorimetry: A general assay for enzyme activity? *Analytical Biochemistry*, 296, 179–187.

Torrisi, M., Scuri, S., Dell'Uomo, A. & Cocchioni, M. (2010) Comparative monitoring by means of diatoms, macroinvertebrates and chemical parameters of an Apennine watercourse of central Italy: The river Tenna. *Ecological Indicators*, 10, 910–913.

Turner, A.P. (2013) Biosensors: Sense and sensibility. *Chemical Society Review*, 42 (8), 3184–96. DOI: 10.1039/c3cs35528d

Tyrell, T. & Law, C.S. (1997) Low nitrate:phosphate ratios in the global ocean. *Nature*, 387, 793–796.

UNEP (2011) *United Nations Environment Programme – World Conservation Monitoring Centre*. [Online] Available from: http://.www.unep-WCMC.org. [Accessed 29th July 2015].

UNEP (2012). *Early Warning Systems: A State of the Art Analysis and Future Directions*. Nairobi, Division of Early Warning and Assessment (DEWA), United Nations Environment Programme (UNEP).

van den Berg, A., van der Wal, P.D., van der Schoot, B.H. & de Rooij, N.F. (1994) Silicon-based chemical sensors and chemical analysis systems. *Sensors and Materials*, 6(1), 23–45.

van der Oost, R., Beyer, J. & Vermeulen, N.P.E. (2003) Fish bioaccumulation and biomarkers in environmental risk assessment: A review. *Environmental Toxicology and Pharmacology*, 13(2), 57–149.

van Emon, J.M. (2006) *Immunoassay and Other Bioanalytical Techniques*. Boca Raton, FL, CRC Press.

Védrine, C., Leclerc, J.C., Durrieu, C. & Tran-Minh, C. (2003) Optical whole-cell biosensor using *Chlorella vulgaris* designed for monitoring herbicides. *Biosensors & Bioelectronics*, 18(4), 457–463.

Vettigera, P., Bruggera, J., Desponta, M., Drechslera, U., Düriga, U., Häberlea, W., Lutwychea, M., Rothuizena, H., Stutza, R., Widmera, R. & Binniga, G. (1999) Ultrahigh density, high-data-rate NEMS-based AFM data storage system. *Microelectronic Engineering*, 46, 11–17.

Wang, J., Chen, L., Mulchandani, A., Mulchandani, P. & Chen, W. (1999) Remote biosensor for in situ monitoring of organophosphate nerve agents. *Electroanalysis*, 11(5), 866–869.

Wang, J., Gu, B., Lee, M.K., Jiang, S. & Xu, Y. (2014) Isotopic evidence for anthropogenic impacts on aquatic food web dynamics and mercury cycling in a subtropical wetland ecosystem in the US. *Science of the Total Environment*, 487, 557–564.

WCMC (2015) *World Conservation Monitoring Centre*. [Online] Available from: http://www.unep-wcmc.org. [Accessed 29th July 2015].

WFD (2000) *Water Framework Directive. A framework for community action in the field of water policy*. 2000/60/EC Directive of the European Parliament and of the Council. [Online] Available from: http://eur-lex.europa.eu/LexUriServ/LexUriServ.do?uri=CELEX:32000L0 060: EN:NOT. [Accessed 29th July 2015].

Whitton, B.A., Say, P.J. & Wehr, J.D. (1981) Use of plants to monitor heavy metals in rivers. In: Say, P.J. & Whitton, B.A. (eds.) *Heavy Metals in Northern England: Environmental and biological aspects*. Durham, Department of Botany, University of Durham. pp. 135–145.

WHO (1996) *Analysis of wastewater for use in agriculture*. Geneva, World Health Organization. [Online] Available from: http://www.who.int/water_sanitation_health/wastewater/labmanual .pdf?ua= 1. [Accessed 28th May 2015].

Wiersma, G.B. (ed.) (2004) *Environmental Monitoring*. Boca Raton, FL, CRC Press.

WMO (2015) *World Meteorological Organization*. [Online] Available from: http://www.wmo. int/pages/index_en.html. [Accessed 29th July 2015].

Wu, Y., Liu, L., Zhan, S., Wang, F. & Zhou, P. (2012) Ultrasensitive aptamer biosensor for arsenic(III) detection in aqueous solution based on surfactant-induced aggregation of gold nanoparticles. *Analyst*, 137, 4171–4178.

Yagi, K. (2007) Applications of whole-cell bacterial sensors in biotechnology and environmental science. *Applied Microbiology and Biotechnology*, 73(6), 1251–1258.

Yang, Y. & Zeyer, J. (2003) Specific detection of Dehalococcoides species by fluorescence in situ hybridization with 16S rRNA-targeted oligonucleotide probes. *Applied and Environmental Microbiology*, 69(5), 2879–83.

Yarsan, E. & Yipel, M. (2013) The important terms of marine pollution "Biomarkers and Biomonitoring, Bioaccumulation, Bioconcentration, Biomagnification". *Journal of Molecular Biomarkers & Diagnosis*, S1, 003. DOI: 10.4172/2155-9929.S1-003.

Zhong, Z., Fritzsche, M., Pieper, S.B., Wood, T.K., Lear, K.L., Dandy, D.S. & Reardon, K.F. (2011) Fiber optic monooxygenase biosensor for toluene concentration measurement in aqueous samples. *Biosensors & Bioelectronics*, 26, 2407–2412.

Zhou, X., Liu, L., Xu, W., Song, B., Sheng, J., He, M. & Shi, H. (2014) A reusable evanescent wave immunosensor for highly sensitive detection of bisphenol A in water samples. *Scientific Reports*, 4, 4572. DOI: 10.1038/srep04572.

Chapter 3

In-situ and real-time measurements in water monitoring

K. Gruiz[1] & É. Fenyvesi[2]

[1]*Department of Applied Biotechnology and Food Science, Budapest University of Technology and Economics, Budapest, Hungary*
[2]*CycloLab Cyclodextrin Research & Development Laboratory Ltd., Budapest, Hungary*

ABSTRACT

In situ and real-time measurements are increasingly used in environmental monitoring and control to enhance the efficacy of decision-making in environmental management. *In situ* measurement techniques enjoy high priority in surface water and wastewater management, particularly in contaminated or otherwise endangered water and land investigations. Real-time measurement methods in particular are employed in water management and global earth monitoring systems. The *in situ* measurements and associated telecommunication, along with the local and worldwide network technologies, make early warning and automation possible. Only *in situ*, real-time measuring methods can fulfill the requirements of environmental regulations worldwide to ensure good quality air and water, healthy agricultural products and a sustainable ecosystem.

In situ sensors that provide real-time output data are a rapidly developing field with hundreds of innovations and applications. The number and quality of the commercial products has also shown a significant increase in recent years.

This chapter categorizes *in situ* and real-time water monitoring measurement techniques as geophysical, geochemical, chemical analytical, biological, ecological and ecotoxicological methods. The fields of application cover surface waters and oceans, drinking water and wastewaters. Rapid methods, field equipment and portable devices are discussed starting from the simplest test papers and visual observations to the most advanced sensor techniques.

The soil and solid waste measurement and test methods are discussed in Chapter 4.

An automated continuous monitoring device for measuring toxic element concentrations in surface-, ground- and wastewaters is also introduced and its use demonstrated.

I INTRODUCTION

Continuous real-time observations and reporting on the status of the environment is the only way to understand and manage regional and global environmental problems. Real-time information on the environment is essential in problem- and site-specific investigations for dynamic decision-making as well as in technology monitoring and process control.

Real-time information may originate from *in situ* detection – when the measurements are realized in the field –, or from remote sensing. On-site or near-site measurements can provide almost real-time data with little delay. *In situ* measurements may be continuous or intermittent, depending on the time requirement of the measurement method and the type of the device used.

1.1 Regional and global monitoring

Real-time serial data show the changes and the trends which are especially useful in environmental monitoring both for endangered and contaminated land as well as for process control of environmental technologies. Information based on real-time measurements provides the best early warning solution and certainly constitutes the only viable approach to more efficient environmental management in the future. Atmosphere and land surface have been observed remotely for many decades, but waters need an urgent change from a data acquisition point of view. Conventional, laboratory-based water analytical techniques currently used hamper fast and effective interventions needed to supply sufficient quantities of high-quality water. There is a huge global interest in efficient water management which is also reflected by WHO regulations and recommendations (GLAAS, 2014), the European WFD (2000) or the CWA (2002) in the US. In addition, the time requirement and cost of the conventional high-sensitivity analytical methods represent obstacles: ever faster and more sensitive and precise methods are required and developed to detect the hundreds of contaminants in environmental samples, and these analytical methods are becoming more and more expensive. It is clear that the strategy of measuring extreme numbers of water samples with extremely sensitive (while still not measuring all of the risky contaminants) analysis methods cannot be followed. New thinking is necessary to optimize the use of measuring capacities: stepwise assessment, cheap, well-designed, selective or generic early warning systems and screening tools, exclusion of the negative samples as soon as possible, and ensuring a good choice of inexpensive, readily available, verified, rapid, and *in situ* analytical and test methods.

Figure 3.1 illustrates the main regional and global observation and monitoring concepts, highlighting the differences in the length of management paths between online/remote monitoring and laboratory-based methods. Other advantages of online/remote monitoring compared to the conventional procedure are summarized below:

– On-line remote monitoring: continuous sampling results in an unlimited sample size and immediate results, without delay (green line in Figure 3.1). The evaluated results can be used both for short-term forecasting and for long-term management, depending on the coupled models.
– Laboratory-based monitoring: intermittent manual sampling and a limited number of samples, labor-intensive laboratory analysis, delayed decision-making and interventions make the corrective actions inefficient (violet line in Figure 3.1).

1.2 Technology monitoring and process control

Technology monitoring and process control in terms of industrial technologies are real-time, online measuring methods applied over a long period of time. Some environmental technologies, e.g. wastewater treatment, have also moved in this

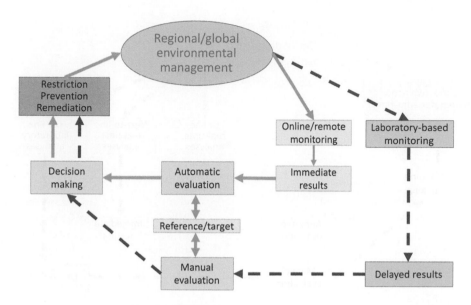

Figure 3.1 Comparison of online/remote monitoring and the laboratory-based methods in regional and global management.

direction, and environmental monitoring stands just before this step. Technology monitoring is *ab ovo* closely linked to process control and regulation, however the basic industrial concepts have been created for closed reactors and physicochemical processes or biotechnologies, while environmental technologies are in use mainly in open reactors and in complex biological-ecological systems. Technology monitoring solutions may be based on online, 'next-to-line' or on off-line measurements. Please see Figure 3.2 for a comparison.

Environmental technology monitoring and process control may increase the efficiency of wastewater or groundwater treatment and soil remediation significantly. The delay in workflow may result in deviations from the optimum and thus result in poor performance. If an environmental technology placed into or in close contact with the environment exhibits poor performance, it endangers the environment, surface waters or groundwater directly.

1.3 Measurement concepts and definitions

A few measurement concepts are described in the following, bearing in mind that the meanings of the used terms are not exactly and consistently defined and applied in the professional literature.

In situ measurements have specific characteristics such as the fact that the sample remains in its original place and in its natural state. It is not separated from the surrounding medium, thus the interactions with its environment are continuous. It also means that the measurement technology should either be employed in the field by placing it there stably or by using mobile, portable devices.

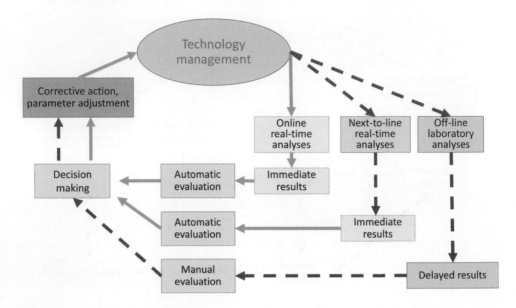

Figure 3.2 Comparison of online, next-to-line and off-line technology monitoring and process control.

It may be extremely important in many environmental studies to keep the sample in continuous interaction with its environment, exposed to air, precipitation and the surrounding abiotic and biotic compartments. An uprooted sample is significantly different from the original one, especially in cases of low redox potential and anaerobic samples or solid matrices. The separation of the sample from the dynamic physical, chemical and biological context and from its natural habitat results in a one-off static situation at the moment of sampling and an entirely different 'new life' following sample removal. Preservation, packaging of samples and shipping into a laboratory cause further multiple uncertainties which cannot be compensated by accurate and costly analyses.

Invasive and non-invasive *in situ* measuring methods differ in terms of the type of signal and the scale of interaction during the placement and operation of the sensors/measuring devices. The scale of interaction of the sample with the measuring device may cover a wide range of

–   analyzers working in deep boreholes – measuring gas, water or solid samples after an invasive preparatory phase;
–   technologies and modern sensors based on *in situ* chemical reactions or very mild interactions;
–   detection of the reflected electromagnetic radiation – a natural interaction without any additional interaction between the object and the sensor.

In situ measurements provide ***real-time information,*** making early warning, process control and long-term environmental management possible.

1.3.1 Some definitions

Real-time measurement means data acquisition and processing during a chemical, physical, biological or other process without delay or asynchronism. A real-time environmental signal is well defined spatially and temporally. It can be a single measurement point or several time points creating a series. Long-term data-series are usually what is meant when using the term *real-time data*. In the context of real-time measurements, all the steps of sensing, detection, data transfer, processing, and evaluation should also be 'real-time', with the acceptance of a minimal delay which is necessary for data transmission and processing.

Sensing and detection of environmental events and changes in quantities or qualities is the basis of environmental monitoring. The term *sensor* is used for a device that measures signals with little interaction and shows the scale of response in the form of an interpretable signal. Another term often used for the activity of placing the sensors or equipment containing sensors is *probe*. An *environmental probe* can be defined as the act of exploring or monitoring the environment with a device or instrument, e.g. a sensor or electrode that can be placed into environments to take and convey measurements.

A *sensor* is a transducer which detects signals from its environment. It detects quantities reflecting interactions, events or changes and provides an electrical or optical signal, e.g. electrical current or voltage. The signal may derive from chemical species present (chemical sensors) or from the activity of a living organism (biosensors). Biosensors can detect the biological response of indigenous organisms (e.g. the chlorophyll fluorescence excitation spectra of algae) or that of a test organism which is built into the 'whole-cell biosensor'. Another way of sensing biological responses is the detection of chemical species which are the product of biological activities, for example switching on genes to adapt or to become resistant to toxicants. Using this sensor concept the biospecific chemical species (genes, enzymes and immunomolecules) can be measured directly as the product of the indigenous biota, or in a simulation test where the environmental sample and test organisms interact. The latter approach ensures uniform, repeatable and comparable results and avoids a negative result if the response of the natural ecosystem is weak or missing.

Sensors are able to detect physical, chemical and biological signals, as follows:

- Physical signals: light, temperature, magnetic fields, gravity, humidity, moisture, vibration, pressure, electrical fields, sound, motion, position, etc.;
- Chemical signals: nutrients and toxic chemical substances, indicator molecules, biomolecules;
- Biochemical and biological signals: metabolic indicators, signal molecules, e.g. hormones, neurotransmitters, specific indicators such as DNA, RNA and several omics (see details in Section 2).

Local and remote sensing is possible by placing the sensor into the location of interest, in other words, into the sample, or by remote application, without direct contact, at various distances (typically on-site, in air or in space), depending on the type and transmission of the emitted signal to be detected. Data may be collected *in situ* or on site, close to the monitoring point and accessed locally or remotely.

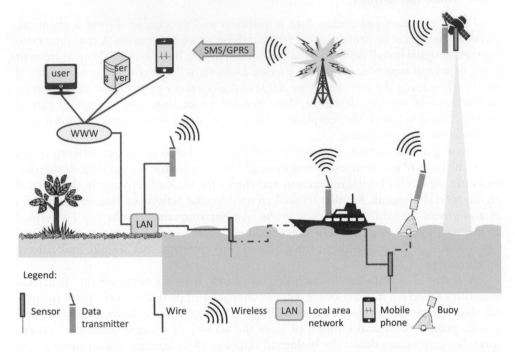

Figure 3.3 Local and remote placing of monitoring system parts: sensing, data logging/communication and workstation.

Remote sensing is the acquisition of information about the environmental object or process without making physical contact with it. Sensors can detect natural signals (passive systems) or the responses on artificially emitted signals.

Remote surveillance is the alternative to reading electronic measuring equipment and collecting the data of sensors in person. It requires access to the collected data from the remotely placed base station via telemetry. One single base station can receive a large amount of data from monitoring systems (Figure 3.3).

1.3.2 Data transmission and processing

Detection by sensors, data recording, transmission, data logging and evaluation can be done *in situ*, on-site or remotely, and all kinds of combinations may be feasible. Remote sensing is essential for non-accessible objects. In this case, primary signals should be able to reach the remotely placed sensor. In other cases, the sensor is placed *in situ* and the data logger and the following step of data processing is carried out remotely. As an alternative, more sensors and the data logger are assembled in a measuring station and the information is forwarded by telecommunication to a remote place for data processing and evaluation. Special combinations occur in professional practices, e.g., an oceanographic device with sensors and data logger can be operated (i) from a ship and the data read either after taking the sensor out or (ii) continuously through a wire, on the ship, or (iii) autonomously once programmed and placed into the proper place and communicate via satellite. Whichever method is applied, the real-time signal characterizes the momentary state of a system or a compartment of the system, and serial signals reflect how a process progresses in time; the only difference is in the mode and place of reading, processing and use of results.

A *data logger* or data recorder is a computer-based, static or programmable data acquisition system, an electronic device that records data over time or in relation to location. It can be either built into the sensor/instrument or external. Electronic data loggers have replaced the former chart recorders in many applications. They are generally deployed and left unattended to record real-time data from the environment or from technologies such as weather stations, water monitoring systems (water level, depth, flow, pH, conductivity, etc.), soil moisture recorders, flow meters, gas pressure, temperature, light intensity recorders, and several other environmental and process monitoring solutions.

The priority activities requiring real-time methods as concerns water bodies are the oceanographic and other surface water monitoring activities, including runoff and wastewater quality control. As regards soil, the groundwater and the natural- and agro-ecosystems are meant to be protected by the integration of these innovative measurement techniques. In addition to generic environmental monitoring, there is a great need for *in situ* and real-time methods for endangered and contaminated land. Real-time data may be decisive in risk management as they can characterize true actual risks and the change trends.

The signals of the sensors can be used not just for monitoring, but also for control and regulation. Several computer-based combined data acquisition and control systems are in operation for industrial, agricultural and environmental technologies as well as for early warning. The control of remote equipment is possible via communication channels with coded signals. The main types of monitoring and control systems are:

- ICS: Industrial Control System for technology control;
- SCADA: Supervisory control and data acquisition from large-scale processes that can include multiple sites and large distances;
- DCS: Distributed control system for a process, wherein control elements are distributed throughout the system. A hierarchy of controllers is connected by communication networks for command and monitoring;
- PLC: Programmable (logic) controller is a computer used for automation of processes.

To exploit the advantages of the *in situ* and real-time information and the connectable dynamic decision-making, the innovative approach is not enough, but innovative tools are also needed. *In situ* and real-time measuring tools are introduced in this chapter through their application. Their advantages and disadvantages will also be discussed to determine the most efficient application of these methodologies in environmental management.

2 *IN SITU* AND REAL-TIME MEASUREMENT TECHNIQUES FOR ASSESSMENT AND MONITORING

In situ environmental investigation and monitoring gives real-time and real-space information about the environment but – of course – it is still loaded with uncertainties based on spatial and seasonal heterogeneities, similarly to most of the techniques applied directly on the environment. The uncertainties and the random errors can be reduced by larger sample sizes and by the elimination of the outliers in a time series. *In situ*, real-time measurements can be used at molecular, microscopic or macroscopic as well as global scales, by means of airborne or satellite coupled sensors in the latter case.

An optical sensor can measure and characterize:

- The DNA-protein interactions by detecting the kinetics of DNA conformational changes;
- The growth of microorganisms or the heart rate of a microscopic insect;
- The movement of an aquatic organism e.g., the opening frequency of a clam;
- The light absorption of local air or the global atmosphere, surface waters and oceans or the terrestrial surface using airborne or spaceborne sensors.

Other types of signals such as radar (synthetic aperture radar = SAR) or hyper-spectral signals, can be logged and converted into images and treated in a similar way to optical sensor data. Computer programs evaluate the arriving data, so the logged number of signals can be increased to extreme scale. The evaluation and interpretation of the large amounts of data generally requires modeling and statistical tools.

In situ applicable non-invasive site assessment tools vary within a wide range from the visual observations of macromorphological characteristics of the ecosystem to the molecular-level biomarkers, and from sensors (including human eye) used in the close environment of the biomarkers to remote sensing with space satellites. A practical combination of *in situ* sensors with remote data collection and processing makes full automation possible: it may change the control and intervention in environmental management to become more efficient in the future.

The advantages and disadvantages of *in situ*, real-time measurements compared to laboratory-based assessments can be summarized as follows:

Advantages

- Data are gathered under ambient conditions;
- The sample is not separated from its environment;
- *In situ* data acquisition is extremely useful for exploratory studies and screening;
- Delineation of contaminated sites is possible;
- Sampling strategy can be modified during field work;
- Research/management strategy can be altered during field work;
- Samples for laboratory analyses can be selected;
- Samples for technological experiments can be selected;
- Real-time data acquisition shows the change trends and allows better estimates;
- Data series from frequent sampling decrease uncertainties and show the trends;
- Long-term data series can serve as basis for statistical evaluation and forecasting;
- The measured values can be supplemented by visual observations, taking photos or videos;
- Sensors and rapid methods may give immediate results on actual risk (bioavailable nutrients and contaminants, toxicity, presence of toxins and pathogens, etc.);
- *In situ* real-time information is directly related to environmental risk; it results in a shortcut in environmental management, and as such avoids (i) the reduction of

the environment to a chemical or biological model; (ii) the re-extrapolation from the results to the real environment – as illustrated by Figure 2.17 in Chapter 2;
– Supports a better understanding of ecosystem complexity;
– Combining *in situ* measurement with large distance data transmission and remote data access will ensure its widespread use.

Disadvantages:

– The sensitivity of the *in situ* methods is often lower compared to the sophisticated laboratory analytical methods;
– Not every type of measurement can be implemented in the field;
– Certain assessment tools are not available in an *in situ* applicable, e.g. portable form;
– Sensors sensitive to contact with solid and biological matter cannot be placed directly into surface waters or soil;
– Part of the *in situ*-applicable sensors should be in direct contact with the environment, which causes deterioration;
– Maintenance and regeneration of the sensors need to be improved.

High-frequency discrete or continuous signals of real-time measuring devices place data in a time dimension, which widens their applicability. Sensors which detect electric signals directly from the analyte have the best accuracy and precision. However, uncertainty due to environmental variability may override this benefit, and therefore harmonizing and optimizing the sampling plan and sensor accuracy is the best strategy. Real-time detection of light or electrons emitted by chemical or biological reactions, although it is less precise, provides more realistic results e.g. nutrient content in water. The latter can be determined by a color reaction with a reagent or microbiota respiration based on CO_2 production.

Types of in situ and real-time measurement methods show great variability: traditional and innovative ideas are used and combined for acquisition and transformation of data into environmental information. Some of the traditional methods such as geophysical assessments are *ab ovo* used *in situ*. Others such as geochemical methods and contaminant analyses, are traditionally carried out in laboratories. These results are loaded with high uncertainty due to sample collection, storage, transport, extraction or other sample preparation methods and with significant delay compared to the date of sampling.

It is important to emphasize that the monitoring itself is a management issue and that individual devices are merely the tools serving the design in line with the scope and concept of the monitoring. After the concept has been laid down, the best fitting tool battery should be assembled and the individual tools selected in harmony with the monitoring requirement and among the tools themselves. The following sections will introduce a number of commercially available measuring equipment and devices, and this overview can help practitioners to make the optimal choice.

2.1 Geochemical and chemical monitoring

Geophysics and hydrogeology traditionally apply *in situ* methods. The measuring devices applied for positioning and for invasive or non-invasive explorations, are all portable devices, often supported by airborne or spaceborne technologies, telecommunication and electronic data storing and processing.

Geochemical and chemical analytical methods are used to describe the chemical composition of the earth's gas, liquid and solid phase compartments and for the identification and measuring of contaminants in the environment.

The theoretical background of the environmental monitoring techniques as well as the methods and devices for general use are discussed in Chapter 4. The *in situ* rapid technologies from the simplest colorimetric test kit applications to *in situ* placed sensor techniques are discussed in this and the next chapter; the methods for water are presented in this Chapter and for soil in Chapter 4.

2.2 Rapid test kits for *in situ* water analysis

A *test kit* is a commercially packaged system of an analytical method's key components used to determine the presence of a specific analyte in a given matrix. Test kits include instructions for their use and are often self-contained, complete analytical systems in easy-to-carry, lightweight boxes. They may require supporting supplies and equipment, which can be the part of the 'box' or these could be available separately. The key components frequently represent proprietary elements or reagents that may be readily prepared by the producer of the kit (AoAC, 1994). Many of these kits fulfill the requirements of relevant standard analytical methods.

Most of the rapid test methods in water analyses are chemical analytical methods, but some tests based on biochemical, enzymatic, immunochemical or DNA techniques are also available in the form of kits and usable in the field with or without mobile laboratories.

2.2.1 Rapid chemical analytical methods and test kits

The *in situ* rapid versions of conventional chemical analyses based on colorimetry apply reagents and indicators for colorimetric evaluation. The simplest and least precise manual analyses apply visual evaluation with test strips or cuvette tests with color cards, cubes or wheels for comparison and reading of the results. More precise titration-based methods need mobile instruments or mobile laboratories. Transportable photometers allow running complex analysis. All the necessary chemicals and tools are assembled into an analytical kit or set available in a handy package. The methods follow international standards and easy-to-understand instructions are added to the kits. The verified products ensure adequate sensitivity and selectivity regarding the analyte, limit or exclude interferences successfully, and can compensate turbidity and color.

Colorimetric analytical kits for waters are provided by AppChem, CHEMetrics, Hach, Lovibond and the Tintometer Group, Macherey-Nagel, Merck Millipore, Systea, Wagtech, Waterworks and many other companies. Table 3.1 summarizes the types, methods, indicators and the detectable concentration ranges of some commercially available colorimetric rapid test kits.

Similar to other titrimetric/colorimetric analytical methods, the corresponding rapid kits are used for targeted assessment, i.e. only known, formerly identified, or otherwise predicted/expected parameters or contaminants.

Table 3.1 Rapid colorimetric test kits for water analysis.

Measured chemical parameter	Method/indicator	Range (mg/L)
Acidity/base capacity	Litmus (pH 5–8), methyl orange (pH 3.1–4.4)	2–7.0 mmol/L H$^+$
Alkalinity	Phenolphthalein (pH 8.3–10.0)	0.2–7.0 mmol/L OH$^-$ or 0–240 CaCO$_3$
Alkalinity – total (pH 5.1; 4.8; 4.5 or 3.7)	Bromocresol green-methyl red and Bromophenol blue	30–500 CaCO$_3$ or 0.2–7.0 mmol/L OH$^-$
Aluminum	Eriochrome cyanine r	0.002–0.25 Al$^+$
Ammonium	Nessler or indophenol blue reagent	0.02–0.4 NH$_4$
Ammonium-N (more ranges)	Salicylate	0.02–50 NH$_3$-N
Anionic surfactants	Cristal violet	0.2–2 LAS
Arsenic	Ag-DDTC = Ag-diethyldithiocarbamate	2–400 ppb
Boron	Carmine	0.2–14 B
Bromine	DPD = N,N-diethyl-p-phenylenediamine sulfate	0.05–4.5 Br$_2$
Calcium	Eriochrome blue, black r	1.5–5.0 Ca
Carbonate hardness	Eriochrome black t or calmagite	0.07–4.0 Ca-Mg
Carbon dioxide	Phenolphthalein	
Chloride (more ranges)	Mercury thiocyanate	0.1–25 & 5–1000 Cl$^-$
Chlorine dioxide	DPD indicator	0.04–5.0 Cl$_2$
Chlorine – free and total	DPD indicator	0.02–10 Cl$_2$
Chromate	Sodium thiosulfate or DPC reagent	0.1–25 CrO$_4^2$
Chromium – hexavalent	1,5-diphenylcarbazide (DPC)	0.01–0.7 Cr^{6+}
Chromium – total	Ox-alkaline hypobromide	0.01–0.7 Cr
Cobalt	PAN indicator (=1-(2-pyridylazo)-2-naphthol)	0.01–2.0 Co
COD high (more ranges)	Reactor digestion, ferroin	100–1500–15,000 COD
COD low (more ranges)	Reactor digestion, ferroin	2.0–40 & 10–150 COD
Copper	Bicinchoninate	0.04–5.0 Cu
Copper	Porphyrin	0.002–0.2 Cu
Cyanide	Pyridine pyrazalone, or p-dimethylamino-benzalrhodanine	0.001–0.25 CN$^-$
Fluoride (SPADNS method)	SPADNS reagent (=4,5 dihydroxyl-3-(p-sulfophenylazo)-2,7-naphthalene-disulfonic acid-Na salt)	0.02–2.0 F$^-$
Hardness	Calmagite colorimetric	0.07–4.0 Ca–Mg
Hydrazine	p-dimethylaminobenzaldehyde	0.004–0.6 ppb
Iodine	DPD indicator t	0.07–7.0 I$_2$
Iron	Ferrozine	0.009–1.4 Fe
Iron ferrous	Phenanthroline 20	0.02–3.0 Fe
Iron total	Phenanthroline 10	0.02–3.0 Fe
Manganese	Periodate	0.2–25 Mn
Manganese	PAN indicator	0.007–0.7 Mn
Molybdenum	Mercaptoacetic acid	0.3–40 Mo
Nitrogen (Ammonium-N, more ranges)	Salicylate	0.02–50 NH$_3$-N
Nickel	PAN indicator	0.007–1.0 Ni
Nitrite	Diazotization	0.003–0.5 N-NO$_2$
Nitrite	Ferrous sulfate	2–250 N-NO$_2$
Nitrate (chromotropic)	Chromotropic acid	0.2–50 N-NO$_3$
Oxygen (dissolved) (Winkler method)	Manganese sulfate and alkaline iodide-azide reagents and thyosulfate titration	1–12 O$_2$

Table 3.2 Rapid colorimetric test kits for water analysis.

Measured chemical parameter	Method/indicator	Range (mg/L)
pH (several ranges)	Test strips for different pH ranges	0–14 pH
Phenols (several ranges)	Sodium peroxidisulfate	0.5–50 phenols
Phenols	4-aminoantipyrine	0.1–3.0 phenols
Ortophosphate	Molybdovanate	1–1000 P-PO$_4$
Total phosphate	Acid persulfate	0–1.5 P-PO$_4$
Silica	Silicomolybdate	1–100 SiO$_2$
Silica ULR rapid	Heteropoly blue rapid liquid	0.003–1.0 SiO$_2$
Sulfate	Sulfate	100–1000 SO$_4^{2-}$
Sulfate	Sulfate	2–70 SO$_4^{2-}$
Sulfide	Methylene blue	0.005–1.0 S^{2-}
Sulfite	Starch	0.1–0.8; 2–100 SO$_3^{2-}$
Total nitrogen (more ranges)	Chromotropic acid	3–150 N
Total phosphorous	Molybdovanadate	0–3.5 P
Total phosphorous	Acid persulfate	1–100 P
Zinc	Zincon	0.01–3.0 Zn

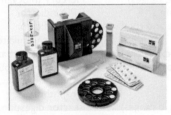

Figure 3.4 Test strips for sensitive pH measurement (Indigo, 2015), color chart for hydrogen sulfide analyses with reagent paper (Hach, 2015), comparator color discs (Comparator disc, 2015) and cube for iron (Comparator cube, 2015).

For arsenic, which is one of the main problems of drinking water supply in many areas of the world, several simple field technologies are available which do not need analytical instruments and results are obtained in 12–30 minutes. Quick test kits are available for average, low and ultra-low arsenic ranges from several companies (Hach, 2015, LaMotte, 2015, Merck, 2015, MN, 2105, ITS, 2015). The color reaction can be evaluated visually using comparative color charts, color discs or color cubes (see Figure 3.4 for some examples). More recent field kits include a digital display of arsenic levels to rule out subjective judgement by the professional who visually detects the difference between the color shades of the strip. One example for digital reading is the *Arsenator* from Wagtech (2015).

Comparative studies support the usability of test kits for field assessments of arsenic. The results suggest that the portable kits can be used to identify water sources with high arsenic concentrations and may provide an important tool for arsenic surveillance and remediation programs (Spear *et al.*, 2006).

Test kits are typically organized into sets by the manufacturers, for example the *ten-parameter test kit* developed for aquacultures (Hach Aquaculture, 2015) to measure

Figure 3.5 Rapid Response Test Kit (Eclox, 2015), Portable Microbiological Water Quality Laboratory (Potatest®, 2015) and Portable Water Quality Laboratory (Potatech®, 2015).

the ten most important parameters for keeping aquacultures well-balanced: acidity, alkalinity, ammonia, carbon dioxide, chloride, dissolved oxygen hardness, nitrite, pH and temperature. Sets for drinking water chemical and microbiological quality are offered by Wagtech (2015) for field analysis. The microbiological set includes an incubator; the chemical set includes the photometer, a compact turbimeter and pocket sensors for pH and conductivity. The mentioned sets are shown in Figure 3.5.

2.2.2 Enzymatic test kits

Enzyme-analytical methods and test kits are customarily used in the food industry for food quality control and are becoming more and more widespread for environmental analytical purposes, too. Enzyme-analytical methods can be characterized by

–   selectivity: enzymes find their target analytes even in complex mixtures;
–   sensitivity: low detection limits;
–   specificity: enzymes react with high certainty with their substrates, i.e. the target analytes (in its equilibrium state an enzyme binds to its substrate with an affinity of 10^5–10^8 M, meaning that the associated complex is 10^5–10^8 times that of the dissociated enzyme and substrate);
–   safety: working under biological conditions, using non-toxic reagents.

Some examples of commercially available enzymatic test kits are introduced below:

–   *Organophosphate/Carbamate* screen kit from Abraxis (Abraxis, Pesticides, 2015) is an *in vitro* enzymatic test used to detect organophosphate and carbamate (OP/C) type insecticides in water and other environmental matrices. It is a qualitative, colorimetric assay based on the inhibitory effect of OP/C on the acetylcholinesterase enzyme. In the case of enzyme inhibition, acetylcholine will not be hydrolyzed, it does not react with 5,5′-dithio-bis(2-nitrobenzoic acid) (DTNB), and fails to produce a yellow color, which is the expected color when no inhibition occurs. The test kit is designed for field use.

- *The organophosphate test kit*, of OP-Stick Sensor is a largely simplified Japanese development for OP/C pesticides. On the stick, two spots can be seen after use, which indicate the presence of the insecticide (yellowish) by contrast with the reference (brown). If no insecticide is present, both spots are brown (OP-Stick, 2015).
- Rapid enzymatic methods with fluorescence detection can quantify fungi – *Mycometer®-test* and bacteria – *Bactiquant®-test* – present in water and air. The technology is based on fluorogenic detection of fungal/bacterial enzyme activities. The sample is contacted with a test solution containing a synthetic substrate, which can be hydrolyzed by the fungal/bacterial enzymes. The hydrolysis product fluoresces upon excitation with ultraviolet light. Fluorescence is measured by a handheld fluorometer after processing for a reaction time at the ambient temperature. Sample preparation and analysis is performed on site within one hour (Mycometer, 2015).
- *Nitrate determination in wastewater* is a newly developed enzymatic method approved by the US EPA (Campbell & Davidson, 2014). Nitrate reductase replaces the cadmium of the traditional nitrate determination method here. In this way cadmium can be eliminated and the substitute is a safe, biodegradable protein. NECi provides easy-to-use test kits for determining nitrate content in any water, soil, plant tissue or livestock feed sample (NECi, 2015).

2.2.3   Immunoanalytical test kits

Immunoanalytical test kits form a special group of rapid methods applicable *in situ* for toxic, mutagenic and reprotoxic contaminants in waters and soils. Immunoassays are based on the very selective and strong binding of an antibody to antigen, i.e. the analyte in this case. The affinity can be characterized by an equilibrium constant of 10^9–10^{12} M, meaning the rate of the associated and dissociated molecules under equilibrium conditions. Several technical solutions have been developed for qualitative and quantitative analyses, for rapid *in situ*/on site analyses such as test strips, tube kits or kits using microplates and readers.

Enzyme-linked immunosorbent assay – ELISA is the immunoanalytical technique on which most of the rapid immunological test kits are based. Its essence is that the analyte from the sample (this is the antigen in the immune reaction) is attached to a solid surface. The antigen-specific antibody is linked to an enzyme. The enzyme-antibody complex is contacted and reacted with the surface bound antigen, and the unbound surplus is washed out. The added substrate of the enzyme produces a measurable color change in proportion to the amount of the bound enzyme.

Magnetic Particle Enzyme Immunoassay (MPEIA) is a relatively new immunoassay method for isolating and measuring antigen-antibody complexes. In the simple magnetic immunoassays (MIA), the antigen is the analyte, and the antibody is labeled by magnetic beads. The complex is formed on the solid-phase surface of the magnetic microparticles. The magnetic bead-linked immunocomplex is then detected by a magnetic reader, measuring the magnetic field change induced by the beads.

In MPEIA, a competitive immunoassay is applied: an enzyme linked analyte-antibody complex is added to the above described reaction mixture. The competition between the analyte in the sample and the enzyme labeled and affixed to the antibody binding sites on the magnetic particles results in an exchange of the labeled and unlabeled analytes. A relatively long reaction time (typically 1–2 hours) is needed to reach equilibrium. At the end of the incubation period, a magnetic field is applied to immobilize the magnetic particles in the test tube. To do so, a magnetic separation rack is used, which allows the separation and immobilization of magnetic particles to the side or the bottom of the test tubes. Racks for special tubes or for normal microtiter plates are available. The unbound reagents can be washed out and the substrate of the enzyme and the chromogen added. The measured color is inversely proportional to the concentration of the analyte in this competitive assay.

Such technology is provided by the test kits of Abraxis (2015), Biosense (2015) for several pesticides, industrial chemicals and estrogens, or the RaPID Assay® (2015) by Modern Water for polycyclic aromatic hydrocarbons (PAHs). The latter is a rapid field testing kit for water and soil, suitable for testing 50 samples at a time within 60 minutes (see also MPEIA Video, 2015).

Immunoassay test kits are provided among others by Hach for the semi-quantitative determination of total petroleum hydrocarbons (TPHs) and polychlorinated biphenyls (PCBs) in soil (see more in Chapter 4) or alachlor and atrazine in water (Hach Immuno, 2015). The rapid test kits include a waterproof pocket colorimeter. We introduce the commercially available products developed for water contaminants:

– *Atrazine immunoassay* reagent set for the determination of atrazine in water. The US EPA Environmental Technology Verification (ETV) Program's Advanced Monitoring Systems (AMS) Center has tested (EPA ETV Atrazine, 2007) the quantitative and the qualitative immunoassays: an ELISA kit (Abraxis Atrazine, 2015), a tube kit (Beacon Atrazine, 2015), as well as the qualitative Watersafe® Pesticide kit (Silver Lake, 2015).
– *Several other pesticides* can be detected by the immunoassay kits (Modern Water, test kits, 2015) listed below:
    o RaPID Assay® tube kits for 2,4-D (2,4-dichlorophenoxyacetic acid), atrazine, triclopyr;
    o EnviroGard® tube kits for triazine;
    o EnviroGard® well kits for isoproturon and triazine;
    o QuickCheck® for chlordane, DDT, isoproturon and triazine (Modern Water, pesticides, 2015).
– Abraxis (Abraxis Pesticides, 2015) provides ELISA kits for the pesticides of 2, 4-D, acetochlor, alachlor, atrazine, azoxystrobin, triazine, carbendazim/benomyl, DDE/DDT, diuron, fluridone, glyphosate, imidacloprid/clothianidin, metolachlor, organophosphate/carbamate (OP/C), penoxsulam, pyraclostrobin, pyrethroids, spinosyn and trifluralin.
– *Microcystins,* the toxic compounds produced by the cyanobacteria species within their cell wall, can also be detected by immunoassay test kits. When the cell dies and disintegrates, microcystins are released into the water, where they have the potential to cause skin rashes, eye irritations, respiratory symptoms, and liver damage for humans, and toxic effects for cohabiting ecosystem members. In 2010 and 2011, six microcystin test kits produced by Abraxis (Abraxis

Microcystin, 2015), Beacon (Beacon Microcystin, 2015) and Zeu-Immunotech (Zeu Microcystin, 2015) were evaluated by the US EPA ETV Program (EPA ETV Microcystins, 2012) in recreational waters. Modern Water, too, produces microcystin immunoassay test kits (EnviroGard® microcytins, 2015).

– *Several biotoxin-specific* immunoassay test kits are available for algal toxins of saxitoxin, domoic acid or octanoic acid accumulated by shellfish. These test kits can also be used for the well-known pathogenic bacterial toxins of anthrax, botulinum, ricin, plague, brucella serving biosafety purposes. The immunoassay kits exist in the form of immunoassay test strips (ADVNT Biotechnologies, 2015, Tetracore, 2015, Zeulab, 2015, Abraxis, 2015), test kits with automated analyser (BioVeris, 2015) or immunoassay test cartridge (Response Biomedical, 2015). The biosensor developments are focused on hand-held tools for rapid and accurate detection of bacterial targets (such as *Bacillus anthracis*) and protein toxins (such as botulinum toxin).

– *Endocrine disrupting chemical compounds* (EDC) are typical water contaminants, which are difficult to analyse and pose an increasing risk for humans and ecosystems. Estrogen ELISA kits have been developed for the detection of estrogenic chemical compounds in water, among others by the Japanese Tokiwa Company (Ecologiena, 2015) and distributed by Abraxis (2015) in the US and Biosense (2015) in Europe for:

  o total estrogen;
  o 17β-estradiol;
  o estrone;
  o ethinyl estradiol.

– *Industrial chemicals* such as surfactants, bisphenol A, triclosan or PCBs can be detected by using rapid immunoassay kits in waters. Several distributors provide ELISA kits such as Abraxis, Biosense and Modern Water. Ecologiena (2015) for example produces kits for:

  o anionic surfactants: linear alkylbenzene sulfonate(LAS) by ELISA kit;
  o nonionic surfactants: alkylphenol ethoxylate (APE) ELISA kit;
  o alkyl ethoxylate (AE) ELISA kit;
  o alkylphenol (AP) ELISA kit;
  o bisphenol A (BPA): super-sensitive ELISA kit.

– Other available ELISA test kits for industrial chemicals:

  o benzo(a)pyrene (B(a)P);
  o coplanar polychlorinated biphenyls (PCBs);
  o PCBs – high chlorination;
  o PCBs – lower chlorination,
  o polybrominated diphenyl ether (PBDE);
  o triclosan.

– *Stress biomarkers* play an important role in the detection of contaminants and early warning. Biosense (2015) has developed a range of new monoclonal and polyclonal antibodies against biomarkers for semi-quantitative rapid detection such as:

  o Vitellogenin (Vtg) and vitellogenin standards from different fish species. Vtg is an egg yolk precursor protein in fish and other egg-laying species. In the presence of estrogenic endocrine disrupting chemicals (EDCs), male fish express the vitellogenin gene in a dose manner, so it is a molecular marker of exposure to estrogenic EDCs.

o Fish *zona radiata* proteins (eggshell proteins, Zrp) are more responsive than vitellogenin for estrogenic effects, so Zrp may function as a more sensitive biomarker compared to vitellogenin.

o Cytochrome P450 1A1 protein is encoded by the CYP1A1 mammal/human gene of the enzyme aryl hydrocarbon hydroxylase (AHH). It is involved in phase I xenobiotic metabolism and is induced by PAHs. The increased amount of gene product indicates the presence of the inductor, the PAH.

o Spiggin, the glue protein of the three-spined stickleback fish species (*Gasterosteus aculeatus*) is produced by the male fish to construct a nest for the eggs. Spiggin production is regulated by androgens. Exposing female fish to androgenic contaminants, the female's kidney also produces spiggin at a contaminant-proportional rate.

– *Metallothioneins* (MT) are specific proteins with the capacity to bind both physiological (zinc, copper, selenium) and xenobiotic (cadmium, mercury, silver, arsenic) metals. MT is a biological indicator for metal stress; its amount is proportional to bioavailable metal exposure.

– *Other stress proteins* such as the so-called heat shock proteins (HSPs) can also be used as indicators for environmental exposures. This family of proteins is produced by cells in response to exposure to stressful conditions. They were first described in relation to heat shock, but currently more proteins are listed which are expressed during other stresses including exposure to cold, UV light, toxic metals, metabolic inhibitors, amino acid analogs, chemotherapeutics, during diseases, or wound healing. The small-size protein ubiquitin which marks proteins for degradation is also classified as a HSP. Detection of the increased production of these HSPs by ELISA or other immunotechniques indicates stress.

– *Gonadotropin-releasing hormone* (GnRH) production and the consequent feminizing effects on male development may be the response to exposure to xenoestrogens such as atrazine, BPA, DDT, dioxin, endosulfan, PBB, PCBs, phthalates, or zeranol. The immunoanalytical detection of GnRH is a possible indication for the presence of xcnoestrogens.

Mercury(II) immunoanalytical test strip can measure Hg(II) between 1–10 mg/L linearly, and its detection limit is 0.23 mg/L. Other metals had a negligible effect on the detection of Hg(II) (Xing *et al.*, 2014). Test kits are mainly used in the initial environmental characterization phase of targeted assessments, but their application in later stages may also be justified, for example for monitoring groundwater or wastewater treatment conditions or for checking whether the treatment is effective. Test kits are not suitable for continuous monitoring; *in situ* or on-line applied sensor-based analytical techniques are recommended instead.

It is important to emphasize that such kits do not substitute the detailed chemical analysis in the positive cases but are able to select negative cases and save a lot of money spent on costly chemical analyses of negative samples. Another advantage is that the test kits provide quick and cost-effective results for the contaminant (e.g. atrazine) levels in positive cases, making the conventional laboratory analysis by gas chromatography/mass spectrometry (GC/MS) easier when the expected concentration range is known.

2.3 Biosensors

The term *biosensor* covers a wide range of equipment which applies biologically active molecules (nucleotides, enzymes, immunomolecules) or living organisms to detect a biologically relevant response to pollutants via electrochemical or optical signals.

One type of biosensor representing the molecular level responses of living systems is the highly selective sensor applied for targeted analyses, namely to find the target molecule in a complex mixture and/or matrix. These sensors follow the concept of traditional chemical analysis: highly selective and sensitive detection of target analytes in the sample. The reaction between the analyte and the sensor-fixed molecules itself is a pure chemical reaction. The traditional chemical analysis generally achieves selectivity not by selective detection, but rather by a selective enrichment of the target analyte during sample preparation. In contrast, biosensors ensure very selective biochemical binding without enrichment, just with the help of the built-in bioactive molecules which mimic biological responses. The built-in molecules may be nucleotides, enzyme-proteins, immunomolecules or engineered molecules with similar roles.

Other types of biosensors representing organism-level responses have broad-spectrum sensitivity, aggregating all exposures present in the sample while taking the biological availability of the analytes into consideration. These types of sensors work with built-in cells or organisms, or such components of the living systems which give the same response for several impacts (such as the luminol or chlorophyll a) so that not only the exposures are aggregated but also the compensatory response of the test organism. This means that the selection of the proper organism assumes the conscious integration of the rules of chemical analysis and ecotoxicology to serve the monitoring concept. Most of these sensors are used for toxicity testing or pathogen detection in waters.

3 *IN SITU* ECOTOXICOLOGY

In situ ecotoxicological methods can measure real-time adverse effects on living organisms in the real environment. They may give highly realistic results, although still loaded with spatial and seasonal heterogeneities and uncertainties. Whole-cell biosensors (see in Section 2.7), are the most advanced *in situ* ecotoxicity measuring devices, whose response is representative of the environment under assessment. Unfortunately, not all organisms can be integrated into sensors.

The toxicity of waters measured by the representative aquatic species is essential information for the regulation of discharges into surface waters from different facilities, e.g. industrial, mining, agricultural or urban areas as well as runoffs and storm sewer systems. Direct toxicity testing is necessary in every case when unknown substances or a mixture of substances are expected in the water, or in the case of a completely unknown situation when non-targeted toxicity screening is necessary. In the US EPA regulation, the National Pollutant Discharge Elimination System (NPDES) includes whole effluent toxicity (WET) testing as a monitoring requirement in the permits the facilities must obtain if they have direct discharge to surface waters (SETAC, 2004).

The test organisms can be the native inhabitants of the aquatic ecosystem or sound/controlled representative ecosystem members.

Figure 3.6 Portable luminometers: Smart Line TL (2015), Lucetta (2015), Clean-Trace™ (2015), System SURE Plus (2015), Junior LB 9509 with and without case (JuniorLB, 2015).

Direct toxicity testing or whole effluent toxicity testing is based on the actual toxicity of all toxic components in contrast with the chemical model-based approach, which measures the concentrations of some supposed contaminants and from these chemical concentrations tries to extrapolate the actual adverse effects and ecotoxicological risks. A whole effluent diluted with the water of the receiving body of water can simulate the real situation and may support the decision on whether or not to allow the discharge into the stream or lake. In the case of significant toxicity, chemical analysis is used for the identification of the responsible contaminants. Freshwater or marine fish, invertebrates, algae and/or macroplants can be used for studying water toxicity.

3.1 Mobile laboratories, rapid toxicity testing, and toxicity test kits

The *mobilisation of laboratory bioassays* is an innovation in environmental toxicology and monitoring. These bioassays can be transformed into rapid on-site test methods, meaning that a short contact of the test organisms with the material is sufficient. If the specific effects measured manifest themselves during the test organisms' growth or propagation, the resulting lengthy time does not allow for the bioassay to be adapted to rapid on-site use. However, if the test is performed in the organisms' non-reproductive phase, the bioassay has a good chance for rapid on-site use. It is worthwhile to work with preserved test organisms which are revitalized in the field as part of the test method. The use of color indicators or other easy-to-detect reagents to the test medium makes the test evaluation easy. The application of portable detectors is also a good option for *in situ* real-time measurements, e.g. colorimeter, densitometer, luminometer (see Figure 3.6). The most advanced solution is the construction of sensors which detect primary signals from selective reactions in a miniaturized system.

3.1.1 General and toxicant-specific testing

Toxicity screening of environmental samples, unlike the detection of particular targeted contaminants, aims to assess the sum of adverse effects. For this purpose, it uses *sensitive and non-selective test organisms* and test end points. Bioluminescent bacteria and their capability to emit light is such a sensitive and generic end point. When the living cell is exposed to toxic substances, the amount of light emitted decreases

proportionally with the toxicity. Microtox® technology is based on this bacterial bioluminescent light emission. It serves as a basis for most of the commercialized whole-cell
toxicity tests. Measuring the difference in light emission between bacterial cultures
unexposed and exposed to toxic substances will therefore indicate the presence of toxicants in the water sample. The gene responsible for light emission can be a 'naturally'
or an artificially (by genetic engineering) built-in element of the living test bacterium.

New developments are steering the engineering of *semi-specific biosensors* which
contain fusions of stress-regulated promoters and reporter genes. These may have
advantages over the generic biosensors due to higher sensitivity and specificity.

Whole-organism-containing biosensors can be created by using the organism's
selective molecular response for a specific contaminant. Such molecular level responses
may come from the genes of adaptive enzymes, stress proteins, immunomolecules or
metabolites responsible for resistance, tolerance or biotransformation (read more in
Charrier *et al.*, 2010). The disadvantage of the limited duration of these kinds of sensors
and their time-consuming development often makes their application unfeasible.

Bioluminescence was probed by a single-photon avalanche diode detector by Elad
et al. (2011) for an agar gel *immobilized recombinant luminobacterium*, which is
sensitive to water pollutants. A flow-through biosensor was constructed in this way
for online continuous water toxicity monitoring.

Jouanneau *et al.* (2012) constructed bioluminescent bacterial biosensors for the
online detection of metals in environmental water samples. They applied freeze-dried
bacteria on a disposable card, which allowed stable detection for 10 days with 3%
reproducibility of the bioluminescence signal both in laboratory conditions and in the
environment. The application of an analytical software made *multidetection of Cd,
As, Hg, and Cu* possible.

A *contaminant-selective, in situ* or remotely readable *Hg biosensor* was developed with a whole-cell system by Goddard *et al.* (2009) who have successfully built
a biosensor containing intact cells to detect both inorganic Hg(II) and methyl-Hg(II).
A hypersensitive Gram-negative mutant was created by directed evolution of MerR
(mercury resistance operon repressor) with subsequent high throughput microplate
screening to increase detection sensitivity. The MerR family is a group of transcriptional
activators, activating the transcription through protein-dependent DNA distortion.
The regulators of Gram-negative mercury resistance (mer) operons were found on
transposable genetic elements (Lund *et al.*, 1986). The Hg(II) biosensor with mutant
MerR can detect Hg(II) at concentrations of 0.1 nM (20 ppt = 0.02 µg/L). The developed Hg(II) biosensor has high specificity, gives a signal only to Hg(II) ions and no
signal with other metals, and is stable for up to 7 days. The prototype has been
completed in the form of a handheld portable detector.

3.1.2 Test kits for general and targeted toxicity

Test kits for on-site aquatic ecotoxicity testing are available with the *Aliivibrio fischeri*
luminobacterium or crustaceans. Rapid on-site screening of water for the presence
of specific bacteria or *bacterial contamination* can also be performed using field kits
based on chemical ATP measurements and luminescence detection with a portable
luminometer.

3.1.2.1 Chemiluminescence in a cell-free system

Chemiluminescence is applied as a test end point based on the inhibitory effect of pollutants on the oxidation reaction of luminol. Toxicants, being free radical scavengers, prevent the reaction leading to chemiluminescence and proportionally reduce the amount of light of the sample compared to the reference (pure water). The percentage inhibition of the light emission is the calculated end point of the test. Some of the products available on the market for rapid *in situ* toxicity assessment based on chemiluminescence are introduced shortly below.

- *Eclox*™ Rapid Response test kit: a qualitative chemiluminescence technology with a luminometer to test toxicity of trace contaminants in water, in the field. Eclox is the abbreviation of Enhanced ChemiLuminescence and OXyradical test. The free radical oxidation of luminol is enhanced by the presence of the oxygen source of horseradish peroxidase (HRP), and the enhancer of 4-iodophenol. Luminescence will be reduced in the presence of toxic substances inhibiting the oxygen-producing HRP enzyme reaction. The results of inhibition rates correlate with other established toxicity tests for several types of contaminant such as metals, antioxidant toxicants, phenols, cyanides, permanganates, pesticides, etc. The chemiluminescence-based rapid toxicity test is unable to identify specific contaminants or their concentrations; it functions instead as a screening tool to quickly determine whether water is potentially toxic (Eclox™, 2015).
- *Chlorophyll fluorescent signals* from photosynthetic enzyme complexes have become one of the most powerful indicators for ecophysiologists in the last few decades. When samples are illuminated by UV light, the intensity of the resultant fluorescence is proportional to the chlorophyll concentration. The major part of the absorbed light energy is used to drive photochemical reactions during photosynthesis. A certain part of the absorbed light is emitted in the form of fluorescence. The presence of electron transport inhibitor toxicants modifies the ratio of absorbed and emitted light energy and the parameters of fluorescence (Boucher *et al.*, 2005). Commercially available apparatuses based on chlorophyll fluorescence are LuminoTox and Robot LuminoTox, produced by the Canadian laboratory Bell Incorporated.
- *LuminoTox* and *Robot LuminoTox* are rapid toxicity detection systems which work as easily as a chemical test: the toxicity result can be seen within less than 15 minutes. Both the handheld model for field analysis and the automated model for online monitoring are available and they measure photosynthetic efficiency by the fluorescence of chlorophyll *a*, an indicator of electron transport efficiency (LuminoTox, 2015; LBi, 2015).

3.1.2.2 Whole-cell toxicity measuring devices

The *in situ* applicable rapid versions of *whole-cell toxicity methods* are designed as portable monitors or are integrated into mobile labs. The whole cells can be single microorganisms (bacteria, algae), eggs of aquatic invertebrates or the mixed microbiota of activated sludge. Most of the methods are based on the luminescent marine bacterium, the *Aliivibrio fischeri* (formerly *Vibrio fischeri*) and the *Microtox*® technology applying it. These devices try to combine the advantages of whole organism toxicity testing and instrumental precision. A few commercial products are listed below:

- *DeltaTox® II* portable toxicity monitor uses Microtox® technology for measuring bioluminescence. It is a simple, rapid, portable water quality test system combining Microtox for acute toxicity and another method for measuring the microbial pollution in the water. Applications include drinking water emergencies and detection of chemical spills entering water systems. Results are available within 5 minutes (DeltaTox II, 2015).

- *Microtox® CTM* is a site-based, broad range, continuous toxicity monitor (CTM). It continuously measures the chemical toxicity of a water source, giving an instant indication of water health. It is a fully automatic instrument that offers a 4-week, autonomous operating cycle and requires a low level of skill for both operation and maintenance (Microtox® CTM, 2015).

- *AppliTOX®* includes a fully automated batch bio-assay using freshly prepared luminescent bacteria based on the standardized laboratory luminescence inhibition test (ISO 11348 – Part 1, 2007) with *Aliivibrio fischeri*. The AppliTOX analyzer system is suitable for ensuring the security of drinking water (intake water, distribution systems), monitoring river water quality (monitoring stations), controlling water recycling of industrial technologies, and monitoring the effluent in wastewater treatment plants (WWTPs) (AppliTOX, 2015).

- *ToxBox* is a toxicity testing box for monitoring toxicity based on bacterial luminescence autonomously, allowing continuous monitoring of rivers, drinking water production or wastewater treatment. With the application of special bacterial strains mutagenicity, biocorrosion and metabolism inhibition can also be monitored. ToxBox is fully autonomous and does not require manual preincubation of the monitor microorganism. Depending on the analysis frequency, ToxBox will operate autonomously for up to four months (ToxBox, 2015).

- *Toxi-chromotest* is a bacterial assay based on the inhibition of the *de novo* synthesis of the inducible enzyme of beta-galactosidase in the *Escherichia coli* K12 OR85 strain. The test applies freeze-dried bacteria, and a rehydration cocktail containing the enzyme beta-galactosidase as inducer. The toxicant-containing sample is added to the revitalized bacterial culture. The toxicants penetrate the cell wall of the bacterium and inhibit the *de novo* synthesis of the beta-galactosidase. The produced amount of the induced enzyme is detected by its reaction with a chromogenic substrate. The greater the toxicity, the lesser the color intensity (Toxi-Chromotest, 2015).

- *Rapidtoxkit* is a rapid, 1-hour toxicity test with larvae of the anostracan crustacean *Thamnocephalus platyurus* for rapid detection of water contamination. The test organism is included in the kit as dormant eggs (cysts) which can easily be hatched on demand to supply the live biota for the assays. This very sensitive sublethal assay is based on the decrease or the absence of ingestion of red indicator microspheres by the test organisms exposed to contaminated waters (Rapidtoxkit, 2015). The colored particulate matter is added to the test after 15 minutes incubation of the test organism in the sample. The control is clean freshwater, wherein the healthy animals take up more microspheres than the stressed ones in the sample. The colored particles can be observed in the digestive tract under a low magnification microscope (e.g. a stereomicroscope).

- *ToxAlarm* toximeter is designed for continuous monitoring of toxicity in drinking or surface waters. It is based on assessing the inhibition of nitrification of activated sludge microorganisms. The conversion of ammonia to nitrate needs oxygen. When the process is inhibited by toxic substances, oxygen consumption decreases. ToxAlarm monitors this oxygen consumption and hence the toxicity. The highly sensitive self-reproducing nitrifying bacterial culture is constantly and independently producing biomass, so enough fresh bacteria are always available for the new measurement. It is characterized by low operational costs since no purchase or external cultivation of bacteria is necessary. The response time is 5–10 minutes (ToxAlarm, 2015).

- *NitriTox* is an online toximeter developed for wastewater treatment plants, especially for the protection of the biology of the nitrification process. Its operation is similar to the previous ToxAlarm equipment, but it is suitable for the continuous toxicity monitoring of wastewater treatment plants. The measurements follow at intervals of less than 5 minutes, thus allowing enough time to introduce countermeasures after the occurrence of pollution. NitriTox offers three warning levels which can be individually set (NitriTox, 2015).

- *TOXcontrol* is a completely automated online toxicity monitoring system. It uses the freshly cultivated test organism of the *Aliivibrio fischeri* bacterium. The luminescence is measured before and after exposure and the inhibition is calculated as a percentage. The automated cultivation of the bacteria occurs inside the instrument and the test method is the online version of the conventionally used ISO standard method. The toxicity information can be sent to a database that is accessible online. Its ability to give an online signal when the water quality has changed allows the operator to take immediate action, for instance to shut down the water intake or stop the water processing. Subsequently, the operator can start a more detailed analysis of the nature of the pollution. Its integration with the monitoring of pathogenic bacteria (BACTcontrol) and algae (ALGcontrol) as well with an optional online solid-phase extraction method makes it possible for the monitoring system to function as a monitoring station. This water quality monitoring system became known from its first application as a biological early warning system in 2004 at a Dutch water intake station along the Rhine River (TOXcontrol, 2015).

- *TOXmini* (2015) is a portable and easy-to-use device for lab and field toxicity testing. It uses *Aliivibrio fischeri* and the same reagents as TOXcontrol.

3.1.2.3   Detection of the presence of microorganisms in waters

A special in situ applicable, real-time microbiological monitor was developed for the detection of the metabolically active acidophilic microorganisms in bioleaching solutions by bioluminescence. The activity of the **sulfide-oxidizing bacteria** is responsible for the production of acidic mine drainage from mines or mine waste disposal sites. The same microbes are responsible for the efficiency of the heap or dump bioleaching technologies applied as metal ore processing (Viedma, 2010).

Pathogenic microorganisms in drinking waters, bathing waters and pools, in surface waters as well in some technological waters need time-consuming and costly inspection when using conventional cultivation-based laboratory methods. This inefficiency is further increased by the probable high number of negative samples. The coliforms in drinking water, the algae in surface waters or the *Legionella* species in bathing water require rapid and automated measuring devices.

Escherichia coli (E. coli) and total coliform detection is essential for good quality drinking water supply all over the world.

– *TECTA™* is a polymer-based optical sensor built into an incubator-analyzer-data logger system. The test utilizes enzyme substrates: beta-galactosidase enzyme for total coliform and β-glucuronidase enzyme for *E. coli*. The enzymes of the bacteria present in the water cleave the substrates, resulting in the release of fluorescent products. The fluorescent molecules are extracted and concentrated within the polymer of the optical sensor, facilitating early and rapid detection by a UV detector (TECTA™, 2015). The equipment can be operated in both manual or automatic modes. A 100–mL water sample is needed and 2–18 hours' cultivation to reach the cell number causing the minimum measurable signal.
– *Colifast ALARM* is an at-line automated remote monitor. The technology uses fluorogenic substrates that are hydrolyzed by the enzymes of coliforms and *E. coli*. An increase in the cell concentration leads to an increase in the proportion of the fluorescent product which is measured by an internal spectrophotometer. It can be operated both manually using intermittent sampling and automatically using periodical or continuous sampling. The 100 mL water samples are automatically collected at programmed intervals. In addition, it measures the turbidity level of the water. The system can automatically send results to the control room or to any remote workplace (Colifast ALARM, 2015).
– *ColiPlate kit* (2015) is a prefabricated 96 well microplate for *E. coli* detection. A convenient test for the quantitative measurement of total coliforms and *E. Coli* bacteria from waters within 24 hours.
– *BACTcontrol* (2015) online monitor is also based on the measuring of fluorescence produced after the bacterial enzyme cleavage of the flouorogenic substrate.
– *ALGcontrol* (2015) equipment offers an online monitoring solution for different kinds of algae through fluorescence detection. It can identify different kinds of algae.

Legionella species are widespread pathogens in waters, which live primarily in cooling towers, swimming pools, domestic water systems and showers, ice making machines, refrigerated cabinets, whirlpool spas, hot springs, and fountains. It is transported by air and vapor from water into the respiratory system, where the bacteria can infect alveolar macrophages. Several rapid kits have been developed for the detection of *Legionella* in waters (Figure 3.7).

– *Legionella* detection (2015) is an on-site applicable rapid test kit using a lateral flow immunochromatographic assay to detect the presence of cell surface antigens from *Legionella pneumophila* serogroup 1 within 30 minutes. The presence of the antigen in the water causes the 'test line' to turn red in color. A 'control line' is included which should always turn red if the test has been performed correctly. It is developed for the rapid analysis of water systems such as cooling towers, hot

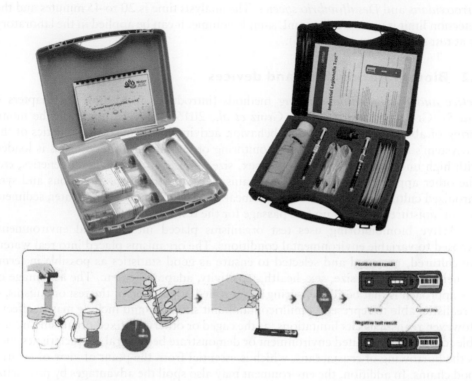

Figure 3.7 *Legionella* kits from Drop Test Kit and the evaluation of the lateral flow immunochromatographic assay (DTK, 2015).

and cold water systems, showers or pools. Several companies produce and sell this test such as for example Accepta, Biótica, Lovibond, etc.

– *Legionella kit* (2015) from Drop Test Kits (DTK Water) is also a rapid immunoanalytical method developed for weekly and monthly analysis of water systems.

– *Legipid* test is a fast detection system with combined magnetic immunocapture and enzyme-immunoassay for the detection of *Legionella* in water. It can simultaneously process up to 40 tests in 1 hour. It is a low-cost mobile device. It detects the amount of free and intact *Legionella species* in water, based on the capture of the bacterium by an interaction that depends on the integrity of the cell envelope, because the recognized element is that in the cell envelope which regulates the infectivity of this bacterium (Legipid, 2015).

There are many other hazardous bacteria living in waters which represent a human health and ecological hazard or may cause technological problems. *AquaScope*® is a fully autonomous biosensor which can be applied for the rapid biomonitoring of specific microorganisms, both in the laboratory as well at the test site. It combines filter cytometry with fluorescence *in situ* hybridisation. It can quantitatively measure the total number of bacterial and yeast cells, numbers of *Escherichia coli*, *Enterococcus species*, *Legionella pneumophila*, aeromonads, pseudomonads, *Thiobacillus ferrooxidans* and *Desulfovibrio species*. The analysis time is 20 to 45 minutes and the detection limit is up to 1 cell per mL sample volume. It can be applied in the laboratory or at test sites (AquaScope, 2015).

3.2 Biomonitoring tools and devices

Active and passive biomonitoring methods (introduced in Volume 2. Chapters 4 and 5) (Gruiz & Molnár, 2015; Gruiz *et al.*, 2015b) may be based on the monitoring of abundance, morphology, behavior, activity, biochemistry or genetics of the ecosystem's native species. Passive monitoring of the inhabiting organisms is loaded with high uncertainty as regards age, sex, size, antecedents, individual genetics, etc. The other approach is to apply test organisms of controlled, homogenous and synchronized cultures in cages or boxes permeable for the monitored air, water, sediment or soil moisture, but with no free passage for the test organisms.

Active biomonitoring uses test organisms placed into the real environment, exposed to variable environmental conditions. The organisms placed into real waters are cultured, prepared and selected to ensure as good statistics as possible in terms of their number, age, size, sex, health, sensitivity, adaptability, etc. The advantage of this approach is that besides ensuring a controlled population of the test organisms, it is realistic, able to represent a multicontaminant situation and include matrix effects. However, realism has its limitations, as the caged or otherwise fixed organisms are not able to avoid the polluted environment or demonstrate behavioral characteristics such as the burrowing of crustaceans, which is essential from the point of view of healthy food chains. In addition, the environment may also spoil the advantages by producing extreme conditions, differing greatly from a normal situation (e.g. high temperature, heavy rain, flood, storms, or other disasters).

Conventionally, the recollected test organisms (active biomonitoring) or the collected natural inhabitants (passive biomonitoring) are investigated in the lab. Some advanced methods provide continuous signals during the stay of the test organisms in the environment at sublethal contaminant concentration ranges. Rapid methods and mobile labs are becoming more and more available for *in situ* investigations of the sampled organisms' morphology and biochemistry.

A *Musselmonitor* (Mosselmonitor®, 2015) is an *in situ* passive biomonitoring method with remote data processing, indicating the frequency of valve opening of mussels in a cage equipped with a motion detector. It works in most cases with the mussels *Dreissena polymorpha* or *Mytilus edulis* and is applied as an early warning tool for the continuous monitoring of drinking waters, surface waters or effluents. The observed and measured end point is the opening of the valve, whose frequency and duration depend on the type and level of contamination. The variations on the normal movement pattern include a more rapid opening and closing of the valves (flapping), keeping the shell closed for a fixed period or opening the shell for shorter time and to a lesser extent. Extreme contamination causes the death of the mussel.

Valve movement is detected and transformed into an electric signal by a microprocessor and the signals are processed by software to get the end point of the movement pattern, which unequivocally indicates the negative effect of the water on the mussel.

The mussel is glued to a platform and the sensors are fixed on each half of the shell (Figure 3.8). The sensors are small coils, one of which generates a magnetic field

Figure 3.8 Mosselmonitor: *in situ* submersible (inside compilation and ready to submerse) and online applicable flow-through versions (Mosselmonitor, 2015).

when current passes through and the current induced in the other coil is measured. The magnitude of this electric current depends on the distance between the two coils. Current is continuously measured. The change in the current is converted into distance and the movement pattern is evaluated as a function of time. The Musselmonitor can be used as an *in situ* placed field unit deployed into surface waters with a locally or remotely arranged data logger, or it can be used in flow-through mode by placing the measurement chamber with the mussels into the side-flow of any water fluxes of a water treatment system (Mosselmonitor®, 2015).

The monitor was applied as far back as 1998 for the automatic water quality monitoring of the Danube River at Bad Abbach (near Regensburg) and Jochenstein (on the German/Austrian border). In this study, a surveillance system was used: when deviations from the normal behavior were recognized, an alarm was triggered and the water was sampled for detailed analysis (IAD, 1999). Several successful applications have been carried out since that time, e.g. for assessing offshore contamination in the Adriatic Sea (Gorbi *et al.*, 2008; Gomiero *et al.*, 2011; Pilot project Mosselmonitor, 2013).

Other caged animals, e.g. daphnia or fish, can be observed by digital video camera. The video image analysis indicates the probabilistic relationship between health and the adversely effected state of the test organism, e.g. abnormally rapid and slow motions or immobility.

A *daphnia toximeter* is an instrument to observe living daphnids in the targeted water from water bodies and water treatment plant intakes of sewers. Its predecessor is the Extended Dynamic Daphnia Test, the oldest *in situ* biomonitoring method, in use from the 1980s. The new development of the company bbe Moldaenke is DaphTox II for the detection of toxic substances in water via computer-assisted digital image analysis. If the change is statistically significant, an alarm is triggered (Figure 3.9). The image analysis covers speed parameters (average speed, speed distribution, distance between the animals), behavioral parameters (swimming height and location, turns and circling movements, curviness) and growth (daphnia size). The system triggers the alarm when more parameters at the same time give characteristic results within a certain period of time (DaphTox II, 2015).

Figure 3.9 AlgaeTox, DaphTox, FishTox: online toxicity assessing equipment from bbe Moldaenke (TOX, 2015).

Changes in ventilatory behavior and certain locomotor activities of fish can be detected by non-invasive electronic sensors in a tank. Fish signals are amplified, filtered, and interfaced to a computer. When a significant number of fish respond simultaneously in an abnormal manner, an alarm is initiated.

A *fish toximeter* can continuously analyze fish behavioral patterns for the detection of toxins in water. It observes fish under the influence of a 'sample' water stream (Figure 3.9). The technique is aided by a digital video camera and continuous computer-assisted image analysis. The measuring system, based on the videos, evaluates the speed, swimming depth, size and the number of fish, and indices are calculated based on the determined values. Animal avoidance behavior can also be observed. If the aggregated 'Toxic Index' exceeds a default criterion for a certain duration, an alarm is triggered (Fish Toximeter, 2015).

Similar measuring systems can be applied for drinking water supply protection, waterway quality analysis and assessment, dam monitoring and for general surveillance. The instrument *ToxProtect 64* has been created especially for drinking water. The evaluation is mainly based on fish activity, swimming on the surface and the escape reaction. In contrast to the more general fish toximeter, the evaluation here is based on interruptions in light barriers by the fish movement. Unacceptable toxicity is associated with a certain level of interruptions per minute and fish. The thresholds for the alarm triggers can be set individually (ToxProtect, 2015).

An *algae toximeter* (2015) continuously monitors water for the presence of toxicity with the help of sensitive green algae. The algal concentration and the photosynthetic activity are measured in the measuring chamber or, alternatively, in the flow-through sample loop. The fluorescence measurement is carried out by the coupled *AlgaeOnlineAnalyser* for online detection of chlorophyll concentration, algae classes and photosynthetic activity. Chlorophyll a is responsible for the fluorescence of algae via excitation by visible light. The presence of other pigments indicates different algae classes. The interaction of these different pigments with chlorophyll-a results in a special excitation spectrum for taxonomic algae classes. The AlgaeOnlineAnalyser (2015) can be switched from continuous to batch mode. The algae are continuously propagated in a separate turbidistatic reactor, producing a well-controlled standard algae culture for the measurement. First the concentration and the activity of the naturally occurring algae are determined in the water, then the standard algal culture is added and the changes observed in the measurement chamber (Figure 3.9).

Bioaccumulation is a very plausible end point for chronic exposures to persistent organics and toxic metals. Some of the accumulator organisms may collect significant amounts from the environment or from food, often without visible health effects. Filter feeders such as bivalves (clams and mussels) tend to concentrate metals in their gills or other organs and tissues. This is because mollusks can limitedly excrete or metabolize pollutants directly and therefore attain higher bioaccumulation or bioconcentration factors compared to other taxonomic groups.

Active biomonitoring with caged mollusks has long been known and practiced (Mussel Watch from the 1970s), however the acquired information is in most cases not proportionate with the extensive workload, the number of problems to be solved and the analysis cost. On the other hand, for some purposes such as long-term monitoring is ideal, as the contamination levels in the mollusks reflect a time-integrated amount and the ecologically relevant bioavailable fraction. The accumulation of filter feeders characterizes water pollution, whereas the sediment-living deposit feeders characterize sediment pollution (Oehlmann & Schulte-Oehlmann, 2002).

The separately grown and then translocated animals are exposed to the contaminated natural waters. They are left unattended in the cages for a certain time. The conventional monitoring method is to chemically analyze the accumulated toxicant in the tissue after retrieval of the mollusks, and the body burden is calculated. A more promising and less time-consuming biomonitoring solution is the investigation of bioaccumulation-specific biomarkers in an animal exposed for a relatively short time. The biomarkers could be metallothioneins, stress/heat shock proteins, several oxidative enzymes including the cytochrome P-450-dependent monooxygenase (MFO) and the flavine-dependent monooxygenase (FMO), monoamine oxidase (MAO), dehydrogenases, peroxidases, etc. The lysosomal stability and membrane integrity may also be characteristic of bioaccumulation. DNA damage in mollusks is detected by the comet assay already after a short time exposure (Steinert *et al.*, 1998).

4 APPLICATION OF *IN SITU* AND REAL-TIME METHODS FOR SURFACE WATERS AND OCEANS

Ensuring water quality, primarily drinking water quality, requires urgent action around the world. Millions of people, mainly children, die every year due to the lack of clean drinking water. Thousands of chemical substances and aggressive microorganisms are contaminating our waters. In order to improve the situation, an exponentially increasing number of measurements would be necessary, which is not feasible due to the lack of the enormous equipment capacity needed for the analysis. The conventional, laboratory-based chemical analytical and microbiological tools are unable to fulfill the requirements of low cost, speed and precision necessary to deal with the large number of samples required. Regulators encourage the development of innovative methods and instruments, aided by frequent data acquisition and getting rapid analytical responses. There is great demand for miniaturized and automated systems, which can function in the long term without significant human workload and costly laboratories.

4.1 Real-time water quality monitoring

Real-time water quality monitoring is an essential need to reduce health and environmental risk. *In situ*, real-time methods are needed in surface water monitoring activities, in oceanography, as well as for runoff and wastewater management. Both research and practice require real-time information on the qualitative and quantitative characteristics of our waters. Changes in flow rates and water levels, temperature, pH, redox potential, nutrient and contaminant concentration may have significant impacts on the aquatic ecosystem. They also largely influence human water uses and health risks. Early warning is essential to prevent damage due to delayed risk reduction measures. Acquiring continuous real-time data may increase the efficiency of risk management of physical, chemical and biological hazards such as algal blooms, which are typically easy to monitor with *in situ* sensors. Most of the *in situ*, real-time measurements and devices have their conventional counterparts for measuring depth, flows, water chemistry, and biology-based physical or chemical signals, but the conventional ones cannot compete with the benefits of the *in situ* placement, the high measurement frequency and the programmable and autonomous versions. Some *in situ* and real-time instruments used in water monitoring are introduced in the following.

A *Conductivity-Temperature-Depth recorder* (CTD) is the basic instrument of all practitioners working in marine and freshwater environments. It may be equipped with sampling rosettes, an additional oxygen sensor, transmissometer and fluorescence detector. A CTD recorder is typically placed by ships into the sea/ocean or other surface waters and is connected by cables to transmit real-time data to the data logging system on board the ship. It is continuously let further down in the water. Depths and intervals of measurements are programmable by the user. Modern equipment has an internal memory and can be powered both by batteries or externally. Designs differ for 600 m depth use with a plastic housing, and at 7,000 or 10,500 m, with a titanium housing (SBE, 2015a).

Submersible multi-channel data loggers, recorders, versatile probes, controllers and sensors are produced for water quality measurement by RBR (2015). The high-precision instruments are recommended for oceanographic, freshwater, groundwater and cryospheric research. The standard data logging instruments range from one to 24 channels, configured as a CTD, conductivity, temperature, depth (pressure) or multi-parameter sensors/recorders.

Real-time water quality assessment of pipe discharges, streams and rivers, lakes, estuaries and other shallow waters can be implemented by the YSI multiprobe instruments, for example the 6600 V2-4. It measures dissolved oxygen, pH and redox potential, turbidity, chlorophyll and blue-green algae. Additional calculated parameters include total dissolved solids, resistivity, and specific conductance. Self-cleaning optical sensors with integrated wipers remove biofouling and maintain high data accuracy. The fluorescence-based blue-green algae sensor enables monitoring of blue-green algae populations where their presence is a concern. The sensors provide early warning of algal bloom, track taste and odor-causing species in drinking water supplies,

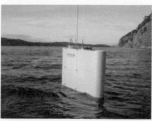

Figure 3.10 Compartments of the SMHI Måseskär buoy: the Sea Tramp, a surface and a subsurface buoy and the SeaMoose (Ocean Origo, 2015).

or conduct ecosystem research (YSI, 2015). The experience of users is that YSI sensors – similar to all water-placed sensors – require rigorous maintenance and frequent calibration. The performance of sensors begins to deteriorate after 2–3 weeks.

Real-time ocean observing systems provide critical information for the study of ecosystems, water quality, and fisheries, as well as data for long-term climate change studies. The Inductive Modem (IM) system for moorings provides reliable, real-time data transmission for up to 100 instruments that can be positioned or repositioned at any depth, in wireless mode. The Inductive Modem Module (IMM) communicates with the buoy controller and with the underwater instruments measuring various combinations of temperature, conductivity, pressure, dissolved oxygen, and data from integrated auxiliary sensors. The data are transmitted from the buoy to the remote receiver (SBE, 2015b).

A *Sea Tramp profiling system* is – similar to the previous SBE system – an autonomous, multi-cycling, data collecting platform designed for unattended marine monitoring and research. It profiles along a guiding wire and performs well also in stratified waters and when equipped with a non-stream-lined payload. Sensors are selected by the operator and may be installed on site (Ocean Origo, 2015). The buoy system (surface and sub-surface buoys) includes data logger, controller and the remote communication unit of SeaMoose, which is a flexible *'meteorological and oceanographic observation system for the environment'*. These parts of the system are shown in Figure 3.10 and the complete system in Figure 3.11. The buoy system also includes a bottom-mounted acoustic doppler current profiler (ADCP) (Teledyne RDI, 2015) with integrated wave measurements for real-time monitoring of coastal currents.

4.1.1 Surface water and oceanographic sensors

Chelsea sensor technology has built several surface water and oceanographic sensors (CTG, 2015) for environmental monitoring of rivers, reservoirs, lakes, and groundwaters. Sensors are provided for *in situ* chlorophyll and algae class studies, dye tracing, oil spill monitoring, airport runoff or water abstraction management and effluent detection. The priority technologies are i) fluorometers for water quality monitoring, ii) compact, multi-parameter monitoring system for oceanography and limnology, iii) sensitive digital infrared turbidity sensor designed for compliancy with ISO 7027:1999 standard iv) submersible bioluminescence sensor which monitors the visible emissions

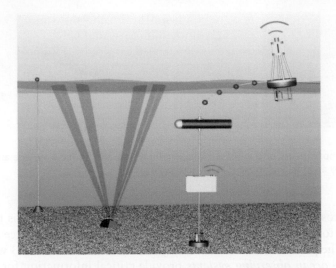

Figure 3.11 The SMHI Måseskär buoy: the complete system (Ocean Origo, 2015).

from bioluminescent organisms in seawater, and v) sensors for the measurement of photosynthetically active radiation (PAR). This company offers a plankton sampler for automated towed and shipborne use, too.

The *autonomous profiling nutrient analyzer* (APNA) is designed to be adaptable for deployment on a wide variety of ocean observation platforms including: shipboard profiling or towed sensor array; fixed-depth or vertical profiling moorings; autonomous underwater vehicles and gliders. The commercially available analyzers can monitor and establish the concentrations and distributions of nutrients and other chemicals – nitrate, nitrite, phosphate, ammonium, silicate and iron – in fresh and marine waters (SubChem, 2015). They are equipped with a 4- or 6-channel analyzer with multichemical capability. It is able to conduct autonomous vertical profiling, continuous underway surveying and intermittent long-term sampling (APNA, 2015).

The *flow-through analyzer system* of NAS-3X (2015) is a robotic analyzer. It is the latest development for high-frequency, time-series determination of nutrient concentrations (nitrate, phosphate, silicate and ammonia) in marine and fresh waters. The NAS-3X is typically deployed unattended for periods of 1 to 3 months, although longer deployments can also be achieved. It has been used near surface in many buoy and riverine applications or moored at depths down to 250 m. It can be applied for early warning of phytoplankton blooms, eutrophication and for the identification of episodic events. Suitable for run-off monitoring and changes in nutrient concentrations.

Satlantic (2015) together with SBE (2015a) developed a wide range of sensors and measuring systems for the study of aquatic environments. They offer real-time *in situ* i) nutrient sensors such as the SUNA V2, Deep SUNA and ISUS V3, ii) *in situ* fluorometric analysis for chlorophyll fluorescence in photosynthetic organisms, iii) radiometers for optical profiling, water color and PAR, and iv) hyperspectral and multispectral radiometers, etc. The company also develops large-scale ocean observatory systems and data extraction tools.

The LISST-100X instrument from Sequoia (2015) is a multi-parameter system for *in situ* observations of particle size distribution and volume concentration. Additionally, it records optical transmission, pressure and temperature.

4.1.2 Global monitoring

Argo is a global array of more than 3,000 free-drifting profiling floats that measures the temperature and salinity of the upper 2000 m of the ocean. This allows systematic, continuous monitoring of the temperature, salinity, and velocity of the upper ocean. Measured data are assimilated in near real-time into computer models and made publicly available within hours after collection. Compared to the traditional ship-based measurements, Argo covers the oceans in their entirety (not only shipping routes), summer and winter period equally, and with a much larger number of real-time measurements (Argo, 2015). It is part of an integrated global ocean observing system (GOOS, 2015), within the global earth observation system of systems (GEOSS, 2015). Figure 3.12 shows the concept and operation of Argo. The floats weigh 25kg, their operating depth is 2000 m and comprise three subsystems: (i) the hydraulics controlling buoyancy adjustment by an inflatable external bladder so the float can surface and dive, (ii) microprocessors dealing with function control and scheduling, and (iii) a data transmission system controlling communication with a satellite. Several types of floats are used in the ARGO project such as:

1 PROVOR and ARVOR floats built by *nke* (2015) and IFREMER (2015);
2 APEX, produced by Teledyne Webb Research Corporation (TWR, 2015);
3 SOLO float designed and built by Scripps Institution of Oceanography (Scripps, 2015) and the SOLO-II float built by MRV Systems (MRV, 2015).

The early projects, such as MAST I and MAST III – the European contribution to the GOOS (2015) – can be considered the predecessors of ARGO. An autonomous *in situ* multidisciplinary ocean observatory was developed within MAST I (BABAS project from 1990 to 1994) and MAST III (YOYO 2001 – Ocean ODYSSEY project from 1998 to 2001). The YOYO is a Eulerian (starts and ends on the same vertex) autonomous multisensor profiler providing time series of parameters continuously over the water column. It is intended for long-term *in situ* monitoring of the ocean, opening a wide range of possible scientific applications ranging from specific process studies to climate monitoring. YOYO was equipped with the *Autonomous Nutrient Analyser in Situ (*ANAIS), a spectrophotometrical instrument providing real-time data on the nutrient status of the ocean water by measuring nitrate, phosphate and silicate. The analyzer is a set of three chemical sensors, a manifold where the reaction takes place, a colorimeter for the analyses, and two clamping plates for fixing the pumps and sealing the manifold. The set is placed in a container together with the bags for the reagents. An IT card system was built in to control the sensors and for data storage and transmission (Jońca *et al.*, 2013).

Advanced sensing for ocean observing systems and the projects of *O-SCOPE* – Ocean-Systems for Chemical, Optical, and Physical Experiments and *MOSEAN* – Multi-disciplinary Ocean Sensors for Environmental Analyses and Networks, were carried out from 1998 to 2008, sponsored by the National Oceanographic Partnership Program (NOPP, 2015). The projects focused on developing and testing

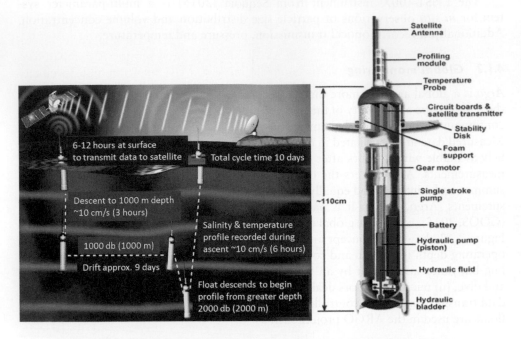

Figure 3.12 Argo operation (2015) and the design of the float (Argofloat, 2015).

new sensors and systems for autonomous, concurrent measurements of biological, chemical, optical, and physical variables from a diverse suite of stationary and mobile ocean platforms. Design considerations encompassed extended open-ocean and coastal deployments, instrument durability, biofouling mitigation, data accuracy and precision, real-time data telemetry, and economy (Dickey *et al.*, 2009a).

O-SCOPE aimed to measure pH, CO_2, partial pressure (p_{CO2}), dissolved inorganic carbon, total alkalinity, dissolved oxygen, water turbidity, chlorophyll, and optical absorption and scattering for the applications of reflectance models for remote sensing of ocean color (Dickey *et al.*, 2009a). O-SCOPE sensors were tested on three deep-sea moorings: (i) about 80 km southeast of Bermuda, (ii) Monterey Bay, California, and (iii) the NOAA Tsunami warning buoy at Ocean Weather Station Papa in the North Pacific.

The *in situ* instrument of a spectrophotometric elemental analysis system (SEAS) autonomously mixes seawater and reagents, and records absorbance at user-defined wavelengths. The precision of the spectrophotometric pH measurement is ±0.001 pH units. A non-dispersive infrared spectrometer has also been developed for measuring the difference in CO_2 partial pressure (Δp_{CO2}) across the air-sea interface (Friedrich *et al.*, 1995). Further developments in this area resulted in a sensor suite for measuring absolute air and sea surface ΔpCO_2 (with an accuracy of ±3 μbars), dissolved oxygen and nitrate concentrations (Johnson & Coletti, 2002).

The O-SCOPE project also focused on the application of bio-optical sensors with improved stability and endurance for operational monitoring. A chlorophyll fluorometer and a multi-angle scattering sensor for measuring the volume scattering function (VSF) were developed for phytoplankton biomass monitoring (Moore *et al.*, 2000). A new modular servo-controlled anti-biofouling shutter system for open-faced optical sensors was tested on the O-SCOPE optical systems (Manov *et al.*, 2004).

The main goal of the MOSEAN project was to test small, lightweight optical and chemical sensors for autonomous deployment on a variety of stationary and mobile platforms. The MOSEAN mooring sites – using the channel relocatable mooring (CHARM) – are located on the Hawaiian HALE-ALOHA, in an open ocean oligotrophic environment ca. 100 km north of Oahu, HI and on the coast in the Santa Barbara Channel. The project was also aimed to develop a near real-time data telemetry system and the mitigation of biofouling.

In summary, the two projects of O-SCOPE and MOSEAN were used to successfully develop and test new, compact, energy-efficient sensors and systems for the autonomous measurement of biological, chemical and optical parameters, in particular chemical sensors, water samplers, and spectrophotometric elemental, pulsed-membrane, colorimetric and microfluidic/fluorometric technologies. Optical technologies such as fluorescence and turbidity meters, multispectral and multi-angle scattering and backscattering sensors, a hyperspectral absorption-attenuation meter, and spectral fluorescence sensors have also been developed. These advances, along with improved water storage and validation techniques, enable accurate *in situ* and remote observations and estimates of a wide variety of biogeochemical parameters, for example inorganic and organic particles such as phytoplankton and hazardous algal blooms (HABs) (Dickey *et al.*, 2009b).

5 APPLICATION OF *IN SITU* REAL-TIME MEASUREMENTS FOR WASTEWATER TREATMENT AND QUALITY CONTROL

The risk of wastewaters on surface waters can be lowered both by process control of the wastewater treatment technology to ensure its optimal functioning, and by product control of the treated wastewater to stop its release when quality does not fulfill the quality objectives.

Monitoring for the purpose of process control in wastewater treatment and for quality control of the inflowing and treated wastewaters still has several shortcomings such as limitations in accurate and frequent sampling and analyses by conventional, laboratory-based measuring methods. Innovative, rapid and cheap *in situ* and real-time methods and devices are needed to drive the monitoring of wastewater treatment efficiency and product quality.

Conventional end points based on standardized methods to follow microbial activity in wastewater treatment plants are dissolved oxygen, oxidation reduction potential (ORP), and solids retention time (SRT) to control the sludge: sludge blanket level and total suspended solids. To control treated water quality the following variables are generally measured: biological oxygen demand (BOD), chemical oxygen demand (COD), total organic carbon (TOC), total suspended solids (TSS), specific organic compounds, e.g. phenols; mineral compounds, e.g. total nitrogen and total phosphorus; pH, residual Cl_2 (after treatment with chlorine chemicals), toxicity and pathogenic microorganisms. These conventional methods are extremely time-consuming and cannot be used for real-time/online monitoring and automation. The measurements are loaded with uncertainties due to sampling problems, conservation, storage, transportation and laboratory analysis – immediate or postponed – depending on capacity.

Water quality is a huge world-wide problem. Residual BOD and nutrients load cause eutrophication in receiving surface waters. The problem of pathogenic microorganisms and micropollutants of emerging concern, typically with long-term human health effects such as endocrine and immune system disruption must be solved.

To improve the situation, innovative methods should be developed and introduced into practice. For example, instead of the 5-day long BOD_5, several rapid methods have been developed and demonstrated such as rapid BOD determination based on respirometry, COD, TOC, fluorescence and UV absorbance (Guwy *et al.*, 1999), or biosensor-based methods (Liu & Mattiasson, 2002). Several online methods have already been applied for controlling the wastewater treatment process such as photometric, colorimetric, or titrimetric methods, ion-specific electrodes, UV spectrometry for organic contaminants, etc. (Vanrolleghem & Lee, 2003). Qualitative or semi-quantitative colorimetric test kits are also available for rapid, on-site (ready-to-use) application. The ISO 17381 (2003) standard establishes criteria for the selection and application of test kit methods in water analysis. Specific sensors, DNA- and immunotechniques are increasingly being developed and routinely applied also in the management of wastewater treatment.

The application of online working sensors would make efficient control of wastewater quality possible and of the treatment process itself. For the efficient utilization of the real-time online signals of the sensors, the whole wastewater treatment and control strategy should be harmonized with a high-level online monitoring. Monitoring should be linked to an automatic control system which is coupled to several technological options. The control system can process the output of the measuring device, select and carry out the proper countermeasure.

Online sensors for temperature, pressure, liquid level, flow of liquid/gas, pH, and conductivity are commonly applied devices in several technologies and these are not discussed here in detail. Biological activities, biodegradation rate, suspended solid, gaseous biodegradation products, dissolved metabolites, and the microorganisms present are typical variables for wastewater treatment, and their conventional laboratory-based analysis methods are extremely time-consuming and costly. A detailed discussion of the innovative methods focuses on the latter subjects.

5.1 Innovative analytical tools for wastewater management

The easiest way of innovation is to modify a conventional method e.g. by size reduction or portable design for on-site applications. A more advanced type of innovative method is based on a new principle, e.g. applying optical sensors or biosensors. Many of these sensors do not give a correct absolute value; they can be used after site-specific calibration of the local wastewater. The best known optical method in this group is the estimation of total suspended solids (TSS) from the results of turbidity or nephelometry, detecting transmission and scattering of light.

Exploitation of the whole UV spectrum and analysis of the spectrum makes it possible to do parallel analyses of several organic pollutants (total organic carbon, phenols, surfactants, etc.). Biosensors are increasingly available and accepted; some are validated for treated wastewater quality monitoring as will be shown in Section 5.3. Immunoenzymatic test kits are available for several micropollutants and electrochemical measurement systems for metals.

Toxicity monitoring is essential both for the undisturbed operation of wastewater treatment plants and the acceptable quality of the product, the treated wastewater. Handheld instruments are in widespread use for measuring the main physico-chemical parameters such as temperature, conductivity, pH, and dissolved oxygen and are often integrated into a multiparameter device. Handheld and portable designs are available for most of the conventional parameters measured. Passive samplers represent a new way and new concept in sampling. The specific materials of the different samplers ensure selective adsorption of specific molecules and micropollutants when the sampler is immersed into a stream. The application of cyclodextrins for selective sorption of certain contaminants is detailed as an example in Chapter 7 Section 3. Short- and long-term applications may refer to instantaneous and aggregated loadings. Their calibration may still be problematic.

The application of models may be a good solution in those cases when the mathematical function between online measurable variables and a difficult-to-measure quality indicator (BOD or nitrifiable nitrogen) is known.

Another conceptual innovation could be (as in other environmental management tasks too) tiered monitoring: the frequent or online monitoring by a qualitative or semiquantitative method as a first tier, and the quantitative analysis of the positive or borderline samples as the second tier.

Emerging measurement techniques have also approved also for online monitoring of wastewater during and after treatment. ISO 15839 (2003) prescribes the test procedures to be used to evaluate the performance of online sensors and devices.

Biomonitoring, early warning systems, bioassays using intact organisms and biomarkers are emerging surface water analytical tools (Allan *et al.*, 2006a and 2006b) which can be, and partly have been, adapted to wastewaters (see the Volume 2 of this series) (Gruiz *et al.*, 2015). The quality of, and the risk due to, the wastewater can be characterized using these biomonitoring results, e.g. the 'no effect dilution' can easily be determined (see Chapter 9 in Volume 2 – Gruiz *et al.*, 2015a).

Online biomonitoring is considered rapid and inexpensive, but currently it has significant limitations (lifetime, reproducibility, etc.) and only a few parameters can be measured online.

Biosensors, introduced in Sections 2.3, can detect chemical substances based on a very selective biocatalytic or bioaffinity response or on the more general, integrating response of living organisms. Many of the techniques developed for surface waters or drinking waters can be applied for wastewaters, and conversely, the methods and tools developed for wastewater can also be applied for contaminated groundwater, leachates and soil waters as well as for the pore water of sediments and saturated soils. The additional problem in their use for soil and sediment is their deployment and protection from the impact of the solid matrix.

Some DNA- and immunotechniques, microanalyzers, online respiration and toxicity measuring methods used for wastewater analysis are introduced briefly in the following section.

5.2   Measuring microbial quality and activity in wastewater treatment plants

Biological wastewater treatment is one of the best known biotechnologies and as biotechnologies in general, it can properly work if both the 'catalyst', the living biomass, and the 'technological parameters', ensuring the conditions for the best functioning of the biomass, are kept at an optimum. Optimization needs continuous feedback on the quality of the wastewater, both the quality (composition) and quantity of the biomass and the value of the technological parameters, e.g. temperature, O_2 supply as well as the level of toxicants which have a potential inhibitory effect on the biomass. The efficiency of the technology determines the quality of the product, i.e. the treated wastewater.

Hundreds of publications deal with the developments of, and the experiences with, the developed innovative wastewater monitoring methods and devices. A practical summary of the applied technologies was published in the form of a guidance document by US EPA in 2013. All the emerging (less than three years after demonstration), innovative (less than five years' application) and established (more than five years' application) methods, including adapted uses, are briefly described (US EPA, 2013). Quevauviller *et al.* (2006) reflect the global concepts on wastewater treatment-related monitoring and control, giving an overview on policies, standard methodologies, reference materials and discussing biosensors and alternative methods.

The innovative measuring methods and devices introduced in the following section are applicable online in wastewater treatment plants and can fulfill the previously listed requirements of biomass and technology. Some of them are already validated and accepted methods, whereas others are still under development.

5.2.1   Whole-cell biosensors for measuring biodegradable organic material content

Biosensors that are able to continuously measure the BOD value are highly desired devices for wastewater treatment. Such equipment is commercially available and a method has recently been standardized in Japan. The whole-cell sensor did not measure the BOD directly, but the end points of respiration and biodegradation rate from which the BOD can be estimated. The same sensors may also function as an alarm system for toxicity based on a massive decrease or full termination of respiration. The measurement is done in a small volume reactor, where the biodegrading biomass is generally fixed on a solid carrier. When the wastewater containing biodegradable organic material meets the metabolically activated microorganisms under aerobic conditions, oxidation-based biodegradation begins. The active biomass in the equipment can be the separately propagated artificial mixture of microorganisms or the local sludge. The microorganisms degrade the nutrient content of the wastewater at the expense of the dissolved oxygen (DO) and produce CO_2 in parallel; both DO and CO_2 can easily be monitored. The cell based BOD-sensors generally consist of more reactors and are equipped with a dissolved oxygen sensor and a thermometer and can be mixed, pumped, heated or cooled. Constant activity of the biomass is critical. Diez-Caballero (2002) successfully applied an independent chemostat (a flow through bioreactor to which a fresh medium is continuously added, in order to keep constant chemical composition and nutrient supply) for propagating and maintaining the activated cells for continuous application in the BOD sensor.

BOD measurement with constant activity biomass can be applied both to characterize the wastewater's biodegradable material content, or the toxicity of the influent waste into the wastewater treatment plant. The conventional BOD_5 testing is not enough for the simulation of the wastewater treatment plants' operation; instead it is *total BOD* which should be measured, including both BOD_5 and the nitrification step. An online total BOD and a rapid BOD analyzer are introduced below:

– *BioMonitor (2015)* is an online BOD analyzer for the determination of biological oxygen demand (total BOD), respiration and toxicity in wastewater. It is a miniature wastewater treatment plant, using its own activated sludge. Measurement of BOD takes place in less than 4 minutes. This speed guarantees that very short peaks can be determined during a daily cycle. Its measurement range is 1–200,000 mg/L BOD, 0–100% toxicity and respiration O_2 mg/L $\times$ min.
– *Quick Scan* BOD analyzer is a small size respirometer that has 4 reactors with a magnetic stirring base. It is ideal for toxicity screening and short-term BOD assessment. This same system can be used for soil and compost testing with the use of special soil/compost columns in place of standard glass reactors (Challenge Technology, 2015).

5.2.2   Online respirometry

Respirometry to obtain estimates of the rates of metabolism of the microorganisms during biodegradation, is a basic measurement in the process control of wastewater treatment. By measuring the rate of respiration (amount of oxygen/volume $\times$ time), a measure of the rate of biodegradation can be obtained. It makes respiration rate an important end point both in process control and toxicity management (Spanjers *et al.*, 1998).

The traditional means of measuring respiration rate is a laboratory method in a closed system based on the decrease of dissolved oxygen content directly in the liquid phase (oxygen electrode) or indirectly in the air space above the liquid (manometric method). Rapid methods need a very small amount of biomass/activated sludge and respond within 5–10 minutes. Multiple electrodes and specific software support the evaluation. Rapid laboratory measurements are useful in influent toxicity control in order to protect biomass from toxic wastes. The results can be used for the determination of the maximal inflow rate of the toxic waste, in other words, the necessary dilution of the waste before releasing it into the wastewater treatment plant.

Laboratory measurements are not suitable for getting real-time feedback from the continuously working biomass and the dynamic wastewater treatment process. Online respirometry applied for process control should be continuous and carried out in real-time. Open respirometry measures dissolved oxygen (DO) in an open flow system. The sensor can be immersed into the main reactor/flow; however, in most cases the location of the sensor is a separate reactor or side flow established for measurement purposes.

The respiration rate measurement in addition to process control can also determine residual BOD in treated wastewaters, the toxicity (Geenens & Thoeye, 1998;

Davies & Murdoch, 2002) and the shock load (Henriques *et al.*, 2007) in a system, as long as the baseline rate has been set for that system. Rapid laboratory and online respirometry are complementary in toxicity management.

Sensors of a respirometer can also be calibrated to measure other gases of concern like carbon monoxide, hydrogen sulfide, and methane.

A classification of online respirometers may occur according to the phase where oxygen is measured, i.e. liquid or gas and the reactor, which can be static or flow-through. In the flow-through system, the flowing phase can be either gas, liquid or both (Vanrolleghem, 2002). Examples of online monitors:

–   *The Ra-BOD* is an online process analyzer for determining short-term BOD in wastewater and surface water and fits into a control strategy for activated sludge plants. *Ra-COMBO* is both for determining BOD or toxicity (AppliTek, 2015).
–   *The Rodtox NG* has been designed for determining BOD and toxicity and to detect acute and chronic toxic effects of incoming wastewater streams of wastewater treatment plants (Kelma, 2015).
–   *STIP TOX* a toxicity monitor which has been designed for continuous toxicity measurement for protection of biological processes from toxic substances (Axon Automation, 2015).
–   *Amtox* uses an immobilized culture of nitrifying bacteria to produce a fast and reliable result. Inhibition is measured by comparing feed and effluent ammonia using a probe technology. Results are displayed continuously via a graphic interface (PPM, 2015).
–   *Online Oxygen Demand Monitor* model ODM–100 is a portable unit for measuring real-time oxygen demand (oxygen uptake rate = OUR) at any point in a wastewater treatment process. It can be used for monitoring in continuous, batch or sequential batch modes.It is supplied with a submersible sewage pump to drop in at any point along the treatment process to allow real-time data at any given location (Challenge Technology, 2015).
–   *Strathtox respirometer* is a 6-cell OUR measurement system for rapid measurement of activated sludge bacterial performance. The corresponding software provides a real-time display of bacterial respiration. Automatic report generation and the calculation of EC_{10}, EC_{20} or EC_{50} values (Sensara, 2015). The software can calculate the following end points: respiration inhibition; nitrification inhibition; short-term BOD, nitrification status, sludge health, OUR and SOUR (specific OUR), and critical oxygen concentration point.
–   *Bio-Scope* is an immersible sensor to see how the bacteria are performing in the wastewater treatment plant in real-time. It gives information on the biodegradation rate profile under real-time conditions and on bacterial health by comparing the current sample with the last tenmeasurements at a specified point. It also provides information on the critical oxygen point for energy optimization, DO and temperature (Sensara, 2015).

5.2.3   Fluorescence in situ hybridization

Fluorescence in situ hybridization (FISH) can be applied for identifying any kind of microorganism or group of microorganisms, even in the smallest amount of a complex mixture. This is why it is ideal for the study of wastewater biomass. Based on the knowledge of concrete DNA or RNA sequences exclusively occurring in the targeted microorganism, genetic engineers are able to prepare a probe. The probe is a polynucleotide sequence complementary to the targeted sequence in the microorganism to be detected or identified. Complementary nucleotides link to each other with high affinity and form a double helix. To find and detect these hybrid nucleotide pairs in a complex mixture, the artificially prepared component of the 'hybrid' nucleotide is labeled (marked). In the case of FISH the label is a fluorescent marker, a covalently attached fluorophore which fluoresces after staining with a specific dye. The fluorescently labeled 16S rRNA (ribosomal RNA) probes are hybridized, stained and observed under an epifluorescent microscope. A validated method exists for filamentous and nitrifying bacteria and for phosphorus accumulating organisms (PAOs).

5.2.4 Quantitative polymerase chain reaction for the quantification of microorganisms

Quantitative polymerase chain reaction (qPCR) technology is able to quantify microorganisms based on their DNA, both in influents and effluents as well as in the sludge of the wastewater treatment plants. PCR is used to amplify specific regions of DNA to be able to selectively detect microorganisms of small copy numbers in mixtures. To use the technique for the identification and quantification of *Escherichia coli* and enterococci in wastewaters, – similar to hybridization – the species-specific DNA sequence of the 16S ribosomal region of the targeted microorganism should be known to be able to synthesize the primer. Real-time quantitative PCR (qPCR) is a relatively rapid method (2–4 hours) for simultaneous DNA quantification and amplification. It uses a hybridization probe and a PCR primer together, to measure the amount of the product in real time. The probe containing a fluorophore is attached to its template in the specific region of the targeted cell, close to the location of the primer linkage. When the polymerase enzyme moves the template DNA forward, it encounters the probe and degrades it. The released fluorophore from the degraded probe is quantifiable by measuring fluorescence, and is proportional with the amount of the target microorganism. A calibration with known cell count of the target microorganism is needed to determine the absolute value of the cell count.

5.2.5 Nicotinamide adenine dinucleotide probes

The presence of the reduced form of nicotinamide adenine dinucleotide (NADH or NADPH) is proportional with the reduction potential of the biomass. Light of 340 nm wavelength induces fluorescence in NADH, and the emitted fluorescence light is detectable at 460 nm; therefore monitoring the level of reducing power is possible by measuring fluorescence at 460 nm. The measurement is done using immersed probes with no sampling or subsequent analysis.

Different NADH probes (molecules emitting light at 340 nm), fluorescent sensors and reporters (manifesting a large change in fluorescence upon NADH binding) have been developed and applied for measuring NAD and NADH ratios as well as NADH outside and inside the cells (Lemke & Schultz, 2011, Zhao *et al.*, 2011 Zhao *et al.*, 2014). By means of the SymBio process (SymBio, 2015, Spellman, 2014) nitrification and denitrification can be quantitatively characterized in one step. NADH is measured in the intracellular pool to quantify the real-time biological activity. Dissolved oxygen (DO) is measured in parallel to control aeration and to precisely adjust the relatively low DO level that is necessary for the simultaneous nitrification in the outer aerobic zone and denitrification in the inner, anoxic zone of the activated sludge flocs (Trivedi, 2009).

5.2.6   Immunosensors and immunoassays

The application of antigen-antibody interaction to detect the presence of toxins in wastewater is based on the specific recognition and binding of a bacterial antigen by antibodies. Immunosensors detect the signals of immunoassays. The artificially labeled antibody is responsible for the detectable signal. Both direct and competitive assays are used, and the label in the practice of wastewater management can be

–   an enzyme such as alkaline phosphatase and horseradish peroxidase in enzyme immunoassay or enzyme-linked immunosorbent assay i.e. ELISA;
–   chemiluminescent (e.g., acridinium ester), or
–   fluorescent (e.g., fluorescein) agents.

Detection of the signal is carried out by spectrophotometric or colorimetric measurement in the enzyme linked immunoassays: the enzyme label on the unbound antibody produces a signal in the presence of a specified substrate that is added to develop the color. When using luminescent or fluorescent labels, luminescence or fluorescence detectors should be applied.

5.2.7   Biological microelectromechanical systems for characterizing microbial activity

Biological micro-electro-mechanical systems (bioMEMS) are miniaturized biosensors used for rapid testing of biomolecules that are indicative of an upset process due to bulking, foaming or disadvantageous changes in the microbial population of the activated sludge. BioMEMS are very similar to lab-on-a-chip (LOC) and micro total analysis systems (μTAS), but strictly for biological applications. The mini mechanical sensors mechanically detect stress or mass through micro- and nano-scale cantilevers or micro- and nano-scale plates or membranes. Bending of the cantilever is caused by the biological process on one side of the cantilever and is measurable either optically or electrically. Electrical and electrochemical detection is possible by amperometric, potentiometric or conductometric sensors, based on the changes in redox, or electric potential as well as impedance caused by the biochemical reactions.

5.2.8   Handheld advanced nucleic acid analyzer for detecting pathogens

The handheld advanced nucleic acid analyzer (HANAA) can be used for real-time detection of pathogens in water and wastewater by relying on a polymerase chain reaction (PCR). This technique allows for a small amount of DNA to be amplified exponentially. Commercially available HANAA are Bio-Seeq (2015) and RAZOR (2015) (originally developed for bioterrorism monitoring purposes). HANAA is a portable/handheld design, otherwise equivalent with a laboratory thermal cycler.

5.3 Toxicity measuring biosensors

The majority of the possible toxicants in wastewaters are not included in monitoring programs so they remain unknown and uncontrolled in many cases, causing significant long-term health and environmental risk due to discharging treated wastewaters into surface waters. Toxicity affects not only receiving waters but the wastewater treatment plants themselves. Incoming toxic wastes lower the treatment capacity of the microbes and to compensate it, the treatment plants must either be larger or use more energy. A rapid, reliable test method for incoming waste increases the stability and efficiency of the wastewater treatment technology. For toxicity measurement some of the equipment uses pure microorganism cultures, activated sludge from a wastewater treatment plant or even GEMs, which recognize the presence of specific environmental pollutants.

5.3.1 Respirometry based toxicity measuring methods

Respirometry is the primary method for the detection of toxicity, as toxicants inhibit the biodegradation capacity of the sludge microorganisms, which is indicated by reduced respiration rate. The oxygen consumption can be measured electrochemically by an oxygen electrode or optically with an optrode, using optical fibers as signal transducer (read more in Quevauviller *et al.*, 2006). Online applicable respirometers have already been introduced in Section 5.2.2. Toxiguard is a commercially available automated respirometer for toxicity evaluation which sounds an alarm when the oxygen concentration is higher (not consumed by the micro-organisms) than a preset value. Toxalarm (2015) uses standard bacterial culture to produce a toxicity result in the water within minutes. Rapid oxygen demand for toxicity (RODTOX) assessment from Kelma (2015) is recommended both for rapid BOD and toxicity.

5.3.2 Microtox and online Microtox

Microtox is based on a bacterial whole-cell biosensor. The measured end point in the patented Microtox® method is the luminescence of the marine bacterium *Aliivibrio fischeri* (strain, NRRL B-11177). Indigenous bioluminescence significantly decreases due to the effect of toxicants. The water samples are added to the standardized bacterial culture and the decrease in light intensity is compared to the negative control. The test is standardized in many countries and also by ISO (2007a,b,c). The ASTM D5660 96 (2014) standard was withdrawn in 2014. The measurement can be carried out online or offline. It can provide near real-time information on water and wastewater toxicity. Microtox CTM (Continuous Toxicity Monitor) is a fully automatic instrument that offers a four-week, autonomous operating cycle and requires a low level of skill for both operation and maintenance. It produces real-time toxicity results without manual intervention except for monthly maintenance (Modern Water, Microtox®, 2015). Laboratory and portable versions are available.

5.3.3 Toxicity testing methods and equipment – commercially available devices

Microtox (2015) is not the only commercially available rapid toxicity test based on *Aliivibrio fischeri* luminescence: ToxAlert® 101 (Merck) and LUMIStox (1999 and 2010) from Hach-Lange are similar in applicability to Microtox as published by

Figure 3.13 The SciTOX electrochemical sensor detects the signal of an artificial electron-acceptor which captures the electrons from the energy-producing process of the microorganisms (SciTox, 2015).

Jennings *et al.* (2001). Methods verified by US EPA ETV are: ToxScreen II (2003) (Checklight Ltd.) and BioTox™ (2003) (from Hidex Oy,). Deltatox® (2015) is a portable device from Modern Water. All luminescence-based rapid toxicity analyzers have similar design: they include a highly sensitive luminometer with built-in software, a freeze-dried bacterial reagent, test control and reconstitution solutions.

Toxicity monitoring in wastewater treatment plants primarily serves the protection of the activated sludge microbes. For the purpose of measuring toxicity, the respiration or other activities of the local sludge microbiota can be used. Alternatively, stable quality pure or mixed microbial cultures can also be applied, similar to those which are used for normal chromotests or luminotests. An abridged list is as follows:

–   The *Eclox*™ *luminometer* – a handheld design – applicable both for the chemiluminescence toxicity test and the luminescent bacteria test to measure luminescent light inhibition of water samples. The Eclox luminometer is designed for field use. Together with the LUMIStherm thermoblock and the corresponding software, *in situ* toxicity measurements are possible (Eclox handheld luminometer, 2015).
–   *SciTOX*™ *ALPHA* is a patented technology for the measurement of toxicity in sewage and in wastewater treatment plants. The complete assay requires only fifteen minutes including sample incubation. It is effective for a wide range of inorganic and organic toxicants. The system uses indigenous bacteria with no importation requirements, and the reagents are simple to prepare (see Figure 3.13). The respiration of the indigenous bacteria is not measured via oxygen consumption, but rather the wireless electrochemical (amperometric) sensor detects the signal of the reduced electron-acceptor of potassium ferricyanide. The artificial electron acceptor, added to the reaction mixture as an indicator, captures the electrons produced by the microorganisms. When the catabolic activity of the microorganism is inhibited by toxicants, the signal of the reduced electron-acceptor is smaller. Due to the high solubility of potassium ferricyanide, the toxicity analysis can be completed in less than 20 minutes, assuming that the inoculum is ready to use and checked. The inoculum is checked by a toxicant of known effect (3,5-dichlorophenol) on wastewater treatment plant microorganisms (SciTox™ ALPHA, 2015).

- *POLYTOX-RES* (2015) is an automated real-time monitoring of the overall toxicity present in water in one hour's time. It works based on respirometry with a dissolved oxygen sensor. The programmable frequency of analysis covers 10 to 20 tests per day. The equipment can be applied as an early warning system and for capturing samples exceeding certain limits for later analysis. Data logging at, and transmission to, remote locations is possible by telemetry.

5.4   Online analyzers and electrodes for the water phase in wastewater treatment

The practice requires reliable, simple and low-maintenance sensors for continuous monitoring and control of the wastewater treatment in order to meet effluent quality objectives. The routine online monitoring of several parameters of the water phase has already been solved and applied in wastewater treatment plants. The parameters monitored are the temperature, pressure, pH and redox potential, conductivity, dissolved oxygen concentration, liquid level, flow rate, nutrients such as NH_4^+, NO_3^-, TOC, BOD, respiration rate, and toxicity. Some other sensors for measuring wastewater parameters or components are still under development or need further development, for example the measurement of chemical oxygen demand (COD) or phosphate concentration. The online applications are based on membrane technologies, UV spectrometry or fluorescence detection and ion-selective sensor techniques. The measuring techniques may be intermittent, continuous flow-through systems (e.g. flow injection analysis, FIA), or sequential injection analysis (SIA). Compared to batch, FIA and SIA have the advantage of small sample size, low reagent use and high sample throughput (Vanrolleghem & Lee, 2003). The processes of nitrification/denitrification and phosphorus removal are critical in wastewater treatment and require intensive and thorough monitoring and control. These innovative methods provide real-time or near real-time monitoring data in the wastewater treatment system, making immediate feedback, timely process adjustments and corrective actions possible if a shock or toxic load occurs.

The most important applications of the online measuring methods and instruments are the monitoring and control of nitrification, denitrification and phosphorus removal. The target molecules of ammonium (converted to ammonia), nitrate, orthophosphate and total phosphorus are analyzed based on colorimetry or ion-selective electrodes. Some commercially available *in situ* real-time measuring devices are briefly introduced below. See references for detailed information.

5.4.1   Real-time measurements based on colorimetry

Colorimetry, the most conventional analysis method, has been developed to make it suitable for real-time application. Some of the commercially available analyzers are listed below:

- *Alert colorimeter* is an analyzer for ammonia, nitrate, nitrite, or phosphorus. A differential technique is applied for compensating fouling and initial sample color (Metrohm-Applikon, 2015a).

- *Trescon analyzer* measures orthophosphate by colorimetry using the vanadate/molybdate method (yellow product), and total phosphorus after a chemical-thermal digestion with the molybdenum blue method (WTW, 2015a).
- *ChemScan* UV process analyzers are online, single or multiple parameter analyzers using full-spectrum, UV-visible detection with chemometric analysis of spectral data. Multiple sample lines allow sampling from several locations to the same analyzer. Nitrate and nitrite are directly analyzed from the spectra of the sample. Ammonia analysis is reagent-assisted using bleach and hydroxide reagents (ASA Analytics, 2015).
- *AMTAX* ammonia analyzer takes samples automatically in every 5 to 120 minutes (arbitrary set). The sample is mixed with sodium hydroxide to convert all ammonium to free ammonia; ammonia gas is then expelled from the sample, redissolved in the indicator reagent and the color measured with a colorimeter (Hach AMTAX, 2015).
- *PHOSPHAX* is a continuous flow analyzer with a five-minute cycle time for ortho-phosphate using the photometric methods with vanado-molydan (Hach PHOPHAX, 2015).
- *NitraVis®-system* is a UV/VIS spectrometer probe for *in situ*, real-time spectral measurement of nitrate concentration without filtering. Turbidity is detected and compensated for. Automatic cleaning is solved with compressed air before each measurement (WTW, 2015b).
- A *NITRATAX* probe is based on UV light absorption. The photometer measures the primary UV 210 beam, and a second beam at 350 nm provides a reference standard. It includes a self-cleaning wiper system (Hach NITRATAX, 2015).

5.4.2 Real-time measurements based on ion-selective electrodes (ISE)

In wastewater treatment, ion-selective electrodes are most frequently used for ammonium and nitrate analysis, but electrodes are also available for sodium, potassium, calcium, chloride and fluoride.

- *Direct immersion ISE sensors*:
  o for ammonium and nitrate: Varion® Plus 700 IQ;
  o for ammonium with potassium compensation: AmmoLyt® Plus;
  o for nitrate with chloride compensation: NitraLyt® Plus. (WTW ISE, 2015).
- *Myratek Sentry C-2 electrode*, based on ISE technology, continually measures ammonia and nitrate levels in the treatment process. Calibration is performed automatically at user-set intervals (Biochem, 2015).
- *AISE sc ISE ammonium probe* provides continuous measurement by direct immersion. Potassium interference is compensated by including a potassium ISE. Optional air cleaning ammonium, nitratesystem may reduce maintenance frequency (Hach ISE, 2015a).
- *AN-ISE sc combination sensor* for ammonium and nitrate provides reliable measured values and considerably reduces maintenance time and costs compared to conventional ISE probes. It is equipped with a cartrical sensor cartridge and automatic cleaning unit with a compressor (Hach ISE, 2015b).

- **ISEMax CAS40** is a potentiometric ion-selective electrode system for the continuous measurement of ammonium and nitrate (Endress, 2015).
- **Alert Ion Analyzer ADI 2003** is a potentiometric ISE either for ammonium, calcium, chloride, fluoride, nitrate, potassium or sodium (Metrohm, 2015b).

5.4.3   Voltammetry for trace metal monitoring

In voltammetry, information about an analyte (toxic metals) is obtained by measuring the current as the potential is varied. The metal ions are drawn onto the working electrode when a specific voltage is applied to the water sample under test. In stripping voltammetry for the plating step, the potential is held at an oxidizing potential, and the oxidized species are stripped from the electrode by sweeping the potential positively. When the stripping voltage is applied, the metals return to the sample solution, generating a small current. Each metal has a specific voltage at which it returns to solution. So the metal is identified by its stripping voltage, and the current generated indicates the concentration of metal in the sample. The stripping step can be either linear, staircase, square wave, or pulse. Some equipment useful in wastewater treatment based on voltammetry are introduced below:

- **PDV 6000 portable analyzer** measures trace metals in water, soil and food. Voltammetry offers a generally accepted alternative to laboratory analysis or automatic samplers in dissolved metal analysis. The sensors provide an easy way to generate and store real-time data, which in turn allows real-time decision-making. The sensor can be used as a stand-alone device or logged to a computer with the VAS software according to purpose. The PDV6000 ultra version is equipped with a standard analytical cell, which can detect a wide range of different metals. An extra accessory is the SV LabCell that allows cathodic stripping voltammetry (CSV) in the same PDV analyzer. The SV LabCell extends the PDV's range of metals to include molybdenum and uranium, and it also gives a better response for nickel, cobalt and chromium at low concentrations. Color or turbidity does not affect the method. Dirty water and soil samples may require simple on-site sample preparations to prevent interferences. Further advantages are: multiple, sequential metal analysis is possible and the detection limits may be below 1 ppb (PDV 6000, 2015).
- **TraceDetect Nano-Band™ explorer** is a metal analyzing system which works with stripping voltammetry and includes the Nano-Band instrument and Explorer software to operate the instrument. It measures trace metals in aqueous solutions: ppm measurements instantly, ppb measurements in seconds, and ppt measurements in minutes. The same equipment can be used for anodic and cathodic stripping voltammetry, cyclic and square wave voltammetry, potentiometric stripping analysis, amperometry, chronocoulometry. The improved version of **Explorer II** is an on-site applicable system with **TriTrode™** electrode technology. This unites all three system electrodes into one piece: the Nano-Band working electrode, the reference electrode and the auxiliary electrode, resulting in simpler maintenance (Nano-Band, 2015).
- **Lead SA1100** (2015) scanning analyzer based on voltammetry, represents a step forward in real-time testing of lead and copper in water. It is a robust, portable instrument, including a disposable electrode which can quickly and accurately detect the presence and concentration of lead and copper.

5.4.4 Real-time measurement of chemical oxygen demand (COD)

Photoelectrochemical oxygen demand (PeCOD™) technology can measure photocurrent charge originating from the oxidative degradation of soluble organic substances. The extent of electron transfer at a TiO_2 nanoporous film electrode is measured during the exhaustive photoelectrocatalytic degradation of organic matter in a thin layer photoelectrochemical cell. It overcomes the problems of other rapid COD determinations, i.e. partial oxidation due to matrix effect. The sensor produces an objective value and no calibration is necessary. The PeCOD method has been validated by comparing it to the standard dichromate method: good agreement was achieved. It is robust, rapid (0.5-5 min/measurement), simple to use, and easily automated. Long sensor life, high sensitivity and a wide linear range characterize the sensor. Real-time soluble COD monitoring enables efficient process control and waste management (Zhao *et al.*, 2004). Some products are commercially available, such as the CONDIACELL.

CONDIACELL (2015) for industrial wastewater treatment uses an electrochemical advanced oxidation process (EAOP®), a technology which is nowadays used for wastewater treatment, too. EAOP® reduces all organic water components by approximately 99%. The high oxidation potential of hydroxyl radicals ensures good efficiency by non-selective oxidation of any kind of organic substances to carbon dioxide. Hydroxyl radicals produce a 'cold incineration' of the organic components. DIACHEM diamond electrodes are used for this rapid electrochemical COD/TOC determination.

5.5 Real-time and online methods for controlling the solid phase in wastewater treatment

In wastewater treatment the most important phase is the solid phase, the activated sludge. Total suspended solids, sludge volume, settling velocity, sludge blanket level, and sludge density, are important technological parameters in controlling the performance of a plant. Laboratory-based methods are extremely time-consuming and thus cause long time delays. Innovative methods are based on *in situ* sensors establishing the way to automation.

A *sludge blanket level* detector integrates ultrasonic absorption and turbidity devices to detect the suspended solids interface as a result of the sudden change in sludge concentration when penetrating into the sludge blanket.

Settling velocity measurement applies a similar approach. In the settlometer of Vanrolleghem *et al.* (1996) the evolution of the blanket height is recorded with a moving optical detection system. From the resulting sludge sedimentation curve, the maximum sedimentation velocity and the sludge volume index can be obtained.

Sludge density can be monitored online by a microwave density analyzer. Solids flowing through pipes cause a phase lag of the microwave. The difference in microwave phase lag between the control wave and the one that passed through the fluid containing sludge is proportional to sludge density. The density meter measures density in electric current, so the signal can be directly applied for process monitoring and control. The measured phase difference is not affected by flow velocity and is resistant to the effects of contamination, scaling, fouling, and gas bubbles (US EPA, 2013).

Floc size and size distribution are the result of a dynamic equilibrium state between formation, transformation and breakage of the microbial aggregates. Floc parameters represent one of the most important parameters for characterization of the process performance and the influence of technological parameters such as substrate loading, sludge age or dissolved oxygen concentration (see more in Govoreanu *et al.*, 2009). To model sludge dynamics and control nitrification/denitrification and settling of the sludge, *in situ* measurement of floc size and distribution is essential. A laser light diffraction technique has been developed for on-line monitoring of the changes in floc structure expressed as a fractal dimension (Guan *et al.* 1998). Biggs & Lant (2000) applied the laser light diffraction technique for direct observation of size distribution. De Clerq *et al.* (2004) successfully applied the focused beam reflectance to measure the floc chord length distribution *in situ* in a secondary clarifier of a wastewater treatment for a wide range of solids concentrations, up to 50 g/l.

6 AUTOMATED INSTRUMENTS FOR CONTINUOUS MONITORING OF TOXIC ELEMENTS IN SURFACE, GROUND- AND WASTEWATER

The assessment of surface waters and effluents of municipal landfills and of abandoned mining sites with associated waste dumps requires continuous control of their toxic metal contents. The most widespread method for determination of metal concentrations in surface, ground and wastewater is via grab sampling and subsequent laboratory analysis. This method is both costly and typically involves a 24 hour turnaround time, which means that pollution events might be missed or detected too late.

Electrolyte-cathode discharge (ELCAD) spectrometry was invented as a direct analytical detection method for dissolved metals in aqueous solutions (Cserfalvi *et al.*, 1993). This analytical method was used to develop an automatic instrument for monitoring heavy metals in wastewaters loaded with high fat emulsion and suspended solid contents like municipal sewage waters mixed with industrial wastewaters (GREENWW1, 2014). Further developments have improved the sensitivity to make the method useful for the measurement of toxic metals in groundwater, contaminated surface water and sediment.

6.1 Principle of ELCAD

The measuring principle is cathode sputtering wherein the cathode is the electrolyte solution (the water sample) to be measured. The main unit is a flow-through cell in which atmospheric direct-current glow discharge plasma is generated emitting light with specific wavelengths which are characteristic of heavy metal contamination. The emitted light is processed by a built-in spectrometer unit.

The resulting optical emission spectrum of this plasma is very simple, containing only the basic atomic lines of the metals and background molecular emissions from the water matrix and the air atmosphere. The atomic lines of metals dissolved in the solution appear immediately in the spectrum emitted by the ELCAD. In this way the concentrations of metals in a sample can be determined within a few minutes. The plasma receives only the cathode-sputtered components and is therefore not disturbed or interfered with the non-sputtered components such as suspended solids and emulsions.

The different systems and designs developed in the past 20 years have recently been reviewed (Jamroz *et al.*, 2012).

Time-programmed sampling of the sewerage water is done by a submerged pump through a coarse pre-filtering (5x5 mm) net. An approximately 20 L sample is pumped to a raw water vessel in the monitor device. An innovative self-cleaning rotating-slit filter (Cserfalvi, 2011) operates in this vessel which removes particles that are larger than 200–300 μm from the analytical sample stream of 10 mL/min pumped in by the monitor unit. The only sample treatment is a controlled acid addition to the sample to solubilize metals present in the form of suspended particles of hydroxides, sulfides and carbonates, or they are partly complexed and bound to the suspended organic particles. The sampling (= measuring) frequency can be as high as 3–6 measurements/hour. It depends mostly on the level of the suspended and organic load of the water stream (flushing of the sample lines).

6.2   *In situ* applications

The ELCAD instrument can be operated on-site, installed in a measuring station or in mobile laboratories. An early industrial demonstration of the performance of the ELCAD method was done on the inflow stream in the North-Pest Wastewater Treatment Plant. It revealed sporadic but high metal pollution peaks occurring at midnight and weekends on mornings (Mezei & Cserfalvi, 2007). After several demonstrations (Figure 3.14) the recent field test now runs on the inlet flow stream of the Municipal Wastewater Treatment Plant of Székesfehérvár, Hungary. ELCAD is able to detect illegal industrial metal releases into the municipal sewerage as well as fluctuations in the metal content of effluents from landfills, abandoned industrial and mining sites. It is capable of monitoring around and above the regulatory limits. It can trigger a regular sampler device for the standard laboratory measurement providing an early warning.

Figure 3.14 Testing the ELCAD instrument in Malta and its application in Bohumin Steelworks (CZ) (Cserfalvi, 2014).

Recent developments (introduction of the capillary technique) resulted in highly improved detection limits (LOD). LOD depends on the chemical nature of the metal. It is as low as 28, 14, 22, 34 and 28 g/L for Zn, Cd, Cu, Ni, Pb, respectively (Cserfalvi & Mezei, 2003). The LOD for some other toxic metals such as Cs, Sr, Ag, Au and Hg 211, 49, 5, 78 and 349μg/L, respectively, were also published (Webb *et al.,* 2005). By applying additives (surfactants or organic acids), further improvement in sensitivity for various metals, for example mercury (2 μg/L), was achieved (Shekhar, 2012; Zhang *et al.,* 2014). These developments are the basis for the construction of a metal monitor for ground and surface waters.

6.3 Advantages and disadvantages

The high-sensitivity laboratory methods (Inductively Coupled Plasma spectrometry – ICP and Atomic Absorption Spectrometry – AAS) can be applied for measuring concentration values of low ppb levels, but they require thorough sample preparation. In most cases the sample has to be colloid-filtered and heavily acidified. Some type of nebulization technology (pneumatic, electrospray, ultrasonic, etc.) is necessary for the introduction of the sample. In addition, these instruments require auxiliary gases (ICP: argon, AAS: acetylene) for the operation. ICP has very significant energy demand with its 1–3 kW high-energy excitation unit. The most important distinction is that these methods cannot be applied for *in situ* monitoring.

Advantages of ELCAD:

- *in situ*/on-site, continuous monitoring technology;
- long-term stability of autonomous operation;
- no specific reagents or rare gases are used, only HCl is added;
- high suspended solid content and even oil and fat emulsion contents are allowed in the sample, because it only measures components dissolved at pH 1.5–1.7;
- low environmental footprint (low power consumption);
- low investment/maintenance cost.

The disadvantages of lower sensitivity compared to the high-sensitivity laboratory methods (ICP and AAS) are overcome by simpler operation, and the provision of real-time results.

7 CONCLUSION

This chapter provided an overview of the *in situ* and real-time monitoring methods useful for the characterization of surface and groundwater as well as of wastewater. The main advantages of the *in situ* real-time measurements, the fast response and undisturbed sample compensate for the usually lower sensitivity compared to the laboratory-based water analytical techniques in assessment, technology monitoring, and in decision-making both at global and regional level. The online real-time analysis in process control, e.g. in wastewater treatment, increases the efficiency of the technology.

The principles of local and remote sensing are discussed showing several examples of geophysical, hydrogeological, geochemical, biochemical, biological, ecological, ecotoxicological, etc. sensing tools.

Real-time water quality monitoring of surface waters and oceans is useful as early warning and makes possible immediate intervention and avoidance of damage to the aquatic ecosystem. The recently developed sensors autonomously measure some biological, chemical and optical parameters. Innovative analytical tools, bioassays, and biosensors are used for measuring the microbial quality and activity, the biodegradable organic compounds, and toxicity in wastewater. The online analyzers based on colorimetry and the ion-selective electrodes provide data for continuous monitoring and control. A recent innovation applying electrolyte cathode discharge (ELCAD) spectrometry is used for continuous control of toxic metal content even in municipal sewage water mixed with industrial wastewaters.

Many of the methodological developments have achieved practical implementation in the form of commercially available mobile equipment, portable or handheld devices. The mobilization of the environmental analytical tool system enables on-site environmental assessment and decision making, which, in turn, results in dynamic and efficient environmental risk management.

In addition to this review of monitoring tools for the aquatic domain, the next chapter covers the methods and tools applicable for soils, primarily for subsurface soils.

REFERENCES

Abraxis (2015) *Abraxis LLC.* [Online] Available from: www.abraxiskits.com and www.abraxiskits.com/products/pesticides/. [Accessed 28th May 2015].

Abraxis, Atrazine (2015) *Atrazine ELISA Kit* from Abraxis. [Online] Available from: www.abraxiskits.com/moreinfo/PN500001USER.pdf. [Accessed 28th May 2015].

Abraxis, Microcystin (2015) *DM Plate and strip test kit.* [Online] Available from: www.abraxiskits.com/products/algal-toxins and www.abraxiskits.com/uploads/products/docfiles/389_Microcystins%20Tube%20Field%20Kit%20R060512.pdf. [Accessed 28th May 2015].

Abraxis, Pesticides (2015) *Pesticides immunoassay kits.* [Online] Available from: www.abraxiskits.com/products/pesticides/. [Accessed 28th May 2015].

ADVNT Biotechnologies (2015) *BADD and Pro Strips™* from ADVNT Biotechnologies. [Online] Available from: www.advnt.org/pro-strips. [Accessed 28th May 2015].

ALGcontrol (2015) *Algae Monitoring System* from microLAN On-line Biomonitoring System. [Online] Available from: www.environmental-expert.com/products/?keyword=ALGcontrol. [Accessed 28th May 2015].

Algae Toximeter (2015) *Online Biomonitoring Using Green Algae* from bbe, Moldaenke. [Online] Available from: http://www.bbe-moldaenke.de/en/products/toxicity/details/toxprotect-64.html. [Accessed 3rd December 2015].

AlgaeOnlineAnalyser (2015) *Algae Monitoring* from bbe, Moldaenke. [Online] Available fromhttp://www.bbe-moldaenke.de/en/products/chlorophyll/details/algaeonlineanalyser.html. [Accessed 3rd December 2015].

Allan, I.J., Vrana, B., Greenwood, R., Mills, G.A., Roig., B. & Gonzalez, C. (2006a) A "toolbox" for biological and chemical monitoring requirements for the European Union's Water Framework Directive. *Talanta*, 69, 302–322.

Allan, I.J., Vrana, B., Greenwood, R., Mills, G.A., Knutsson, J., Holmberg, A., Guigues, N., Fouillac, A-M. & Laschi, S. (2006b) Strategic monitoring for the European water framework directive. *TrAC Trends in Analytical Chemistry*, 25(7), 704–715.

AoAC (1994) *Test Kit Definitions and Modifications Guideline* from AOAC Res. Inst. [Online] Available from: www.aoac.org/imis15_prod/AOAC_Docs/RI/RI_Resources/0217_Appendix_18.pdf. [Accessed 28th May 2015].

APNA (2015) *Autonomous Profiling Nutrient Analyzer*. [Online] Available from: www.subchem.com/SubChem%20FLYERS%202011.pdf. [Accessed 14th March 2015].

AppliTek (2015) *Online analysers – RA-BOD.* [Online] Available from: www.applitek.com/en/offer/analyzers/water-quality/environmental-compliance/ra-bod. [Accessed 24th March 2015].

AppliTOX (2015) *Toxicity Analyzers for Water Quality* from AppliTek. [Online] Available from: www.applitek.com/en/offer/analyzers/water-quality/toxicity/applitox/. [Accessed 28th May 2015].

AquaScope (2015) *Fully autonomous biosensor* from Biotrach V.B. [Online] Available from: www.environmental-expert.com/products/aquascope-fully-autonomous-biosensor-394077. [Accessed 28th May 2015].

Argo (2015) *Argo Project* part of the integrated global observation strategy. [Online] Available from: http://www.argo.net and http://www.argo.ucsd.edu. [Accessed 14th March 2015].

ASA Analytics (2015) *ChemScan® Analyzers* from ASA Analytics. [Online] Available from: www.asaanalytics.com/chemscan.php and www.asaanalytics.com/pdfs/2150N%20Effluent%20RecSpec.pdf. [Accessed 24th March 2015].

ASTM D5660 96 (2014) *Standard Test Method for Assessing the Microbial Detoxification of Chemically Contaminated Water and Soil Using a Toxicity Test with a Luminescent Marine Bacterium* (Withdrawn 2014). [Online] Available from: http://www.astm.org/Standards/D5660.htm. [Accessed 24th March 2015].

Axon Automation (2015) *STIP TOX Toxicity Monitor.* [Online] Available from: http://www.axonautomation.ca/E+H%20Literature/TIs/menu/docs/TIs/Analysis/STIP/TI832 CAE_STIP_TOX_TI.pdf. [Accessed 24th March 2015].

BACTcontrol (2015) *Automatic Online Measuring Device for Bacteria Pollution* from microLAN On-line Biomonitoring System. [Online] Available from: www.environmental-expert.com/products/brand-bactcontrol. [Accessed 28th May 2015].

Beacon Microcystin (2015) *Microcystin Plate Kit and Microcystin Tube Kit* from Beacon Analytical Systems. [Online] Available from: http://www.beaconkits.com/welcome/products-page/algal-toxins/microcystin/microcystin-tube-kit. [Accessed 28th May 2015].

Beacon, Atrazine (2015) *Atrazine Tube Kit* from Beacon Analytical Systems. [Online] Available from: www.beaconkits.com/welcome/products-page/crop-protection-chemicals/atrazine/atrazine-tube-kit. [Accessed 28th May 2015].

Biggs, C. A. & Lant, P. A. (2000) Activated sludge flocculation: on-line determination of floc size and the effect of shear. *Water Research* 34 (9), 2542–2550.

Biochem (2015) *Model C-2-Sentry Ammonium & Nitrate Analyzer* by Biochem Technology Inc. [Online] Available from: www.environmental-expert.com/products/myratek-model-c-2-sentry-ammonium-nitrate-analyzer-129013. [Accessed 24th March 2015].

BioMonitor (2015) *Model BOD and Toxicity* by LAR Process Analysers AG. [Online] Available from: www.lar.com/products/bod-toxicity/biomonitor.html. [Accessed 24th August 2015].

Bio-Seeq (2015) Biological agents detection from *Smiths Detection-Edgewood* (SDE). [Online] Available from: www.smithsdetection.com. [Accessed 24th March 2015].

Biosense (2015) *Biosense Laboratories* AS. [Online] Available from: www.biosense.com. [Accessed 28th May 2015].

BioTox® (2003) *Hidex Oy BioTox rapid toxicity testing method.* Environmental Technology Verification Report. ETV Advanced Monitoring Systems Center. [Online] Available from: www.epa.gov/etv/pubs/01_vr_biotox.pdf. [Accessed 24th March 2015].

BioVeris (2015) *BioVeris diagnostics*. [Online] Available from: http://bioveris.com/. [Accessed 28th May 2015].

Boucher, N., Lorrain, L., Rouette, M-E., Perron, E., Déziel, N., Tessier, L. & Bellemare, F. (2005) *Rapid Testing of Toxic Chemicals*. [Online] Available from: www.americanlaboratory.com/914-Application-Notes/19161-Rapid-Testing-of-Toxic-Chemicals. [Accessed 28th May 2015].

Campbell, E.R. & Davidson A-M. (2014) Enzymatic Method for Determining Nitrate in Wastewater Reaches First Step Toward EPA Approval. *Environmental Protection*. [Online] Available from: https://eponline.com/articles/2014/05/05/enzyme-based-nitrate-analysis.aspx. [Accessed 28th May 2015].

Challenge Technology (2015) *Quick Scan BOD Analyzer and Online Oxygen Demand Monitor* by Challenge Technology. [Online] Available from: www.copybook.com/environmental/challenge_technology#home. [Accessed 28th May 2015].

Charrier, T., Durand, M.J., Affi, M., Jouanneau, S., Gezekel, H. & Thouand, G. (2010) Bacterial bioluminescent biosensor characterisation for online monitoring of heavy metals pollutions in wastewater treatment plant effluents. In: Serra, P.A. (ed.) *Biosensors*. ISBN 978-953-7619-99-2, 302. DOI: 10.5772/7210.

Colifast ALARM (2015) *Rapid, on-line microbial water analysis*. [Online] Available from: http://www.colifast.no/products/alarm/. [Accessed 28th May 2015].

ColiPlate kit (2015) *E. coli detection on microplate* by EBPI. [Online] Available from: http://www.ebpi.ca/index.php?option=com_content&view=article&id=91&Itemid=81. [Accessed 28th May 2015].

CONDIACELL (2015) *Condiacell for Industrial Wastewater Treatment* by CONDIAS GmbH. [Online] Available from: http://condias.de/en/products/condiacell/condiacell.html. [Accessed 28th May 2015].

Cserfalvi, T. (2011) Apparatus for fluid sample filtration. *Hungarian Patent Application* P-1100411 27.07.2011, *International Patent Application* WO 2013/ 014475 A1 31. 01.2013.

Cserfalvi, T. & Mezei, P. (2003) Subnanogram sensitive multimetal detector with atmospheric electrolyte cathode glow discharge. *Journal of Analytical Atomic Spectrometry*, 18, 596–602.

Cserfalvi, T., Mezei, P. & Apai, P. (1993) Emission studies on a glow discharge in atmospheric pressure air using water as a cathode. *Journal of Physics D: Applied Physics*, 26, 2184–2188.

CTG (2015) *Environmental sensors* – Chelsea Technologies Group Ltd. [Online] Available from: http://www.chelsea.co.uk. [Accessed 14th March 2015].

CWA (2002) *Clean Water Act*. [Online] Available from: http://www.epw.senate.gov/water.pdf and http://www2.epa.gov/laws-regulations/summary-clean-water-act. [Accessed 24th March 2015].

DaphTox II (2015) *Biomonitoring using Daphnia*. bbe, Moldaenke. [Online] Available from: http://www.bbe-moldaenke.de/en/products/toxicity/details/daphtox-II.htmlr. [Accessed 3rd December 2015].

Davies, P. S. & Murdoch, F. (2002) The increasing importance of assessing toxicity in determining sludge health and management policy. *Measurement and Control*, 35 (8), 238–242. DOI: 10.1177/002029400203500804.

De Clercq, B., Lant, P.A. & Vanrolleghem, P.A. (2004) Focused beam reflectance technique for in situ particle sizing in wastewater treatment settling tanks. *Journal of Chemical Technology and Biotechnology*, 79(6), 610–618.

DeltaTox II (2015) *Portable toxicity testing* from Modern Water. [Online] Available from: www.modernwater.com/assets/downloads/Brochures/MW_Toxicity_Brochure_Low_res.pdf. [Accessed 28th May 2015].

Deltatox® (2015) *DeltaTox® II User's manual.* [Online] Available from: http://www.sdix.com/uploadedFiles/Content/Products/Water_Quality_Tests/3099915%20DeltaToxII%20User%20Manual-1.pdf. [Accessed 24th March 2015].

Dickey, T., Bates, N., Byrne, R.H., Chang, G., Chavez, F.P., Feely, R.A., Hanson, A.K., Karl, D.M., Manov, D., Moore, C., Sabine, C.L. & Wanninkhof, R. (2009a) The NOPP O-SCOPE and MOSEAN Projects: Advanced sensing for ocean observing systems. *Oceanography,* 22 (2), 168–181. DOI: 10.5670/oceanog.2009.47.

Dickey, T., Hanson, A., Karl, D. & Moore, C. (2009b) *Multi-disciplinary Ocean Sensors for Environmental Analyses and Networks (MOSEAN).* [Online] Available from: http://www.nopp.org/wp-content/uploads/project-reports-cdrom/reports/06dickey.pdf. [Accessed 14th March 2015].

Diez-Caballero, T.J., Rodriguez Albalat, G., Ferrer, C., Espinas Marti, E., Montoro Rodriguez, S., Erchov, V., Mendoza Plaza, A. & Diez-Caballero, T.G. (2002) Method for continuous monitoring of chemical substances in fluids. *US Patent* US 20020137093 A1. [Online] Available from: www.google.com/patents/US20020137093. [Accessed 14th April 2015].

Eclox™ (2015) *Rapid response water toxicity test kit* by Hach. [Online] Available from: www.hach.com/hach-eclox-rapid-response-water-test-kit/product?id=7640273470. [Accessed 28th May 2015].

Eclox™ handheld luminometer (2015) *Eclox™ Rapid Response Test Kit including handheld luminometer* by Hach. [Online] Available from: www.hach.com/asset-get.download.jsa?id=7639983069. [Accessed 28th May 2015].

Ecologiena (2015) *Ecologiena Series for Environmental Pollutants* from Tokiwa Chemical Industries Co. [Online] Available from: www.ecologiena.jp/index-e.html. [Accessed 28th May 2015].

Elad, T., Almog, R., Yagur-Kroll, S., Levkov, K., Melamed, S., Shacham-Diamand, Y. & Belkin, S. (2011) Online monitoring of water toxicity by use of bioluminescent reporter bacterial biochips. *Environmental Science & Technology,* 45 (19), 8536–8544.

Endress (2015) *ISEMax CAS40 for ammonium and nitrate* from Endress. [Online] Available from: http://portal.endress.com/wa002/ProductContainerGUI/hu/en/CAS40?tab=documents&filter=010.030. [Accessed 14th April 2015].

EnviroGard® Microcystins (2015) *Tube kits for microcystins.* [Online] Available from: www.modernwater.com/assets/downloads/PesticideDetectionProducts_01.03.12.pdf. [Accessed 28th May 2015].

EPA ETV, Atrazine (2007) *Immunoassay Test Kits for Atrazine.* [Online] Available from: http://nepis.epa.gov/Adobe/PDF/P10012ZS.pdf. [Accessed 28th May 2015].

EPA ETV, Microcystins (2015) *Immunoassay Test Kits for Microcystins.* [Online] Available from: http://nepis.epa.gov/Adobe/PDF/P100EEXL.pdf. [Accessed 28th May 2015].

Fish Toximeter (2015) *Online toxicity monitoring by fish* from bbe Moldaenke. [Online] Available from: http://www.bbe-moldaenke.de/en/products/toxicity/details/fish-toximeter.html. [Accessed 3rd December 2015].

Friedrich, G.E., Brewer, P.G., Herlien, R. & Chavez, F.P. (1995) Measurement of sea surface partial pressure of CO_2 from a moored buoy. *Deep-Sea Research I,* 42 (1), 175–1186.

Geenens, D. & Thoeye, C. (1998) The use of an online respirometer for the screening of toxicity in the Antwerp WWTP catchment area. *Water Science and Technology,* 37, 213–218.

GEOSS (2015) Global Earth Observation System or Systems. [Online] Available from: www.geoportal.org/web/guest/geo_home_stp. [Accessed 14th March 2015].

GLAAS (2014) *Investigating Water and Sanitation: increasing access, reducing inequalities.* UN-Water Global Analysis and Assessment of Sanitation and Drinking-Water. [Online] Available from: http://apps.who.int/iris/bitstream/10665/139735/1/9789241508087_eng.pdf?ua=1. [Accessed 24th March 2015].

Goddard, W.A., Ferkel, C. & Ferkel, M. (2009) Highly Sensitive Micro-Biosensors for in situ Monitoring of Mercury (II) Contaminants through Genetic-evolution and Computer modeling of Metal-binding Proteins, *Annual DOE-ERSP PI Meeting, April 22, 2009, Environmental Remediation Sciences Program.* [Online] Available from: http://esd.lbl.gov/research/projects/ersp/generalinfo/pi_meetings/PI_mtg_09/present/Wednesday/Goddard_DOE%20PI%20Hg%20biosensor.pdf. [Accessed 14th December 2014].

Gomiero, A., Da Ros, L., Nasci, C., Meneghetti, F., Spagnolo, A. & Fabi, G. (2011) Integrated use of biomarkers in the mussel *Mytilus galloprovincialis* for assessing off-shore gas platforms in the Adriatic Sea: Results of a two-year biomonitoring program. *Marine Pollution Bulletin*, 62, 2483–2495.

GOOS (2015) *Global Ocean Observing System.* [Online] Available from: www.ioc-goos.org. [Accessed 14th March 2015].

Gorbi, S., Virno Lamberti, C., Notti, A., Benedetti, M., Fattorini, D., Moltedo, G. & Regoli, F. (2008) An ecotoxicological protocol with caged mussels, *Mytilus galloprovincialis*, for monitoring the impact of an offshore platform in the Adriatic Sea. *Marine Environmental Research*, 65, 34–49.

Govoreanu, R., Saveyn, H., Van der Meeren, P., Nopens, I. & Vanrolleghem, P.A. (2009) A methodological approach for direct quantification of the activated sludge floc size distribution by using different techniques. *Water Science & Technology*, 60 (7), 1857–1867.

GREENWW1 (2014) *Development of a cost-effective, low-maintenance, online instrument to detect heavy metal concentrations in wastewaters.* [Online] Available from: www.greenwwater.eu/hu/node/3. [Accessed 14th December 2014].

Gruiz, K. & Molnár, M. (2015) Aquatic toxicology. In: Gruiz, K., Meggyes, T. & Fenyvesi, E. (eds.) Engineering tools for environmental risk management. Volume 2. *Environmental Toxicology*. Boca Raton, Fl., CRC Press. pp. 171–228.

Gruiz, K., Meggyes, T. & Fenyvesi, E. (eds.) (2015) Engineering tools for environmental risk management. Volume 2. *Environmental Toxicology*. Boca Raton, Fl., CRC Press.

Gruiz, K., Hajdu, Cs. & Meggyes, T. (2015a) Data evaluation and interpretation in environmental toxicology. In: Gruiz, K., Meggyes, T. & Fenyvesi, E. (eds.) Engineering tools for environmental risk management. Volume 2. *Environmental Toxicology*. Boca Raton, Fl., CRC Press. pp. 445–544.

Gruiz, K., Molnár, M., Feigl, V., Hajdu, Cs., Nagy, Zs. M., Klebercz, O., Fekete-kertész, I., Ujaczki, É. & Tolner, M. (2015b) Terrestrial toxicology. In: Gruiz, K., Meggyes, T. & Fenyvesi, E. (eds.) (2015) Engineering Tools for Environmental Management, Volume 2. *Environmental Toxicology*. Boca Raton, Fl., CRC Press. pp. 229–310.

Guan, J., Waite, T.D. & Amal, R. (1998) Rapid structure characterization of bacterial aggregates. *Environmental Science & Technology*, 32, 3735–3742.

Guwy, A.J., Farley, L.A., Cunnah, P., Hawkes, F.R., Hawkes, D., Chase, M. & Buckland, H. (1999) An automated instrument for monitoring oxygen demand in polluted waters. *Water Research*, 33 (14), 3142–3148.

Hach (2015) *Arsenic test kit.* [Online] Available from: www.hach.com/arsenic-low-range-test-kit/product?id=7640217303&callback=pf. [Accessed 28th May 2015].

Hach AMTAX (2015) *Amtax-sc-ammonium-analyzer.* [Online] Available from: www.hach.com/amtax-sc-ammonium-analyzer-0-05-20mg-l-nh4-n-115v-5m-filter-probe-sc/product?id=7640092926. [Accessed 24th March 2015].

Hach Aquaculture (2015) *Ten-Parameter Test Kit, Model FF-2.* [Online] Available from: www.hach.com/ten-parameter-test-kit-model-ff-2/product-downloads?id=7640218453. [Accessed 28th May 2015].

Hach ISE (2015a) *AISE sc ISE ammonium probe.* [Online] Available from: www.hach.com/aise-sc-ise-ammonium-probe/product-details?id=14667082652. [Accessed 24th March 2015].

Hach Immuno (2015) *Immunoassay for atrazin, PCB and TPH.* [Online] Available from: http://in.hach.com/test-kits/immunoassay-test-kits/family?productCategoryId=22216767297. [Accessed 28th May 2015].

Hach ISE (2015b) *AN-ISE sc combination sensor for ammonium and nitrate.* [Online] Available from: www.hach.com/an-ise-sc-combination-sensor-for-ammonium-and-nitrate/product?id=9296230750. [Accessed 24th March 2015].

Hach NITRATAX (2015) *Nitratax-sc-nitrate-sensors.* [Online] Available from: www.hach.com/nitrate-sensors/nitratax-sc-nitrate-sensors/family-downloads?product CategoryId=22219796950. [Accessed 24th March 2015].

Hach PHOPHAX (2015) *Phosphax-sc-phosphate-analyzer.* [Online] Available from: www.hach.com/phosphax-sc-phosphate-analyzer-0-05-15mg-l-po4-p-115v-5m-filter-probe-sc/product?id=7640094090. [Accessed 24th March 2015].

Henriques, I.D.S., Kelly, II, R.T., Dauphinais, J.L. & Love, N.G. (2007) Activated sludge inhibition by chemical stressors: A comprehensive study. *Water Environment Research*, 79 (9), 940-951.

IAD (1999) Biomonitoring of Danube Water Quality. *Danube News*, 1999 (1).

IFREMER (2015) *Institut Français de Recherche pour l'Exploitation de la Mer.* [Online] Available from: http://wwz.ifremer.fr. [Accessed 14th March 2015].

ISO 11348-1 (2007) *Water quality* – Determination of the inhibitory effect of water samples on the light emission of *Vibrio fischeri*. [Online] Available from: www.iso.org/iso/catalogue_detail.htm?csnumber=40516. [Accessed 28th May 2015].

ISO 11348-1 (2007a) *Water quality* – Determination of the inhibitory effect of water samples on the light emission of *Vibrio fischeri* (Luminescent bacteria test) – Part 1: Method using freshly prepared bacteria. [Online] Available from: www.iso.org/iso/catalogue_detail.htm?csnumber=40516. [Accessed 24th March 2015].

ISO 11348-2 (2007b) *Water quality* – Determination of the inhibitory effect of water samples on the light emission of *Vibrio fischeri* (Luminescent bacteria test) – Part 2: Method using liquid-dried bacteria. [Online] Available from: www.iso.org/iso/catalogue_detail.htm?csnumber=40517. [Accessed 24th March 2015].

ISO 11348-3 (2007c) *Water quality* – Determination of the inhibitory effect of water samples on the light emission of *Vibrio fischeri* (Luminescent bacteria test) – Part 3: Method using freeze-dried bacteria. [Online] Available from: www.iso.org/iso/catalogue_detail.htm?csnumber=40518. [Accessed 24th March 2015].

ISO 15839 (2003) *Water quality* – On-line sensors/analysing equipment for water – Specifications and performance tests. [Online] Available from: www.iso.org/iso/catalogue_detail.htm?csnumber=28740. [Accessed 24th April 2015].

ISO 17381 (2003) *Water quality* – Selection and application of ready-to-use test kit methods in water analysis. [Online] Available from: www.iso.org/iso/catalogue_detail.htm?csnumber=30626 and https://www.iso.org/obp/ui/#iso:std:iso:17381:ed-1:v1:en. [Accessed 24th March 2015].

ISO 7027 (1999) *Water quality* – Determination of turbidity. [Online] Available from: www.iso.org/iso/catalogue_detail.htm?csnumber=30123. [Accessed 14th March 2015].

ITS (2015) *Arsenic Low-range, Ultra low- and Econo Quick IITM* from Industrial Test Systems. [Online] Available from: www.sensafe.com/?s=Quick+II and http://www.sensafe.com/arsenic-kits/. [Accessed 28th May 2015].

Jamroz, P., Greda, K. & Pohl, P. (2012) Development of direct current atmospheric pressure glow discharges generated in contact with flowing electrolyte solutions for elemental analysis by optical emission spectrometry. *Trends in Analytical Chemistry*, 41, 105–121.

Jennings, V.L., Rayner-Brandes, M.H. & Bird, D.J. (2001) Assessing chemical toxicity with the bioluminescent photobacterium (*Vibrio fischeri*): a comparison of three commercial systems. *Water Research*, 35(14), 3448–3456.

Johnson, K.S. & Coletti, L.J. (2002) *In situ* ultraviolet spectrophotometry for high resolution and long term monitoring of nitrate, bromide, and bisulfide in the ocean. *Deep-Sea Research I*, 49 (1), 291–1305.

Jońca, J., Comtat, M. & Garçon, V. (2013) In situ phosphate monitoring in seawater: today and tomorrow. In: Mason, A. & Mukhopadhyay S. (eds.) (2013) Smart Sensors for Real-Time Water Quality Monitoring. *Smart Sensors, Measurement and Instrumentation*, 4, 25–44.

Jouanneau, S., Durand, M.J. & Thouand, G. (2012) Online detection of metals in environmental samples: comparing two concepts of bioluminescent bacterial biosensors. *Environmental Science & Technology*, 46(21), 11979–11987.

Kelma (2015) *Rodtox NG Online BOD & toxicity analysers. [Online] Available from:* www.kelma.com/product/71/rodtox-ng. [Accessed 24th March 2015].

LaMotte (2015) *Arsenic Test Kit* from La Motte. [Online] Available from: http://www.lamotte.com/en/; http://www.lamotte.com/en/browse/4053-02.html. [Accessed 28th May 2015].

LBi (2015) *Lab Bell Inc.*, Canada. [Online] Available from: www.lab-bell.com/index.php. [Accessed 28th May 2015].

Legionella detection (2015) *Legionella Detection Test Kit – Rapid Test Method* from the company Accepta. [Online] Available from: http://www.accepta.com/water-testing-water-quality-analysis-equipment/legionella-testing-monitoring-equipment/62-legionella-test-kit-rapid-test-method. [Accessed 28th May 2015].

Legionella kit (2015) *Universal Rapid Legionella Kit* from droptestkits (DTK Water). [Online] Available from: http://www.droptestkits.com/test-kits/legionella-compliance/. [Accessed 28th May 2015].

Legipid (2015) *Legionella Fast Detection Kit* from Bioquímica Analítica, S.L. [Online] Available from: www.biotica.es/en/Products/Legipid%C2%AE%20Legionella%20Fast%20Detection%20Kit. [Accessed 28th May 2015].

Lemke, E.A. & Schultz, C. (2011) Principles for designing fluorescent sensors and reporters. *Nature Chemical Biology*, 7, 480–483. DOI: 10.1038/nchembio.620.

Liu, J. & Mattiasson, B. (2002) Microbial BOD sensors for wastewater analysis. *Water Research*, 36(15), 3786–3802.

LuminoTox (2015) *LuminoTox, Robot LuminoTox* from LBi. [Online] Available from: http://www.environmental-expert.com/products/luminotox-robot-luminotox-22996. [Accessed 28th May 2015].

LUMIStox (1999) *LUMIStox 300 LPV321* – User Manual of Hach-Lange. [Online] Available from: www.manualsdir.com/manuals/333014/hach-lange-lumistox-300-lpv321-user-manual.html. [Accessed 24th March 2015].

LUMIStox (2010) *Joint Verification of the Hach-Lange GmbH LUMIStox 300 Bench Top Luminometer and ECLOX Handheld Luminometer*. US EPA ETV and ETV Canada. [Online] Available from: http://nepis.epa.gov/Adobe/PDF/P100ELTP.pdf [Accessed 24th May 2015].

Lund, P.A., Ford, S.J. & Brown, N.L. (1986) Transcriptional regulation of the mercury resistance genes of transposon Tn501. *Journal of General Microbiology*, 132, 465–480.

Manov, D.V., Chang, G.C. & Dickey, T.D. (2004) Methods for reducing biofouling of moored optical sensors. *Journal of Atmospheric and Oceanic Technology*, 21, 958–968.

Merck (2015) *As MQuant™* from Merck Millipore. [Online] Available from: www.merckmillipore.com/FI/en. [Accessed 28th May 2015].

Metrohm (2015b) *Alert Ion Analyzer ADI 2003* from Metrohm-Applikon. [Online] Available from: www.metrohm-applikon.com/Products/ADI2003.html. [Accessed 24th March 2015].

Metrohm-Applikon (2015a) *Alert Colorimeter* – On-Line Analyzers Metrohm Applikon. [Online] Available from: www.metrohm-applikon.com/Downloads/APPK_ALERT_2011_FC_LR.pdf. [Accessed 24th March 2015].

Mezei, P. & Cserfalvi, T. (2007) Electrolyte cathode atmospheric glow discharges for direct solution analysis. *Applied Spectroscopy Reviews*, 42, 573–604.

Microtox® (2015) *About Microtox by Aquatox Research.* [Online] Available from: www.aquatoxresearch.com/microtox.html. [Accessed 28th May 2015].

Microtox® CTM (2015) *Continuous Toxicity Monitor* by Modern Water. [Online] Available from: www.modernwater.com/assets/downloads/Brochures/MW_Toxicity_Brochure_Low_res.pdf. [Accessed 28th May 2015].

MN (2015) *QUANTOFIX® test strips* from Machrey-Nagel. [Online] Available from: http://www.mn-net.com/tabid/10304/tabid/4928/default.aspx. [Accessed 28th May 2015].

Modern Water, test kits (2015) Available from: http://www.modernwater.com/monitoring/environment/test-kits. [Accessed 28th May 2015].

Modern Water, Microtox® (2015) *Microtox® CTM.* [Online] Available from: www.modernwater.com/monitoring/toxicity/on-line-microtox-ctm. [Accessed 24th March 2015].

Modern Water, pesticides (2015) QuickChek® *Testing Products for Pesticides/Herbicides/Fungicides in Water.* [Online] Available from: www.modernwater.com/assets/downloads/PesticideDetectionProducts_01.03.12.pdf. [Accessed 28th May 2015].

Moore, C., Twardowski, M.S. & Zaneveld, J.R.V. (2000) The ECO VSF – A multi-angle scattering sensor for determination of the volume scattering function in the backward direction. In: *Proceedings of Ocean Optics XV. October 16–20, Monaco.*

Mosselmonitor® (2015) *Mosselmonitor.* [Online] Available from: http://www.mosselmonitor.nl/html/Engels/general%20information.html. [Accessed 28th May 2015].

MPElA Video (2015) *Magnetic Particles ELISA Kit.* [Online] Available from: https://www.youtube.com/watch?v=2XYPyURZCqw. [Accessed 28th May 2015].

MRV (2015) *MRV Systems. Marine Robotic vehicles.* [Online] Available from: www.mrvsys.com. [Accessed 14th March 2015].

Mycometer (2015) *Rapid fungi and bacteria detection by Mycometer® and Bactiquant® tests.* [Online] Available from: http://www.mycometer.com/products/. [Accessed 28th May 2015].

Nano-Band (2015) *TraceDetect Nano-Band™ ExplorerII fpr voltammetric metal analysis* from EVISA: European Virtual Institute for Speciation Analysis. [Online] Available from: *www.speciation.net / Database / Instruments / TraceDetect / NanoBand-Explorer-II-; i2564.* [Accessed 28th May 2015].

NAS-3x (2015) *In situ Nutrient Analyzer* by EnviroTech. [Online] Available from: http://www. hydroacoustics.com/library/catalogos/Envirotech/nas3x.pdf. [Accessed 14th March 2015].

NECi (2015) *Enzymatic Nitrate Quantification.* [Online] Available from: www.nitrate.com. [Accessed 28th May 2015].

NitriTox (2015) *Online Toximeter for waste water treatment plants* from LAR LAR Process Analysers AG. [Online] Available from: www.lar.com/products/bod-toxicity/nitritox.html. [Accessed 28th May 2015].

nke (2015) *nke-INSTRUMENTATION for in situ measurements.* [Online] Available from: www.nke-instrumentation.com. [Accessed 14th March 2015].

NOPP (2015) *National Oceanographic Partnership Program.* [Online] Available from: http://www.nopp.org/ [Accessed 14th April 2015].

Ocean Origo (2015) *SMHI Måseskär buoy.* [Online] Available from: http://www.oceanorigo.com/seatramp_buoysystems.asp. [Accessed 14th March 2015].

Oehlmann, J. & Schulte-Oehlman, U. (2002) Molluscs as bioindicators. In: Markert, B.A., Breure, A.M. & Zechmeister, H.G. (eds.) *Bioindicators and biomonitors.* Amsterdam, Elsevier. pp. 577–635.

OP-Stick (2015) *OP-Stick Sensor for OP and carbamate – Agri-Stick* from AR Brown. [Online] Available from: www.arbrown.com/english/products/agri_stick/. [Accessed 28th May 2015].

PDV 6000 (2015) *Portable metal monitor* by Modern Water. [Online] Available from: www.modernwater.com/monitoring/trace-metals/portable-and-laboratory-pdv. [Accessed 28th May 2015].

Pilot project Mosselmonitor (2013) *Biomonitoring of coastal marine waters subject to anthro pogenic use: development and application of the biosensor Mosselmonitor®*. Action 4.4 MSP Pilot project – ARPA Molise Final Report. [Online] Available from: www.shape-ipaproject.eu/download/listbox/WP4%20action%204.4/Molise%20Pilot%20Project%20Final%20report.pdf. [Accessed 18th July 2015].

POLYTOX-RES (2015) *Automated Real Time Monitoring Device* by Biosensores, S.L. [Online] Available from: http://www.biosensores.com/EN/polytox-res.php. [Accessed 28th May 2015].

PPM (2015) *Pollution & Process* Monitoirng, Toxicity Technology, Amtox. [Online] Available from: http://www.pollution-ppm.co.uk/pdf/Amtox%20lab.pdf. [Accessed 24th March 2015].

Quevauviller, P., Thomasand, O. & van der Beken, A. (eds.) (2006) *Wastewater Quality Monitoring and Treatment*. Hoboken, NJ., John Wiley & Sons, Ltd. [Online] Available from: www.academia.edu/319633/Wastewater_Quality_Monitoring_and_Treatment. [Accessed 24th March 2015].

RaPID Assay (2015) *Rapid field enzyme immunoassay method* from Modern Water. [Online] Available from: www.modernwater.com/assets/downloads/Factsheets/MW_Factsheet_Rapid %20Assay_PAH_highres.pdf. [Accessed 28th May 2015].

Rapidtoxkit (2015) *Thamnocephalus platyurus rapid toxicity test* from Environmental Bio-Detection. [Online] Available from: www.ebpi.ca/index.php?option=com_content&view =article&id=59&Itemid=69. [Accessed 28th May 2015].

RAZOR (2015) *RAZOR® and RAZOR EX instrument and pouch support*. [Online] Available from: http://biofiredefense.com/support/razor-support. [Accessed 24th March 2015].

RBR (2015) *RBR Products*. [Online] Available from: www.rbr-global.com/products/ctd-a-multi-channel-loggers. [Accessed 14th March 2015].

Response Biomedical (2015) *RAMP® test kits* from Response Biomedical. [Online] Available from: http://responsebio.com/. [Accessed 24th October 2015].

Lead SA1100 (2015) *Scanning Analyzer* from Palintest by DelAgua Group. [Online] Available from: www.palintest.com/products/sa1100-scanning-analyzer/. [Accessed 28th May 2015].

Satlantic (2015) *Advanced ocean technology*. [Online] Available from: http://satlantic.com. [Accessed 14th March 2015].

SBE (2015a) *SBE 19plus V2 SeaCAT Profiler CTD* by Sea-bird Electronics. [Online] Available from: http://www.seabird.com/sbe19plusv2-seacat-ctd. [Accessed 14th March 2015].

SBE (2015b) *Real-Time Ocean Observing Systems* by Sea-bird Electronics. [Online] Available from: *http://www.seabird.com/real-time-ocean-observing-systems*. [Accessed 14th March 2015].

SciTOX ALPHA (2015) *Rapid wastewater toxicity measurement* by SciTox Ltd. [Online] Available from: http://www.scitox.com/products.php. [Accessed 28th May 2015].

Scripps (2015) *Scripps Institution of Oceanography*. [Online] Available from: https://scripps. ucsd.edu. [Accessed 14th March 2015].

Sensara (2015) *Strathtox Respirometer and Bio-Scope*. [Online] Available from: www.sensaratech.com/productos.php. [Accessed 28th May 2015].

Sequoia (2015) *Products and Research for Environmental Science*. [Online] Available from: *http://www.sequoiasci.com/products*. [Accessed 14th March 2015].

SETAC (2004) *Technical issue paper: Whole effluent toxicity testing*. Society of Environmental Toxicology and Chemistry. Pensacola FL, USA, SETAC.

Shekhar, R. (2012) Improvement of sensitivity of electrolyte cathode discharge atomic emission spectrometry (ELCAD-AES) for mercury using acetic acid medium. *Talanta*, 93, 32–36.

Silver Lake (2015) *Watersafe® from Silver Lake research.* [Online] Available from: www. silverlakeresearch.com and www. discovertesting.com/articles/14,1.html and www.discovertesting.com/. [Accessed 28th May 2015].

Spanjers, H., Vanrolleghem, P.A., Olsson, G. & Dold, P.L. (1998) *Respirometry in Control of the Activated Sludge Process: Principles.* Scientific and Technical Reports, IAWQ Task Group on Respirometry in Control of the Activated Sludge Process.

Spear, J.M., Zhou, J.M.,. Cole, C.A. & Xie, Y.F. (2006) Evaluation of arsenic field test kits for drinking water analysis. *Journal – American Water Works Association*, 98(12), 97–105.

Spellman, F.R. (2014) Energy conservation methods and sustainability. In: *Handbook of Water and Wastewater Treatment Plant Operations.* Boca Raton, Fl, CRC Press. pp. 86–87.

Steinert, S.A., Steib-Montee, R., Leather, J.M. & Chadwick, D.B. (1998) DNA damage in mussels at sites in San Diego Bay. *Mutation Research*, 399, 65–85.

Sub Chem (2015) *Website of SubChem Systems, Inc.* [Online] Available from: www.subchem.com. [Accessed 14th March 2015].

SymBio (2015) *SymBio process.* [Online] Available from: www.wateronline.com/doc/symbio-process-0001. [Accessed 24th March 2015].

TECTA™ (2015) *Automated rapid microbial detection systems* from ENDETEC Veolia. [Online] Available from: http://www.endetec.com/en/products/tecta/. [Accessed 28th May 2015].

Teledyne RDI (2015) *Marine measurements.* [Online] Available from: http://www.rdinstruments.com/mm_main.aspx. [Accessed 14th March 2015].

Tetracore (2015) *BioThreat Alert® kits of lateral flow immunoassay test strips, BioThreat Alert® Reader and ELISA Kits.* [Online] Available from: http://www.tetracore.com/products. [Accessed 28th May 2015].

ToxAlarm (2015) *Online Toximeter for Drinking Water and Surface Water Monitoring* from LAR Process Analysers AG. [Online] Available from: www.lar.com/products/bod-toxicity/toxalarm.html. [Accessed 28th May 2015].

Toxalarm (2015) *Toxicity analyser for surface water and drinking water.* [Online] Available from: www.envirotech-online.com/news/water-wastewater/9/lar_process_analysers_ag/toxicity_analyser_for_surface_water_drinking_water/32954. [Accessed 24th March 2015].

ToxBox (2015) *Toxicity Testing Box* from Biotrack B.V. [Online] Available from: http://www.biotrack.nl/products/toxbox.html. [Accessed 24th March 2015].

TOXcontrol (2015) *On-line Toxicity Monitoring System* by microLAN On-line Biomonitoring Systems. [Online] Available from: http://toxcontrol.com/products/toxcontrol-line-biomonitor-system/. [Accessed 28th May 2015].

Toxi-Chromotest (2015) *Toxi-Chromotest* from Environmental Bio-Detection. [Online] Available from: www.ebpi.ca/index.php?option=com_content&view=article&id=5&Itemid=53. [Accessed 28th May 2015].

TOXmini (2015) *Portable toxicity testing device* from microLAN. [Online] Available from: http:// www.environmental-expert.com/products/?keyword=Toxmini. [Accessed 28th May 2015].

ToxProtect (2015) *Online instrument for drinking water protection.* [Online] Available from: http://www.bbe-moldaenke.de/en/products/toxicity/details/toxprotect-64.html [Accessed 3rd December 2015].

ToxScreen II (2003) *Environmental Technology Verification Report.* Checklight Ltd. Toxscreen-II rapid toxicity testing system. Advanced Monitoring Systems Center. [Online] Available from: http://www.epa.gov/etv/pubs/01_vr_toxscreen.pdf. [Accessed 24th March 2015].

Trivedi, H.K. (2009) Simultaneous Nitrification and Denitrification (SymBio®Process). In: Wang, L.K., Shammas, N.K. & Hung, Y.T. (eds.) *Advanced Biological Treatment Processes. Handbook of Environmental Engineering Series.* NY, USA, Humana Press. pp. 185–208. DOI: 10.1007/978–1–60327–170–7.

TWR (2015) *Teledyne Webb Research Corporation*. [Online] Available from: www.webbresearch.com. [Accessed 24th March 2015].

US EPA (2013) *Emerging Technologies for Wastewater Treatment and In-Plant Wet Weather Management*. Process Monitoring Technologies. [Online] Available from: http://nepis.epa.gov/Exe/ZyPDF.cgi/P100ILDC.PDF?Dockey=P100ILDC.PDF. [Accessed 20th February 2016].

Vanrolleghem, P.A. (2002) *Principles of Respirometry in Activated Sludge Wastewater Treatment*. Department of Applied Mathematics, Biometrics and Process Control. [Online] Available from: http://modeleau.fsg.ulaval.ca/fileadmin/modeleau/documents/Publications/pvr402.pdf. [Accessed 24th March 2015].

Vanrolleghem, P.A. & Lee, D.S. (2003) Online monitoring equipment for wastewater treatment processes: state of the art. *Water Science and Technology*, 47(2), 1–34.

Vanrolleghem, P.A., Van der Schueren, D., Krikilion, G., Grijspeerdt, K., Willems, P. & Verstraete, W. (1996) Online quantification of settling properties with in-sensor-experiments in an automated settlometer. *Water Science & Technology*, 33(1), 37–51.

Viedma, P. (2010) Method for detecting presence of acidophilic microorganisms in bioleaching solution. *US Patent 7,851,177*.

Wagtech (2015) *Arsenator: Digital test kit* from Wagtech. [Online] Available from: www.wagtech.co.uk/products/water-and-environmental/water-test-kits/arsenator%C2%AE-digital-arsenic-test-kit. [Accessed 28th May 2015].

Webb, M.R., Andrade, F.J., Gamez, G., McCrindle, R. & Hieftje, G.M. (2005) Spectroscopic and electrical studies of a solution-cathode glow discharge. *Journal of Analytical Atomic Spectrometry*, 20, 1218–1225.

WFD (2000) *Water Framework Directive*. Directive 2000/60/EC of the European Parliament and of the Council of 23 October 2000 establishing a framework for Community action in the field of water policy. [Online] Available from: http://eur-lex.europa.eu/LexUriServ/LexUriServ.do?uri=CELEX:32000L0060:en:HTML and http://ec.europa.eu/environment/water/water-framework/info/intro_en.htm. [Accessed 24th March 2015].

WTW (2015a) *TresCon® analyzer modules for online orthophosphate and P total measurement*. [Online] Available from: www.wtw.de/en/products/online/phosphate/tresconr-analyzer.html. [Accessed 24th March 2015].

WTW (2015b) *NitraVis spectral sensor* from WTW. [Online] Available from: www.wtw.de/en/products/online/nitrogen/uv-visuv-sensors.html. [Accessed 24th March 2015].

WTW ISE (2015) *ISE sensors for ammonium and nitrate* from WTW. [Online] Available from: http://www.wtw.de/en/products/online/nitrogen/ise-sensors.html. [Accessed 24th March 2015].

Xing, X., Liu, L., Zhang, X., Kuang, H. & Xua, C. (2014) Colorimetric detection of mercury based on a strip sensor. *Analytical Methods*, 6, 6247–6253. DOI: 10.1039/C3AY42002G.

YSI (2015) *6600 V2 Multiparameter sondes*. [Online] Available from: https://www.ysi.com/600OMS-V2. [Accessed 14th March 2015].

Zeulab (2015) *Okatest, domotest, saxitest, in bivalves and microcystest in water*. [Online] Available from: http://www.zeulab.com/products.html/toxins.html. [Accessed 28th May 2015].

Zeu Microcystin (2015) *MicroCystest Plate Kit* from Zeu-Immunotec. [Online] Available from: http://www.zeulab.com/products.html/toxins/117-microcystest-kit.html. [Accessed 28th May 2015].

Zhang, Z., Wang, Z., Li, Q., Zou, H. & Shi, Y. (2014) Determination of trace heavy metals in environmental and biological samples by solution cathode glow discharge-atomic emission spectrometry and addition of ionic surfactants for improved sensitivity. *Talanta*, 119, 613–619.

Zhao, H., Jiang, D., Zhang, S., Catterall, K. & Richard, J.R. (2004) Development of a direct photoelectrochemical method for determination of chemical oxygen demand. *Analytical Chemistry*, 76(1), 155–160. [Online] Available from: http://pubs.acs.org/doi/abs/10.1021/ac0302298. [Accessed 14th May 2015].

Zhao, Y., Jin, J., Hu, Q., Zhou, H.-M., Yi, J., Yu, Z., Xu, L., Wang, X., Yang, Y. & Loscalzo, J. (2011) *Genetically encoded fluorescent sensors for intracellular NADH detection.* [Online] Available from: http://dx.doi.org/10.1016/j.cmet.2011.09.004. [Accessed 14th March 2015].

Zhao, Y, Yang, Y. & Loscalzo, J. (2014) Real time Measurement of Metabolic States in Living Cells using Genetically-encoded NADH Sensors. *Methods in Enzymology*, 542, 349–367. DOI: 10.1016/B978–0–12–416618–9.00018–2.

Photos:

Argo float design (2015) [Online] Available from: www.argo.ucsd.edu/float_design.html. [Accessed 14th March 2015].

Argo operation (2015) *Scheme of the operation.* [Online] Available from: www.argo.ucsd.edu/operation_park_profile.jpg. [Accessed 14th March 2015].

Clean-Trace™ (2015) *3M™ Clean-Trace™ NG Luminometer* from 3M. [Online] Available from: Solutions.3m.com/wps/portal/3M/en_US/Microbiology/FoodSafety/product-information/product-catalog/?PC_Z7_RJH9U523003DC023S7P92O3O87000000_nid=0S2MVTKC2GbeJ9C52DTHJWgl. [Accessed 28th May 2015].

Comparator cube (2015) *Iron color disc test kit, Model IR–18B* from Hach. Available from: www.hach.com/iron-cube–0–10–mg–l–fe/product?id=7640212241&callback=qs. [Accessed 28th May 2015].

Comparator disc (2015) *Comparator discs for water analyses* from Lovibond. [Online] Available from: www.lovibondwater.com/News/balancing-the-art-of-water-testing.aspx. [Accessed 28th May 2015].

Cserfalvi, T. (2014) *ELCAD wastewater metal monitor.* [Online] Available from: www.gold-aquapacific.com/ELCAD.html. [Accessed 14th December 2014].

DTK (2015) Rapid *Legionella* test kit. [Online] Available from: www.droptestkits.com/microbiology/universal-rapid-legionella-test-kit/. [Accessed 20th July 2015].

Eclox (2015) *Eclox Rapid Response Test Kit from Hach.* [Online] Available from: www.hach.com/hach-eclox-rapid-response-water-test-kit/product?id=7640273470. [Accessed 28th May 2015].

Hach (2015) *Color chart for hydrogen sulfide test kit, Model HS-C* by Hach. [Online] Available from: www.hach.com/hydrogen-sulfide-test-kit-model-hs-c/product?id=7640219546. [Accessed 28th May 2015].

Indigo (2015) *Test strips for pH measurement from Indigo Instrument.* [Online] Available from: http://www.indigo.com/test_strips/ph-litmus/#.VbIPdflW-88. [Accessed 28th May 2015].

Junior LB (2015) *Junior LB 9509 Portable Tube Luminometer* from Berthold [Online] Available from: www.berthold.com/en/bio/portable_tube_luminometer_Junior_LB9509. In the courtesy of Berthold Technologies GmbH. [Accessed 14th March 2015].

Lucetta (2015) Lucetta luminometer from Lonza. [Online] Available from: www.lonza.com/products-services/bio-research/cell-culture-products/mycoplasma-detection-and-removal/lucetta-luminometer.aspx. [Accessed 14th March 2015].

Mosselmonitor (2015) [Online] Available from: www.mermayde.nl/mosmonfunc.html; www.mosselmonitor.nl/html/Nederlands/detectie%20limieten.html; www.mermayde.nl/mosmonoptions.html. In the courtesy of AquaDect (Brouwershaven, The Netherlands) [Accessed 28th May 2015].

Ocean Origo (2015) *Compartments of the SMHI Måseskär buoy and the SeaTramp™ buoy systems*. [Online] Available from: http://www.oceanorigo.com. Products. [Accessed 14th March 2015].

Potatech® (2015) *Portable Water Quality Laboratory* from Wagtech. [Online] Available from: www.palintest.com/products/potatech-intermediate-portable-water-quality-laboratory. [Accessed 28th May 2015].

Potatest® (2015) *Portable Microbiological Water Quality Laboratory* from Wagtech. [Online] Available from: www.palintest.com/products/potatest-rapid-response-portable-water-quality-laboratory. [Accessed 28th May 2015].

SciTox (2015) *Alpha for toxicity measurement in wastewater*. [Online] Available from: http://www.scitox.com/products.php. [Accessed 28th May 2015].

System SURE Plus (2015) *SURE Plus portable luminometer* from Analytika High Quality Scientific Equipment. [Online] Available from: www.hygiena.com/systemsure-healthcare.html. [Accessed 14th September 2015].

Smart Line TL (2015) *Portable tube luminometer – Smart Line TL* from Titertek Berthold. [Online] Available from: http://www.titertek-berthold.com/smartline-tube-luminometer_19.html. [Accessed 28th May 2015].

TOX (2015) *Detection of Water Contamination* by bbe Moldaenke. [Online] Available from: http://www.bbe-moldaenke.de/en/products/toxicity.html. In the courtesy of bbe Moldaenke GmbH. [Accessed 3rd December 2015].

Chapter 4

In-situ and real-time measurements for effective soil and contaminated site management

K. Gruiz,[1] É. Fenyvesi,[2] M. Molnár[1],V. Feigl,[1]
E. Vaszita[1] & M. Tolner[1]

[1]*Department of Applied Biotechnology and Food Science, Budapest University of Technology and Economics, Budapest, Hungary*
[2]*CycloLab Cyclodextrin Research & Development Laboratory Ltd, Budapest, Hungary*

ABSTRACT

In situ measurement techniques have become indispensable tools in soil and contaminated site investigation because they enable the collection of real-time information. This provides increasingly efficient management of soil, groundwater and contaminated sites. Early warning and rapid intervention, e.g. risk reduction of endangered or damaged soil and groundwater, can be promptly implemented when up-to-date information on the state and activities of the environment's living and non-living parts is available.

In situ soil investigations can equally be applied to small sites, watersheds or global areas. Sampling, sensing, data collection and the evaluation methods that fit best to spatial and temporal requirements can be chosen from a rich reservoir of methods.

This chapter gives an overview of *in situ* methods and tools for surveying and monitoring soils and contaminated sites. Special measurement methods and devices, including sensors and rapid test kits, and portable and handheld field equipment, are introduced in detail. The detection of organic and inorganic contaminants, as well as measuring adverse effects, are discussed. In addition to methodical considerations, commercially available tools and devices are also described.

I INTRODUCTION

Spatial heterogeneities and the dynamic nature of soil processes make soil monitoring a demanding task. In addition to the two or three physical phases of the soil matrix that strive for an equilibrium – which in itself is difficult to comprehend – soil monitoring also covers the biota and uses all the methods and tools of atmospheric and aquatic monitoring. Soil can be characterized by the interactions between physical phases, chemical species and their interaction with the biota. Bioavailability of nutrients and toxicants and the physico-chemical and biological changes in reaction to external and internal conditions (temperature, humidity, pH, redox potential, etc.) have significant influence on soil properties and functions.

A survey can characterize the effect of weather conditions, natural and anthropogenic land uses and the processes occurring on the soil surface and in the subsurface. In addition to the complex and unclear baseline, contaminants and deteriorating effects create new interactions and shifts in the equilibrium. In conventional site investigation, sampling, packaging and delivery to the laboratory for analysis result in significant delay. It may be that the situation has completely changed by the time the final analysis results are available and since the sampling date. Thus the application of risk management measures based on the delayed information may not be relevant for the actual situation.

In situ **site investigation** provides a better fit to the specifics of the site and greater flexibility in field work compared to laboratory-based solutions. Systematic planning greatly increases the efficiency of the site investigation and characterization. The concept of *in situ* site assessment differs from that of the conventional procedure in terms of analysis, decision making and control/corrective actions. Figure 4.1 illustrates three typical cases: i) remote sensing for site monitoring; ii) *in situ* measurements providing real-time data; and iii) laboratory-based analyses.

–   Remote sensing enables continuous measurement and arbitrarily frequent sampling. The evaluation can be made immediately and the results are available without delay. Consequently, immediate or even automatic decision making is possible. The decision may require further investigation or risk reduction. Default algorithms or individual solutions (green line) can be used to specify the necessary intervention.
–   *In situ* measurements can use automatic or manual sampling of arbitrary frequency. The result with little or no delay makes immediate decisions and interventions possible (violet and green lines).

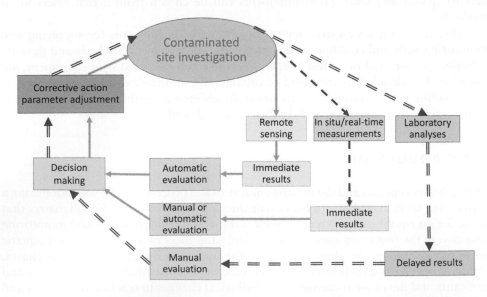

Figure 4.1 Comparison of laboratory analyses, *in situ* assessment and remote sensing in site investigation.

– Site investigation based on laboratory analysis works with infrequent manual sample collection, packaging & transport, with delayed laboratory analysis, resulting in further delay in getting the results and making decisions. It is typical that one workflow cannot satisfy the aim of the assessors: based on the first round of information, a second and third round should in most cases be planned and executed at different time points (double violet lines).

2   *IN SITU* AND REAL-TIME MEASUREMENT TECHNIQUES FOR SOIL AND CONTAMINATED LAND

In situ site investigation provides real-time and real-space information, but it is accompanied by significant uncertainties due to spatial and temporal (i.e. seasonal) heterogeneities typical for soils. Information acquired *in situ* can be best utilized for:

– The assessment and monitoring of the surface and subsurface as well as the terrestrial ecosystem including soil biota;
– Contaminated land investigation;
– The follow-up of *in situ* soil and groundwater remediation technologies.

Soils and sediments accumulate soil nutrients and contaminants due to their high sorption capacity and the partition of elements and, inorganic and organic compounds between the liquid and solid soil phases. The high sorption capacity of the solid phase is beneficial from the point of view of groundwater and soil water. It is also beneficial for the protection of soil-living organisms from contaminants. However, stressors or harmful chemicals in the solid environmental phases create a long-term contaminant source which may lead to chronic risks and chemical time bombs. It represents a latent risk, which may remain undetected under normal circumstances for a while, but which may suddenly emerge due to physical, chemical or biological mobilizing processes or as the consequence of the deterioration of soil sorption capacity.

In situ measurements are based on the signals of sensors placed either remotely (on or near the soil surface) or into deeper layers for detecting physical, chemical or biological signals and changes in the gaseous, liquid or solid phases. Sensors can be placed into the three-phase soil (also called unsaturated soil or the vadose soil zone) or in the two-phase (saturated) soil, i.e. under the water table. Thus the sample is minimally disturbed and maintained in its environment. Other types of *in situ* measurements are based on the removal of the sample (soil gas, groundwater, soil moisture or soil solid). On-site analyses of these separated/removed samples performed in mobile laboratories or by portable/handheld devices may increase the efficiency of site investigation significantly compared to laboratory-based methods. Decontextualized samples of course cannot give information on the context and do not reflect realistically those transport and biological processes which are linked to the surroundings. The sensors emplaced in the soil can be operated continuously or intermittently and arranged on-line or off-line. The separated/removed samples are analyzed typically intermittently and off-line. A sample-flow e.g. a side stream from extracted groundwater can be measured continuously and on-line.

A sensor emplaced either a few centimeters or several hundred meters deep can:

- Detect or measure the signal inside the soil;
- Forward the signal to a data logger on the surface via wire or telecommunication (telecommunication works only within short distances inside the soil);
- Automatically forward the data/information to the central laboratory for processing.
- As an alternative, it is possible to separate the meter on the surface from the sensor and to connect it intermittently to the stably emplaced sensors. Such a meter can serve as a multisensory system.

Another scheme moves the sample to the surface and analyses it on site. Continuous sampling of the gas and liquid phases generates a flux that can be channeled into a flow-through measuring device. Discrete, removed samples can be measured by the rapid and mobilized versions of the laboratory methods, including physico-chemical analysis and biological or toxicological tests.

Emplacement of subsurface sensors and samplers is a special task which, depending on the depths and the targeted sample's physical state, may be done from the surface by burying, pushing (direct push), or by using existing wells. Pushing of sensors or samplers may be executed by hand to a maximum of a few meters, or by power tools to greater depths.

Sampling is a crucial step in soil investigation. The spatial and temporal variations in soils are extremely high and one has to decide first if the aim is to gain a characteristic value (e.g. concentration) at a certain point (defined in space and time) or a spatial/temporal average (a representative mean). After gaining this information, one can create the sampling strategy and plan the best sampling method. Time series, cyclic variables, and gradual changes may increase the statistical quality of the results. The type of statistics for the evaluation of the results should be harmonized with the sampling plan.

Conventional soil sampling is highly destructive. Some soil characteristics, soil structure, aggregate structure, pore distribution and other micro-scale structures and micro-scale habitats can be investigated and characterised exclusively in undisturbed soil. Quasi-undisturbed soil can be acquired by core sampling and careful handling of the core so as not to lose the original structure and, if required, to keep the original microbiota in the soil. The context – the former contacts and interactions of the removed soil volume with its surroundings – are terminated by its removal from the original place, so the situation can never be the same as it was before removal. The investigation of a certain soil volume together with the surrounding matrix is possible by using non-invasive sampling and sensing techniques, with very little artificial impact and without sample removal. This kind of sampling covers the following.

- The use of absolutely non-invasive, remote or proximal sensors that detect magnetic, electromagnetic, seismic or radio waves.
- Minimally invasive electric or optical microsensors, which can be placed into the soil without significant disturbance to the investigated soil.

If larger sensors or measuring devices are deployed in the soil, the investigators should expect disturbance by the deployment and significant interactions between the soil and the measuring device.

Table 4.1 In situ site investigation: sampling and detection.

Soil phase	Remote sensing	Sensor deployed by direct push	Sensor deployed into wells	Sensor combined with deployed sampler	Sensors placed in gas or groundwater side stream	Analyses/tests on removed samples	Other methods/analysis
Soil gas	Transported from soil to air	In the vadose zone	In gas exhaust wells	Passive and active samplers	In gas exhaust, side stream or recycled gas flow	Gas exhaust by pumping	Transportable device or in box, mobile or remote laboratory
Soil moisture	Content in surface soil	In combination with passive samplers	Equilibrium vapors in wells	Passive and active samplers and lysimeters	Vapor exhaust or side stream	Vapor exhaust by pumping; solid soil sampling	On-site applied rapid tests or laboratory moisture analysis
Solid soil unsaturated	Surface characteristics	For equilibrium moisture characterization	In vapor exhaust wells	Passive gas and moisture samplers	In gas or vapor exhaust, side stream or recycled gas flow	Gas and vapor exhaust by pumps; disturbed/undisturbed solid soil sampling	On-site applied rapid tests, transportable devices, mobile or remote laboratory analysis of phases or the soil as a whole
Solid soil saturated	Not possible	For soil characterization	Equilibrium water	Passive and active water collectors and lysimeters	Extracted, side stream or recycled ground water	Water sampling by pumps solid sampling by core sampler	On-site rapid methods, mobile or remote laboratory analysis of the phases or the whole
Ground water	Not possible	For groundwater characterization	For groundwater characteristics	Passive and active water collectors	Extracted, side stream or recycled groundwater	Surface or submersed pump, flow split	On-site applied rapid methods, mobile or remote laboratory analysis
Biota	On the surface	Presence, activities, products in the vicinity of the sensor	Presence, activities products in the ground water	Optical sensors and image analysis, e.g. pitfalls with camera; biosensors for products	Presence of activities and products of living organisms in flow-through cells measured by biosensors	Presence of organisms, activities or products in water and soil, box, mobile or remote laboratory	Assessment of the biota on the site or in laboratory from samples

2.1 Soil investigation in endangered and contaminated land

Soil survey is the process of soil characterization regarding its formation, geological and geochemical properties, and structure. The results enable classification of the soil and preparation of soil maps covering soil type, coverage, land uses, etc. Contaminated land management may rely on soil maps – for example the diffuse pollution of watersheds – when the large size of the area and no need for high resolution justify it. Smaller sites polluted from point sources as a rule need higher resolution assessment.

Site investigation is generally a simple or tiered one-off process, while site monitoring is based on a time series of measurement results. The measuring principles may be the same but the equipment and its installation and design may differ – e.g. mobile, handheld or stationary. In addition to soil characteristics and contaminant content, remediation technology monitoring focuses on technological parameters such as aeration rate; groundwater level; soil gas and groundwater fluxes; amount/concentration of additives, and their fate and behavior in soil.

2.1.1 Contaminated site investigation

In situ measurements for contaminated site investigation – as explained in Figure 4.1 – are crucial in acquiring immediate information to allow a dynamic site assessment, and decision making. The results of *in situ* and on-site measurements – using *in situ* emplaced sensors, portable devices and mobile laboratories – can be considered as real-time information and applied to on-site decision making as part of a dynamic work strategy. *In situ* measuring methods are used either for one-off site investigation, or continuous environmental monitoring, or for remediation technology monitoring. The difference in these applications is generally not in the sensor or the analytical technique but in the mode of emplacement into the soil and the implementation and maintenance of the measuring system.

2.1.2 Soil remediation process control

In situ and, if possible, online monitoring of soil remediation technologies typically apply to small contaminated sites. The measuring methods applied for assessing soil phases are similar to those used for air and water. However, the emplacement of the sensor or the monitoring device into soil needs additional technologies and the interpretation of data is different. Its specificity lies in the dominance of the solid phase in the soil and the difficulty of accessing groundwater and soil air. The measuring and sampling devices may need to be deployed at different depths depending of the location of contaminants, the water table, or the relevant redox potential.

Remote sensing of the soil surface (see also Chapter 2) and real-time – even online – sensing of groundwater or soil gas in wells, or in extracted groundwater or soil gas flows, are becoming increasingly widespread in soil remediation practice. Direct push technologies for the subsurface deployment of sensors are also emerging techniques in soil engineering.

Real-time and online monitoring can be applied either for *ex situ* or *in situ* soil remediation, for soil characterization and process control. *Ex situ* soil remediation is executed on excavated soil after placing it into a closed or open reactor or forming a

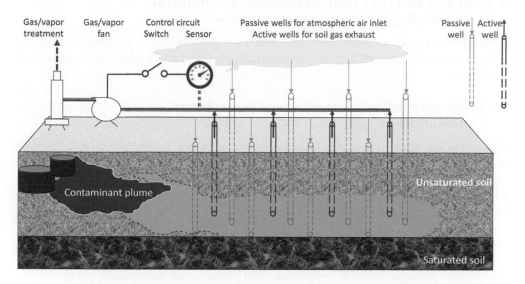

| Gas/vapor treatment | Gas/vapor fan | Control circuit Switch Sensor | Passive wells for atmospheric air inlet Active wells for soil gas exhaust | Passive well | Active well |

Figure 4.2   Monitoring of *in situ* bioventing of the three-phase soil. CO_2 and/or O_2 contents are measured in the exhausted soil air. Volatile contaminant content can also be monitored in this way.

simple heap from the soil material. The treated soil remains in place when applying *in situ* soil remediation, or to be exact, the solid phase of the soil remains in place, the gas and liquid phases can be removed, their flow can be controlled and partly or fully recycled. Solid movement and mixing is possible in *ex situ* treatment systems. The stationary or moving nature of the gas, liquid or solid in the course of remediation may determine their sampling. Samplers can be placed into the main flow or in a side stream in a fixed or removable mode. Design of the sampler can be passive, static or flow-through.

Online monitoring of contaminated soil remediation can be applied to the gas and vapor phase in three-phase soils, and to the water phase in two-phase soils. A schematic example is introduced in Figure 4.2, which shows the monitoring of *in situ* bioventing. Gaseous microbiological products and volatile contaminants can be analyzed from the extracted soil gas/vapor. Information on the solid phase is typically acquired indirectly via the gas or liquid phase results, based on estimated partitions between phases. Solid phase characteristics significantly influence the results of the analyses of the gas and liquid phases, so the measurements should be calibrated to the analyzed soil type. The equilibrium and the kinetics of the transport processes are primarily influenced by the sorption capacity of the solid phase. The same signal size may indicate different moisture or contaminant content in a sandy or a loamy soil.

Sensor technologies are the best methods for process monitoring during soil remediation – similar to other technological processes. Some remediation technologies, e.g. those based on natural attenuation, may take years, so monitoring should also be performed for years. Sensors can be emplaced and operated with minimal supervision and used in long-term subsurface applications.

2.2 *In situ* and real-time geotechnical, chemical and biological soil characterization

Measurement methods applied to soil – either for general characterization or for contaminated soil – are classified as geotechnical, chemical and biological/toxicological methods.

2.2.1 *Geophysical soil properties and geotechnical soil investigations*

Conventional geophysical and hydrogeological methods for the characterization of soil physical properties such as particle density; shape; size and size distribution; pore volume; and specific surface area are limitedly applicable *in situ*, so the more complex soil structure; soil water content and water characteristics; and mechanical and hydrologic behavior are measured instead.

Soils structure is a commonly used term for overall physical soil characterization. It refers to soil resistivity to environmental and anthropogenic stresses, sustainable soil uses, and remediation possibilities of soil and groundwater. In addition to the above-mentioned physico-chemical properties, soil structure is determined by secondary aggregate formation and stability and by several biological factors. Soil structure is a dynamic phenomenon and greatly influences function and activity, nutrient storage and cycling, as well as water-holding capacity of the soil. This explains why soil structure is the target of many conventional and innovative soil characterizing methods and models.

The composition and the ratio of clay and silt particles, and the organic matter content together with the effects of soil-living micro-organisms, plants and soil-dwelling animals determine the structure of microaggregates and aggregates, as well as other morphological and hydrologic properties of the soil such as bulk density, porosity, pore size distribution, and its resulting water retention and hydraulic conductivity. These characteristics can be best measured in undisturbed soil, and with *in situ* methods in particular.

Mechanical stress has a deteriorating effect on soil. Stress resistance, or the strength of the soils, mainly depends on their physical characteristics such as particle size distribution, bulk density, pore size distribution, pore continuity, and water content. This is influenced by the type of clay minerals, organic matter/humic substances which soil contains and the stabilizing effect of plant and tree roots. The stress is proportional to the pore air and pore water pressure. The indicators for mechanical stress are bulk density; soil consistency (liquid limit, plastic limit); penetration resistance; compressibility; shear strength and tensile strength; and the rheological properties (typical for colloidal materials) characterized by the parameters of shear modulus, plastic viscosity and yield stress. Several *in situ* methods are available for measuring static and dynamic stress in soil. The response on a vibratory load measured by pressure cells, seismographs or a miniaturized shear cell and several other innovative methods are coupled to direct push technology. The interactive nature of direct push technologies enables stress mapping of subsurface soils.

Soil water content is an essential characteristic in soil that influences soil structure, and determines water potentials and fluxes; water cycling; solute transport; the element cycles; and the life and activity of the soil living biota. Soil water fluxes, water retention, hydraulic conductivity of the saturated and unsaturated soil depend on soil hydraulic resistivity and capillary structure.

Groundwater levels and subsurface water flows determine the water quantity and quality at local and watershed scales. Water flux in soil is closely related to nutrient and contaminant transport by infiltration, evaporation, capillary rise from the water table and the groundwater flow. Water movements in saturated and unsaturated soils are primarily determined by the hydraulic conductivity (a function of soil permeability and fluid density). Under saturated conditions the flow can be considered as steady-state, whereas under unsaturated soil conditions as transient. Advection, dispersion and other transport processes are described by equations and transport models based on these characteristic (see Radcliff & Simunek, 2012).

Non-invasive technologies are widespread for profiling and mapping the soil sub-surface using magnetism; the frequency or time domains of different electromagnetic waves; resistivity; radars; seismic reflections; or microgravity for indicating areas of less dense materials and flow paths. Two-dimensional and three-dimensional imaging techniques are highly adaptive and some of them can be applied *in situ*.

Direct push technologies are used for moderately invasive exploration and sampling of the subsurface and groundwater (discussed in detail in this and Chapter 5).

2.2.1.1   In situ *applicable geotechnical sensors for subsurface soil*

The ideal sensors are miniaturized, smart sensors: a chemical/biochemical laboratory on a chip for *in situ* monitoring and characterization of the vadose zone, the ground-water or the saturated soil zone. The purpose of using sensors may be environmental monitoring or technology monitoring – typically before and during the application of soil remediation. Development of such sensors has been ongoing for at least thirty years, but the requirements and the predicted developments (DOE, 2001) have still not been fulfilled. The ideal sensors are injectable or emplaceable by new, slightly invasive deployment methods, e.g. microinjection. In addition to the new sensors, new evaluation and interpretation methods are necessary to create useful site-specific information from the measured data.

Geotechnical sensors can provide information about the physical properties of the subsurface environment such as density, thickness, resistivity and microgravity of layers of soil or sediment. Sensors can provide information about stratigraphy, estimate depth to groundwater, or approximate hydraulic conductivity, etc. Pore pressure transducers, geophones, accelerometers, settlement monitors, inclinometers, etc. are commonly used in geotechnical fieldwork.

Cone penetrometer technology (CPT) and direct push technologies are closely connected. Cone penetration testing applies a hydraulic ram against the static reaction force (i.e. the 10–30 tons mass of a vehicle) to advance steel rods into soil. CPT directly measures the force needed to push through different subsurface soil layers. The electric CPT cone measures the tip resistance (against pressure), the local sleeve friction and the deviation from the vertical axis in two directions. Pressure sensors apply pressure to the soil or sediment. The resulting resistance to the probe is measured to provide information about physical properties of the material which the probe is pushed through. Pressure sensors can measure lithostatic pressure (cone penetrometer) and hydrostatic pressure (pore pressure transducer).

The output signal from the cone is voltage. The analogue signal is transferred from the cone to the surface by cable (after amplification) to the data logger where it is converted into a digital signal, which is then processed by a computer program. Cones typically have a $10-15\,cm^2$ cross-sectional surface. Measurements can be continuous or intermittent.

CPT is often combined either with sensors and analytical devices or samplers, typically with a battery of geotechnical sensors for measuring tip resistance, hydraulic conductivity, sleeve friction, DC resistivity, and pore pressure. Acquiring several measurement end points makes it possible to cross-check different geotechnical data. The different end points can be measured continuously and simultaneously.

Percussion hammers (PH) combine the static reaction force (generated by the mass of the vehicle) with a percussion hammer to advance steel rods and either a sampler or an analytical device. A hammer ensures quicker progression with less effort and better access to more remote places. A PH device is much smaller than a CPT, and handheld versions are also available. Normal hammers can also be equipped with probe rods by using adapters, so the same sensors and devices can be used as for CPT. Percussion hammer systems are capable of directional drilling into the subsurface at up to 37 degrees.

Detailed information on direct push tools can be found on the website of US EPA (Clue-in DP, 2015). Trading companies which market direct push platforms and equipment such as Geoprobe (2015), Fugro (2015) or Gouda Geo (2015) also provide information on measurement principles and applicability and installation of the equipment.

A piezocone penetration test (PCPT) is an electric CPT cone equipped with a pressure sensor for measuring pore pressure *in situ*. PCPT measures the transient pore pressure generated during penetration as well as hydrostatic pore pressure. The dissipation of the transient pore pressure is measured by interrupting the penetration. The *in situ* coefficient of consolidation can be estimated from the result of the dissipation measurement.

Seismic cone penetration testing (SCPT and SCPTU) applies seismic sensors to detect reflected and refracted acoustic energy and geophone arrays to determine the time of travel. The angles of reflection and refraction are determined from arrival information. The position of geologic units, layers of different density or buried objects can be determined from this information. SCPT may be used to measure *in situ* compression and shear wave velocity profiles for layers of known depth and thickness. The P-wave (primary) and S-wave (secondary) velocities (calculated from arrival time) are directly related to the soil elastic constants: Poisson's ratio, shear modulus, bulk modulus, and Young's modulus, which are important parameters in foundation design.

A magnetometer cone measures the strength and/or direction of the magnetic field. The earth's magnetic field varies spatially due to heterogeneities of rocks and the presence of ferromagnetic materials causing, thus detectable disturbance in the magnetic field. Magnetic anomaly is measured by a magnetometer. The electronically amplified signal is collected by a data logger and forwarded to a computer for processing. Magnetometers are typically used for detecting mineral geological structures; buried or submerged objects, unexploded bombs and detecting the position of power cables, the depth of foundation piles, and the length of sheet piles.

An electric vane tester can be used to measure the shear strength of a soil. It is a commonly-used method to determine the structural strength of soil. The electric vane tester is applicable for measuring undrained and remoulded shear strengths of saturated cohesive soils. It also gives a good indication of the over-consolidation ratio, the stress-strain relationship and the sensitivity of cohesive soils. The vane tester is generally combined with a CPT or PCPT. After pushing the rod supporting the vane to the test depth, the vane is rotated from the surface at a prescribed rate and maximum torque required to reach soil failure is measured. At least three complete turns are necessary for measuring the residual torque (Pagani, 2015; Gouda Geo VT, 2015).

Geoprobe uses the name *Direct Image*® *technologies* and provides the following sensors for soil mechanics and physics tests: electrical conductivity sensor, membrane interface hydraulic profiling tool, hydraulic profiling tool, cone penetrometer, and pneumatic slug test kit. For the detection of volatile contaminants the geotechnical tools are complemented by membrane interface probe (MIP), and special low level MIP if necessary (Direct Image, 2015).

2.2.1.2 Subsurface exploration by direct push technologies

In addition to geotechnical sensor-based techniques, several other measuring methods have become available by attaching sensors and samplers to the direct push equipment to collect geophysical and geochemical information.

Miniature video imaging tools can be coupled to direct push probes to characterize lithologic properties and fracture patterns or simply to visually inspect CPT and the attached probes or samplers.

A *wireless sensor network* is more suitable for long-term subsurface monitoring, in contrast to currently-used solutions that forward data from the geotechnical instruments to local data loggers by electric wires/cables. A wireless network is not restricted to a one-off contaminated site investigation or the follow-up monitoring of remediation but can be used for long-term monitoring based on locally or remotely collected and processed data. Wireless systems enable a large number of sensors to be linked to a data logger, and the measured data can be collected and processed remotely. As the signals of sensors at greater depths cannot directly be transmitted to remote receivers, cables are used to connect the subsurface instruments to a transmitter which is placed on or near the surface and forwards the signal to the receiver. The distance between the transmitter and receiver specifies whether the communication is performed by SMS or Internet. Wireless data acquisition may be 25–30% cheaper than cable-based systems and can be applied to long-term subsurface monitoring and geotechnical health inspection of buried pipelines or subsurface structures.

Direct push technology uses a truck-mounted or anchored cone penetrometer system to directly push the instrumented probe into the ground for soil physico-chemical characterization and contaminant analysis. The truck may be a CPT penetrometer rig based on a truck chassis. In this arrangement the penetrometer pusher is installed in the truck's center of gravity. The truck's own large weight is able to provide the required penetrative force (Gouda, 2015). Before starting the test, it is imperative to level the truck horizontally. The CPT functionalities are hydraulically powered, operated by a hydraulic pump. It is equipped with an operating lever for the up-and-down movement of the penetrometer pusher. The system's electronics control the entire CTP and the combined sensors and devices. The system is known as SCAPS: Site Characterization and Analysis Penetrometer System. In addition to a wide selection of sensors and samplers attached directly to the CPT, portable and transportable instrumentation may also be included (A.P. van den Berg, 2015).

The *Geoprobe (2015) anchor system* can substitute the heavy trucks that provide suitable reaction force for pushing with a direct push machine. The three-, five- or seven-anchors system holds down a foot anchor bridge, and the probe foot slides underneath the foot anchor bridge where it is held during pushing. The anchors are screwed into the ground before testing and screwed out after testing. Soil type and the anchor depth are important characteristics of the system (Geoprobe Anchor, 2015).

Contaminant-specific or contaminant group-specific sensors such as MIP, LIF, FFD, XRF, LIBS sensors can be attached to the probe and thermal desorption or Hydrosparge™ samplers. These VOC samplers consist of a nose cone with a sampling chamber that can be opened to collect a soil sample. The sample is heated in the chamber and VOCs are volatilized and transported by a carrier gas flow to the surface, where they are analyzed by a portable mass spectrometer. The hydrosparge sampler inserts a sparge into the groundwater using helium gas, then purges the VOCs from the water and transports them to the portable mass spectrometer on the surface.

2.2.2   Chemical soil properties and in situ analysis methods

In situ chemical analysis of soil and groundwater can detect and identify natural structural and functional components or stressors/contaminants in soil and groundwater. *In situ* methods can be based on the mobile version of conventional chemical analyses, or electrochemical sensors. Sensors in direct contact with soil are deployed/pushed into the soil, while remote or proximate sensors detect the signals without direct contact with soil. The modifications of the conventional chemical analytical methods include: complete mobile laboratories, portable equipment, handheld instruments and rapid, easy-to-use analytical kits.

Non-invasive sampling and testing can be performed by remote sensing, practically non-invasive ones by miniaturized sensors, and minimally invasive ones by conventional sensors. *In situ* applied direct push measuring technologies (MIP, etc.) and direct push sampling are less invasive than conventional drilling boreholes and taking soil cores or disturbed soil samples.

The basic geochemistry and chemistry of soil is rather complex in itself, given that the sources, forms, mobility, and bioavailability of elements and compounds – as well as the chemical processes, material transports and reactions between the various forms – lead to a continuously changing dynamic system that is hard to map. Contaminants or technological additives may change and further complicate the original situation and the interactions of molecules with each other and with structural building blocks or biological systems. These changes may influence soil structure and function significantly.

The organomineral complex of the soil matrix is the sum of two complex systems. The inorganic particles in soil are made of rocks that have broken down: small inorganic molecules and ions, and natural and contaminating metals and semi-metals. The organic matter of the soil comprises living and non-living organic matter, such as humus, consisting of several fractions with different molecular masses and chemical properties. Organic *soil contaminants* also tend to be attached to this part of the soil.

Information on the carbon forms and the carbon cycle is necessary to follow global carbon cycles and carbon sequestration, humus quantity and quality, which determine soil behavior and functions. The organic material content of the soil is generally measured as total organic carbon (TOC) directly – e.g. after complete or incomplete combustion – as well as by approximation methods such as UV and IR spectroscopy. The optical sensor for laser-induced fluorescence (LIF) is mainly applied to organic contaminants such as mono- and polycyclic aromatics and NAPLs. The reduced size of mass spectrometers enables their field use by constructing portable – or at least transportable – versions. Unfortunately, these simplified methods cannot reflect the complex chemical composition of the soils' organic matter and humus content. On the other hand, the rapid methods make it possible to follow up relative changes and detect unacceptable differences compared to a relevant reference.

Rapid, semi-quantitative analytical kits based on colorimetry may apply vials or multi-well plates (for liquid phase reactions between the sample and the reagent) or strips (to be immersed into the sample or place the sample on the strip infused with the reagent) and visual or colorimetric evaluation. Such kits are available for many of the soil and groundwater components and contaminants. One typical group of these rapid methods measures environmental parameters such as pH and dissolved oxygen in groundwater; another group of methods plays a role in assessing microbiological carbon, nitrogen, sulfur, and phosphorus cycles. The third group consists of toxic metals and organic contaminants. Kits are available for rapid biochemical, enzyme and immunological analysis, providing *in situ*/on site and close to real-time information. Most of these kits do not provide precise analytical results, but they make the exclusion of negative cases possible in a cost-efficient way. The positive ones should be further analyzed by more precise analytical methods. Some specific enzyme, immune, and DNA methods are introduced in Section 6 of this chapter.

The chemical composition of surface minerals can be identified by remote investigation based on satellite or airborne information. The information can be used for mapping and differentiating the mineral composition of large areas by multispectral and hyperspectral analyses.

Tracing radioactivity is a typical *in situ* assessment task. The connecting data processing may take place *in situ* or remotely. Airborne radiometric data (e.g. gamma-ray flux) can be combined with digital maps for the identification of the location of radioactive pollution (leaking storage bunkers, reactors, or illegal use) or other radioactive sources.

Elements, including toxic metals, can be detected and quantified by the hand-held, multi-element XRF (X-ray fluorescence) device (US EPA Method 6200, 2007). It is ideal for an *in situ* assessment of contaminated sites and the application of the Triad approach for exploration of contamination, and for mapping or delineating contaminated areas. Its precision when measuring soil without sample preparation does not match that of sophisticated laboratory equipment, but the advantages of rapidity, repeatability and high measurement frequency overweigh any loss in precision. These capabilities provide greater advantage in the case of soil heterogeneities than does laboratory preciseness. The greatest benefit is that its field use allows the application of the Triad approach. Section 9 of this chapter describes an application at a mining site.

Soil and sediment contaminants can be analyzed *in situ* based on their partition between gaseous, liquid and solid phases of soil or sediment. Knowing that every chemical substance is partitioned between the environment's physical phases, the detection of certain contaminants in the gas or water phase indicates their presence in the solid phases of soils, sediments or wastes. The mobile environmental phases have an averaging role, eliminating small-scale spatial heterogeneities of the solid phase. Indirect results of soil gas and soil solution are indicative for soil solid. The precise concentration can be determined when the contaminant and its partition are known. The soil type also influences the calibration of the instruments analyzing soil gas or liquid phases.

Portable instruments (UV, VIS, IR, NIR, GC, MS and GC/MS) are used for *in situ*/on-site detection and analysis of organic pollutants, including volatile organics (VOCs), hydrocarbons, chlorinated hydrocarbons, explosives and several other pollution indicators. Portable instruments can measure discrete or continuous gas/vapor, and liquid or solid samples from the soil surface, the vadose and the saturated soil.

The combination of drilling rigs and *direct push equipment* with contaminant-specific sensors enables *in situ* measurement in undisturbed or minimally disturbed samples. For example, a cone penetrometer combined with a photoionization detector can measure VOCs at different depths of the soil without separating/removing the solid soil sample from its context, which is a prerequisite for gaseous phase analysis.

Volatile soil contaminants causing air pollution can be monitored by *open path technologies* that measure the concentrations of chemicals along an open path in the air. A concentrated beam of electromagnetic energy is emitted to the air, and its interactions with the chemical components of the atmosphere can be followed by LIDAR (Light Detection and Ranging), FTIR (Fourier Transform Infrared Spectroscopy), Raman spectroscopy or TDL (Tunable Diode Laser) absorption spectroscopy.

Sensor technologies have been developed especially for soil and sediments based on physical, chemical, biochemical, immunological, DNA, etc. interactions (see the description of the principles in Chapters 2 and 3). Special attention is attributed to the protection of sensors from the soil environment and their deployment in the solid matrix, sometimes as deep as 100 meters or more. The selection of the housing materials, the deployment technology and long-term protection against chemical and microbial corrosion are prerequisite of their safe and long-term application. Sensors provide data in real time, can be operated remotely, and may even acquire data continuously, so they are also able to detect short-term and unexpected changes. Sensors can collect a large amount of data, and data loggers and processors forward them in suitable forms directly to the user. Sensors placed directly in a well can reduce monitoring costs significantly. Multi-sensor networks are ideal for identifying trends in time and space. Sensors are discussed in detail in Sections 3; 4.1; 4.4 and 5.4 of this chapter, while only some sensor types – those suitable for the detection of chemical and biological stressors in soil gas and groundwater – will be briefly discussed here.

Ion selective electrodes are the oldest and most widespread type of sensor. As an example, there is the potentiometric sensor system based on potassium ion-selective electrodes which was developed by Lemos *et al.* (2004) for agricultural purposes. Another example is the nitrate-selective sensor for groundwater monitoring which has high importance for the monitoring of water gases (drinking water toxicity) and for the protection of surface waters from too much nitrogen (eutrophication). In the course of soil bioremediation of saturated soils polluted with hydrocarbons, the problem may be the opposite: nitrate is often the bottleneck of anoxic/anaerobic biodegradation (at 450–250 mV redox potential), where nitrate plays a duplicate role as (i) an alternative electron acceptor in bacterial respiration and (ii) an essential building block for bacterial growth. Nitrate sensors measure dissolved nitrogen by direct potentiometry. The reference electrode is immersed into a solution of a constant nitrate concentration within the sensor housing, and the ionic charge transfer between the solution and the sensing electrode is measured and converted into a concentration. This concentration is representative of the dissolved nitrate concentration of the analyzed aqueous solution. The reference electrode is built into the housing of the electrode. Several companies produce such nitrate sensors generally combined with pH, temperature, other ions and redox potential sensors (Catalog, Envco, 2015).

Several developments are available for soil gas and groundwater analyses, some of which are listed below (Ho *et al.*, 2005).

Detection of radiation. Several principles have been used and various types of equipment developed for the detection of different radiations, such as PIN diodes, thermoluminescent dosimeters, gamma detectors for isotope identification, neutron generators for nuclear material detection, Geiger counters, and CZTs. CZT, the cadmium, zinc, telluride detector is an inexpensive and sensitive detector for measuring radiation levels of gamma and neutron radiation. It is a compound semi-conductor detector that uses a wide band-gap and produces a current flow under the influence of a gate voltage upon exposure to high-energy radiation (Amptek, 2015).

Detection of metals. Nanoelectrode arrays and LIBs are the two most popular sensors for this purpose. Nanoelectrode arrays are extremely small-size arrays that measure dissolved metals. The electrodes produce current due to the application of an electrical potential.

Laser-induced breakdown spectroscopy (LIB) is applicable for the rapid analysis of metals and other inorganic components in drinking waters, surface waters, groundwater and soils. Its principle is that the laser rapidly heats a small portion of soil or water thereby generating plasma from the material in the laser beam's focal point. The plasma is then analyzed by spectroscopy.

Volatile organic contaminants can be analyzed by a wide range of sensors such as the microChemLab™ (a miniature gas chromatography system) and the chemiresistor array (to discriminate chemical classes) (Sandia micro, 2015).

Fiber-optic chemical sensors (FOCS) can be used for hydrophobic organics: an evanescent wave interacts with the matrix to be analyzed, within the penetration depth. Chemical species are preferentially concentrated from the matrix into the evanescent interaction zone. After this selective concentration near-infrared (NIR) spectroscopy is used for quantitative measurement, combined with multivariate statistical analysis.

Grating light reflection spectroscopy (GLRS) is an emerging technique for spectroscopic analysis and sensing. A transmission diffraction grating is placed in contact with the sample to be analyzed, and an incident light beam is directed onto the grating. At certain angles of incidence, some of the diffracted orders are transformed from traveling waves to evanescent waves. In combination with electrochemical modulation of the grating, the technique was applied to the detection of trace amounts of aromatic hydrocarbons (Kelly *et al.*, 2000).

Micro Hound is a complete analytical system applying diffusion-based separation for chemical pre-concentration and a miniature ion mobility spectrometer (IMS) for analysis. It has been developed for semi-volatiles, pesticides and halogenated semi-volatiles in gases and vapors (Micro Hound, 2015).

Surface acoustic wave sensors (SAW) are chemical sensor arrays measuring a decrease in the active resonant frequency of the chemicals that is related to trace mass loading on the active surface. They have been used for chemical speciation and quantification of vapors, but several other applications of SAW also exist, such as measuring ferromagnetic elements, humidity, viscosity and biological materials. Equipment and application are detailed in Section 3 of this chapter.

2.2.3   In situ *biological soil characterization and toxicity measuring methods*

Many of the biological properties, activities or products of soil-living fauna, flora and microbiota members can be measured to characterize the biological status of soil, as well as the effects of nutrients and contaminants in it.

Soil biodiversity and function, as well as their changes, may reflect the healthy or deteriorated condition of the soil, but they cannot characterize the complete soil ecosystem, the density and distribution of all species. There are two concepts to assess soil biodiversity:

1.   The first concept aims to get as close as possible to a correct species diversity value by traditional field observations or by innovative metagenomic approximation. Conventional field methods can assess macroscopic organisms; airborne or space-borne remote sensing can describe plants; and the metagenome or any of the omics can characterize the diversity of soil micro-organisms.

2.   The other concept focuses on some well-selected indicator species. A wide selection of responses – from gene level, through biochemical, physiological, and behavioral, to community level – can be applied to soil characterization. Many of these biological responses can be measured *in situ* using traditional field tests, mobile versions of laboratory tests, or innovative biosensors and microprobes.

In situ biological and toxicological assessments play an increasingly important role in soil risk characterization and management. Most soil contaminants are reactive, volatile and capable of being degraded and/or dissipate during transportation and storage. Any change in the pH and redox potential between the time of sampling and testing may significantly influence the effect of the soil on living organisms. Thus, *in situ* testing of biological effects is extremely important. The information on actual effects characterizes the bioavailable fraction of the chemical substance. Information obtained promptly on soil or groundwater toxicity enables *in situ* decision making during pollutant screening and mapping.

The same *sensors* can measure biological soil activity and toxicity, and sensors from waste water practice can be applied to groundwater to measure growth, respiration, metabolic activities, as well as their inhibition (see Chapter 3). When sensors are used in two-phase soils and sediments, they must be protected from solid intrusion. The characterization of three-phase soils is a more complex task, since free water cannot ensure the contact with sensors placed *in situ*. Thus gaseous phase metabolites are measured by gas/vapor sensors, and the solid phase components by sensors which do not need direct contact, e.g. those which detect electromagnetic waves or fields. Microsensors placed into the biofilm and sensors combined with gels; porous or capillary sorbers; lysimeters; or other special tools can ensure contact via collected water in three-phase soil. Mildly invasive methods may apply heat to evaporate volatiles and semi-volatiles or solvents (typically water) to mobilize the water-soluble and biologically available analytes.

Many *field applicable ecotoxicity measuring methods* are the modifications of original laboratory soil tests. They are extremely useful in quickly mapping the extent of bioavailable pollution at contaminated sites. These types of tests should be low-cost and rapid methods, enabling immediate decision for the next step of the site investigation. Rapid test kits use conserved test organisms and early responses, thus ensuring good reproducibility and a short response time. One example of commercially available mobile tests for field use is the Microtox®, which applies the *Vibrio fischeri* luminescent bacterium for toxicity measurement. It is a sensitive method with average time and cost requirements. Its environmental relevance may be low as it uses a marine species, but, due to its sensitivity, it can indicate the presence of toxic pollutants.

Species density and diversity are conventionally assessed *in situ*. The assessment includes direct observation and visual counting of species density and calculation of species distribution. The following conventional methods can characterize the healthy or degraded state of an ecosystem and measure adverse effects: field micro- and mesocosms; cotton strip for cellulase activity; litter bag for litter and organic waste decomposition in soil; pitfalls for trapping certain species; bait lamina for nutrient consumption intensity; avoidance tests for monitoring the escape of individuals or populations from the environment; and the observation of caged test organisms (see Gruiz *et al.*, 2015).

Conventional ecosystem assessments and ecotoxicity test methods can be applied to macroplants and the members of the meso- and macrofauna, either in themselves or in combination with molecular methods or sensor techniques. The density and diversity of the soil microflora can be best characterized by molecular methods and microsensors, detecting colors (direct optical sensing) or biomolecules (chemical sensing) and processing data to identify individual or community fingerprints.

Resistance may be triggered by soil contaminants in many species. The processes behind this trigger is the evolution of new genes or an increase in the copy number of already existing relevant genes in the metagenome. Indication of the presence and frequency of such genes may also serve as basis for *in situ* toxicity tests.

Functional tests characterize the activities of the soil microbiota. Molecular techniques are generally used to analyze the enviromics: the genome (DNA/RNA), and the products of genes or the activities of these products, typically enzymes. The molecules should reflect the interaction between the environment and the inhabiting community. Microbial energy production; respiration; nitrification or denitrification; and biodegradation or biosynthesis are governed by enzymes such as oxidases, glucosidase, saccharase, xilanase, proteases, lipases, urease, amidases, esterases, cellulases, chitinase, and arylsulfatase. Their presence, as well as a change in their volumes, may indicate stress as a result of soil deterioration or contamination. That is why they can serve as end points in monitoring and early warning.

Biosensors and microprobes can be used for continuous soil and groundwater monitoring. These tools provide real-time information on the state of the environment via the response of living organisms. In addition to plant and animal species, the microbial communities are the most important components of the soil ecosystem. Their diversity cannot be characterized by conventional methods based on counting (suitable for macroplants and animals) due to the limitations of their laboratory cultivation and microscopic identification. The characterization of soil microbial diversity by innovative, possibly *in situ*, methods may therefore play an increasing role in soil and contaminated site management. Both microbiological communities and individual species can give characteristic information on soil health or deterioration and on actual adverse effects.

In situ, real-time measurement methods using biosensors and microprobes can observe details of the microbial community in action, given that the measurements are carried out at the sub-millimeter scale. Microprobes have been developed to characterize temperature, pH and the redox system of the soil. Nitrification can be monitored by a combined oxygen and nitrogen oxide microprobe. The combination of oligonucleotides and microelectrodes may result in methods that can indicate bacterial processes such as sulphate reduction. Genetic bioindicators (i.e. bioreporter genes) create detectable products such as light emission due to the green fluorescent protein (GFP) or the natural or cloned lux gene of bacterial strains or higher organisms (see Chapter 2, Sections 4.4 and 5.2.3).

Biosensors for soil microbiota assessment are introduced in the followings.

BioSAW sensors are selective for biological stressors and pathogens based on picograms of protein detection. A biologically-active layer can be placed between the interdigitated electrodes which contain an immobilized pathogen-selective antibody. Bonds between the analyte antigen and the sensor-bound antibody cause a mass loading on the electrode surface. A similar design with DNA probes can sense messenger RNA (mRNA) in samples.

The *fatty acid methyl ester* (FAME) analyzer is another selective and sensitive biosensing device which can measure fatty acids of microbial origin in quantities of a few nanograms. The ratios of FAMEs can be used to distinguish bacteria according to gram-type, genera, or species level. The analyzer can detect toxic pathogens, food contaminants, and biological warfare agents, but it requires manual sampling.

The *μProLab acoustic sensor* has been developed for measuring picograms of proteins and peptides even after 1,000-fold pre-concentration using programmable switchable polymers and electrokinetic trapping. Its main application is the fingerprint identification of micro-organisms, typically pathogens.

Hyperspectral imaging, using multivariate analysis, can be applied at the micro- and the macroscale for soil analysis and quantitative species mapping (Ho *et al.*, 2005).

Hyperspectral microscopy works with quantitative spectral analysis of nanoscale materials that are imaged with the help of a microscope. CytoViva applies a darkfield-based microscope system. It is a spectral analysis of nanoscale samples which may be isolated or integrated in cells, tissues or other – non-living – matrices (CytoViva, 2015).

Hyperspectral microsensors, using defined wavelengths for excitation, allow spatial localization of pigments and mapping microbial communities in solid matrices. This rapid, non-invasive, long-term method can be applied *in situ* and provides a high degree of species resolution. Biomass is quantified by deconvolution into single-cell spectra. Hyperspectral microsensors can distinguish with high precision the distribution of benthic species groups such as diatoms, green algae, flagellates, cyanobacteria and various anoxygenic phototrophs in sediments (Chennu, 2010; Nielsen *et al.*, 2015).

The application of biosensors for research on benthic micro-organisms and phyto-benthos has been reported in several publications: a study of benthic boundary layers (Kühl and Revsbech, 2001), quantification of ice algal communities (Glud *et al.*, 2002), and the effect of oil contamination on the microbial mat (Benthien *et al.*, 2004).

3 *IN SITU* SOIL GAS AND VAPOR ANALYSES: SENSORS AND SAMPLERS

Understanding, monitoring, and predicting contaminant fate and transport in the unsaturated soil zone needs to be developed, as was suggested by the US Department of Energy (DOE, 2001). In contrast to groundwater, which has priority due to its importance in the drinking water supply, the vadose zone has been neglected by environmental engineers for a long time. Only agriculture has shown interest for many years. The transport of gases and vapors is playing increasingly important role in the remediation of three-phase soil.

Natural and contaminating gases and vapors in the soil are partitioned between the three physical phases, so their presence can be detected in the gaseous, liquid or solid phase. Real-time and online methods focus on gaseous and liquid phases, and some dynamic *in situ* measurements techniques also cover vapor desorption from the solid phase.

The gaseous phase of a contaminated soil significantly differs from that of non-contaminated soil due to the presence of the contaminant gases, vapors and metabolites produced by the micro-organisms interacting with the contaminants and transforming or degrading them.

Remediation technologies that apply soil gas and vapor exhaust recycling, can be monitored by placing an online measuring device into the gas flow to measure gaseous-form natural products (typically of biological origin) or contaminating/toxic gases and vapors. These gases and vapors may derive directly from soil contaminants, from biological processes, or from technological additives. The gases which are measurable online are: CH_4, Cl_2, ClO_2, CO, CO_2, H_2, H_2S, LPG (liquefied petroleum gas), natural gases, NH_3, NO, NO_x, O_2, O_3, and SO_2. The principles of the measurement methods can be catalytic combustion, other traditional detection methods such as UV, IR, FTIR, FID, PID, MS or sensor techniques – such as a hot-wire semi-conductor, piezoelectric sensor – or other sensor-based and electrochemical methods, e.g. controlled potential electrolysis or galvanic cell.

The detection of *volatile organic soil contaminants* such as volatile hydro-carbons and chlorinated volatile hydrocarbons (VOCs) needs contaminant-specific sensors or arrays. Some typical volatile soil contaminants of industrial and waste origin are: acetone, benzene, butadiene camphor, chlorobenzenes, decane, 1,4-dichlorobenzene, ethanol, ethylbenzene, formaldehyde, hexane, d-limonene, methylene chloride, naphthalene, octane, pentane, phthalates, styrene, tetrachloroethene, toluene, trichloroethanes, trichloroethene, trimethylbenzenes, and xylenes.

Vapor-detecting devices are deployed from the surface either manually by direct push or, for larger depths, through a direct push platform attaching the device to a cone penetrometer. Other detectors are used on the surface and the vapor sample is forwarded to the surface from the deep layers by a heated carrier gas.

Devices interfaced to a cone penetrometer are: acoustic wave detectors, conductivity detectors, infrared spectrometers, photoionization detectors, Raman or laser-induced fluorescence detectors. The latter two can apply substance-specific wavelengths such as the nitrogen laser for polycyclic aromatic hydrocarbons with three rings (diesel, heating oil, creosotes) or lower wavelengths (higher energy) lasers such as the neodymium laser for one- and two-ring lighter hydrocarbons (kerosene, jet fuel). The same detectors can also be applied for surface use. The use of these detectors requires the same standardized methods and quality assurance as laboratory analyses.

Gas chromatography (GC) in vapor analysis has long been the priority method. Combined with subsurface sampling techniques and the application of a MIP, GC became a popular method in mobile laboratories is used for direct push explorations. A special easy-to-transport GC instrument, the Geoprobe™ GC, has been designed to meet the needs of Geoprobe™ operators (SRI instruments, 2015). It can perform a total VOCs analysis of the purge gas during drilling and identify specific contaminants in near real-time. It is equipped with PID (photo ionization detector), and FID/DELCD (flame ionization detector/dry electrolytic conductivity detector) combination detectors (sequential arrangement), with a gas sampling valve, a compressor and an analogy output. FID responds to all hydrocarbons, while DELCD only to chlorinated or brominated compounds (vinyl chloride, DCE, TCE, PCE, and others).

Gas chromatograph with a mass spectrometer for the detection of the components of volatile soil contaminants is also a popular combination for environmental monitoring. Contaminant components are separated by a GC column and passed into the mass spectrometer (MS) through a membrane interface that is only permeable for volatile organic compounds. The miniaturized, field-portable versions opened new horizons for *in situ* real-time data acquisition by GC-MS. Both the chromatograph and the mass spectrometer are so small that they fit in the palm of the hand. Two commercial products are introduced here: the Hapsite™ and the TRIDION™ systems.

– *The Hapsite™* portable system can be used for quick identification of volatile chemicals and for characterizing contaminated sites (Figure 4.3, left). It is a quadrupole GC/MS for compound identification and quantification. In conjunction with a headspace equilibrium sampling accessory, the instrument has the capability to analyze water and soil samples. The dimensions of the equipment are $45 \times 42 \times 17$ cm and it weighs 16 kg (Hapsite™, 2015).

Figure 4.3 The Hapsite™ (2015) and the TRIDION™ (2015).

– *The TRIDION^{TM}-9* is also a portable GC-TMS and weighs 14.5 kg (with battery) with a size of 38 × 39 × 23 cm (Figure 4.3, right). It is fast, reliable, and easy to use and operates with a touch screen. The system includes a low thermal mass capillary gas chromatograph with high-speed temperature programming and a miniaturized toroidal ion trap mass spectrometer (TMS) with a mass range between 50 and 500 Daltons. Samples are injected using a CUSTODION® solid-phase microextraction fiber syringe or a needle trap (CUSTODION-NT). The GC-TMS can be accomplished by the *FUZION^{TM}* sample preparation module. TRIDION-9 is used for volatiles and semi-volatiles, explosives, and several hazardous, contaminating substances. It is ready for use within five minutes and can analyze 12 samples per hour (TRIDION^{TM}, 2015).

The development and the results from exhaustive performance testing of the new TRIDION instrument with the toroidal ion trap mass spectrometer are introduced to validate its robustness and ability to identify targeted and unknown chemicals.

Besides GC-MS, *innovative sensor techniques* have been developed, such as microrespiration tubes detecting single volatile compounds (Kaufmann *et al.*, 2005); metal oxide-based olfactory sensors, also named electronic noses (De Cesare *et al.*, 2011; Bruins *et al.*, 2009); and nanoparticle-structured sensing arrays (Han *et al.*, 2005). VOC fingerprints may play an important role in future microbial ecology research.

The electronic nose is a Surface Acoustic Wave (SAW) vapor detector which can be used as single detector, or coupled to a gas chromatograph. The application of the SAW resonator sensing element provides a highly increased sensitivity compared to conventional SAW sensors, and its combination with a GC results in high specificity at the part per trillion level in near real-time.

The portable GC-SAW version can simultaneously detect and quantify multiple chemical vapors within a single environmental sample. It is useful for trace detection of chemical contaminants, toxic agents or explosives in the field (GC-SAW, 2015). Two practical applications of this sensitive field technique to detect volatile chemical species are the analysis of volatile soil biomarkers and the detection of explosives at contaminated and abandoned sites.

Volatile biomarkers of microbiological or plant origin represent an interesting new field in soil VOC analysis. Mycorrhizal plant roots emit significant amounts of volatiles, as the interaction between root and mycorrhizal micro-organisms is mediated by VOCs. Similar to human diagnosis or the monitoring of the composting technology, the volatile biomarkers (soil volatilomics) can characterize the microbiological structure and the health or deterioration of soils (Insam & Seewald, 2010). VOC production depends on the soil-specific community composition, on nutrient and oxygen availability, and on the physiological state of the community. Under microaerobic and anaerobic conditions homofermentative processes result in a larger amount and variety of VOCs. Production and release of VOCs differ from each other because microbes live and produce VOCs in water-based biofilms, from where the release of VOCs is determined by actual partitioning. A single micro-organism species can produce 15–20 different detectable VOCs, while microbial communities produce 50–100, or even several hundreds. Most of these VOCs have not been identified to date. Among the identified VOCs are low molecular weight alcohols, aldehydes and ketones, methyl ketones, esters, acids, amines, branched-chain alcohols, straight chain aldehydes, alkyls, oximes, phenols, heterocyclic compounds, terpenes, isoprene, monoterpenes, sesquiterpenes e.g. trichodiene, furfural and similar furan compounds. Some identified species are: cyclohexanone; 1-octene-3-ol; 2-heptanone; 4-allylanisole; 3-methyl-1-butanol; propanoic and butanoic acid; limonene; geosmin; and fungistatic VOCs. Several VOCs and gaseous species are the products of biodegradation of organic compounds and contaminants.

Field explosive detection is a standard assessment task at abandoned military and mining sites. *Enhanced Spectrometry (2015)* developed *portable Raman-luminescent spectrometers* and analyzers for a broad range of applications, e.g. for narcotics and explosives. Its main advantages are that it is a rapid, handheld detector for field operation, and it gives fast results.

3.1   Chemiresistor

The chemiresistor contains a special polymer-carbon particle mixture. The dissolved form is dried onto wire-like electrodes on a specially designed integrated circuit. VOCs absorb into the polymer, causing it to swell and, as a consequence, change the electrical resistance. The process is reversible: polymers will shrink once the chemical is removed (Sandia chemiresistor, 2015).

The sensors are calibrated to provide 'training sets' for pattern recognition of various chemicals and chemical mixtures. The array has also been tested in the field during soil remediation by soil venting and air sparging.

The Sandia National Laboratories' Directed Research and Development (Sandia LDRD, 2015) project can be mentioned as an example here, as this company developed and applied microchemical sensors for *in situ* monitoring of subsurface volatile contaminants. Their microchemical sensor employs an array of chemiresistors in stainless steel housing.

3.2   MIP, Membrane Interface Probe

A *MIP* is a semi-quantitative, field-screening probe for the detection of volatile organic compounds (VOCs) in soil and sediment. The MIP technology is capable of sampling VOCs and some semi-VOCs from subsurface soil in the vadose and saturated zones. It uses heat to volatilize and mobilize soil contaminants for sampling. The soil and/or the groundwater is heated up in the surroundings of the MIP, and the volatilized VOCs diffuse (passively pass) through the semi-permeable membrane of the probe into the carrier gas which then transports the vapors to the surface for analysis by a gas chromatograph equipped with a photoionization detector (PID), flame ionization detector (FID), dry electrolytic conductivity detector (DELCD) or other contaminant-specific detectors, often a mass spectrometer (see Table 4.2).

MIP is typically used for *in situ* determination of volatile or semi-volatile contaminants such as hydrocarbons or solvents. The rapid identification and location of subsurface contaminants enables the application of the Triad approach during site investigation and subsequent remediation. The source zone identification is a crucial point in soil and groundwater remediation and is generally loaded with extremely high uncertainty. MIP was successfully applied for dense non-aqueous phase liquid (DNAPL) and light non-aqueous phase liquid (LNAPL) plume identification and for screening of chlorinated hydrocarbons (CHCs); aromatic hydrocarbons of benzene-toluene-ethylbenzene-xylenes (BTEX); perchloroethylene (PCE); trichloroethylene (TCE) and their biodegradation products; and several other volatile organic compounds (VOCs), in both saturated and unsaturated zones (see also Chapter 5).

3.3   Detectors for volatile soil contaminants

A wide variety of *in situ* analytical solutions are available for soil organic gas and vapor analyses, from the simplest colorimetric tube methods to sophisticated sensor technologies.

Single detectors, or a combination of detectors, for selective specification of gases and vapors of contaminants can be used for *in situ* gas or volatile compounds analyses in soil. Part of these detectors can be coupled to MIP for underground application; others are used as independent probes for analyzing soil gas and volatile organic content. Table 4.2 summarizes the frequently used detector types. Section 4.4 describes some commercially available *in situ*-applicable, portable/handheld soil gas and vapor analyzing equipment and devices.

Photoionization detectors (PID) are in the group of efficient and inexpensive broadband detectors suitable for analysis of the total amount of many gas and vapor components, without identifying them specifically. Among its conventional applications, the handheld version has been available for a long time and used widely for soil organic volatiles detection. The small-size, handheld versions are highly suitable for *in situ* soil gas/vapor detection and for the monitoring of changes in the field or

Table 4.2   Detectors for VOCs and semi-VOCs.

Detector type	Contaminants the detector is applicable for
PID = Photoionization Detector	Hydrocarbons and chlorinated VOCs with ionization potential
FID = Flame Ionization Detector	Hydrocarbons, chlorinated VOCs and all types of organic pollutants
TCD = Thermal Conductivity Detector or Katharometer	All organic and inorganic compounds with thermal conductivity different from helium.
ECD = Electron Capture Detector	Chlorinated VOCs, other halogenated organic compounds (PCB, DDT), nitro compounds, organometals
DELCD = Dry Electrolytic Conductivity Detector	Chlorinated VOCs, electronegative molecules such as oxygen or halogens
FID/DELCD combination	Selective differentiation between halogenated and non-halogenated organic compounds
XSD = Halogen-Specific Detector	Chlorinated VOCs
IR and FTIR spectrometer	Gases, VOCs and any targeted chemical species
NDIR = Non-Dispersive Infra-Red spectrometer	CO and CO_2
CLD = Chemi-Luminescence detector	NO
SAW = Surface Acoustic Wave detector	VOCs
LIF = Laser-Induced Fluorescence	PAHs
FFD = Fuel Fluorescence Detector	Hydrocarbons

on sites of remediation technology applications. The PID technology measures total organic volatiles, including all chemical species yielding a positively-charged ion and a free electron. The incoming gas/vapor molecules are subjected to ultraviolet (UV) radiation and each molecule is transformed into a charged ion pair, creating a current between the two electrodes.

Different *UV light sources* can be built into the detectors. Ionizing energies of 9.5, 10.2, 10.6 or 11.7 electron volts (eV) (HNU, 2015) can ionize large-molecule hydrocarbons but perform differently for ionizable, smaller organic molecules and halogenated hydrocarbons. The 10.2 eV probe, for example, is the most useful for environmental monitoring because it can detect the highest number of VOCs. PID is non-selective for organic compounds: it detects VOCs without identifying the composition of the vapor. The equipment can be calibrated for several known organic chemicals, but if the contaminant is a mixture, the result will never reflect the real composition, instead giving an equivalent value, e.g. the total measured contaminant amount as if it were benzene, when calibrated against benzene. Additional laboratory analysis and contaminants identification are necessary for monitoring other than known VOCs.

Flame ionization detectors (FID) are often applied for the quick detection of organic and inorganic VOCs. Many of the VOC detectors can be configured as FID, as PID, or as dual detection technologies (FID and PID) to offer fast response time and more flexible analytical capabilities. FID detects all kinds of combustible organic carbon, while the DELCD is able to detect organic-bonded chlorine. The two together make possible highly selective detection of halogenated and non-halogenated compounds from the contaminant mix.

The *thermal conductivity detector* (TCD) is a bulk property detector sensing changes in the thermal conductivity of the column effluent. The signal is compared to the reference thermal conductivity of the carrier gas by feeding it into a Wheatstone bridge circuit which produces a measurable voltage change proportional to the difference between the signal and the reference. Most analyzed compounds have a thermal conductivity much lower than that of the common carrier gases helium or hydrogen, so a detectable signal is produced every time an organic compound passes the column.

The *dry electrolytic conductivity detector* (DELCD) is selective for chlorinated and brominated molecules, and its detection limit is in the parts per billion range. While traditional ELCD uses a solvent electrolyte, DELCD detects the analyte in the gaseous phase. The small ceramic tube reactor within the detector is heated up to 1,000 degrees Celsius, where the chlorinated and brominated chemical species will react with oxygen and produce ClO_2 and BrO_2 which are detected by the conductivity measuring electrode. DELCD is often used in combination with PID as the detectors of gas chromatographs. The combination of PID and DELCD enables deciding if hydrocarbon peaks detected by the FID belong to halogenated or non-halogenated components of a mixture.

An *electron capture detector* (ECD) uses a radioactive beta particle (= electron) emitter for the detection of highly electronegative compounds such as chlorinated or brominated VOCs in conjunction with the 'makeup gas', which is nitrogen in this case. When the electrons of the beta emitter collide with the nitrogen molecules several free electrons result, as nitrogen is easy to ionize (by removing an electrode from its molecule). Nitrogen gives a high basic signal, which decreases when an electron-absorbing analyte (e.g. a chlorinated molecule) captures the electrons. Out of the chlorinated compounds ECD is sensitive to organometallic compounds, nitriles, and nitro compounds.

Infrared spectroscopes (IR) have widespread use both as point and open-path sensors, and as sensor systems. IR, NIR, FTIR and NDIR are often differentiated. Some simple IR devices contain selective optical filters, e.g. specific filters and individual detectors for methane, carbon dioxide and petroleum hydrocarbons. Thus, these three species can be measured in parallel. Others, using FTIR and computer aided mathematical tools for data analysis, can identify hundreds of chemical species. Some commercially available equipment are based on IR detection:

- NIR uses near-IR wave bands from the spectrum, between 4,000 and 14,000 cm^{-1} wave numbers. NIR can penetrate much further into a sample than mid-IR radiation. Therefore, near-infrared spectroscopy is used for bulk materials with little, or no, sample preparation. Unfortunately, the spectrum is much weaker since NIR is based on molecular overtone and combination vibrations.
- A non-dispersive infrared (NDIR) detector is non-dispersive in the optical sense, since the infrared energy is allowed to pass through the chamber containing gas/vapor without deformation.
- Fourier transform infrared spectroscopy (FTIR) applies a mathematical process to convert raw spectral data into targeted spectra. FTIR is applied as a multi-component open path gas analyzer for pollutants diffusing from the soil surface or subsurface to the air, such as acids, alcohols, aldehydes, aromatics, CFCs, combustion gases, fluorocarbons, greenhouse gases and hydrocarbons.
- IR imaging is based on infrared detectors, which are thermal- or photodetectors. As gases have their own characteristic absorption spectrum, detectors can be adjusted to the region of the spectrum specific to the targeted analyte. VOCs and gases of soil origin can be identified in the near-surface atmosphere this way. The thermal imager allows the vapors and gases to be visualized as smoke on the LCD screen.

Cavity ring-down spectroscopy (CRDS) is a highly sensitive optical spectroscopic technique that enables measurement of absolute optical extinction by samples that scatter and absorb light. Near-infrared laser is used to illuminate a high-finesse optical cavity, which in its simplest form consists of two highly reflective mirrors, resulting in a path length of many kilometers by bouncing the laser between the mirrors. When the laser is in resonance with a cavity mode, intensity builds up in the cavity due to constructive interference. Measurement of light intensity decay starts after the laser is turned off. Decay is faster if there is gas or vapor in the cavity. The difference in the so-called 'ring down time' is proportional to the amount of the (near-infrared) light-absorbing molecules. The composition of gas – or rather the concentration of any chemical species – can be determined by measuring the height of specific absorption peaks. By tuning the laser to different wavelengths, the characteristic absorption spectrum can be taken up. The extraordinary low drift of these instruments means they can operate for months without recalibration. The field-deployed version is rugged and insensitive to changes in ambient temperature, pressure or vibration (CRDS, 2015).

The real-time signals of *in situ*-placed measuring devices can be used for early warning (open path and imaging systems) or for control and regulation of technological parameters. For example, in a bioventilation-based technology, excessive CO_2 or insufficient O_2 in the extracted soil gas indicates the need for higher rate aeration, so the air flow should be increased. Bioremediation based on alternating aerobic and anaerobic phases can be controlled by continuous monitoring of the gaseous microbial products such as CO_2, NH_3, H_2S or CH_4. The reduction in contaminant concentration can be followed by using contaminant-specific detectors.

Laser-induced fluorescence (LIF) systems use ultraviolet light to induce the fluorescence of polycyclic aromatic hydrocarbons (PAHs). The UV light is emitted from a nitrogen laser through a sapphire window into the soil. The UV light induces the fluorescence of PAHs. The fluorescence signal is detected or transmitted to the surface via a fiber optic cable.

The fuel fluorescence detector (FFD) applies a mercury lamp as light source but otherwise works similarly to LIF. FFD can be configured to target detection of a number of different hydrocarbon contaminants (see Chapter 5 for more details).

Surface acoustic wave detectors (SAW) are a class of microelectromechanical systems (MEMS). The method is based on the modulation of the surface acoustic wave by the chemical species coming from the sample directly, or through a gas chromatograph. The analyte may change the amplitude, phase or frequency of the surface wave and cause a time delay between the input and output of the measurable electrical signals. In the analysis of VOCs thin film polymers are applied across the delay line. An array of sensors with different polymeric coatings (lab-on-a-chip) can be used for the selective detection of a large range of gases and vapors. The resolution may be reduced to parts per trillion. It is a very sensitive and selective detector system, applicable for soil gases and VOCs. Electric noses apply the same principle.

3.4   Handheld devices for *in situ* soil gas and vapor analysis

Gas and vapor analyzing sensors/detectors can either be attached or unattached to active or passive samplers. The samplers themselves may ensure discrete or continuous gas/vapor sampling for the detector which can be placed on the surface or downhole.

A wide selection of devices is available: from very simple colorimetric single-gas analytical kits to portable multigas detectors based on PID, IR, FTIR or UV detection and produced by a number of manufacturers.

3.4.1   Colorimetric gas/vapor kit for soil

Colorimetric gas/vapor detection tubes can be used for monitoring already identified or predictable chemical species. A piston hand pump makes on-the-spot measurements possible, and the detection tubes work without the need for calibration. Tubes for the following gases and vapors are available: acetone, amines, ammonia, benzene, butane, CO_2, CO, Cl_2, ClO_2, diesel, jet fuel, ethanol, formaldehyde, gasoline, hydrocarbons, HCl, HCN, HF, H_2S, methyl ethyl ketone, mercaptans, methyl bromide, NO_2, NO_x, O_3, phenol, phosphine, SO_2, toluene, trichloroethylene, vinyl chloride, water vapor and xylene (Colorimetric tubes, 2015).

3.4.2 PID-based portable, handheld detectors for soil VOCs

- The *MiniRAE 3000 and ppbRAE 3000* are specialized for soil gas/vapor and remediation monitoring. They both work with real-time wireless data transmission. The sensor provides a three-second response time and a measurement range of up to 15,000 ppm (MiniRAE), or from 1 ppb to 10,000 ppm (ppbRAE) with good linearity. Humidity compensation is solved with integral humidity and temperature sensors. MiniRAE provides calibration for more than 200 compounds through integrated correction factors (RAE, 2015).
- The *UltraRAE 3000* is a handheld device. It is a benzene- and compound-specific VOC monitor with a PID range of 50 ppb to 10,000 ppm in VOC mode, and 50 ppb to 200 ppm in benzene-specific mode. It is suitable for pre-screening of TPH, diesel and jet fuel contaminated sites, and contamination spills, as well as the processes of natural attenuation and remediation technologies (UltraRAE, 2015).
- The *Model 102 Snap-On PID*^{TM} photoionization analyzer is a handheld tool for the measurement of organic and inorganic species that can be ionized by a UV lamp. Model 102+ is an upgraded model with eliminated moisture sensitivity. It can be connected to up to 4 sensors, a pump and a data logger (HNU, 2015).
- The *Model 112 continuous VOC analyzer* has either a 9.5 or a 10.6 eV lamp, an optional 4–20 mA output, a single programmable setpoint, configuration software, and an industrial enclosure if necessary. This model can be used not only with a PID but with a FID detector, too (HNU, 2015).

3.4.3 FID-based equipment for soil VOCs

- The *TVA2020* portable vapor analyzer's key features include portability, hydrogen refill assembly and Bluetooth^{TM} connectivity (TVA2020, 2015).
- The most widespread use of FID is its coupling to a gas chromatograph, for the selective separation of the volatile components. GC-FID analyses may also be a part of *in situ* site assessment and monitoring by using portable gas chromatographs, mini gas chromatographs or mobile laboratories equipped with GC-FID. The rugged field design and the reduction in size – both of the gas chromatograph and FID detector – are necessary elements to achieve an easily portable version. Mobile equipment is offered by several manufacturers and vendors, from Germany to Indonesia. Examples are listed below.

  o Geoprobe^{TM} GC designed for *in situ* subsurface exploration of contaminated soil and groundwater and coupled among others to FID (Geoprobe^{TM} GC, 2015).
  o Meta GC-FID/4HU (Portable GC-FID, meta, 2015).
  o Explorer Portable Gas Chromatograph (Explorer, 2015).
  o Mini GC-FID developed by Acquisition Solutions[k] for C1–C6 short chain hydrocarbons: analysis time 90 seconds, size: $30 \times 20 \times 12$ cm, weight: 5 kg (Mini GC-FID, 2015).
  o Model 8807 Portable Gas Chromatograph by PCF Elettronica (Model 8807, 2015).
  o Portable GC by PT Amerta Labindo Utama (Portable GC, Indonesia, 2015).
  o A small FID detector was developed as early as in 2005 (Deng et al., 2005).

–   The hypersmall size development of Cbana Laboratories for NASA is a micro-gas
    chromatograph with a micro-flame ionization detector and thermal conductivity
    detector (TCD). It is just $20 \times 20 \times 10$ cm in size. It has been designed both for
    terrestrial and space applications for onsite gas/vapor analysis. It is applicable
    for air quality monitoring; natural gas analysis; chemical spill monitoring; indus-
    trial toxic chemical detection; chemical warfare agents and other chemical threat
    detection; petroleum/biodiesel analysis; etc. (NASA, MicroFID, 2015).

3.4.4   IR detection-based field equipment for soil gas and VOC analysis

–   The *Cerex Company* supplies the Shepherd FTIR for 385 gas and volatile chemical
    species (Cerex, FTIR, 2015), and the Hound series: the Hound, Mini Hound and
    Micro Hound. The Hounds use the technique of differential optical absorption
    spectroscopy (DOAS) with UV light absorption. They measure the absorption
    spectra (instead of light intensity at a single wavelength only) and separate the gas
    and volatile species from each other, and from scattering, based on their individual
    absorption spectra (Cerex, Hound, 2015).
–   *Gasmet products* are *in situ* and portable multigas meters based on FTIR (Gasmet
    in situ, 2015). Gasmet™ DX4015 from the portable series includes a built-in
    pump, others need an additional portable sampling system (Gasmet, portable,
    2105).

3.4.5   Combined PID and IR detection of soil VOCs

The *Ecoprobe 5* is a handheld device – shown in Figure 4.4 – providing selective
infrared analysis of methane, petroleum hydrocarbons (TPH) and CO_2, combined
with ultra-sensitive PID analysis of total organic compounds. It is complemented with
O_2, atmospheric pressure, sampling vacuum and soil temperature readings, and a fully
integrated automatic GPS data logger (Ecoprobe 5, 2015). The combined equipment
is suitable for application on the soil surface for the identification of pollutants evap-
orating from soil as well as for downhole application at up to 6 meters' depth in the
vadose soil zone. A bell helps to exclude external air exhaust at surface detection.

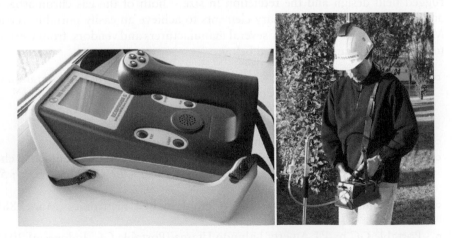

Figure 4.4 Ecoprobe 5 equipment and its application in the field (Ecoprobe 5, 2015).

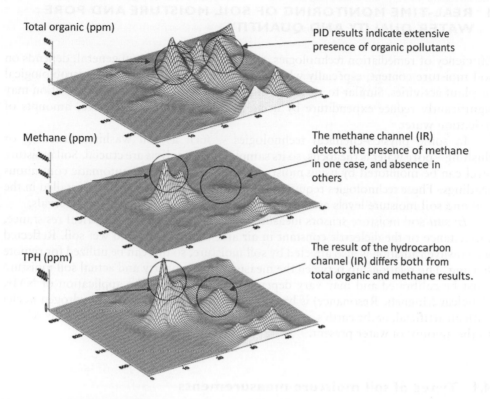

Total organic (ppm)

PID results indicate extensive presence of organic pollutants

Methane (ppm)

The methane channel (IR) detects the presence of methane in one case, and absence in others

TPH (ppm)

The result of the hydrocarbon channel (IR) differs both from total organic and methane results.

Figure 4.5 Graphical visualization of the results: spatial distribution and the result of PID and IR analysis (Ecoprobe 5, 2015).

Small wells used for downhole sampling are sealed from the top so vertical average samples can be collected and analyzed. Soil gas and vapor are sampled by a built-in pump. Vertically sectioned sampling makes measurement in discrete depths possible. A special sampler is used for that purpose with an openable and closeable tip. The closed sampler is driven into the soil, opened for soil vapor extraction and measurement and closed again before pushing down on the drive to a deeper position.

The data from the sensors can be downloaded and evaluated by the relevant software, which provides quantitative results in a table and visualized, three-dimensional images. Figure 4.5 illustrates three of Ecoprobe 5's end products: the PID result covering all organic carbon (excluding methane), the methane result, and the hydrocarbon result of the IR analysis. A comparative evaluation of the results enables decisions about whether the contamination is of petroleum origin, and whether methanogenic bacteria have had the opportunity to start anaerobic hydrocarbon substrate utilization. As this example shows, mapping with the help of Ecoprobe is just the first step in the assessment of a site with unknown contaminants. It can identify hot spots, and monitor spontaneous changes, new releases and the changes of contaminants already identified. However, the identification of the contaminant should be based on an additional contaminant-selective analysis.

4   REAL-TIME MONITORING OF SOIL MOISTURE AND PORE WATER QUALITY AND QUANTITY

Efficiency of remediation technologies, similar to soil function in general, depends on soil moisture content, especially when the technologies are based on microbiological or plant activities. Similar to agricultural practice, soil moisture-based irrigation may significantly reduce expenditure by using sufficient, but not excessive, amounts of irrigation water.

In some physico-chemical technologies – such as soil washing, leaching or flushing – pore water is the focus, so its sampling and analysis are crucial. Soil moisture level can be monitored by using moisture-sensitive probes and automatic continuous readings. These technologies require the equipment to be permanently installed in the soil and soil moisture levels are then automatically measured at regular intervals.

In situ soil moisture sensors measure the difference between electrical resistance, capacitance or the dielectric constant in air and water, or dry and wet soil. Reflected microwave radiation is also affected by soil moisture, so this can be utilized for remote sensing. The relationship between the measured soil property and actual soil moisture must be calibrated and may vary depending on soil type. The application of NMR (Nuclear Magnetic Resonance) is based on the interaction of water's hydrogen nuclei with an artificial, or the earth's, magnetic field and the detected signal is directly related to the amount of water present.

4.1   Types of soil moisture measurements

Soil moisture content is measured by tension or by volume. Methods based on tension use tensiometers which measure suction with the help of porous media and special detectors. Volumetric methods may be based on measuring the soil's dielectric constant, neutron moderation, magnetic resonance or heat dissipation.

4.1.1   Porous media instruments

– *Tensiometers* are water-filled tubes with a porous ceramic tip at the bottom and a vacuum gauge at the top. Soil moisture content depends on soil type. Soil moisture tension (expressed in bars) may be a good indicator of moisture content, and it is proportional to the necessary suction capacity of plants to draw water from the soil. The method does not require calibration. If the water level in the tube drops below a certain level, the reading is not correct, and the tube must be refilled.
– *Gauge-type* tensiometers include a permanently-attached pressure gauge. The gauge can be read directly or replaced with a pressure transducer to enable data logging.
– *Electrical resistance* gypsum blocks are another way to measure soil moisture content by its tension. Two electrodes – coupled to a resistance meter – are placed into the porous material, and when the gypsum block becomes wet, the resistance decreases. The resulting electrical signal should be calibrated with the real soil moisture content. Gypsum blocks should be replaced yearly due to deterioration.

- Instead of gypsum, other porous materials such as granular matrix materials (for soils with high water content), swelling polymeric gels, or porous ceramics can also be used to analyze soil moisture.
- Thermal heat sensors to measure soil matric potential use the heat pulse concept to determine the water content of a ceramic body, which in turn is in equilibrium with the water tension of the surrounding soil (see also Section 4.1.4).

4.1.2   Volumetric soil moisture sensors

Volumetric moisture probes work on the principle that the volumetric soil moisture content ratio ($q_v = (V_w/V_s) \times 100\%$) is related to the apparent dielectric constant (e) of the soil, with a linear correlation between q_v and the square root of e. This relationship has been shown to be valid for many different soil types. Time domain reflectometry and transmissometry are the two main methods in volumetric analysis. Both methods are useful to monitor infiltration and desiccation, and upward soil water transport by capillary forces in soils, in both treated contaminated soils and in landfill covers. A calibration equation must be created and further field calibration may be necessary.

- ***Time domain reflectometry*** (TDR) sends an electrical (voltage) signal through two or three steel rods (waveguides) placed in the soil and measures the return signals in the rods (Figure 4.6). The speed with which the electromagnetic pulse travels down the waveguides depends on the dielectric constant = permittivity (K_a) of the surrounding soil. The theoretical background is that the surrounding material causes a part of the energy to be reflected through the waveguide. The higher the water content, the higher the dielectric constant ($K_{a,air} = 1$, $K_{a,soil} = 2$–5, $K_{a,water} = 80$) and the lower the speed. Thus, wet soil slows the signal down stronger that dry soil. When the pulse reaches the end of the waveguide (typically between 10 and

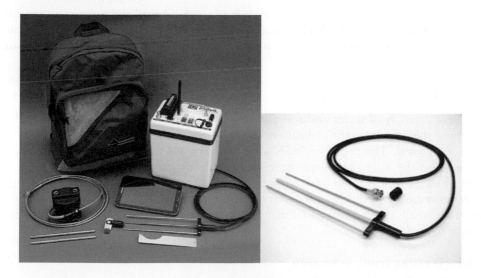

Figure 4.6   MiniTrase Kit portable soil moisture measuring instrument using Time Domain Reflectometry (TDR). The kit and the TDR sensor (Photos: MiniTrase Kit, 2015).

30 cm), all the remaining energy in the pulse is reflected from the end point of the waveguide. Intensity and occurrence in the travel time of the reflected signal is detected and analyzed by the instruments cabled to the waveguides and the dielectric constant is calculated. TDR provides accurate readings of soil water content. It needs little maintenance, but interpretation of data may require special calibration, depending on soil characteristics. A suitable sensor arrangement can provide information on spatial distribution. Figure 4.6 shows the portable MiniTrase Kit and the TDR sensor (Soilmoisture, 2015).

– *Time domain transmissometry (TDT)* observes the pulse at the opposite end of the transmitter, so the time requirement of the one-way propagation is measured. It means that there is no reflection and no signals are superimposed on the transmitted signal.

The advantages of dielectric soil moisture sensors are:

– continuous measurement;
– good repeatability;
– high sensitivity.

Disadvantages are:

– the need for a calibration equation and difficulty in developing this equation;
– relatively small zone of influence;
– that soil salinity may affect probe reading;
– that air gaps surrounding the sensor may spoil contact (careful emplacement is necessary) and influence signal size.

4.1.3   Combined methods

The methods in most cases are a combination of tension-based and volumetric sensors. For example, the electrodes built into a gypsum block may be combined with a water suction device, a porous material with known water-retention properties (gypsum block), and a volumetric sensor.

4.1.4   Heat dissipation sensors

The heat capacity of water and soil differ significantly. The consequence is that when applying the same amount of heat energy, a dry soil will reach a higher temperature than a wet soil. The probe includes a heat source and a temperature sensor and is placed into the soil. The temperature sensor records the peak temperature after the heat shock. The heat input and the peak temperature are used to calculate volumetric water content. Calibration is necessary with the relevant soil type. The measurement sphere is small (about 1 cm) and it may ensure high spatial resolution. Time series data can be logged.

4.1.5   Wetting front detection

All types of devices are suitable for the detection of a *wetting front* in soil due to the difference between wet and dry soil. Wetting front detection represents a simplified application of moisture sensors. The information needed is that the wetting front has arrived at a specified depth in the soil. The signal can be used just for warning, or for controlling irrigation: when the specified change in the signal appears, the sensor switches off, and the electric signal terminates irrigation. Wetting front detection may provide benefits for sampling soil water especially from the wetting front, whose nutrient and salt contents significantly differ from the subsequent, vertically moving, infiltrated waters.

Soil moisture content can be measured by several conventional and innovative *in situ* methods. A comprehensive overview on the topic was given by Charlesworth (2005) and the products have been tested and published by several researchers and practitioners (Vaz *et al.*, 2013; Kargas & Soulis, 2012). Some products are listed below.

–   The **SOILSPEC tensiometer** consists of a tensiometer tube and a vacuum gauge. The scale of vacuum, i.e. the suction can be read from the gauge and the reading can be manually noted or stored by the probe (along with the tube ID and date) and downloaded to a computer (SoilSpec, 2015).

–   Various **UMS tensiometers** are available on the market. The T4 precision tensiometer was developed for outdoor monitoring projects, where the ceramic cup is only filled with water for highest accuracy and year-round operation; the T5 is a very small tensiometer specially made for measurements in soil columns, in small lysimeters or pots; the T8 is designed for long-term monitoring projects; and the TS1 is a self-refilling tensiometer for ecological impact, water balance and transport studies (UMS, 2015b).

–   The **ML3 ThetaProbe** soil moisture sensor is based on TDR and is a Delta-T device providing built-in temperature measurement, compensation for the effect of temperature, and calibration for soil salinity (Delta-T, 2015).

–   The **EnviroPro soil probe's** moisture readings are also compensated for the effects of temperature and can be compensated for errors caused by electrical conductivity. The internal electronics are encapsulated in an epoxy housing and thus protected against moisture and chemicals. No maintenance is necessary (Envirotek, 2015).

–   The **EnviroSCAN** probe, a Sentek Probe, is developed for the monitoring of water and salinity at multiple depths in a soil profile. Several sensors can be built into one probe, e.g. moisture, temperature, pH, salinity, humidity. The electronics of the drill and drop device are protected by a sturdy housing, which is useful when it needs to be moved between sites (Sentek, 2015).

–   **Decagon** moisture sensors are calibrated in a variety of soil types and are insensitive to soil salinity and soil texture in typical soils. Decagon data loggers are available but the sensors are compatible with other standard data loggers, too (Decagon, 2015).

–   The **EQ3 equitensiometer** is a combined device from Delta-T which uses the TDR-based ML3 ThetaProbe (see above) embedded into a porous material, serving as the equilibrium body. Its matric potential equilibrates to that of the surrounding soil and is measured directly by the ThetaProbe. The matric potential of the equilibrium body can be converted into the matric potential of the soil (EQ3, 2015).

–   The **EQ15 equitensiometer** for soil water potential measurement is a Japanese development, similar to the Delta-T product. It is a maintenance-free device with a wide measurement range (0–1500 kPa) and long term stability. It is not affected by over-range values, and is applicable for a wide range of soil types and conditions. Data recording is performed either by a data logger or via the display of a simple voltmeter (Ecomatik, 2016).

- The *CS 229 heat dissipation matric water potential sensor* uses the heat dissipation method to indirectly measure soil water matric potential. A heating element is placed into the cylindrical, porous ceramic body. The temperature increase is measured after heating the body for a certain duration. Temperature increase is related to the thermal conductivity of the porous material, which in turn depends on the amount of water present in the ceramic. A calibration for the soil being tested and the moisture content range is needed. The sensor is capable of reading the moisture content from saturation to air dry state (Campbell Scientific, 2015).
- *Gopher and MicroGopher* soil moisture profilers measure the dielectric constant of wet soil to determine the moisture content of the soil. (The dielectric constant of dry soil is typically 4, and that of water is 80.) The device is portable and therefore should not be placed into soil permanently, but instead deployed just for the measurement. Variations in electrical conductivity of the soil moisture due to salinity have practically no effect on the measurement result. The microprocessor-controlled measurement system consists of the sensor, a data logger recording information for download to a computer at a later time, an LCD dot matrix display for display of graphs and information, and a 16-key keypad for operator interface (Gopher, 2015).
- The *green light red light (GLRL) device* consists of a string of capacitance sensors inserted in an access tube at depths of 10, 20, 30 and 50 cm. Calibration should be made and the full point or field capacity is automatically set at 100%. Based on this calibration one can set the desired soil type-dependent moisture content. The device lights red when above and green when below the set moisture content. The sensor can be used either in portable mode, using a handheld reader, or permanently installed for automatic time series logging through the use of the Odyssey logger (GLRL, 2015).
- The *ECH2O* is a fiberglass printed circuit board inserted into the soil. Copper traces embedded in the fiberglass generate an electromagnetic field which varies with the surrounding soil dielectric. The probe measures voltage which is calibrated against volumetric water content (ECH2O, 2015).
- The *WET sensor* measures water content (W), electrical conductivity (E), and temperature (T) (WET, 2015).
- *Aquaflex methodology* applies time domain transmission (TDT) to measure the soil dielectric constant. An electrical pulse is sent along a 3-meter long transmission line, and the electrical field generated around the transmission line interacts with the surrounding medium. The speed and shape of this pulse is affected by the dielectric properties of the medium. The probes can be installed horizontally or diagonally into a trench (Aquaflex, 2015).

The probes in online applications are attached to a data logger, and the data measured automatically are forwarded by wire, radio or cell phone telemetry to the receiving stations. Local closed systems, mobile phone or Internet-based clouds can be used for recording and analysis.

4.1.6 Nuclear Magnetic Resonance (NMR) to measure soil waters in situ

NMR occurs when a nucleus (usually hydrogen, but any nucleus that has a non-zero spin works) is placed into a magnetic field and is 'swept' by a radio frequency wave that causes the nuclei to 'flip'. This causes the radio frequency energy to be absorbed, and this is what is measured.

Magnetic resonance imaging (MRI) is a complex application of NMR in which the geometric sources of the resonances are detected and deconvoluted by Fourier transform analysis.

Electron spin resonance (ESR) is also a resonance phenomenon, except in this case it is the spin of an unpaired electron that is in resonance, rather than a nuclear spin.

Earth's field NMR (EFNMR), also called *GeoMagnetic Resonance (GMR),* uses the Earth's magnetic field instead of the (much stronger) magnetic field generated by an NMR system. In the geomagnetic field, hydrogen nuclei in groundwater emit a measurable NMR signal when they are energized at a specific resonance frequency.

NMR-based methods are suitable for measuring soil moisture content and for distinguishing between soil water strongly bound in small soil pores and mobile water in large pores. Soil type can also be characterized based on the pore size.

Slim-line logging NMR measurement was reported by Perlo *et al.* (2013). This NMR tool is suitable for measuring water content in the vadose zone due to the kit's relatively small size. It is based on cylindrical permanent magnets of 20 cm length and 5 cm diameter, and has a penetration depth of about 2 cm measured from the surface. After being optimized by maximizing the signal-to-noise ratio, it became a useful tool for *in situ* soil moisture analysis (Perlo *et al.*, 2013).

The developments resulted in a portable, *in situ* MRI for studying deeper soil layers or the root zone. The down-borehole NMR logging tools for normal or direct push wells were applied to ground water investigations by Walsh *et al.* (2013). The new NMR logging tool was tested in the field and proved to be able to provide reliable, direct, and high-resolution information for total water content (total porosity in the saturated zone or moisture content in the unsaturated zone), and estimates of relative pore-size distribution (bound vs mobile water content) and hydraulic conductivity.

The application of *NMR for root imaging* was published as early as in 1986 (Rogers & Bottomley, 1986). From that time many developments were published for its application to the root zone to study water content and plant uptake, typically for trees (Kimura *et al.*, 2011 and Jones *et al.*, 2012).

Jones *et al.* (2012) present the design of the Tree Hugger, an MRI system for the *in situ* study of living trees in the forest. It is a transportable (55 kg), 1.1 MHz, ^{1}H MRI system, able to achieve access down to 2 m depths.

Several companies offer NMR and MRI tools for soil geophysics, mainly for measuring soil properties based on water content. The *in situ*-applicable instruments, Javelin and Discus/Dart, are provided by Vista Clara (2015) for soil moisture and groundwater characterization.

– The *Discus* (2015) is suitable for direct characterization of soil moisture in the vadose zone based on the interaction between water's NMR active ^{1}H nuclei and a static magnetic field. The NMR system operates with flat, discus-like sensors; the magnet interacts with water molecules in the soil and the sensor measures the NMR signals of the water molecules. The signal amplitude quantifies soil moisture content and it is able to differentiate between bound and mobile waters in unsaturated soils, and thaw water in permafrost. The decay time of the signal reflects soil texture, permeability and pore size: the decay time of a silty soil is 0.01 s, for example, while that of a sandy soil is 0.3 s. The device can be applied for compaction monitoring and analysis, infiltration assessment and monitoring, carbon cycling, and generic agricultural characterization of soils. It can establish a soil moisture profile by acquiring data at discrete depth intervals.

– The *Dart* (2015) is a close relative of Discus, able to distinguish water bound in small pores as distinct from mobile water in large pores, thus providing classification of soil type and prediction of dynamic groundwater flow behavior. Unlike the flat Discus, the Dart has a rod-form design and can be deployed in 2 cm PVC pipes or auger holes. The Dart is especially useful for monitoring laboratory and field micro- and mesocosms.

– The *Javelin* (2015) operates in slim boreholes, projecting a magnetic field several inches beyond the borehole and creating a cylindrical-shaped sensitive region from which the groundwater NMR signal can be detected. The sensitive region is located beyond the region disturbed by drilling or pushing, within the representative subsurface formation of rocks and sediments. The detected signals reflect the quantity of groundwater as well as the hydrogeological properties of the formation. The amplitude of the signal indicates the total amount of water. The decay behavior of the signal over time (decay time) conveys information about the pore environment. Water in low-permeability silts and clays exhibits a short decay time, and a long one indicates high-permeability sands and gravels. Detection and quantification of groundwater; measuring bound and mobile water content; pore size; and permeability at successive depth intervals enables the characterization of the soil profile. Hydrocarbon-type contaminants and fluid diffusion can also be quantified.

Geomagnetic resonance (GMR, 2015) sensors enable the detection of weak NMR signals produced by the nuclei of soil water in the earth's magnetic field at a specific resonance frequency. Protons resonate in the earth's magnetic field at audio frequencies of around 2 kHz (compared to the 900 MHz of high field NMR) and generate very weak signals. The magnetic resonance sounding (MRS) method is used for the detection of these weak NMR signals. Large wire loops are arranged across the surface of a groundwater investigation site to measure the response of the groundwater. The magnetic field energizes the hydrogen nuclei in groundwater at their resonance frequency (1–3 kHz). After a delay the surface loop(s) switch to receive mode and record the NMR signal generated by the energized groundwater. This NMR signal provides information about the abundance of water and also the size of the pore spaces in which the groundwater resides in the saturated soil. Different layers can also be studied by changing the amount of the energy delivered by the surface coils.

4.1.7 Rapid test kits

Soil moisture content is generally measured by sensors calibrated to moisture content, but more precise field applicable test kits are also available.

– The *HydroSCOUT* (2015) test kit is based on the standardized quantitative calcium hydride reaction (US EPA standard) and produces a result within 10 minutes. The results correlate with the standard oven drying laboratory method for determining soil moisture content.

4.2 Lysimeters and wireless lysimeters for online soil water monitoring and sampling

Lysimeters are containers isolating a disturbed or non-disturbed soil column from the surrounding soil. They have a sampling system at the bottom for collecting leachates, i.e. percolating water moving from the surface down to the bottom gravimetrically. Lysimeters may simulate situations above the groundwater level (three-phase soil with soil moisture) without connection to groundwater or in part under groundwater level with connection to groundwater. In the latter, the transport routes are completed by capillary water seeping upwards from the water table.

Lysimeters in various sizes and installations are versatile tools in soil management mainly for investigating seepages, infiltrates, leachates, capillary fringes and all types of pore water, as well as dissolved elements/ions, nutrients and contaminants and their transport in the vadose soil zone. As described in Gruiz *et al.* (2015), subsurface and surface lysimeters are equally applicable *in situ,* in the field. Lysimeters may contain undisturbed soil cores or can be filled with excavated soil or any other material. Subsurface models contain the undisturbed indigenous soil profile and are exposed to the complexity of natural impacts so that these *in situ* lysimeters simulate the actual field conditions. The ones placed on the surface are exposed to the same weather conditions, but segregated from the subsurface soil matrix (isolated core or a mixed, rearranged or otherwise manipulated aliquot). They duly simulate vertical fluxes and the situation of natural leaching by precipitation, e.g. in the case of disposed waste, without the horizontal transports via neighboring soil volumes.

UMS (2015a) manufactures complete lysimeters equipped with built-in measuring devices that provide information for water management in environmental, agricultural and scientific fields. Their products are as follows.

– Meteo-lysimeters that can measure water balance parameters and weather data.
– Hydro-lysimeters are for field-identical measurement of precipitation as well as determination of true evapotranspiration and leachate rate.
– Agro-lysimeters are for field-identical measurements for sustainable agriculture and long-term groundwater protection. They work by the visualization of the availability of water and fertilizers, and measuring precipitation, evapotranspiration and leachate.
– The Smart Field lysimeter is a weighable and tension-controlled field lysimeter, combined with a tensiometer, a soil moisture sensor, a data logger and powered by solar panels. It is especially useful for water balance and evapotranspiration determination.
– The Science-lysimeter is a modular precision measurement system for scientific studies and research in soils. 'Laboratory precision under field conditions' is the vendor's interpretation of its capabilities (UMS, 2015a).

Compared to the sophisticated lysimeters mentioned, much simpler passive equipment is also used for soil water and leachate investigation and sampling, which contains just a suction cup or plate, a sampling bottle and a buffer vessel connected to a vacuum pump. Some of them are introduced below.

– The traditional porous ceramic suction cup sampler, which has been used for at least a hundred years. In spite of its low cost and easy usability, its application is limited by disadvantages such as (i) inability to characterize the solute flux and amount and, (ii) lack of representative samples because the waters of different origins cannot be sampled proportionally due to a non-continuous vacuum during sampling.
– The application of a low vacuum (low suction) to a suction cap may sample waters, which would not be subject to natural leaching.
– The zero-tension lysimeter collects freely draining leachate.
– The zero-tension pan can collect water only from a soil matrix with higher than zero tension.
– Equilibrium tension lysimeters maintain equilibrium between the suction applied to the leachate collection system and soil matrix potential by setting the rate of applied suction.

4.2.1 Lysimeters for soil investigation

Gravity pan lysimeters are sampling lysimeters placed underground for collecting leachates without exerting suction, thus they only collect water when the subsoil (above the sampler) enables free drainage. They are refilled with soil material and supplied with a drainage layer in the bottom and with a sampler for moving leachate to the surface.

Suction lysimeters apply the above assemblage – supplemented with a sampling plate under suction as shown in Figure 4.7. This type of lysimeter can be used for vadose-zone studies, when the soil is not saturated at the depths of the lysimeter's bottom. The tensiometer applied in parallel provides information on the changes of moisture content in the soil during leachate collection. Sampling plates placed on the bottom of the lysimeter are specifically designed to be buried beneath soil columns to collect soil water. The material of these plates can be porous plastic films or blocks, sintered metals, ceramics or glass. The plate is supplied with a rubber backing and a stainless steel outflow stem on the bottom side (Sampling plate, SM, 2015; Sampling plate, EK, 2015).

Tension lysimeters are specially designed suction cups (i.e. tensiometers) for collecting soil solution. A vacuum is applied to the porous ceramic cup and the solution in the soil in close contact with the cup is pulled into the cup, from where the sample is removed after a certain time. Several problems arise during the interpretation of the results because of the applied sampling time and pressure (suction), the origin and type of the collected water (peak water flows can be characterized by larger amounts and dilutions compared to dry periods), the nature of the contaminants, e.g. volatile or showing a delay due to ion exchange. All in all it is not always clear what the collected sample represents. Their advantage is that they can be placed in any depths

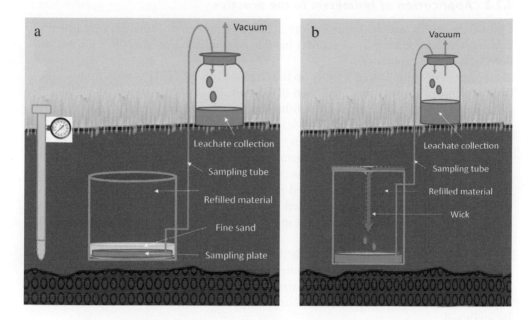

Figure 4.7 Refilled lysimeters: a) with a sampling plate, and b) a wick for sampling.

compared to zero-tension or suction lysimeters and they cause less disturbance in the soil during emplacement. They can follow the changes (degradation, transformation) of soil contaminants in the vadose zone.

Passive capillary wick samplers apply a hanging wick to extract soil solution from the soil column above the container (Figure 4.7). Wick-type samplers are cheap and reliable instruments for measuring groundwater fluxes and nutrient or contaminant concentrations in the vadose zone when their design fits soil properties and climatic conditions, as demonstrated by a numerical analysis (Gee *et al.*, 2005; Mertens *et al.*, 2007). The material of the wick is generally fiberglass or rock wool.

Wireless lysimeters equipped with soil moisture sensors and coupled with a distributed wireless sensor network (WSN) have been proved by Kim *et al.* (2010) to be suitable for real-time online monitoring of drainage water. They published an account of the installation of twelve passive capillary samplers (PSAPSs) sensing stations across an agricultural field at 90 cm below the soil surface. In addition to the twelve stations measuring the amount of drainage water, a weather station was included in the WSN. Sensory data were periodically sampled and transmitted by Bluetooth wireless radio communication to a base station, from where the data were forwarded via the Internet. A developed and published web-linked WSN system provided convenient remote online access to monitor drainage water flux and field conditions during the whole of the growing season without the need for time-consuming field operations. Such wireless sensor networks can be coupled with remote control and irrigation regulation in a system called agromotic.

4.2.2 Application of lysimeters in the practice

Lysimeters can be widely applied in the management of environmental contamination and hazard, and hundreds of reports have been published on their use for real-time detection of contaminants.

Kram (1998) published a study on the combination of PCAPSs with hydrocarbon-specific sensors for real-time detection of DNAPL.

Virtanen *et al.* (2013) used monolithic (undisturbed) lysimeters for monitoring the oxidation of sulfidic materials upon drainage of acid sulfate (AS) soils, which causes episodic hazards to aquatic ecosystems worldwide.

Wang *et al.* (2012) measured nitrate leaching in lysimeters and compared the results with those obtained by porous ceramic cups. There were differences between the cumulative leaching loss measured in the lysimeter drainage and the values estimated from the cups. It was concluded that suction cups were inappropriate for the determination of cumulative leaching in silt loam, but provided useful data in sandy loam soil.

Herbicide transport was studied by several lysimeter-type samplers by Peranginangin *et al.* (2009). They studied pesticide leaching through the vadose zone via preferential flow paths and emphasized the importance of selecting porous or other capillary materials for the samplers. These materials must not retain the chemicals to be analyzed.

The authors of this chapter applied field lysimeters for the study of several environmental technologies such as the following.

– Long-term behavior of sulfidic mine waste at variable precipitate amounts: monitoring leachate chemistry and microbiology.
– Chemical stabilization of toxic-metal-contaminated soil by various stabilizing additives: monitoring the leachate.
– Studying permeable reactive barriers prepared from waste materials to retain toxic metals: monitoring the leachate.
– Water infiltration and capillary transport in the cover layer of a red mud tailings pond: the lysimeters were equipped with self-developed moisture, elecrical conductivity and temperature sensors.

The company UMS established a *lysimeter mesocosm station* – with 48 lysimeters, each 2 m^3 and a service cellar – to close the gap between field and greenhouse experiments by determining processes in naturally embedded soils (UMS Lysimeter station, 2015). The filled lysimeters were equipped with sensors at five depths. Tensiometers, TDR probes, temperature probes, pore water samplers, empty tubes for a camera, and soil gas samplers were available for the experiments.

4.2.3 Capillary water absorbers

Capillary water absorbers produce suction (due to lower water potential compared to soil) and collect water and dissolved chemicals through a porous membrane. The membrane is brought into hydraulic contact and the absorber gets into equilibrium with the surrounding soil. The water content of the absorber can be measured *in situ* by moisture sensors. The dissolved chemicals can be analyzed *in situ* using contaminant-specific sensors or in the laboratory after removal from the absorber.

   Passive capillary samplers (PCAPS) may be the wick pan type of passive samplers, applying negative tension (suction) to soil water. These types of water samplers act as a hanging water column, developing a flux-dependent suction. The length and diameter of the wicks are adjusted according to the equation of Knutson and Selker (1994) to achieve the closest possible match for the expected pressure/flux condition. They are more efficient compared to other passive samplers and enable the collection both of water and dissolved chemicals (Brown *et al.*, 1986; Holder *et al.*, 1991; Boll *et al.*, 1992; Louie *et al.*, 2000; Tuller & Islam, 2005). Passive capillary wick-type lysimeters may use several different types of materials with large suction capacity, not only ceramics but other sintered materials, polyethylene or nylon, silicon carbide or borosilicate glass.

4.2.4 Commercial passive capillary lysimeters

There are a number of passive capillary lysimeters and samplers on offer, and the following is a selection.

– *Drain gauge G3 and G2* passive capillary lysimeters are manufactured by Decagon to determine the volume of water and chemicals draining from the vadose zone into groundwater. The 1.5-meter tall drain gauge is buried directly in the ground to measure the flow rate in unsaturated soils and collect soil water samples for chemical analysis. Water samples can be collected through a surface port for analysis of chemicals, fertilizers, and other contaminants (Decagon PCAS, 2015).
– *Monoflex lysimeters* are designed for permanent subsurface installation. They are closed tubular devices with a porous ceramic filter element at one end. Monoflex lysimeters are provided with two ports: one to allow application of a vacuum or pressure, the other to allow delivery of collected water samples to the surface. Contaminants may partly be sorbed by the ceramic cup, so the concentration of toxic metals in the sample is typically 10% less than the actual amount in the soil water. The same retention percentage of chlorinated hydrocarbons can reach as much as 90% (Monoflex, 2015).
– *Sampling lysimeters* by ICT International include simple pan lysimeters; drain gauges made of PVC and stainless steel; large-volume and slim-tube samplers; pressure vacuum samplers; and deep profile samplers. The company offers a 'ready-to-go' lysimeter which is instrumented with soil moisture probes and matrix potential probes which are connected to a data logger and a compact climate station (ICT, 2015).

4.3 Water and contaminant mass and flux measurements

The measurement of nutrient or contaminant concentration in groundwater is not sufficient for the calculation of risk, as water and contaminant mass and flux should also be determined and the exposure of the environment and humans calculated from these data.

   Water mass and flux at the surface and subsurface are equally essential information for calculating loads or modeling material transports of nutrients or contaminants.

Water and water-dependent material fluxes in the soil are determined by the transport processes of lateral advection in a porous material due to pressure difference in the groundwater and by the gravitational transport of the pore fluid in the vadose zone. These two main water transport processes are supplemented with the diffusion of dissolved substances and their partitions between the physical phases. Water transport combined with substance-dependent partition results in selective retention of the components. The complex transport forms in the various pore sizes and capillaries, including the capillary fringe, may further complicate mapping/modeling water and solute transports in soil.

4.3.1   Water and contaminant flux in groundwater

Measuring flux in a porous medium is extremely difficult, so water fluxes are generally estimated from locally measured water potential values and the hydraulic properties of the porous subsoil by applying Darcy's law:

$$Q = -\frac{kA\Delta p}{\mu L}$$

where:
Q: flow rate or flux (volume per time), m³/s
k: permeability of the fluid, m²
A: cross-sectional area, m²
Δp: pressure difference, Pa
μ: viscosity, Pa·s
L: length, m.

The negative sign indicates that flow direction is opposite to pressure increase.

A summary list of water and contaminant mass flux measuring methods (ITRC, 2010) is given below.

–   The *transect method* uses estimates of groundwater contaminant concentration and groundwater velocity at a series of monitoring points across a plume.
–   The *well capture method* calculates the mass flux (mass per time) by measuring the concentration and the flow rate of the well. This approach assumes that the well or well system fully captures the horizontal and vertical extent of the contaminant plume.
–   A *passive flux meter* consists of a permeable sorbent infused with soluble tracers packed in a nylon mesh tube and placed in a well for a given period. Dissolved contaminants from the groundwater flow will be sorbed on the sorbent and the soluble tracers leached out. The mass flux can be calculated from the sorbed contaminant mass (g) and the estimated groundwater flux (m³) during the measurement time (h).
–   *Transects based on isocontours* can be used for contaminant mass flux calculation by multiplying the velocity at the isocontour, cross-sectional area and fluid density.
–   *Solute transport models*, such as MODFLOW or REMChlor, or those for modelling monitored natural attenuation (MNA), e.g. BIOSCREEN, BIOBALANCE or BIOCHLOR can calculate not only concentration but also flux as output information.

Whichever method is applied for mass flux calculation, a pumping test may provide useful information. A pumping test is a field experiment in which one of the wells is pumped at a controlled rate and the response is measured in the surrounding wells. The measured responses, i.e. the water levels in the wells, are used to determine the hydraulic properties of the aquifer: transmissivity, hydraulic conductivity (horizontal and vertical) and storativity (storage coefficient).

–   The *Sorbisense fluxsampler* can be applied for measuring water and contaminant flux as well as groundwater flow direction in groundwater wells. The fluxsampler serves for the direct monitoring of the Darcy flux and flow direction by a simple, hydraulically permeable polypropylene cartridge filled with soluble tracer salt. The tracer salt is environmentally friendly and dissolves proportionally to the volume of water passing through the cartridge. The flux can be determined from the emplacement duration and the residual salt mass. It can be combined with special sorbents (SorbiCell) for various contaminants (see also passive sampling in Section 8.3). The method enables measuring average flow volume over an extended time period (Sorbisense, 2015).

4.3.2   Water and contaminant flux in the vadose soil zone

–   *Passive capillary samplers* can directly measure water flux in the vadose zone. A specially designed filled funnel and an *in situ* conical collector ensure suction. The funnel is filled with local soil and a fiberglass wick, and is equipped with a heat pulse sensor which determines water flux from the temperature response following the application of a heat pulse. The flux of water carries proportionate heat, so the temperature response can be calibrated for water flux (Gee *et al.*, 2002; Gee *et al.* 2004).
–   *Lysimeters combined with a balance* have been developed by Meissner and Seyfarth (2004). The lysimeters are supplemented with a weighing function to measure the soil water balance of the soil monoliths in the lysimeters with high precision (± 30 g). The actual evapotranspiration can be determined from the results. The collected sample can be analyzed for nutrients or contaminants by complementing the weighing with sampling.
–   *Weighable groundwater lysimeters* have been developed in addition to weighable gravitational lysimeters by Meissner and Seyfarth (2004) for measuring the volume of water flown through the lysimeter during a certain period of time. Instead of a simple collecting tank, the developer applied a balancing tank with readable water levels.
–   *Fiber-optic sensors* have been applied for measuring soil water content (Alessi and Prunty, 1986; Garrido, *et al.*, 1999; Ghodrati *et al.*, 2000 and 2001).
–   *Solute transport in porous media* can be characterized via fluorescent tracers (Ghodrati, *1999*). This is a promising application for the vadose zone.

4.4   *In situ* characterization of polluted groundwater and the control of remediation

Remediation technology typically applies groundwater extraction, directed underground circulation or other directed flow types or recirculation of the extracted

Figure 4.8 TROLL® 9500 sensor system; flow-through chamber; field installation; measuring in a groundwater well (TROLL 9500, 2015).

groundwater. Many of the methods and sensors/instruments used for surface water or wastewater monitoring can also be applied to groundwater.

Online probes for measuring level, temperature, conductivity, dissolved oxygen (DO), pH, redox potential and nutrient content, various contaminant concentrations or adverse effects, i.e. toxicity in groundwater, may be the same as those for waters in general. The difference is that the devices used for groundwater are generally designed for more demanding environments and are equipped with abrasion-resistant foils which withstand fouling, high solid/sediment loads, and rapid flow rates.

A wide supply of sensors for measuring basic groundwater characteristics is available on the market. Multiparametric or 'smart' sensor-systems can measure a set of parameters. Some sensors from the selection are listed below.

– The *Aquistar® TempHion*™ smart sensor is a submersible water quality sensor and data logger capable of measuring pH, specific ions, redox, and temperature (*TempHion*™, 2015).
– The *INW multiparameter smart sensor* measures and records water level, pressure, temperature, pH, redox, conductivity and ions, e.g. chloride, bromide and nitrate (Multiparameter sensor, 2015)
– The *TROLL® 9500* multiparameter instrument, equipped with several sensors, is recommended for spot monitoring, low-flow groundwater sampling, soil and groundwater remediation and mine water monitoring (In-Situ, 2015). The sensors – recommended especially for the monitoring of soil remediation – make it possible to carry out real-time online monitoring of several chemical and biological soil treatment processes such as *in situ* chemical oxidation (ISCO), *in situ* chemical reduction (ISCR), biosparging (biodegradation in the saturated soil zone), and air sparging (*in situ* air stripping). The instrument shown in Figure 4.8 supports real-time measurement of all known performance indicators by specific sensors: dissolved oxygen (DO) and rigged DO (RDO) conductivity, pH, redox potential, temperature, barometric pressure and site-specific contaminants. With a S2XP restrictor it can log water level. Real-time information enables the dynamic work strategy and the Triad approach to be applied to site remediation (see also Section 2). The design allows the deployment of the probes in harsh

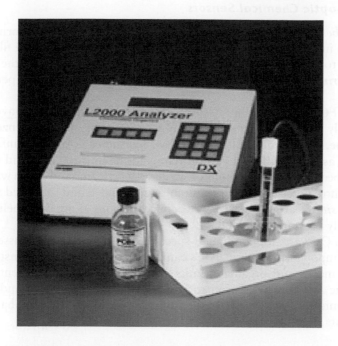

Figure 4.9 Field applicable measuring device with ion-specific electrode (L2000DX, 2015).

conditions: the corrosion-resistant housing is suitable for soil and groundwater remediation applications. It can be linked to a telemetry system for remote access and/or an automatic data collecting and reporting system.

–   PCB and other chlorinated contaminants can be analyzed quantitatively by the electrode-based ***L2000DX analyzer system*** (see Figure 4.9). The field applicable analyzer uses an ion-specific electrode to quantify the chlorinated contaminant, while contaminant identification is an additional task. The analyzer is preprogrammed with conversion factors for all major PCBs or chlorinated pesticides and solvents already identified (L2000DX, 2015). It applies a sodium-based reagent to remove the chloride from the PCB backbone and the chlorine-specific sensor for measuring the concentration of the chloride ion.

General water chemistry and contaminant content can be monitored in groundwater by methods similar to those described in Chapter 3. The samplers or sensors are placed into the groundwater flow (main flow or by-pass) when the groundwater is extracted, recycled or its flow is directed correspondingly. These processes are typical parts of the soil/groundwater remediation technologies, so they are useful both for contaminated site/soil assessment and technology monitoring. Nutrients and contaminants in the water can be measured by the mobile/transportable versions of conventional chemical, biochemical (enzymological, immunological or DNA) or biological (bioassay) methods or sensors.

4.4.1   Fiber-optic Chemical Sensors

Fiber-optic chemical sensors (FOCS) transport light to provide information about contaminants in the environment surrounding the sensor. The optical fiber may only be a conduit to transport the signal to the detector, e.g. placed on the soil surface, known as *extrinsic FOCS*. The laser-induced fluorescence (LIF) cone penetrometer is one example.

Intrinsic FOCS use the fiber itself as a detector. Intrinsic FOCS have been developed primarily to detect volatile petroleum constituents such as benzene, toluene, methylbenzene, and xylene (BTEX) as well as chlorinated volatile organic compounds (VOCs) such as TCE, PCE and carbon tetrachloride in water, air, and soil gas. The sensors have been developed to monitor waste water streams and storm water run-offs, to be placed into monitoring wells (both into unsaturated and saturated soil zones) or to provide *in situ* measurements of VOC concentrations along gas pipelines. Intrinsic FOCS typically measure total VOC.

Intrinsic FOCS are not compound-specific, as they respond to classes of VOC or semi-VOC compounds. Extrinsic FOCs can be chemical-specific. For instance, Raman is specific to metals and organic chemicals. LIBS is specific to elements and LIF to monoaromatic and polycyclic aromatic hydrocarbons. The Raman and LIBS instruments are semiquantitative to fully quantitative, but they generally have ppm level detection limits. SERS is capable of parts per billion-level analyses.

4.4.2   Laser-induced fluorescence

Hydrocarbons – the most common soil contaminants, including gasoline, kerosene, diesel fuel, jet fuel, lubricating and hydraulic oils, tars and asphalts – all contain PAHs, which fluoresce when irradiated by ultraviolet light.

Laser-induced fluorescence (LIF) is the best *in situ* field screening method for the qualitative and semiquantitative characterization of the distribution and delineation of soil contaminants containing PAH, typically non-aqueous phase hydrocarbons in groundwater, in saturated subsurface soils and in capillary fringes. Monoaromatics can also be detected by a UV LED (ultraviolet light emitting diode). The LIF sensors can be built into handheld devices to investigate the soil surface or deployed on cone penetrometers (CPT) or percussion direct push rigs to measure underground contaminants. In these cases the UV light source is placed in the cone itself. LIF sensors are not generally designed to detect dissolved-phase contaminants.

Fuel fluorescence detection (FFD) is suitable for differentiation between hydrocarbons. Because different types of PAHs fluoresce at different wavelengths, each has its own fluorescence signature. Using an instrument that measures the intensity and wavelength of the fluoresced hydrocarbon enables the assessment of the hydrocarbons present. This makes UV-induced fluorescence a useful technology to characterise subsurface and groundwater hydrocarbon contamination. There are various types of LIF (LIF, Clue-in, 2015) such as the LIF sensor of the site characterization and analysis penetrometer system (SCAPS-LIF, 2015), the rapid optical screening tool (ROST, 2015, Bujewski & Rutherford, 1997), the ultraviolet optical screening tool (UVOST, 2015), and the tar-specific green optical screening tool (TarGOST, 2015).

Vertek's in situ FFD measures fluorescence produced by aromatic hydrocarbons when excited by ultraviolet (UV) light. This tool is designed to screen and define subsurface plumes and provide discrete depths for soil and groundwater sampling. Vertek's FFD significantly reduces the time required to detect and characterize hydrocarbon fuels and volatile organic compounds (VOCs). The FFD probe system generates minimal investigation-derived waste, thus reducing containment and disposal costs (FFD, Vertek, 2015). There are two options with this system.

- High sensitivity option: by using photo multiplier tubes and traditional UV mercury lamps, the high sensitivity probe identifies low levels of contamination more readily.
- Continuous measurement: the Vertek's FFD provides a continuous, real-time output of fluorescence through the entire exploration. This information can be viewed graphically in real time using Vertek's Data Acquisition System (DAS) as the probe is advanced. Thus a more complete characterization of the plume's extent is obtained than in traditional discrete sampling techniques.

MIP, the membrane interface probe, has been discussed in the gas and vapor analyses in Section 3.2. It is a special sampling device that eliminates the disadvantages of groundwater sampling by pumps and lifts and separates volatiles after having brought up the sample volume. Volatile contaminants may be lost during groundwater sampling. This problem can be avoided by the deployment of substance-specific sensors in the groundwater and by *in situ* quantitative purging of the volatile components from the groundwater. This can be done by a MIP that is actually a heated semipermeable membrane, an interface between the soil matrix and the detector. The volatile molecules moving through the membrane are carried to the detector by a neutral carrier gas. The sensors are generally placed on the soil surface, and the carrier gas of constant temperature moves the vapors to the uphole detectors.

5   *IN SITU* CHEMICAL ANALYSIS OF METALS IN SURFACE AND SUBSURFACE SOIL

In situ analysis of natural or contaminating elements – similar to soil gas, vapor and water analyses – is possible on the soil surface, close to the surface or beneath it by direct push equipment or devices placed into wells.

5.1   X-ray fluorescence detection for surface and subsurface soil analysis

Portable *energy-dispersive X-ray fluorescence* (EDXRF) analyzers, commonly known as XRF analyzers, can quickly and non-destructively determine the elemental composition of metal products and precious metal samples; rocks; ores; soil; painted surfaces, including wood, concrete, plaster, drywall, and other building materials; plastics and consumer goods; dust; and airborne particles collected on filters.

EDXRF analysis is based on the characteristic X-rays emitted by atoms. The element content can be identified and the concentration of each element determined by measuring the peak energies of X-rays emitted by a solid sample exposed to the radiation source, and the elements present in the sample can be identified.

XRF can detect the presence of polychlorinated biphenyls (PCBs) in various media (soil, paint, personal protective equipment, liquids, and oils) by measuring the total chlorine concentration.

An XRF sensor can be employed with a cone penetrometer for real-time, *in situ* field screening of heavy metals in subsurface soils. The XRF sensor system uses an X-ray source located in the probe to bombard the soil sample with X-rays. The emitted fluorescence is detected in the subsurface, and the data are transferred by wire to a data logger.

XRF may have difficulty in detecting very small concentrations of some metals, but these detection limits are generally far below regulatory limits.

Handheld battery-operated XRFs are useful to detect (toxic) metals and waste containing toxic metals on the soil surface. The handheld device performs *in situ* real-time analyses to identify and quantify up to 40 elements including some light elements such as Mg. XRF is a field screening device. Its readings need to be recalibrated by laboratory analysis results at regular intervals. It is ideal for mapping and delineating pollution and finding hot spots on large areas. Some commercially available devices are shown in Figure 4.10 and listed below.

– The *S1 TITAN*^{TM}, a tube-based XRF analyzer with ergonomic pistol grip and trigger, is designed for all-day use. The color touch screen LCD is easily seen in all lighting conditions. With a weight of just 1.5 kg it is among the lightest on the market, according to the manufacturer. It can be configured with either the performance-based silicon drift detector (SDD) or the economical silicon PIN diode (Si-PIN) detector. The fast SDD operates at high count rates and provides excellent precision in short measurement times. In addition, it allows for measurements of light elements such as magnesium, aluminum, and silicon. The Si-PIN detector is an option when detection of light elements is not required. It is characterized by good precision and accuracy but slightly longer measurement times (Titan, 2015).
– The *DELTA* Handheld XRF analyzers are suitable for identifying metal contents which do not meet the regulatory requirement of USEPA 6200, ISO/DIS13196

Figure 4.10 X-Met 8000 (Oxford), Delta (Olympus), S1 Titan (Bruker) and Niton XL3t (Thermo Analytics) handheld XRF devices.

and other SOPs. The handheld device is recommended for soil and sediment (in the field as well as for bagged soil, and soil or sediment cores) and for screening surfaces (solids and fluids) and dust (dust wipes and filters) primarily for Pb, Cd, Cr, As, and Hg. The DELTA series analyzers are configured with miniature X-ray tubes; Si-PIN detectors or highly advanced silicon drift detectors (SDD); specialized filters; and multi-beam optimization (Delta XRF, 2015).

- The **X-MET8000** Series XRF analyzers use the optimized combination of a high performance X-ray tube and a large-area silicon-drift detector (SDD) and include three models to various analysis needs and budgets. Along with the X-MET8000 Expert, the new models X-MET8000 Optimum and X-MET8000 Smart enable reliable, uninterrupted metals analysis, all day long, and in the harshest environments (X-Met 8000, 2015).
- The SPECTRO xSORT family of handheld energy dispersive X-ray fluorescence (EDXRF) spectrometers supplies elemental testing and spectrochemical analysis of solid materials in widely varying conditions. The manufacturer emphasizes repeatability and the laboratory-quality of the results (SPECTRO, 2015).
- The **Niton XL3t** XRF analyzer is also a handheld instrument recommended for mining and exploration, ore grade control, geochemical mapping and environmental analysis. Its prime tasks include soil and sediment testing; soil contamination mapping and delineation; soil remediation; and solid waste testing and control, as well as dust and consumer goods analysis (Niton, 2015). Several devices are available for a variety of purposes, and the newest GOLDD technology covers geometric advantage, optimized excitation, and a large drift detector. Each version displays actual metal values in seconds and can be adjusted to comply with local regulations. The devices offer point-and-shoot simplicity; they are sealed against moisture and dust, and are equipped with a color touch-screen display with daylight-readable icons. The additional software allows the generation of custom reports, printed certificates of analysis, or to remotely monitor and operate the instrument hands-free from the computer (Niton, 2015).

The above portable equipment can be characterized by several benefits and limitations.

- **Benefits:** *in situ* use; non-invasive analysis, without or with simple sample preparation; rapid measurement; immediate results; high sampling density; easy operation; minimal training required; user-friendly interface; and a minimum 10–12-hour battery life. Most of them provide powerful data management and simple data download, an extensive library and an integrated GPS and camera for accurate measurement positioning. They are suitable for a wide range of applications: mineral confirmation and mine mapping; compliance testing; contaminated site mapping; hot spot identification; soil analysis; monitoring of soil removal and remediation; and waste disposal testing (RoHS, 2003), etc.
- **Limitations:** sample heterogeneity and humidity influence the result; matrix interference and chemical interference (overlapping emission lines) may occur; confirmatory laboratory analysis may be necessary; limitations in light metals analysis; and they measure total element content independent from mobility and bioavailability.

5.2 Laser-induced breakdown spectroscopy

Laser-induced breakdown spectroscopy (LIBS) is a type of atomic emission spectroscopy for solid environmental samples (soil) and wastes. A highly energetic laser pulse is used for excitation. The laser is focused to form plasma, which atomizes and excites the sample. It is suitable for gas, liquid and solid phase samples, so it is an ideal tool for soil analyses.

The basis of its function is that the characteristic frequencies of the plasma emission spectrum of the elements in the excited sample are detected by a sensitive detector and evaluated by applying software. The results may show the relative abundance of known elements or the presence of contaminants. LIBS is similar to other laser-based analytical techniques such as vibrational spectroscopy, Raman spectroscopy, or the laser-induced fluorescence (LIF) technology – and manufacturers often combine these techniques in a single instrument. Compared to XRF, the portable LIBS (5–6 kg) is more sensitive, faster and can detect a wider range of elements (particularly the light elements). It is also less hazardous, given that it does not use ionizing radiation to excite the sample, as XRF does.

LIBS just needs optical access to the targeted object, nothing else. The detector can be placed very close to the excited sample, even coupled to an optical microscope, or adjusted to larger distances. It can be employed for remote analyses when coupled to appropriate telescopic apparatus.

LIBS's detection limit is a function of the plasma excitation temperature, the light collection window, and the line strength of the viewed transition. High resolution is extremely important in soil, which has a very complex matrix with several elements.

Kram *et al.* (2000) reported the downhole application of LIBS. Ismaël *et al.* (2011) used LIBS for optimizing sampling operations by preparing a LIBS map directly on site, providing valuable information on subsurface lead and copper distribution.

The U.S. Department of Energy (DOE) National Energy Technology Laboratory (NETL) has developed an integrated laser-induced breakdown spectroscopy (LIBS) and cone penetrometer technology (CPT) system to analyze metals in subsurface soils at a 5 m depth. LIBS and the optical fiber were integrated in CPT, but LIBS can also be deployed in already existing (small diameter) wells or holes in the shallow subsurface (Cohen & Saggese, 2000).

Some commercially available equipment are listed below.

– The *portable LIBS* is for rapid, real-time environmental investigations and does not require sample preparation. Its detection level ranges from parts per million to parts per picogram across the 200–980 nm wavelength range, with ~0.1 nm optical resolution. The standard device includes 2 m optical fiber (Ocean Optics, LIBS, 2015).
– The *PPO-LIBS* (2015) spectrometer system uses a high-efficiency transmission grating combined with a special optical design maximizing the number of photons that reach the ICCD (intensified charge-coupled device) detector. Its rugged structure makes it suitable for field use and remote detection.
– The *ARYELLE 150* (2015) is a compact high-resolution echelle spectrometer for LIBS. It is fiber-coupled to a CCD image detector. Its small size ($12 \times 17 \times 8.6$ cm)

and small mass (2 kg without the detector) makes it particularly suitable for use with portable devices.
– The *mPulse*[TM] (2015) is a hand-held portable metal analyzer developed for alloy sorting purposes, but it is recommended for environmental purposes to identify pollution in soil by metals and from mining.

5.3   DGT method for measuring metal concentration and plant uptake in contaminated and remediated soil

The *in situ* device, based on measuring the diffusive gradients in thin films (DGT), was developed between 1990 and 1995 (Davison *et al.*, 1994; Davison & Zhang, 1994; Zhang *et al.*, 1995a and 1995b). It has become a popular tool in the last ten years for measuring 'labile' or 'mobile' metals, not only in water but also in sediments and soils (DGT Research, 2003; Davison & Zhang, 2012) to estimate the bioavailable fractions in soil and sediment (see also Hajdu & Gruiz, 2015).

DGT directly measures the mean flux of mobile/labile species (including free and kinetically labile metal species) to the device during deployment. The measured concentration represents the supply of metal to any sink, be it DGT or an organism that comes from both diffusion in solution and release from the solid phase. It has been proved by comparative studies that the measured concentration is suitable to measure approximate bioaccessibility of metals, so it can be a simple surrogate measurement of the biologically potentially effective concentration (Søndergaard *et al.*, 2014). It supports well the presently accepted Free-Ion Activity Model (FIAM), which stipulates that the biological response of organisms to metals in water-based systems is proportional to the free-ion activity of metals and not to their total or dissolved concentrations (Zhang *et al.*, 2002; Gimpel *et al.*, 2002). That is why DGT uptake highly correlates with plant uptake, as shown by Davison *et al.* (2000a & 2000b).

Zhang *et al.* (2001) investigated copper-accumulator plants in 29 soils and found a high correlation between the DGT measured concentration of copper (c_{DGT}) and the effective concentration taken up by the plant (c_e). The c_e value is resulted from the concentration existing in the soil water and supplied by the solid. Nolan *et al.* (2005) compared chemical speciation, DGT, extraction, and isotopic dilution techniques and found that cadmium and zinc concentrations measured by DGT highly agree with wheat uptake. The explanation may be that the kinetics of cadmium and zinc release from the solid phase is similar or identical to the kinetics of metal supply to plants. This assumption was supported by Zhang *et al.* (2004), who performed measurements on different soil types and found that the plant available cadmium and zinc concentration in clayey soils was equal to the concentrations in the soil solution. However, prior to using the DTG device for sandy or other low sorption capacity soils, the kinetics of metal release from solid to soil solution has to be explored.

DGT can be applied *in situ* but not online, because the DTG probe should be removed from the soil or sediment for further analysis of the absorbed metals. It can be considered as a sampler, providing samples containing bioavailable metals for chemical or toxicological studies. Chen *et al.* (2013) combined the use of DGT with *in situ* XRF measurement in the DGT.

5.4   Biosensors for metals

Biosensors for toxic metal analysis in soil and groundwater function by sensing the interaction between a metal in the environment and a biomolecule or living cell. Biomolecules selectively interacting with certain metals typically are enzymes; chelating molecules; regulating proteins; immune molecules; participants of the natural nutrient uptake and molecules responsible for resistance mechanisms. Those organisms causing interactions are dominantly micro-organisms or plants (Lasat, 2002; Verma & Singh, 2005; Belkin, 2006; Kahru *et al.*, 2008; Olaniran *et al.*, 2013). It is important to note that the sample matrix significantly affects the interaction of biosensors with real samples, so the predicted values can deviate from the measured ones to a large extent. Therefore, calibration to the matrix is an important step in sensor use for environmental samples.

5.4.1   Biosensors based on metal–protein interactions

Several proteins – both enzymes and non-enzymes – can be selectively bound to metals and serve as a basis for analysis and sensor design. Enzymes built into biosensors may be inhibited or activated by their bonds with certain metals.

Enzyme inhibition-based biosensors use urease, pyruvate oxidase or other oxidases and dehydrogenases. Such electrodes can be regenerated by EDTA or dithiothreitol, and a buffer containing Mg^{2+} and thiamine-pyrophosphate was used for pyruvate oxidase.

Enzyme activation-based sensors such as the alkaline phosphatase synthetase or glutamine synthetase activation-based sensors measure the increased enzyme activity which is the result of a metal functioning as a cofactor to the enzymes. Zinc acts as a cofactor for alcohol dehydrogenases, carbonic anhydrases or DNA polymerases; copper for cytochrome oxidase; magnesium for glucose 6-phosphatase, hexokinase or DNA polymerase; Mn for arginase; Mo for nitrate reductase and nitrogenase, and Ni for urease.

Non-enzyme type proteins such as the chelating proteins can selectively bind to metals. Both naturally occurring metal-binding proteins and, more specifically, engineered molecules have been built into sensors. The fusion of metallothionein of a *Synechococcus sp.* with glutathione-S-transferase is one example for biosensor development in the analysis of Zn^{2+}, Cd^{2+}, Cu^{2+} and Hg^{2+}. A synthetic phytochelatin was built into a capacitive transduction biosensor for sensing Zn^{2+}, Cd^{2+}, Pb^{2+} and Hg^{2+}.

Immunoassay-based metal sensors are characterized by high sensitivity, selectivity and species-specificity and are applicable to any pollutant for which a suitable antibody can be generated (Verma & Singh, 2005).

Regulatory proteins can be very selectively bound to metals such as CueR, a member of the MerR family of transcriptional activators in *Escherichia coli*. The CueR-dependent regulation of RNA polymerase transcription showed 10^{-21} mol sensitivity to Cu^{2+} (Changela *et al.*, 2003).

DNAzymes (Chapter 2) on gold nanoparticles results in a high-activity and -sensitivity biosensor such as the DNAzyme for the detection of Pb^{2+} (Liu & Lu, 2003).

5.4.2 *Whole-cell tests and sensors for measuring bioavailable toxic metals*

Bioavailability of toxicants is proportional to their adverse effects and risks. Bioavailability is responsible for the actual effect and reflects the current risk of a toxic metal posed to a certain organism. It differs from water availability, as it depends not only on the environmental conditions but also on the properties of the interacting biological system. Bioavailability is a useful piece of information in assessing toxicity and risk of metals on living organisms in environmental matrices.

Genetically modified bacteria can be used as whole-cell biosensors for the detection of the bioavailable metal fraction (and other inorganic and organic compounds, including nutrients and environmental toxicants) in various environmental matrices such as soil, sediments, water, and leachates. The genetically modified bacteria contain a built-in regulator gene for sensing the contaminant and making the reporter gene produce a detectable signal (see also Chapter 2). The presence of the specific toxic metal triggers the regulator gene to switch on the transcription of the reporter gene and the production of the enzyme luciferase, which causes the bacteria to emit light. The quantification of the emitted bioluminescence can be recorded by a luminometer and compared to an internal calibration realized with known concentration of metal ions (Corbisier *et al.*, 1999; Ivask *et al.*, 2009). Bacterial sensors specific for Zn, Cd, Cu, Cr, Pb (Corbisier *et al.*, 1999) and Ni (Tibazarwa *et al.*, 2001) have been fully characterized in terms of specificity, detection limits and selectivity. A published development is the luminescence response of *E. coli* bearing a construct of two plasmids in which the metal-inducible zntA and copA promoters from *E. coli* were fused to a promoterless *Vibrio fischeri* luxCDABE operon. The genetically manipulated *E. coli* was studied as a function of the concentration of several toxic metals (Riether *et al.*, 2001). The very first cell constructed for biosensing was the green fluorescent, protein-based bacterial biosensor with an *E. coli* strain as a host cell. It is based on the expression of green fluorescent protein under the control of the cad promoter and the cadC gene of *Staphylococcus aureus* plasmid pI258. The sensor *E. coli* mainly responded to Cd(II), Pb(II), and Sb(III). Lateron, similar gene combinations have been used for several fluorescence-based biosensors to measure bioavailable toxic agents in the environment (Li *et al.*, 2008).

More information on heavy metal bioavailability and toxicity and their measurement by whole-cell sensor technology is available in Turpeinen (2002) and Belkin (2006). Strosnider (2003) published an interview on bioavailable arsenic detection by bacterial biosensors.

6 ON-SITE APPLICABLE RAPID METHODS AND COMMERCIAL KITS FOR SOIL ANALYSES

Colorimetric reagents and indicators represent the oldest and most traditional branch of analytical chemistry. Their revival as rapid environmental test methods can be explained by the urgent need for rapid, cost-effective, real-time – or near real-time – analytical methods, applicable in the course of fieldwork. Contaminated site investigators have to collect information on site about the presence of possible sources of contaminants and delineate the boundaries of the contaminated area. The managers of contaminated sites have to carry out as much efficient site investigation as possible by applying a tiered assessment strategy and the Triad approach. Field-applicable, rapid analytical methods are commercialized as kits, marketed on their ease of handling, and low, or no, need for a laboratory. Test kits offer ready-to-use formulation, packing and dosing, ensuring laboratory quality from the start to finish. The kit includes everything that is necessary for the successful implementation of the test for the desired size and purity. Generally, easily understandable, step-by-step instructions are also presented for less experienced personnel.

Although most environmental test kits are designed for field use, many of them are used in a sample trailer, mobile laboratory, or other fixed location. This is mainly true for the quantitative methods requiring colorimeters or other equipment. The most demanding kits are the immunoassay kits, the enzyme assays and the miniaturized portable PCR and other molecular test systems. These may require more instrumentation, refrigeration, a controlled climate, and other conditions. The EPA has approved immunoassay methods for a number of contaminants (see also Chapter 3), and enzyme assays are also available for food and health applications, but only a few manufacturers produce field-applicable kits for environmental purposes (see also Clue-in, immunoassay, 2015). Some field-applicable environmental test kits are briefly introduced below, starting from the simplest colorimetric method and running to the field versions of DNA techniques such as PCR.

6.1 Commercially available soil testing methods and products

A wide selection of colorimetric kits is available from the largest manufacturers and distributors, such as Aqualytic, Fisher Scientific, Cole Parmer, Hach, Merck Millipore, and Systea. These kits are primarily designed for water and wastewater but can equally well be applied to groundwater, various soil waters, leachates and soil extracts. When using the aquatic test kits for soil extracts, the evaluator should know how to carry out the extraction; which pH, redox potential or salt concentration to use; and how to interpret the results. The aquatic extracts in such a simplified model are treated as the available fraction of the nutrients or contaminants. This approach does not take into account the dynamic nature of solid-water interactions and the biological impacts on nutrient and contaminant mobility.

6.1.1 Basic soil characteristics

Some manufacturers of rapid test kits focus on soil in particular, two such manufacturers are Eijkelkamp and ModernWater, which offer kits for general agricultural and environmental purposes.

-   The *soil test kits* of Eijkelkamp (2015) provide simplified methods of determining available nutrients in agricultural soils. Rapid, colorimetric chemical tests use standardized reagents and the developed color is visually compared to color charts. A simple aqueous soil extract results in the available nutrients. Chloride can be extracted by demineralized water, and pH is also measured.
-   The *Palintest* (2015) soil test kit, similar to the previous ones, has been designed for agricultural and environmental use, and can measure a complete range of

soil macro- and micronutrients. It measures nitrogen, phosphorous, potassium, magnesium, calcium and sulfur, aluminum, copper, iron, manganese, chloride, conductivity, salinity and lime requirement of the soil. A photometer is included in the set for colorimetry, and a pocket sensor for soil pH, conductivity and salinity.

6.1.2 Contaminant-specific rapid tests

Field-applicable forms of laboratory analysis use physical, chemical, enzyme- and immunoanalytical methods as kits, or kit sets. Some commercially available kits for rapid field use are introduced in the following list.

- The *PetroFLAG* (2015) method has been developed for total petroleum hydrocarbons (TPH) analysis in soil, based on turbidimetric screening. The field test kit for a broad-range hydrocarbon analysis works with special reagents: an extractant (methanol) and the aqueous emulsifier development solution (diglyme: diglycol methyl ether). The high-range extractant dissolves, and the developer precipitates, the hydrocarbon from the supernatant in suspended form. The turbidity of the suspension is measured as an end point. The analysis range is 10–2,000 ppm total hydrocarbons, without specifying the components of the mixture. As a screening tool it is fast and cheap: suitable for the exclusion of negative samples from a second tier assessment (PetroFLAG, 2015).
- The *CLOR-N-SOIL 50 PCB* screening kit for polychlorinated biphenyl (PCB) field detection is based on colorimetric titration with a test time of only ten minutes. The kit can test any type of soil, including sand, topsoil, sediment, and clay. As with other screening tests, the decision on the necessary following steps can be made in the field based on the results. Its main advantage is that negative samples can be screened at low cost (CLOR-N-SOIL, 2015).

Immunoassays are widely used in rapid soil analysis based on the selectivity and sensitivity of the immune reaction. Immunoassays allow matrix-independent contaminant and toxin analysis in complicated matrices such as soils, sediments and solid waste samples. Most commercially available kits apply a competitive immunoassay method for the targeted analyte. This method applies a labelled analyte bound to an analyte-specific antibody that is immobilized on a membrane or other solid surface. When the sample contains the targeted analyte, an exchange reaction occurs, and a part of the labelled and immobilized molecule is replaced by the sample analyte, proportionally decreasing the measurable (label-specific) end point.

- The *Modern Water* company (2015) has a wide variety of test kits, which include the following products: EnSys™; RaPID Assay®; EnviroGard® and QuickChek™. The analyses of soil samples require prior extraction from the soil, for which a soil extraction kit is provided.
  - The Ensys™ methods are rapid field or laboratory colorimetric tests for the analysis of explosives such as TNT (trinitrotoluene) and RDX (Research Department explosive, cyclotrimethylenetrinitramine) in soil.
  - The RaPID Assay® enzyme immunoassay is a field and laboratory method for the analysis of TPH, BTEX, pesticides (PCP, atrazine, 2,4-D), PAH,

carcinogenic PAHs (CPAHs) and PCB in soil and water. It applies a new, antibody-coated magnetic particle technology to detect the target analyte.

o EnviroGard® is a rapid field or laboratory immunoassay method for the analysis of pesticides (chlordane, DDT, triazine), microcystins (algal bloom toxins), other algal toxins and PCBs in soil and water. It uses an antibody-coated test tube or plate technology to detect the analyte of concern.

o The QuickChek™ system is a rapid enzyme immunoassay method for the detection of sulfate reducing bacteria (SRB). A characteristic enzyme of SRBs, the adenosine-5′-phosphosulfonate reductase is detected by fixed antibodies.

– **Hach immunoassay** (2015) test kits are available for alachlor, atrazine, TPH and PCB. They are equipped with a Pocket Colorimeter II (PCII) instrument and a soil extraction device.

– The **flow assay and sensing system** (FAST) is a rapid, portable, fluorometric assay system (FAST, 2015). The FAST 6000 can perform six simultaneous assays for six different analytes on the same sample (Figure 4.11). The time requirement for a measurement is two minutes.

– **Delfia® labelling** (europium chelate-labelled antibody), combined with time resolved fluorescence intensity measuring immunoassay, significantly increases the signal-to-noise ratio (Delfia TFR, 2015).

– **Enzyme assays** are less popular in environmental analysis, but the effect of carbamate and organophosphate pesticides can be traced by measuring the inhibition of the enzyme acetylcholinesterase (AChE) on a substrate, acetylthiocholine (ACE).

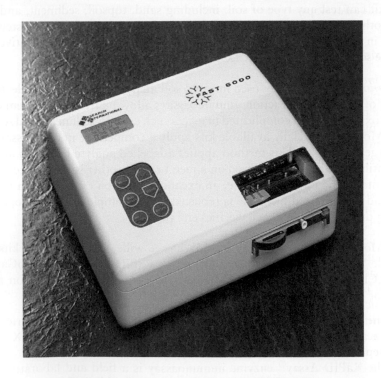

Figure 4.11 Fluorometric assay system (FAST, 2015).

The hydrolysis of ACE by AChE can be detected with the help of the 5,5′-dithiobis-(2-nitrobenzoic acid) reagent, resulting in a yellow color. If AChE is inhibited by organophosphates/carbamates, the color intensity is reduced compared to a control. The intensity of the yellow is measured by a photometer or estimated by visual comparison. As the enzyme/substrate reaction – and so the inhibition – is accompanied by a change in pH, it also can be measured as an indicative end point (ACHE, 2015).

6.1.3 Tests for detecting and measuring toxicity

Rapid toxicity testing kits may also use enzyme systems isolated from bacteria, or without isolation in entire organisms such as freshwater crustaceans, bacteria, or algae. The enzyme systems are linked to fluorescent markers that emit light if the system is functioning, but the light turns off when the toxins inhibit enzyme function. Several toxins such as botulinum, cyanide, ricin, thallium sulfate, and nerve agents can be detected in drinking water, soil and sediments.

– The *Toxi-ChromoPad*[TM] (2015) kit is a bioassay screening tool used to determine acute toxicity in soil, sediment, sludge or other solid waste material directly, without extraction. The Toxi-ChromoPad[TM] bioassay uses a special *E. coli* of wide-range sensitivity. The assay allows the bacteria to grow in direct contact with the toxicants in the sample. The end point of the assay is a color reaction. If the sample is toxic, no color will develop; if the sample is non-toxic, a distinctive blue color develops around the sample.

6.2 Rapid, on-site applicable PCR systems

The *real-time microchip PCR system* for portable plant disease diagnosis is designed for the rapid and accurate detection of plant pathogens in the field. It is crucial to prevent the proliferation of infected crops. The polymerase chain reaction (PCR) process is the most reliable and accepted method for plant pathogen diagnosis; however, current conventional PCR machines are not portable and require additional post-processing steps to detect the amplified DNA (amplicon) of pathogens. Real-time PCR can directly quantify the amplicon during the DNA amplification without the need for post-processing, and thus it is more suitable for field operations. However, this still takes time and requires large instruments that are costly and not portable. Microchip PCR systems have emerged in the past decade to miniaturize conventional PCR systems and to reduce operation time and cost. Real-time microchip PCR systems have also emerged but, unfortunately, all real-time microchip PCR systems declared as portable require various auxiliary instruments. In the following, a real-time microchip PCR system will be described that can diagnose plant diseases. It comprises a PCR reaction chamber microchip with integrated thin-film heater, a compact fluorescence detector to detect amplified DNA, a microcontroller to control the entire thermocycling operation with data acquisition capability, and a battery. The entire system is $25 \times 16 \times 8$ cm in size and weighs 843 g. The disposable microchip requires only an 8-μl sample volume and a single PCR run consumes 110 mAh power. A DNA extraction protocol was also developed for field operations. The developed real-time microchip PCR system and the DNA extraction protocol were used to successfully detect six different fungal and bacterial plant pathogens to a detection limit of 5 ng/8 µl sample (Koo *et al.*, 2013).

– **R.A.P.I.D.®**, the 'ruggedized advanced pathogen identification device', is a portable real-time PCR system designed for the identification of biological agents. Because of its rugged design, reliability and accuracy, it has become the standard for the U.S. Department of Defense (DoD) and other militaries around the world. The R.A.P.I.D. is part of mobile analytical laboratories and field hospitals. It includes a PCR cycler and software which simplifies use and evaluation (RAPID, 2015).
– The *Palm PCR G1-12* is a portable high-speed mini PCR system with three speed levels. It can produce 30 cycles in 18–20 minutes in 'turbo fast' mode (Ahram, 2015).

7   MEASURING THE RESPONSE OF LIVING ORGANISMS – SOIL BIOLOGY AND TOXICOLOGY

Some *in situ* biological and toxicological methods using whole cells and living organisms are briefly discussed here, emphasizing *in situ* methods which produce real-time signals. The topic of soil toxicity testing is discussed in detail in Volume 2 of this book series (Gruiz *et al.*, 2015).

7.1   *In situ* rapid tests based on microbial response

Microorganisms are ready to respond immediately to the occurring stress; thus many of the rapid methods are based on the response of indigenous or specific test species or of the soil microbiota. Some commercially available methods in the form of mobile laboratories of field kits are introduced below. Kits have the advantage of controlled test organisms and auxiliaries, as well as precise instructions.

7.1.1   Toxicity measuring methods

– *Microtox®* (2015), for measuring soil toxicity *in situ*, includes a highly sensitive luminometer with built-in software. It applies a freeze-dried bacterial reagent, and ready-made test control and reconstitution solutions. Microtox is mainly used for groundwater, pore water, soil extract and leachate. Its use for solid soil is limited for technical reasons because the solid particles absorb a large and hard-to-determine part of the luminescent light to be measured by the photodetector. When testing whole soil in a suspension, the test organism is in direct contact with the soil. Bioavailability of the contaminants can be increased by direct contact between sorbed contaminants and the test organism. There are methods of working with carefully selected diluents and reference soils which are comparatively assessed to compensate for light absorption by the solid phase or to calculate the dilution-proportionate solid absorption from the zero time absorption value and use this for compensating the measured results. Ashworth *et al.* (2010) proposed a new method using a modified double-cuvette for correcting the effects of color and turbidity on bacterial light output measurements in the Microtox bioassay.

- *LUMIStox* (1999), from the Hach Lange company, is verified by US EPA ETV (LumiStox, 2010) for groundwater or soil leachate, but many users apply it for soil and sediment direct contact tests, too.
- The *BioTox*^TM (2015) luminometric toxicity test is a plate luminometer.
- The *ToxTrak* (2015) toxicity test kit uses a colorimetric method to determine toxicity with a spectrophotometer or color disc comparator. Percent inhibition either on plant biomass or freeze-dried bacteria can be measured. It is an inexpensive alternative that has results comparable to respirometric methods that measure dissolved oxygen consumption (ToxTrak, 2015)
- The *ECLOX* (2015) *rapid response* test kit, in addition to chemiluminescence toxicity screening, allows measurement of pH, color, total dissolved solid, chlorine and pesticides.
- The *flash test* is an improved version of the *BioTox*^TM (2015) test, designed for solid samples. The signal is continuously recorded for 30 seconds in kinetic mode. The maximum signal received immediately after dispensing (I_{peak}) is compared to the signal after a 30-second incubation period (I_{30s}). Light intensity decrease can generally be observed after a few seconds of incubation; however, some chemicals need longer contact times to become effective. Thus, kinetic data from samples after 15 or 30 minutes give an additional dimension for obtaining reliable results.
- *Deltatox*® (2015) is a portable rapid toxicity monitor by Modern Water. Soil samples need preceding sample preparation.
- *MicroBio Tests Inc.* (2015) produces simple, low-cost field applicable kits for soil with dormant or immobilized test species which can be reactivated at the time of performance of the bioassays. The available test organisms of diverse phylogenetic groups correspond to standard toxicity tests prescribed by environmental legislation at national and international level. Kit-type products are available for toxicity and bacterial contamination screening, packed in a luminescence measurement case.

7.1.2 Soil respiration-based method

Soil microbiological communities sense, and respond to, chemical stress and the presence of toxic substances (e.g. pesticides). Responses may appear at each level: from molecular to community level. Behind the resultant changes in population density and species diversity, rearrangement of the metagenome can be detected. Metabolic changes such as decreased respiration, substrate utilization or mineralization can be monitored at the level of genes, enzymes, or enzyme products. Soil respiration is generally measured by measuring the final product, i.e. CO_2. The cellulase enzyme activity of the detritus (community biodegrading organic waste in the soil) may be followed by a simple cotton strip assay detecting its biodegrading activity using the soil litter bag test. Soil microcosms – small-scale tools simulating the soil's biological state and activity – indicate toxicity with decreased rates of respiration, mineralization and other activities. Terrestrial microcosms for substrate-induced respiration technique (SIR) are systems well known for the dynamic testing of the health and adaptability of the soil microbiota. Its essence is that the respiration rate is measured in the soil before and after the addition of an easy-to-use substrate, e.g. glucose, glutamic acid, mannitol and/or amino acids. Alternatively, site-specific biodegradable substrates can also be used. The response to the increased amount of substrate/nutrient is measured via respiration, i.e. CO_2 production or O_2 consumption. The results obtained by SIR are used both for biomass estimation and toxicity assessment (Gruiz *et al.*, 2015a).

Kaur *et al.* (2015) modernized the conventional SIR method by the application of a non-dispersive infrared (NDIR) CO_2 sensor, and applied the method for assessing and monitoring soil toxicity in the course of soil remediation. They deployed an inexpensive NDIR CO_2 sensor to successfully differentiate the control and a diesel-contaminated soil in terms of CO_2 emission after glucose addition. This *in situ* applicable sensing system provides information on soil microbial activity, an indicator of soil quality. This sensor-linked microbiological method is suitable for the initial screening of contaminated sites and for mapping soil adaptability and health at a relatively low cost.

Commercially available systems for measuring soil respiration, and SIR systems, are suitable for field measurement in mobile laboratories or on-site platforms. This simple equipment can measure soil respiration in the soil and evaluate it in comparison to reference soil. The changes in soil respiration can also be measured after the addition of substrates or toxic agents. The substrate can be a readily mineralizable sugar or one of the contaminants included in the assessment concept. It is typically used to measure the aerobic respiration of chemoorganotrophic organisms by measuring CO_2 production, O_2 consumption via direct CO_2 analysis, or sequestration in alkali. Closed-bottom systems measure pressure changes with parallel CO_2 sequestration. Alternative respiration forms can also be monitored by product analysis or pressure changes. Optimal size and soil-air ratio can minimize the time requirement to a few hours.

Soil respirometry applications are the following: short- or long-term respiration/biodegradation tests under aerobic, anoxic or anaerobic conditions; biodegradation of certain materials e.g. plastics; treatability of contaminated soil by biodegradation-based technologies; and testing the effect of bioactivators and enhancing additives. A selection of commercially available soil respirometer systems will be described below.

- The **Quick Scan** soil and compost analyzer by Challenge Technology measures oxygen uptake rates or other gas production from biological activity under user-defined conditions (Quick Scan, 2015).
- The *OxiTop*® automated closed-bottle system is for measuring basal and induced respiration. Basal respiration refers to soil respiration without the addition of organic matter. Substrate-induced respiration (SIR) refers to soil respiration after the addition of an easy-to-metabolize carbon source for micro-organisms. It can be adapted to any other user-specified purpose such as stress response, adaptation, treatability or toxicity tests (OxiTop, 2015).
- The **Q-Box SR1LP** package is an *in situ*-applicable, rapid method, measuring soil respiration rates with a field-portable gas exchange system. It uses a chamber placed over the undisturbed soil surface, or soil samples that are placed in a flow-through chamber. The box system in a weatherproof case includes a NDIR CO_2 analyzer; a relative humidity and temperature sensor; a gas pump and mass flow monitor; a soil moisture probe; a soil chamber with collar; a flow-through chamber; a six channel data interface, and data acquisition software (Q-Box, 2015).

7.1.3   Physiological profile of soil microbiota

The community-level physiological profile and the metabolic activity of soil microbiota are complex indicators that reflect the health and metabolic changes of the hidden microbial community. The microbial community response can be considered as an early warning of adverse changes related to higher ecosystem members. The physiological profile covers the utilization of a set of suitably selected substrates in ready-made microplates (e.g. EcoPlate™) or the compilation of targeted, problem- or site-specific substrate assemblies (see more in Chapter 2).

The *Biolog® EcoPlate™* was created specifically for community analysis and microbial ecological studies. It provides replicate sets of tests. The wells of the multi-well plate contain 31 typical substrates (carbon sources) in three replicates. Inoculation of the plate with a mixture of micro-organisms (e.g. those within a soil suspension) develops a community metabolic fingerprint or physiological profile after incubation. This profile can distinguish temporal changes in the microbial community. It can be used to characterize normal soil population or to follow the impact of environmental conditions or chemical stress. The adaptability and biodegradative potential of the soil microbiota can be tested, or the ready-made EcoPlates can be used as a toxicity bioassay (Biolog, 2015).

Species density and diversity changes are evaluated from long-term serial data, which are generally not available. Some indicator organisms, such as earthworms or Collembola, are selected and monitored for the follow-up on community- or ecosystem-level changes. An example of European soils is introduced in Section 7.2. Even very simple tools, such as pitfall traps for collecting insects, can be used to acquire the required information for a generic survey of contaminated sites, landfills or areas next to a reference site. Abundance, diversity and bioaccumulation of the animals that have fallen into the pit can be determined in comparison to the reference.

7.2   Terrestrial vegetation – indicators and monitoring tools

Terrestrial plants are in intimate contact with soil through their roots, which are not simply fixed in the soil: the root exudates and the cooperating root-microbiota play an active role in the extraction of nutrients from the soil. Native vegetation; species diversity; plant structural, functional and compositional attributes; production; and (nutrient) contents and bioaccumulation, as well as invasive species and pests, are all suitable indicators for environmental monitoring. More refined plant responses are relevant for detoxication or defense mechanisms, phytosensing of pathogens, genetic regulatory elements such as adaptive enzymes, xenobiotics-responsive or ozone-induced omics (Gruiz *et al.*, 2015a). Some of the same indicators can be used for agricultural purposes to observe vegetation conditions.

The plant-related indicators vary from (i) conventional density, diversity and composition assessments, through (ii) sensor-based chemical detection of genetic, enzymatic or other molecular markers and accumulated contaminants, to (iii) hyperspectral remote sensing of diversity; chlorophyll or carbon content; and carbon flux for global scale environmental – e.g. climate change – modeling. The choice of the best fitting site- or problem-specific indicator(s) is critical in the design of a cost-effective, reliable and repeatable monitoring method (Asner & Martin, 2009; Turner *et al.*, 2003;

Ustin & Gamon, 2010). The selection of the best fitting tool is primarily influenced by the scale of the assessment and the indicators to be monitored. The statistical method should also fit to the methodology to achieve good statistical power.

Wang *et al.* (2010) recently published an overview on remote sensing technologies, instruments and techniques. In the summary of their paper on the assessment of vegetation condition, Lawley *et al.* (2016) listed 33 extensive remote-sensing projects in Australia for several vegetation indicators. These included land cover and its changes (forest cover, non-forest cover, sparse perennial vegetation, regrowth, clearance), differentiated forest characteristics (constant forest, constant non-forest, hardwood and softwood plantations, environmental planting, new forests and plantations), and a density index of forest vegetation communities and vegetation. Further remotely monitored indicators are the extent and distribution of vegetation types; fractional ground cover; persistent green vegetation; photosynthetic and non-photosynthetic vegetation; and biomass. In addition to forest, special attention is paid to pastoral lease vegetation; rangeland; floodplains; riverine vegetation; salinity effects on vegetation; and soil characteristics. Concrete characteristics monitored remotely are vegetation cover, height, stem density, primary productivity; biophysical properties such as leaf area index; absorbed photosynthetically active radiation; biochemical properties including chlorophyll and other leaf pigments or moisture content; and temporal dynamics of vegetation. Canopy health and insect damage have been successfully derived from hyperspectral imagery (e.g. Stone *et al.*, 2013 or Song *et al.*, 2013).

7.2.1   Remote sensing

Information based on remote sensing data is often validated or supported by site-based assessments. As remote sensing provides additional information compared to the conventional ground-based assessments, the integration of the two is highly desired. Lawley *et al.* (2016) discuss the uses, advantages and the obstacles of the integration of site-based and remote sensing data. The typical uses of site-based data are the assessment of the accuracy of classifications from remote sensing analysis and calibration and validation of remote sensing data. The advantages – that large areas could be characterized based on long-term continuous remote sensing data sets – cannot be appreciated as long as the correlation between ground-based and remote data is poor, as it is in many cases. The main obstacles to integrability at present are differences in data quality, and the lack of spatial and temporal fit of data sets (Reinke & Jones, 2006). Further problems may be the lack of reference points for data alignment, poor correspondence between the size of the ground-based plots and image samples, statistically unsuitable numbers of plots and matching dates of data acquisition, and inconsistent sets of guidelines. An interesting and promising impact of the widespread research into, and application of, remote sensing is that – in addition to the calibration and validation of remotely acquired data – the use of the same, or similar, optical methods (multiband and hyperspectral optical sensors) has begun for proximal sensing of the plants in high resolution. The aim is to provide information on key vegetation parameters by field optical measurements to characterize leaf and canopy biochemistry, vegetation stress and seasonal dynamics (Ustin *et al.*, 2009). EUROSPEC identified the main requirements of field sensors to promote the development of new, dedicated instrumentation for continuous field monitoring of the vegetation/canopy. EUROSPEC stated that 'there is a clear need to establish a global network of sites with standardized and coordinated *in situ* spectral measurements to facilitate the integration of remotely sensed data towards improving the global monitoring of the carbon cycle. In addition, such a network is also needed to calibrate and validate satellite data products, and to resolve and avoid problems that appear when inferring ecosystem properties directly from satellite data.' (Porcar-Castell *et al.*, 2015).

7.2.2 Proximal sensing

Some custom-made developments and commercially available sensors are introduced below.

–   *RGB (red-green-blue) cameras* can be used for the monitoring of plants and canopy characteristics by means of very accurate color image acquisitions. An RGB camera delivers the three basic color components (red, green, and blue) on three different wires. This type of camera often uses three independent sensors to acquire the three color signals.

–   The *PSR+3500 field spectroradiometer* (UV/VIS/NIR) for 350–2500 nm and the PSR-1100 for 320–1100 nm range are applied to canopy studies, crop yield forecasting, vegetation identification, and crop condition assessment. The *PSR-1100F* includes a range of accessories for field and laboratory use, including a leaf clip attachment to a tungsten halogen contact probe, specifically designed for leaf reflectance measurements with a built-in white plate, small spot, and low power illumination (PSR+ and PSR-1100, 2015).

–   *Cropscan* Multispectral Radiometers (MSR) can assess various factors of plant health and yield. Upward- and downward-facing sensors and three models are available, which are:

    o   MSR5, a 5-band LANDSAT Thematic Mapper compatible model (460–1750 nm)
    o   MSR87, an 8 narrowband wavelength model (460–810 mm)
    o   MSR16R, a model with up to a maximum of 16 sensor bands in the 450–1750 nm range.

Measured data can be used to characterize/estimate plant growth, canopy color, biochemical content, crop yield, crop quality, and leaf area index, as well as quality loss due to disease, insect infestation, air pollution, nutrient deficiencies, and chemical phytotoxicity (Cropscan, 2015).

–   Vegetation interacts with solar radiation in a specific way and different plant constituents can cause variations across the spectrum. Thus, the spectrum provides information about plant health, water content, environmental stress, etc. Spectral information is expressed as indices e.g. the *PRI* (photochemical reflectance index) and the *NDVI* (normalized difference vegetation index). PRI and NDVI are directly calculated by multispectral sensors of *Skye sensor* (SKR-1800 or 1860 sensor series) and by Decagon (SRS series) (Skye sensor, 2015; Decagon, NDVI, 2015).

- *Ocean Optics* commercial hyperspectral sensors (Ocean Optics, plants, 2015) have been used for several field applicable spectrometer systems such as the *UNIEDI System*, a temperature-controlled spectrometer system for continuous and unattended measurements of canopy spectral radiance and reflectance developed by the University of Edinburgh (Drolet *et al.*, 2014); the *multiplexer radiometer irradiometer* (MRI), developed by the Remote Sensing of Environmental Dynamics Laboratory, University of Milan, Italy and deployed for weeks to months (Cogliati *et al.*, 2015); the *hyperspectral irradiometer* (HSI) developed by the previous university group from Milan (Meroni *et al.*, 2011 and Rossini *et al.*, 2014); and the *AMSPEC II*, an automated, multi-angular radiometer instrument for tower-based observation of canopy reflectance. A Spanish research consortium modified the original instrument to study the relationships between canopy reflectance and plant-physiological processes from multi-angular observations, thereby facilitating a comprehensive modeling of the bidirectional reflectance distribution of the canopy. A webcam permits simultaneous monitoring of phenological changes over time (Hilker *et al.*, 2010).

- The *S-fluor box* is a custom-made device recently developed in collaboration with the Forschungszentrum Jülich GmbH based on an MRI design. An overall compact design and the integration of the cooling system within the instrument box (Cogliati *et al.*, 2015) are the prime technical improvements.

7.3 Indicator species

Our knowledge of terrestrial bioindicators and their application for environmental management is not as clear as that of aquatic bioindicators, in spite of extensive research and the extensive information available. What has become clear is that bioindicators are taxa or functional groups which mirror the state of the environment, acting either as early warning indicators of any change to the local environment (environmental indicator), or monitors of a specific ecosystem stress (ecological indicator), or indicators of taxonomic diversity at a site (biodiversity indicator) (McGeoch, 2007). The objective of the activity, the targeted results, and its spatial and temporal scope greatly determine the selection of terrestrial bioindicators in a study or a management task. Soil microbiota, terrestrial vegetation and soil-dwelling invertebrates are the most frequently used indicator organisms.

Soil microbiota – single species and whole communities – are the preferred bioindicators in the rapid activity and toxicity tests, as shown in Sections 5.4, 6.2 and 7.

Vegetation, i.e. the typical terrestrial land cover, can be considered as the most important generic indicator of the terrestrial ecosystem health and life conditions.

Terrestrial invertebrates are frequently-used indicators, and standard methods using them are also formulated for ecological assessment or environmental purposes, but a general concept for their use is still lacking (see also Gerlach *et al.*, 2013).

7.3.1 Terrestrial vegetation

The vegetation of macroplants can be monitored by ground-based or remote sensing. Individual terrestrial plant species could also be studied as ecosystem indicators or community indicators due to their sensitivity, resistance or accumulation. Plants respond to nutrient availability, fertility, waterlogging, and soil chemical properties, e.g. pH and soil structural characteristics (compaction, etc.). Measurement and test methods can be developed based on these responses.

Phytoindicators can be applied at every level (molecular to ecosystem). Forest is used as an indicator of land productivity and air pollution: evergreen forest and grassland are sensitive to the amount of precipitation, and grass types may reflect the soil type (sandy, loamy, etc.), soil acidity or geochemical content, e.g. sulphur or contaminants. Contaminant-resistant, bioaccumulator, hyperaccumulator plant species can function as early warning indicators; *in situ* screening techniques and remote sensing may be extremely important tools to detect these species and their contaminant content without complicated and time-consuming laboratory analysis. Metallophytes can also indicate the presence of specific minerals.

Mosses (bryophytes): these non-vascular, flowerless plants are highly sensitive indicators and popular organisms for monitoring bioaccumulation. Since they have no roots, the chemical species they accumulate should mainly derive from atmospheric depositions. In addition to the monitoring of naturally occurring mosses and peat profiles, active biomonitoring techniques have been developed, such as moss transplants or moss bags, mainly for urban areas where indigenous mosses may be absent. Water bears, or moss piglets – the ancient taxon of tardigrades – which feed on plant cells and small invertebrates, are extremely resistant to environmental conditions but sensitive to contaminants via food. They are common worldwide and are reliable indicators of atmospheric deposition and soil surface quality.

7.3.2 Soil invertebrates – soil-dwelling worms and insects

Environmental and ecological bioindicators can be classified based on their response to environmental stress, e.g. the presence of contaminants with adverse effects:

– 'Detectors' are naturally-occurring sensitive species which respond to environmental stress with a decrease in their numbers or activity;
– 'Exploiters' take advantage of changes or stress by becoming more abundant in response to environmental conditions and chemicals stress (disadvantaging others);
– 'Accumulators' take up chemical species such as persistent organic chemicals and metals and concentrate them in their body organs or tissue, depending on the mechanisms.

Invertebrates represent great diversity and abundance, which is why they can be used as biodiversity bioindicators. They are sensitive to several soil-borne stresses, and their mobility may result in immediate avoidance. Small size, short generation time, close contacts with soil and cost-efficiency in their testing are all characteristics that make them suitable bioindicators not only for environmental management, but also for

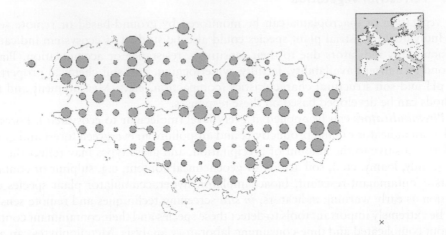

Figure 4.12 Earthworm abundance: monitoring data of the network in Brittany, France. Legend: green circles in decreasing size: 1000, 100, 50, 10 and 1 animals/m² (Jeffery *et al.,* 2010).

landscape ecology and conservation. The proper collation of indicators and the compilation of indicator batteries should follow some basic principles, for example, the bioindicators should belong to several taxa and different trophic levels; random variables in abundance should be known and considered; the linear relationship between a measured end point (abundance, activity) and the scale of effect should be within a broad range; and the selected bioindicator should be identified and familiar to the user (see also Samways *et al.,* 2010).

Some results of the European soil monitoring network are introduced below. Yearly changes and long-term trends can be recorded and compared to climate and land use parameters (Cluzeau *et al.,* 2012, also cited in Jeffery *et al.,* 2010). Earthworm abundance in Brittany, France is shown in Figure 4.12. The European distribution of Collembola (springtails) based on average national data can be seen in Figure 4.13 (Jeffery *et al.,* 2010). The data on abundance or diversity can be mapped to establish any correlation with regional chemical substance production and use, as well as land cover and land use patterns, to support environmental management and decision making.

The presence and density of species is assessed in practice by counting the organisms or by detecting any of the indicators at the levels of community, food chain, food web, population or organism. The indicator may be the behavior, the morphology or metabolic, biochemical, chemical and genetic markers which clearly show the presence and density of the species or higher taxa. Even individual animals can be monitored by using microchip implants. The evaluation in most diversity assessments needs empirical calibration, based on knowledge on the relation between diversity and health of the environment. Simple, but illustrative, radar-type histograms visualize the patterns of taxa distribution. Figure 4.14 shows the impairment scale of 25 different bioindicators (abundance, community composition, functional and land use parameters) relevant for 4 taxa (nematodes, bacteria, earthworms and enchytraeids) (Jeffery *et al.,* 2010).

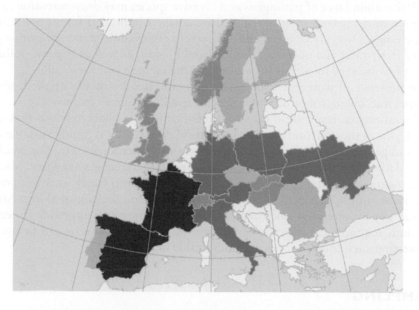

Figure 4.13 The abundance of Collembola in Europe based on national averages. Blue: 488–610; bluish green: 368–488; dark green: 246–367; green: 125–245; yellow: 1–124 animals/m² (Jeffery *et al.*, 2010).

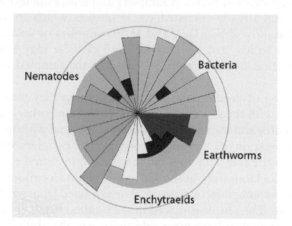

Figure 4.14 Abundance of four groups of soil-dwelling organisms: bacteria, nematodes, enchytraeids and earthworms shown on a radar-type histogram of 25 bioindicators. The concentric rings correspond to the scale of impairment given in percentage of reference: outer: 100%, grey: 75%, black 50% (Jeffery *et al.*, 2010).

In situ diversity assessment means field work: collection and identification of the individuals of one or several species. Species-specific traps may help in the collection of organisms, e.g. soil-living arthropods or insects. The consumption of species-specific food may indicate their overall activity.

Relative abundance of pathogens and invasive species may draw attention to pests. Lichens (comprising both fungi and algae) are sensitive indicators for air pollution and climate change. Diversity of algae and the presence of cyanobacteria are primary indicators of freshwater and ocean nutrient (N and P) load, as well as organic pollutants. Both can be observed by airborne or satellite sensing based on their color.

The presence of dangerous species can be indicated by *in situ* applied, sensitive biosensors and microprobes based on species-specific enzymes and toxins.

Mineralization, respiration rate, and nitrification are characteristics that can be sensed *in situ* in the environment, but their absolute intensity is highly variable due to spatial and geochemical heterogeneity, nutrient supply, aerobicity, etc. The results can be utilized for monitoring and forecast when a reliable reference is available. Microbial activity, together with the relevant environmental parameters such as redox potential and nutrient supply, can provide useful information for decision making. Multiple sensors at key locations and their results may be an innovative approach in real-time online data acquisition. Efficiency of monitoring and early warning can be increased by converting such complex data sets into usable information.

8 SAMPLING

Sampling and samplers are basic tools in environmental assessment and monitoring, and are of major importance for *in situ* and real-time site assessment and soil analysis. There is no sharp delineation between samplers and sensors/detectors in many cases. Some of the equipment already discussed, e.g. lysimeters, which are samplers for water and solutes of the shallow subsurface, and MIPs, samplers of volatile components of soil vapor, water or solids in deeper layers. Both of these may integrate the sensor for immediate detection of the analyte or transport the sampled material to uphole detectors.

Some sensors do not need separate samplers because they measure signals without being in contact with the material to be analyzed. Some others should be in direct contact with the matrix and the analyte, but detect the signal without destructive interaction with, or segregation of, the sample. Some other samplers separate aliquots of fluids or solids from the soil, and then the sample is transferred to, and analyzed on, the surface in a mobile laboratory at the site or at a remote location. Fluid samples can be acquired by passive samplers, or exhausted or extracted from soil by intermittent or continuous pumping. The sensor is placed in the resulted batch or stream of the fluid. Samplers can be designed to have great selectivity, e.g. the ability to sample *all* contaminants separated from the matrix; or only organic contaminants or contaminants separated from each other; or to sample only one specific contaminant. DGT (Section 5.3) is an example of a selective sampler, in which the sampled material can be analyzed at the site in a mobile laboratory. Selective molecular sensors react only with the analyte, while other less specific sensors may only react with ions or electrochemically active species, etc. Sampling and analysis are hardly separable in these cases.

Sensors placed *in situ* may produce results with or without a very short delay: the greater the time requirement of sampling and sample transference, the longer the delay. Destructive methods and sample removal may significantly reduce the assessment results' environmental realism. The dilemma of sample size, point or collective sampling, timing and placing should be solved by a proper statistical method both for sampling planning and results evaluating.

Sampling for site characterization may be carried out at the soil surface in the upper few centimeters. Sampling may cover deeper layers employing direct push technology or boreholes with a combined soil gas, groundwater and soil sample collection from the same borehole.

8.1 Sampling soil means sampling a process

Before soil sampling planning commences the conceptual model of the site or the problem should be developed by identifying sources, transport pathways, and the site's uses and users, i.e. the receptors. The aim of monitoring also determines the sampling strategy. The aim of monitoring may be (i) reference monitoring just for the background characteristics, (ii) to assess contaminant transport, (iii) the relationship of the soil with surface water, (iv) compliance monitoring, primarily drinking water wells and agricultural land, or (v) to monitor the process and the outcome of remediation.

When planning soil sampling one must realize that physical, chemical and biological processes occur continuously in the soil, and the direction of change depends on the relation of the equilibrium to the actual state. The processes show changes in space and time. The transport of soil fluids (vapors and solutes); the partition of molecules between phases; heat dispersion and distribution; ion exchange; and chemical reaction do not only change in time but also show wide spatial variations.

Any time point can be interpreted as a 'state'. Multiparametric and multivariate spatial changes imply spatial differences in soil waters that are determined by spatial differences in soil types associated with the topography, vegetation, precipitation, evapotranspiration, etc. However, temporal changes may be significant due to seasonal and weather changes. Certain processes are dominated by spatial, others by temporal, changes and it is typical that both spatial and temporal differences have great influence on the soil's actual state. The treatment of soil may make the situation even more complicated and difficult to understand. Another problem is the temporal and spatial resolution of changes and their harmonization with the sampling and monitoring method.

How can this situation be approached? One solution is to collect a representative average sample, representative for the state or for the treatment. If the differences between two time or space points are smaller than the natural variance, one considers the sample 'random'. Another good solution is sampling in time series or spatial series. If the process being modelled is based on serial data, it can be compared to cyclic variables, fluctuation patterns or just simple transport regularities. A third solution for the assessment of complex and unclear multiparametric processes is to perform simulation tests or treatment experiments. Under controlled conditions, only one or a few parameters are varied in these experiments instead of having a large number of reference samples. A step forward is to find the mode of action, the effect mechanisms, instead of conducting several randomized experiments. A knowledge of biogeochemical causes and mechanisms can help in understanding the response of the soil and the soil ecosystem.

As mentioned above, the scale of change is a crucial piece of information. Changes of processes and functions at community or ecosystem level over large areas occur over several years and decades, with remote sensing and mapping being the main monitoring tool. Processes at small scale in the soil pores or biofilms occur in a few seconds or less. Microelectrodes or X-ray images can detect these changes and the results are true for the tiny space and for the short time only, because different values can be measured a few millimeters away. Sampling strategy and statistics should be harmonized with the aim of the assessment/monitoring.

After this general overview of sampling we look at sampling strategies and the tools relevant for soil gas/vapor, soil solute and soil solid.

8.2   Soil gas sampling

Soil gas sampling may aim to collect surface or subsurface gas and vapor samples from the vadose zone. The samplers are employed at the soil surface to collect and measure gases and vapors emitted from the soil to the atmosphere, or are deployed some meters deep into shallow wells or directly pushed or emplaced into the soil. The sampler can sorb gas/vapor molecules passively, or soil gas can be exhausted by ventilation or vacuum pumps. Larger depths can be reached through wells or by direct-push technologies which permit real-time chemical monitoring of soil gases and vapors by analytical sensors in conjunction with the direct push tip. Soil gas sampling in this case can be performed using a commercial unit such as Geoprobe® vapor sampling system or a specially designed filter probe attached to a standard penetrometer tip. The latter consists of a filter probe module located immediately behind the penetrometer tip to collect soil gas samples at discrete depth intervals during CPT advancement. This system has the advantage of collecting soil gas samples at multiple depth increments while its geotechnical sensors simultaneously track soil behavior types.

Professionals have to reckon with the consequences of the application of a vacuum, which initiates a higher rate, or another direction, of soil gas transport than is normal: it may cause changes in contaminant concentration and in vapor partition. Opening and purging the sampler as well as duration of sample collection are all important design parameters.

The simplest samplers are manual soil gas probes for shallow depths and loose soil. The sampler consists of a probe tip and holes above for soil gas access. The sampler can be joined to a hand pump. The pump forwards the sampled gas flow into a gasometer and the analytical device directly, or into a bag, a bottle or canister for off-site analysis. Hand-pushed probes and pumps are commercially available from several vendors. The canisters can be used both for manually or machine-pushed samplers.

AMS offers hand-pushed gas samplers for a one-off sample, and stainless steel vapor implants for continuous soil gas sampling. Special tools are available (i) for soil gas: this is the originally developed kit; (ii) for hydrocarbon vapors: a dedicated gas vapor tip with an 'umbrella', ensuring an easy passage for gas entering the collection system; and (iii) the retract-A-tip, which can collect soil vapor samples at discrete depths by opening the tip for sample extraction, closing and removing, then sampling at another depth (AMS, 2015).

Samplers in conjunction with direct push technologies are classified as discrete-interval samplers, continuous and permanent samplers. The discrete-interval sampler consists of a sample chamber of a vacuum pump. Continuous sampling tools are driven in sniffing mode, i.e. collecting the samples as the tool is being driven.

They can be used as discrete samplers or multiple-depth samplers by interrupting pushing and opening of the sampler at a certain depth. Permanent samplers are deployed for long time durations (e.g. in vadose zone wells) to ensure continuous or frequent soil gas sampling.

Geoprobe also applies post run tubing systems and a vacuum/volume system, which is the combination of a vacuum pump and a tank or permanent implant. Permanent implants are anchored at shallow depths and applied both for gas/vapor and groundwater sampling.

Passive soil gas samplers can be used for source monitoring, vapor intrusion detection and remediation technology monitoring. They may be a cost-effective, soft (minimally invasive) solution, producing time-integrated samples before reaching an average saturation in the sample. The sample does not show spatial or time variabilities, rather it represents an equalized load in contrast to active samplers, which collect momentary, non-equilibrium state, disturbed samples. Manual devices and reusable sorbents ensure good environmental efficiency compared to the direct push technologies and large drilling rigs. When using sorbent-based soil gas samplers, gas/vapor partition between soil gas and the sorbent, as well as the competition between the soil solid and the sorbent, is worth considering. Thus, the proper selection of the sorbent material (selective or non-selective for the analytes) and the sampler's housing (non-sorbing for the analyte) may play an important role. Adherence to ASTM's standard D7758-11 (2011) *'Standard Practice for Passive Soil Gas Sampling in the Vadose Zone for Source Identification, Spatial Variability Assessment, Monitoring, and Vapor Intrusion Evaluation'* ensures the uniform application of such samplers. Its main disadvantage is that the result cannot be expressed in terms of concentration (contaminant g/soil gas volume). Therefore, the amount sorbed by passive samplers should be converted to soil gas or groundwater concentrations measured by conventional methods.

– *Passive soil gas sampling* technologies are provided by Beacon (2015) sample collection kits. The samplers are selective for individual halogenated compounds; petroleum blends; BTEX and PAHs; ketones; alcohols; explosives; pesticides; chemical warfare agents; and mercury. Such samplers are recommended for screening to exclude negative samples as soon as possible, and continue the detailed assessment on the positive ones. Thus the expensive methods are applied only to the positive samples.

– *Solid-phase microextraction* (SPME) is a solventless sample extraction technique. In SPME, a polymer-coated fused fiber is used for selective sorption ('extraction') of analytes, from where it will be desorbed directly to a chromatography column. Analytes are concentrated on the fiber and they can be rapidly delivered to the column (e.g. by a transportable GC-MS). Detection limits are low and the resolution is good, and it provides linear results for wide analyte concentration ranges in the sample as long as equilibrium is reached. However, this is questionable in the case of field samplers emplaced directly into soil gas for a short time. SPME can be used for semiquantitative analysis under controlled conditions or just for detection of contaminants under field conditions. SPME field samplers are small and extremely efficient in extracting and transporting volatile and semi-volatile compounds from field samples to the analytical device (portable models used on the field or stable types in the laboratory). The sampler can be reused 50–100 times and the fiber disposed of. The three most popular fibers are: general-purpose polydimethyl-siloxane (PDMS) fiber, PDMS/carboxen fiber for trace levels of volatiles, and PDMS/divinylbenzene (DVB) fiber for semi-volatiles and large-molecule volatiles (Pawliszyn, 2009).

– *Supelco products* are marketed by Sigma-Aldrich, providing ready-made samplers equipped with PDMS, DVB and carboxen fibers, as well as customized SPME samplers provided with other coatings (Supelco SPME, 2015).

– *Custodion SPME* is suitable for semi-quantitative analysis of liquid samples, headspace above liquid or solid samples and gaseous samples. TRIDION™ provides a conventional trap (Custodion CT) for the collection of gas samples followed by gas exhaust or thermal desorption. Their accelerated diffusion sampler (ADS) serves for sampling contaminants on solid surfaces. Volatiles and gases in the trap have more chance to reach equilibrium and provide quantitative results. The Custodion needle trap (NT) is especially designed for sampling and performing quantitative analyses of gaseous samples (Custodion, 2015).

8.3 Soil solution sampling

Soil solution is the aqueous phase of the soil containing the solute, i.e. dissolved inorganic and organic substances. The solutes can be dissolved in water bound strongly to the colloid surface, in free waters in the macropores and in immobile water in micropores. The solutes distribute and seek equilibrium among these water forms, which differ in composition and activity.

Sampling aspires to acquire the unaltered soil solution from unsaturated or saturated soil. A water sample from the saturated soil represents the flowing solute striving for equilibrium with the contacting solid and the more immobile solute forms. The spatial differences in the solute sample from the saturated soil are slurred compared to the solid phase. The sample aggregates losses and increments of different solutes on its way through solid matrices.

Waters of the *vadose zone* show different type of variabilities. In the absence of free water the more strongly bound soil moisture is sampled, which shows a wide range of concentration as a function of soil moisture content. Free water moves gravitationally and is present when excess rain or irrigation water flows downwards. Percolation and leaching are the typical processes in this case, and they can transport solutes into certain depths: in a healthy state to the root zone, in other cases to the groundwater. Another transport direction is the groundwater moving to the unsaturated soil, driven by capillary forces. The layer where this process is typical is called capillary fringe. It may reach the root zone, ensuring water and nutrient supply for the vegetation during drought. Sampling soil solution from the vadose zone involves lysimeters, vacuum or buried samplers and/or sensors.

Sampling water from the *saturated zone* requires the establishment of wells or access tubes and emplacement of the sampler or sensor. Burying the samplers and/or sensors is also a feasible option.

Conventional laboratory-based solute sampling methods from soil are (i) extraction from soil (involving the problem of saturation, equilibrium and separation), (ii) column displacement of, and leaching of, the soil, and (iii) pressure or vacuum extraction or centrifugation. Some of these can be adapted to field application, e.g. field lysimeters for leachate collection and vacuum and/or porous material (with high capillary suction power) for pore water collection.

Sections 4.1 and 4.2 on soil moisture determination introduce the tools, which instead of determining the composition, determine the moisture content of soil by measuring the tension or the dielectric constant; heat dissipation; or NMR spectrum. Specific sensors built into the samplers can detect targeted analytes or adverse effects (immunoanalytical and whole-cell sensors) *in situ*. However, suction cups, lysimeters and passive capillary lysimeters can be applied to take samples for detailed moisture composition analysis on site or off site (Soilmoisture, 2015).

8.3.1   Soil solution sampling from unsaturated soil

Unsaturated soil sampling includes block or monolith lysimeters, zero-tension lysimeters and porous cup vacuum samplers. *Monolithic* lysimeters may contain refilled or undisturbed soil and work either with or without using a vacuum. The leachate is collected and analyzed. The environmental realism of undisturbed blocks with vegetation is much better than that of the refilled ones, but sidewall leakage may impair the results. *Zero-tension* lysimeters are good examples for sampling solutes moving with water downwards. There are several designs available such as pan, trough, funnel or plate installed from a trench. Sample collection vessels are emplaced into a deeper trench, and one version works under tension. *Porous cap* vacuum lysimeters are the most widespread tools and are prepared from porous materials used as a filter cartridge, typically made of ceramic, and apply a vacuum or pressure for solution collection. They can be used for intermittent or continuous sampling. The porous material can sorb selectively dissolved components from the water or can contaminate the sample. The leachate can be analyzed either *in situ*, on-site or in the laboratory.

In addition to the mentioned lysimeters, several *capillary water absorbers*, passive capillary samplers and passive capillary lysimeters are available for collecting both soil moisture or pore water based on samplers prepared of capillary sorbents in the form of cups, plates or wicks (see more in Chapter 2).

Solid-phase microextraction can be applied for sampling dissolved substances in soil solution or from the vapor above groundwater (see details on gas/vapor sampling in Section 8.1). When analyzing soil moisture, the SPME sampler should be in contact with the moisture collected by lysimeters or capillary samplers. In saturated soil, pore water can be reached more easily – even directly – in multipurpose wells.

8.3.2   Soil solution sampling from saturated soil

Pore water or groundwater can be sampled from wells or from direct push systems with or without removing the sample intermittently or continuously. Conventional groundwater sampling from wells seems simple and is in widespread use, but it is still loaded with several conceptual and practical problems which should be considered and solved in order to acquire a sample representative in terms of quality and quantity. The main problems originate from the following: (i) permeability of the adjacent strata is widely variable, so the water level in a well is practically a flow-weighted average, (ii) the depth of the non-aqueous phase is often unknown, and (iii) the length of the screened interval must be decided in advance. Borehole design can be long screened, short screened, nested, clustered, simple rod or dual tube, as well as multilevel. Shorter screen lengths offer sampling at precise depth intervals (especially important for contaminant plumes or dense non-aqueous phase liquids – DNAPL) and less opportunity for mixing water from layers of different permeability. Purging before sampling is also a crucial step. In practice, fixed-volume (e.g. x3 the well volume) and low-flow purging (for groundwater containing volatiles) or zero purge (using passive samplers) are applied.

8.3.2.1 Passive samplers

Passive samplers work without pumping. They can be conventional bailers, or sorbent- or diffusion-based passive sampling devices for retrieving discrete water samples. Bailers retrieve water from the mixture in the well, and passive samplers from the vicinity of the strategically emplaced sampling tool. New types of bailers are biodegradable, disposable or reusable plastic samplers. Sorbisense is a sorbent-based passive sampler for specific water-dissolved components.

– *BioBailers*^{TM} and *EcoBailers* are standard disposable bailers used for groundwater bore sampling with the added benefit of being biodegradable when landfilled. BioBailers are made of PVC or HDPE with a small amount of an engineered additive Ecopure^{TM} to make them biodegradable. PVC has a higher specific density of 1.3 which allows these bailers to sink in the well faster than HDPE bailers. Therefore most applications do not require weighting PVC bailers. Special bailers are suitable for hydrocarbon (non-aqueous liquid phase) sampling (BioBailers, 2015; EcoBailer Pro, 2015).
– The *HydraSleeve*^{TM} is made of light plastic and used with a stainless steel weight (HydraSleeve, 2015). The sample should be transferred into a transportable container.
– The *SNAP Sampler*^{TM} is a double-ended bottle, which is closed while submerged in the well, opened during sampling and closed again after being filled. Samples can be transported in their sampler to the laboratory (SNAP Sampler, 2015).
– The *SorbiCell* is a porous cartridge with various contaminant-specific sorbent fillings especially developed for groundwater solutes. When the cartridge is placed into groundwater (into wells), the sorbent adsorbs the targeted compounds from the water passing through. Contaminant flux can be calculated from the sorbed contaminant mass and the groundwater flux. It is measured by the same cartridge containing an environmentally neutral tracer that dissolves proportional to the passing water. The sorbent and the salt are isolated by a permeable layer. The cartridge is shown in Figure 4.15, before and after use. After deployment for a certain period the cartridge is taken out of the well and its content analyzed in the laboratory. The method enables measuring the average concentration over an extended time scale (SorbiCell, 2015). SorbiCells are especially suitable for measuring several hundred chemicals (about 300 chemicals are specified by the developer) from the following chemical groups: inorganic chemicals such as nitrate and phosphorous; metals, including toxic metals; volatile (e.g. chlorinated and brominated) organic compounds; a wide range of pesticides; petroleum hydrocarbons; and polycyclic aromatic hydrocarbons. Sampling time is typically 1–3 months, but concentrated waters may need only a few minutes. The cartridge is used as a passive

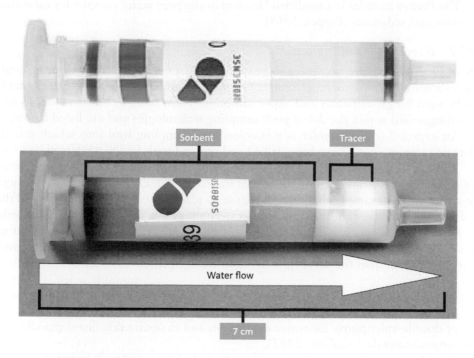

Figure 4.15 SorbiCell cartridge before and after use for contaminated groundwater monitoring (Sorbisense, 2015).

sampler placed in a water flow or into a well. The adsorbed amount is flow-related, and a time-averaged concentration can be calculated from the analysis results.

– **Diffusion samplers** are covered by a permeable membrane or gel that enables selected chemicals (by size and/or electrical charge) to establish an equilibrium between the soil solution and the liquid inside the sampler. Some development and design for diffusion samplers were published by US EPA Clue-in (ITRC, 2005).

– **Dialysis membrane** samplers have been successfully tested for several nutrients, and organic and inorganic toxicants. The regenerated-cellulose dialysis membrane tube is filled with deionized water. The outer layer is a protective LDPE mesh. Ions and smaller molecules diffuse through the membrane from the groundwater into the deionized water in the sampler. The driving force for diffusion is the concentration difference, so it stops after reaching equilibrium. The published technical designs are for the regenerated-cellulose membrane tubes and bags, the nylon-screen passive diffusion samplers and the peeper sampler.

– The **Equilibrator®** passive diffusion sampler is available in pre-filled and self-filled forms. The latter should first be filled with deionized water by the user, then attached to a suspension tether and lowered into the saturated zone in the well screen. While being deployed, contaminant (e.g. VOC) molecules diffuse through the semi-permeable membrane until their concentration in the sampler is in equilibrium with the surrounding groundwater (Equilibrator, 2015).

–   The *Peeper sampler* is a modified Hesslein *in situ* pore water sampler for saturated soils and sediments (Peeper, 2015).

8.3.2.2   Direct push sampling systems

Direct push systems are becoming increasingly widespread and accepted for groundwater sampling. Both cone penetrometer testing (CPT) and percussion hammer systems can be used for groundwater and vapor sampling. Three basic technical solutions can be distinguished within the direct push sampling technologies and are listed below.

An *exposed-screen sampler* is a stainless steel sampling tool into which several inlets or sampling ports have been drilled and covered with fine-mesh screen.

A *sealed-screen sampler* is driven to the desired depth where the protective outer rod is withdrawn, thus exposing the screen to groundwater. The water flows through the screen into the sample chamber. O-ring seals placed between the drive tip and the tool body help ensure that the sampler is watertight as it is driven to the target depth.

Open-hole sampling is conducted by advancing drive rods with a drive point to the desired sampling depth. After reaching the sampling depth, the rods are withdrawn to separate them from the drive tip and allow water to enter.

–   The *Condition Monitoring Technologies* (CMT) and the *Waterloo multilevel samplers* are sealed-screen type samplers, using – depending on depths – a peristaltic or double-valve pump for water extraction, and to obtain detailed depth discrete groundwater data (Solinst, 2015).
–   *Dual tube* sampling uses two sets of probe rods. One set of rods is driven into the ground as an outer casing. These rods receive the driving force from the hammer and provide a sealed hole from which continuous soil samples may be recovered without the threat of cross-contamination. The second, smaller, set of rods is placed inside the outer casing. The smaller rods hold a sample liner in place as the outer casing is driven one sampling interval. The small rods are then retracted to retrieve the filled liner (Geoprobe dual tube, 2015).

Several other advantages of direct push technologies are described in Section 2.2 and Chapter 5. One disadvantage from a sampling perspective is that the sample may be turbid. Turbidity disturbs not only the analysis, but the suspended fine particulate solid can sorb the contaminants and bias the analytical result. To exclude turbidity, samplers with ceramic or other porous tips are used, through which the water sample is drawn, and so the suspended matter is filtered out.

–   The *BAT® sampler system*, contains one such filter tip: a MkIII. The filter tip has a body of high-strength thermoplastic and a filter made of porous polyethylene, or a replaceable ceramic filter, or has a stainless steel body with a replaceable porous polyethylene filter. This sampler excludes not only turbidity but also seals the sample hermetically in order to retain volatiles by septa. The tip is installed at the sampling depth and fixed there. The sampler housing is equipped with a holder for a double-ended needle. The evacuated sample tube is lowered down through the extension pipe, from where it drops (by gravity) on the double-ended needle which penetrates both the septum of the sample tube and septum in the filter tip, establishing a leak-proof connection between them. So the vacuum in the sample tube can soak up the sample of water and vapors together from the BAT filter tip, which is under groundwater pressure. Both septa automatically reseal after lifting the sample tube (BAT, 2015). This sampler can be used both for screening and long-term monitoring.

8.3.2.3   Sealing the borehole with a liner

In addition to the quick closure, sealing the borehole with a liner provides several technological solutions such as these listed by the Flexible Liner Underground Technologies company (FLUTe™, 2015):

– Multi-level groundwater sampling;
– Mapping hydraulic head distribution;
– Landfill monitoring;
– Vadose zone pore fluid sampling;
– Mapping the subsurface flow;
– Mapping contaminant distributions;
– Locating NAPL (non-aqueous phase liquids) in sediment and fractured rock.

8.3.2.4   In situ *applied sensors*

Sampling by sensor probes is the most advanced methodology in soil and groundwater monitoring. Sensors that operate *in situ*, online or in-line, provide real-time data and represent a combination of samplers and detectors. The emplacement of *in situ* sensors does not damage or significantly disturb either fluids or solids. Remote sensors do not come into contact with the sample at all.

Soil sensors are placed into a specific location, and therefore the spatial domain is unchanged during measurement. The sampling/measurement pattern can be one-off, regular, intermittent or continuous. In groundwater or soil gas/vapor the temporal domain associates with a spatial domain, and thus the measured data are determined by the flux and the partitioning of the analyte between physical phases. Sensors have been developed for characterizing a soil's physical, chemical and biological properties. Sensors are used in the analysis of:

– Soil gases and vapors of water and volatile contaminants;
– Organic and inorganic solutes;
– Soil solid: structure and composition, including contaminants;
– Biological entities: presence, conditions and responses.

The numbers in the summary Table 4.3 refer to the sections of Chapters, where sensors are introduced. The classification of sensors has not been clarified, e.g. biosensors are actually chemical sensors (based on chemical reactions between two molecules), but if the molecules used are of biological origin, the sensor is referred to as 'biosensor'. The table uses the same nomenclature. Sensors developed for a certain environmental sample type are applicable to others, e.g. those offered for groundwater are generally suitable for surface waters, runoffs and leachates and also for waste waters.

Table 4.3 Field applicable soil sensors – the section of this book where the type of sensor is introduced

Compartment Method	Gas/vapor	Soil moisture	Soil waters, solution/solute	Soil solid	Soil biota	Soil surface
Physical		4.1	4.3	2.3		4
Chemical	3	4.1	4.4	4.4	7	
Biological			5.4		7	4

8.3.3   Soil sampling

Sampling with hand tools is commonly undertaken to obtain soil samples at, or near, the surface by digging holes 2–20 cm in diameter and a few meters deep. Drilling and direct push techniques enable sampling at great depths, and mapping the subsurface geotechnically, hydrogeologically or chemically. Removed soil samples can be categorized as 'disturbed' (i.e. nothing retained of the original structure) or 'undisturbed' (i.e. the original structure is more or less intact, close to undisturbed). Sensors deployed *in situ*, especially microsensors, eliminate destructive intrusions and ensure close to undisturbed sampling. Remotely placed sensors do not make contact with the sample, but the surface only can be monitored.

8.3.3.1   Hand tools

– *Shovels and scoops* are general-purpose hand tools for taking disturbed samples from the surface of the soil, or near the surface. Stainless steel tools are resistant to caustic and corrosive soil materials.
– *Soil augers* produce disturbed soil samples generally from soil layers near the surface, but augers can bore to greater depths too, e.g. to ensure access for core samplers. They are made of stainless steel and several types of augers are available for different types of soils.
– *Soil core samplers* collect virtually undisturbed soil cores for soil profiling and environmental investigations. Modern samplers collect the soil core in removable liners and can be opened for visual observation of the core.
– *Soil probes* are generally used to collect smaller diameter samples than soil core samplers on, or near, the surface. The undisturbed soil allows the description of the soil profile and other characteristics. The core sample should be slightly smaller than the inner diameter of the probe body to ensure easy removal of the core from the sampler. The core samples can be collected directly into a liner (special design) for later detailed analysis. Several types of tips are available for most of the probes, of which the most frequently used are often arranged into a set as part of soil sampling kits.

8.3.3.2   Power tools

Power tools are needed for greater depths and generally for sampling groundwater. Some conventional tools and developments are described as follows.

- *Conventional drilling techniques* for sampling, using power tools, are listed below.

  o *Augering*: the auger is screwed into the ground then lifted out. Soil is retained on the blades of the auger from where it can be collected for analysis.
  o *The split-spoon sampler* is a hollow tube split into halves lengthwise. A drive shoe is attached to the bottom end and it is driven into the ground with a hammer.
  o *The Shelby tube* is a thin-walled tube with a cutting edge for taking samples for geotechnical investigations.
  o *Piston samplers* are used to collect relatively undisturbed soil samples without generating soil cuttings. The samplers are pushed into the bottom of the borehole by the piston, which remains at the surface while the tube slides past it. The soil sample tube ends are sealed with Teflon® tape and plastic caps to keep the sample secure for later environmental analysis.
  o *The Pitcher^{TM} barrel sampler* is similar to piston samplers, but without a piston. There are pressure-relief holes near the top of the sampler to prevent the build-up of water or air under pressure above the soil sample.

- *Hollow-stem augers and cutter heads* are powerful drills which can auger under complicated field situations, e.g. through hard layers and rocks. They are ideal for drilling boreholes for lithologic description, sample collection for chemical analysis, and conventional well installation for groundwater monitoring.
- The Central Mine Equipment company's *continuous soil samplers* work in conjunction with the hollow-stem auger drilling process, but the important difference is that the sample tube does not rotate with the auger. This enables undisturbed, representative collection of core samples.
- *Direct push technologies* e.g. the dual tube soil sampling system as described in Section 8.3.2.2.

Some of the companies developing and marketing samplers and sampling systems are: Geoprobe Systems, AMS, Eijkelkamp, Mateco, Pagani, Environmental Remediation Equipment Inc. (ERE), Cornelsen Umwelttechnologie GmbH, Peterson Environmental, Inc., Argus-Hazco, Challenge Technology, Geotechnical Services, Inc., and Solinst Canada.

9 FIELD PORTABLE EQUIPMENT FOR ASSESSING METALS

Lead, arsenic, cadmium, chromium and other toxic metals contaminate sites of former industrial and agricultural activity around the world. This pollution may be the unintended by-product of industrial/mining processes, or it may have been dumped deliberately in violation of environmental regulations. Portable X-ray fluorescence (XRF) measuring equipment is an efficient *in situ* assessment tool used in the environmental management of metal-contaminated sites.

X-ray fluorescence (XRF) spectrometry has evolved into an analytical technique used extensively in the last decades. It can provide quantitative analytical data for academic and industrial use in laboratory, plant or field environments. Nowadays a portable X-ray fluorescence technique is available, which is especially suited for *in situ* analysis of various sample types. The portable XRF device is used for the assessment of contaminated land; metal and alloy analysis; analysis of surfaces; coatings and paints; workplace monitoring; archaeological and geochemical investigations; and many other applications.

9.1 Principle of X-ray fluorescence

The XRF method is based on interactions between electron beams or X-rays and the sample. Major and trace elements in materials can be analyzed by XRF because of the behavior of atoms when they interact with X-ray. When the X-ray hits a material, it can either be absorbed by the atoms or scattered through the material. When the X-ray hits an atom, electrons are ejected from the inner shells creating vacancies and making the atom instable. During stabilization of the atom, electrons jump from the outer shell to the inner one and give off the energy between the two shells in the form of a secondary or fluorescent X-ray. Each element has specific energy levels and produces a unique set of X-ray that allows the identification of the element content in a material or environmental sample without extraction of the elements for analysis. This non-destructive measurement of the elemental composition offers multi-element capability.

Portable XRF equipment uses either a miniature X-ray tube or a sealed radioactive source to excite the sample with X-ray photons. The radioisotope excitation source never requires replacement, unlike conventional X-ray generators. An X-ray tube allows for maximum optimization of excitation that cannot be afforded by radioisotope-based devices. This in turn translates into the best possible analytical sensitivity and limits of detection. On the other hand, X-ray tube-based analyzers require more power to operate, are electronically more complex than their radioisotope counterparts, and are somewhat larger and heavier (Potts & West, 2008). Due to the hazards and required preventive measures for the use of radioactive sources, X-ray tubes are used exclusively in handheld devices.

9.2 Field portable XRF device (PXRF)

Field portable X-ray fluorescence measuring devices (PXRF) allow rapid and cost-effective screening of toxic metals – e.g. lead, arsenic, cadmium, and chromium – and a standard analytical range of up to 30 elements from magnesium to uranium in the soil by *in situ* measurement (Figure 4.16). They also provide off-site, prepared-sample analysis of materials in the field with an accuracy challenging that of standard laboratory performance. PXRF provides a time- and cost-effective approach for the analysis of a variety of environmental samples: aerosols, water, sediment, soils or solid wastes.

The PXRF is lightweight and easy to operate with an integrated touch screen display, and there is the possibility of remote operation, plus a custom report-generation capability from a Windows^{TM}-based PC.

The bulk sample mode provides an analysis of the chemical composition of the soil, sediment, and other bulky, homogeneous samples. The pre-set factory calibration enables the simultaneous analysis of up to 25 elements in any bulk material, with no requirement for on-site calibrations or standards. Whether testing is performed *in situ*

Figure 4.16   Portable XRF equipment, menu on the display, *in situ* field use and the results on the screen (Thermo Scientific, 2015).

or off-site, the software automatically compensates for matrix variations. The typical testing time is in the range of 60–90 seconds.

In situ testing with the PXRF placed directly on the ground or on top of bagged samples allows the user to collect a large number of data points within a short time, delineating contamination patterns and achieving a more economical site remediation.

Field portable X-ray fluorescence (PXRF) continues to gain acceptance as a complement to traditional laboratory testing of metal-contaminated soil. The quality of data produced by PXRF varies with site conditions, soil composition and sample preparation. Quality assurance protocols for the field method usually require that a number of field samples be split and sent to a laboratory for confirmatory analysis. This confirmatory analysis (5% of the samples) can provide valuable information on the effectiveness of the field methodology (Sarkadi *et al.*, 2009).

9.3   *In situ* application of PXRF

PXRF may play a practical role in the quantification of the risk of contaminated sites. The device is suitable for source and transport pathway identification, source delineation, and it shortens the time requirement for the preparatory work of planning and executing risk reduction. Field-portable devices allow assessment of large sites, including water catchments, and enable on-site monitoring of polluted sites and remediation technologies (Tolner *et al.*, 2010). Mapping of multi-elemental contaminants is feasible with the application of the XRF device.

Even in cases where laboratory analysis is required, field XRF can be used to rapidly pre-screen samples (undisturbed soil samples at their original place or homogenized samples through the plastic sample bag) to obtain the optimal utility from the laboratory sampling effort (Shefsky, 1995; Spittler, 1995; Swift, 1995).

Negative or other inconsistent analytical results from samples collected, homogenized and delivered may surprise the assessor and necessitate repeated sampling. The use of XRF devices during sampling can make the collection of soil, sediment or solid waste for bioassaying, microcosm studies or technological experiments more targeted than in conventional technologies.

For the efficient removal of contaminated soil, sediment or waste, one has to continuously control the quality of the excavated material and take precautions not to remove excess soil or leave a contaminated proportion in the site.

As for field use, it is worth mentioning that the handheld device requires little or no sample preparation, but any large or non-representative debris (rocks, pebbles, leaves, vegetation, roots, concrete, etc.) on the soil surface should be removed before placing the device directly on the surface. Also, the soil surface must be as smooth as possible so that the probe window has good contact with the surface. A further requirement is that the soil or sediment is not saturated with water (US EPA Method 6200, 2007).

9.4   Uncertainty of *in situ* PXRF measurements

Accuracy of the *in situ* method is element- and matrix-dependent; it is influenced by site-specific conditions, particle size distribution, sample moisture, sample preparation time, and analysis time. The field portable XRF instruments perform properly for soils with moisture contents of up to 10%. Due to the heterogeneous nature of soil samples, *in situ* analysis can provide screening-type data. For on-site measurements the sample should be homogenized before or after drying, dried (if necessary), and ground before analysis. The error of *in situ* measurements decreases with the measurement time, up to a maximum of 90 seconds (Tolner *et al.*, 2008).

9.5   Advantages and disadvantages of *in situ* PXRF measurements

The main advantages of *in situ* PXRF measurements are:

– Rapid and simultaneous analysis of several elements
– As many measuring spots as necessary (no limitation)
– Low-cost measurement and immediate data provision
– Good analytical precision
– Little or no sample preparation.

Disadvantages of *in situ* PXRF measurements are:

– Lower accuracy compared to laboratory analytical measurements
– Uncertainty of the measurement results if soil moisture content exceeds 10–15%
– Detection limits are not low enough to quantify some elements at typical background concentrations in soils (e.g. cadmium)
– Relatively high equipment cost.

In summary, *in situ* PXRF provides rapid, low-cost measurement of toxic metals in the soil, with minimal, or no, sample preparation. Although *in situ* measurements with prepared samples are not as accurate as compared to off-site measurements, the main advantage of the *in situ* application of the portable device is that the sample number is practically unlimited by time and cost. As many as 100–150 measurements can be performed in a day and the sampling strategy can be continuously modified to fulfill site-specific needs.

The accuracy of the *in situ* method depends on site-specific conditions of contaminant particle size and distribution. Accuracy can be assessed in the field by comparison to the prepared sample using the XRF method.

REFERENCES

A.P. van den Berg (2015) *CPT Track-Truck*. [Online] Available from: www.apvandenberg.com/files/.fileserver/043/080.pdf. [Accessed 28th May 2015].

ACHE (2015) *Acetylcholinesterase activity assay kit*. [Online] Available from: http://www.sigmaaldrich.com/life-science/cell-biology/cell-biology-products.html?TablePage=112668420. [Accessed 14th August 2015].

Ahram (2015) *Palm PCR F1-12* from Ahram Biosystems. [Online] Available from: http://www.ahrambio.com/products_palmpcr_F1-12.html. [Accessed 14th August 2015].

Alessi, R.S. & Prunty, L. (1986) Soil water determination using fiber optics. *Soil Science Society of America Journal*, 50, 860–863.

Amptek (2015) *The cadmium, zinc, telluride detector*. [Online] Available from: http://www.amptek.com/pdf/cztapp1.pdf and http://www.amptek.com). [Accessed 14th August 2015].

AMS (2015) *Soil gas sampler*. [Online] Available from: http://www.ams-samplers.com/itemgroup.cfm?CNum=88&catCNum=3. [Accessed 14th August 2015].

Aquaflex (2015) *Aquaflex soil moisture sensors* from Watermetrics NZ. [Online] Available from: http://www.watermetrics.co.nz/products/soil-moisture-probe. [Accessed 28th May 2015].

ARYELLE 150 (2015) *LIBS* from LTB Lasertechnik Berlin GmbH. [Online] Available from: http://www.ltb-berlin.de/Spectrometers.93.0.html?&L=1. [Accessed 14th August 2015].

Ashworth, J., Nijenhuis, E., Glowacka, B., Tran, L. & Schenk-Watt, L. (2010) Turbidity and Color Correction in the Microtox Bioassay. *The Open Environmental Pollution & Toxicology Journal*, 2, 1–7.

Asner, G.P. & Martin R.E. (2009) Airborne spectranomics: mapping canopy chemical and taxonomic diversity in tropical forests. *Frontiers in Ecology and the Environment*, 7, 269–276.

ASTM Standard D7758-11 (2011) *Standard Practice for Passive Soil Gas Sampling in the Vadose Zone for Source Identification, Spatial Variability Assessment, Monitoring, and Vapor Intrusion Evaluations*. DOI: 10.1520/D7758-11. [Online] Available from: http://www.astm.org/Standards/D7758.htm. [Accessed 14th August 2015].

BAT (2015) *BAT groundwater monitoring system*. [Online] Available from: http://www.bat-gms.com/bat-mk-iii-filter-tips.asp. [Accessed 14th August 2015].

Beacon (2015) *Advanced sorbent sampling techniques*. [Online] Available from: www.beacon-usa.com. [Accessed 14th August 2015].

Belkin, S. (2006) Genetically engineered micro-organisms for pollution monitoring. In: Twardowska, I., Allen, H.E., Häggblom, M.M. & Stefaniak, S. (eds.) *Soil and Water Pollution Monitoring, Protection and Remediation*. Springer. pp. 3–23.

Benthien, M., Wieland, A., Garcia de Oteyza, T., Grimalt, J.O. & Kühl, M. (2004) Oil-contamination effects on a hypersaline microbial mat community (Camargue, France) as studied with microsensors and geochemical analysis. *Ophelia*, 58, 135–150.

BioBailers™ (2015) *Biodegradable disposable groundwater bailers*. [Online] Available from: http://www.biobailer.com/biobailer-groundwater-sampling-bailer. [Accessed 14th August 2015].

Biolog (2015) *EcoPlate™, microbial community analysis* by Biolog Inc. [Online] Available from: http://www.biolog.com/pdf/milit/00A_012_EcoPlate_Sell_Sheet.pdf. [Accessed 14th August 2015].

BioTox™ (2003) *BioTox luminometric toxicity tests and BioTox flash test* by Aboatox Oy. [Online] Available from: http://www.aboatox.com/?page_id=22#toxkit. [Accessed 28th May 2015].

Boll, J., Steenhuis, T.S. & Selker, J.S. (1992) Fiberglass wicks for sampling of water and solutes in the vadose zone. *Soil Science Society of America Journal*, 56, 701–707.

Brown, K.W., Thomas, J.C. & Holder, M.W. (1986) *Development of a capillary wick unsaturated zone water sampler*. Coop. Agreement CR812316-01-0, Las Vegas, NV, US EPA Environmental Monitoring Systems Laboratory.

Bruins, M., Bos, A., Petit, P.L.C., Eadie, K., Rog, A., Bos, R., van Ramshorst, G.H. & van-Belkum, A. (2009) Device-independent, real-time identification of bacterial pathogens with a metal oxide-based olfactory sensor. *European Journal of Clinical Microbiology & Infectious Diseases*, 28, 775–780.

Bujewski, G. & Rutherford, B. (1997) *The Rapid Optical Screening Tool (ROST^{TM}) Laser-Induced Fluorescence (LIF) System for Screening of Petroleum Hydrocarbons in Subsurface Soils*. Innovative Technology Verification Report. EPA 600-R-97-020. [Online] Available from: http://nepis.epa.gov/Exe/ZyPDF.cgi/30003J7F.PDF?Dockey=30003J7F.PDF. [Accessed 14th August 2015].

Campbell Scientific (2015) *Heat Dissipation Matric Potential Sensor*. [Online] Available from: www.campbellsci.com/229-l and https://s.campbellsci.com/documents/us/manuals/229.pdf. [Accessed 28th May 2015].

Catalog, Envco (2015) *Water and soil sensors*. [Online] Available from: http://envco.co.nz/catalog/water. [Accessed 14th August 2015].

Cerex, FTIR (2015) *Shepherd FTIR*. [Online] Available from: http://cerexms.com/shepherd-ftir. [Accessed 14th August 2015].

Cerex, Hound (2015) *Hound Point Gas Detector*. [Online] Available from: http://cerexms.com/hound. [Accessed 14th August 2015].

Changela, A., Chen, K., Xue Y., Holschen, J., Outten, C.E., Halloran, T.V. & Mondragon, A. (2003) Molecular basis of metal ion selectivity and zeptomolar sensitivity by CueR. *Science*, 301, 1383–1386.

Charlesworth, P. (2005) *Soil Water Monitoring* – An Information Package. Technical Reports. Irrigation Insights No. 1. CSIRO/CRC Irrigation Futures, Land & Water Australia. [Online] Available from: http://www.irrigationfutures.org.au/news.asp?catID=9&ID=137. [Accessed 28th May 2015].

Chen, Z., Williams, P.N. & Zhang, H. (2013) Rapid and nondestructive measurement of labile Mn, Cu, Zn, Pb and As in DGT by using field portable-XRF. *Environ. Science Process Impact*, 15(9), 1768–1774.

Chennu, A. (2010) *Shedding light on life: Optical technology in biological studies*. Marie Curie Conference, ESOF, 1–2 July 2010, Turin, Italy. Poster section. [Online] Available from: http://www.eu-sensenet.net/sensenet/sites/sensenet/files/images/arjun_esof2010_shedding_light_on_life.pdf. [Accessed 14th August 2015].

CLOR-N-SOIL (2015) *Field Test Kit for PCB Screening of Soil at 50 ppm*, by Dexil. [Online] Available from: http://www.dexsil.com/products/detail.php?product_id=4. [Accessed 14th August 2015].

Clue-in (2015) *High-Resolution Site Characterization (HRSC)*. [Online] Available from: https://clu-in.org/characterization/technologies/hrsc/. [Accessed 14th August 2015].

Clue-in DP (2015) *Direct-push platforms*. Contaminated site clean-up information. [Online] Available from: www.clu-in.org/characterization/technologies/dpp.cfm. [Accessed 28th May 2015].

Clue-in, immunoassay (2015) *Immunoassay and enzymatic assays*. [Online] Available from: https://clu-in.org/characterization/technologies/immunoassay.cfm. [Accessed 14th August 2015].

Cluzeau, D., Guernion, M., Chaussod, R., Martin-Laurent, F., Villenave, C., Cortet, J., Ruiz-Camacho, N., Pernin, C., Mateille, T., Philippot, L., Bellido, A., Rougé, L., Arrouays, D., Bispo, A. & Pérès, G. (2012) Integration of biodiversity in soil quality monitoring: Baselines for microbial and soil fauna parameters for different land-use types. *European Journal of Soil Biology*, 49, 63–72.

Cogliati, S., Rossini, M., Julitta, T., Meroni, M., Schickling, A., Burkart, A., Pinto, F., Rascher, U. & Colombo, R. (2015) Continuous and long-term measurements of reflectance and sun-induced chlorophyll fluorescence by using novel automated field spectroscopy systems. *Remote Sensing of Environment*, 164, 270–281. DOI: 10.1016/j.rse. 2015.03.027.

Cohen, K. & Saggese, S. (2000) Fiber Optic/Cone Penetrometer System Used for Heavy Metal Detection. *TechTrends*, 39, 1. [Online] Available from: http://www.epa.gov/tio/download/ newsltrs/tt1100.pdf. [Accessed 14th August 2015].

Colorimetric tubes (2015) *Colorimetric gas detection tubes and piston hand pump* by Enviroequioment Inc. [Online] Available from: http://www.enviroequipment.com/products/ piston.html. [Accessed 14th August 2015].

Corbisier, P., Lelie, D., Borremans, B., Provoost, A., Lorenzo, V., Brown, N.L., Lloyd, J.R., Hobman, J.L., Csöregi, E., Johansson, G. & Mattiasson, B. (1999) Whole cell- and protein-based biosensors for the detection of bioavailable heavy metals in environmental samples. *Analytica Chimica Acta*, 387, 235–244.

Cropscan (2015) *Multispectral Radiometers*. [Online] Available from: http://www.cropscan. com/msr.html. [Accessed 14th August 2015].

CRDS (2015) *Cavity ring-down spectroscopy* by Picarro. [Online] Available from: http://www.picarro.com/technology/cavity_ring_down_spectroscopy. [Accessed 14th August 2015].

Custodion™ (2015) *SPME sampler and traps* by Torion, by Perkin-Elmer. [Online] Available from: http://torion.com/products/custodion.html. [Accessed 14th August 2015].

Cytoviva (2015) *Hyperspectral microscopy*. [Online] Available from: www.cytoviva.com/ products/hyperspectral-imaging-2/hyperspectral-imaging. [Accessed 14th August 2015].

Dart (2015) *Soil moisture NMR for rapid soil and geotechnical application* from Vista-Clara. [Online] Available from: www.vista-clara.com/wp-content/uploads/2015/03/Dart-one-sheet-v2d.pdf. [Accessed 28th May 2015].

Davison, W., Hooda, P. Zhang, H. & Edwards, A.C. (2000a) DGT measured fluxes as surrogates for uptake of metals by plants. *Advances in Environmental Research*, 3, 550–555.

Davison, W., Fones, G., Harper, M., Teasdale, P. & Zhang, H. (2000b) Dialysis, DET and DGT: *in situ* diffusional techniques for studying water, sediments and soils. In: Buffle, J. & Horvai, G. (eds.) *In situ chemical analysis in aquatic systems*. Wiley. pp. 495–569.

Davison, W. & Zhang, H. (1994) *In situ* speciation measurements of trace components in natural waters using thin-film gels. *Nature*, 367, 545–548. DOI: 10.1038/367546a0.

Davison, W. & Zhang, H. (2012) Progress in understanding the use of diffusive gradients in thin films (DGT) back to basics. *Environmental Chemistry*, 2012, 911–13.

Davison, W., Zhang, H. & Grime, G.W. (1994) Performance characteristics of gel probes used for measuring the chemistry of pore waters. *Environmental Science & Technology*, 28, 1623–1632.

Decagon (2015) *Decagon probes*. [Online] Available from: www.decagon.com/products/soils/ volumetric-water-content-sensors. [Accessed 28th May 2015].

Decagon, NDVI (2015) [Online] *Normalized Difference Vegetation Index*. [Online] Available from: http://www.decagon.com/education/ndvi-and-pri-measurement-theory-methods-and-applications. [Accessed 14th August 2015].

Decagon PCAS (2015) *Drain Gauge G3 Passive Capillary Lysimeter*. [Online] Available from: www.decagon.com/products/hydrology/lysimeters/drain-gauge-g3-passive-capillary-lysimeter. [Accessed 14th August 2015].

De Cesare, F., Di Mattia, E., Pantalei, S., Zampetti, E., Vinciguerra, V., Canganella, F. & Macagnano, A. (2011) Use of electronic nose technology to measure soil microbial activity through biogenic volatile organic compounds and gases release. *Soil Biology and Biochemistry*, 43(10), 2094–2107.

Delfia TFR (2015) *Immunoassay endpoint: time resolved fluoresce intensity*. [Online] Available from: www.blossombio.com/pdf/products/BRO_DELFIATRFTechnology.pdf. [Accessed 14th August 2015].

Delta XRF (2015) *Handheld XRF device* by Olympus. [Online] Available from: www.olympus-ims.com. [Accessed 14th August 2015].

Delta-T (2015) *ML3 ThetaProbe*. [Online] Available from: http://www.delta-t.co.uk/product-category.asp?div=Soil%20Science&cat=Soil%20Moisture. [Accessed 28th May 2015].

Deltatox® (2015) *DeltaTox® II User's manual*. [Online] Available from: www.sdix.com/uploadedFiles/Content/Products/Water_Quality_Tests/3099915%20DeltaToxII%20User%20Manual-1.pdf. [Accessed 24th March 2015].

Deng, C., Yang, X., Li, N., Huang, Y. & Zhang, X. (2005) A Novel Miniaturized Flame Ionization Detector for Portable Gas Chromatography. *Journal of Chromatographic Science*, 43, 355–357.

DGT Research (2003) *DGT – for measurements in waters, soils and sediments*. DGT Research Ltd: Lancaster, UK. [Online] Available from: http://www.dgtresearch.com/dgtresearch/dgtresearch.pdf. [Accessed 14th August 2015].

Direct Image (2015) *Direct Image® Products* from Geoprobe. [Online] Available from: http://geoprobe.com/geoprobe-systems-direct-image-products. [Accessed 14th August 2015].

Discus (2015) *Soil moisture NMR for rapid soil and geotechnical application* from Vista-Clara. [Online] Available from: www.vista-clara.com/instruments/discus/discus-features. [Accessed 28th May 2015].

DOE (2001) *A National Roadmap for Vadose Zone Science and Technology – US DOE/ID-10871*. [Online] Available from: www.pc-progress.com/Documents/RVGenugten/2001_DOE_National_Roadmap.pdf. [Accessed 28th May 2015].

Drolet, G., Wade, T., Nichol, C.J., MacLellan, C., Levula, J., Porcar-Castell, A., Nikinmaa, E. & Vesala, T. (2014) A temperature-controlled spectrometer system for continuous and unattended measurements of canopy spectral radiance and reflectance, *International Journal of Remote Sensing*, 35, 1769–1785. DOI: 10.1080/01431161.2014.882035.

ECH2O (2015) *Volumetric Water Content Sensors* from Decagon. [Online] Available from: http://www.decagon.com/products/soils/volumetric-water-content-sensors & http://www.decagon.com/education/soil-moisture-install-video. [Accessed 28th May 2015].

ECLOX (2015) *Rapid response test kit* by Hach. [Online] Available from: http://hachhst.com/products/portable-rapid-response-testing-systems. [Accessed 14th August 2015].

EcoBailer Pro (2015) *Disposable bailers for discrete sampling*. [Online] Available from: http://www.rshydro.co.uk/ETP-Disposable-Bailers-c-204.html. [Accessed 14th August 2015].

Ecoprobe 5 (2015) *Handheld multiple device* from RS Dynamics. [Online] Available from: www.rsdynamics.com/main.php3?s1=products&s2=soilcont&s3=aplication. [Accessed 14th August 2015].

Ecomatik (2016) *Equitensiometers*. [Online] Available from: http://www.ecomatik.de/pdf/catalog.pdf. [Accessed 13th March 2016].

Eijkelkamp (2015) *Soil test kit for macronutrients & pH*. [Online] Available from: https://en.eijkelkamp.com/products/field-measurement-equipment/soil-test-kit-for-macronutrients-ph.html. [Accessed 14th August 2015].

Enhanced Spectrometry (2015) *Handheld Raman detector for explosives and narcotics*. [Online] Available from: http://enspectr.com. [Accessed 14th August 2015].

Envirotek (2015) *EnviroPro probe*. [Online] Available from: http://www.envirotek.com.au/products/enviropro-capacitance-probe.html. [Accessed 28th May 2015].

EQ3 (2015) *Equitensiometer* from Delta-T. [Online] Available from: http://www.delta-t.co.uk/product-display.asp?id=EQ3%20Product&div=Soil%20Science. [Accessed 28th May 2015].

Equilibrator (2015) *Passive diffusion samplers* by EON. [Online] Available from: http://store. eonpro.com/store/p/1788-EON-Equilibrator-Passive-Diffusion-Samplers.aspx. [Accessed 14th August 2015].

Explorer (2015) *Explorer Portable Gas Chromatograph* by Inficon. [Online] Available from: http://products.inficon.com/en-us/Product/Detail/Explorer-Portable-Gas-Chromatograph?pa th=Products%2Fpg-ChemicalDetection. [Accessed 14th August 2015].

FAST (2015) *Flow Assay and Sensing System* by Research International. [Online] Available from: www.resrchintl.com/Custom_Solutions.html & www.resrchintl.com/Research_Development. html#sthash.P95cscPP.dpuf. [Accessed 14th August 2015].

FFD, Vertek (2015) *Fuel Fluorescence Detector* from Vertek. [Online] Available from: http:// www.environmental-expert.com/products/vertek-model-ffd-in-situ-fuel-fluorescence-detector-323697. [Accessed 14th August 2015].

FLUTe™ (2015) *Flexible liner underground technologies.* [Online] Available from: http://www. flut.com/About/about.html. [Accessed 14th August 2015].

Fugro (2015) *Geotechnical site investigation.* [Online] Available from: www.fugro.com/ our-expertise/our-services/geotechnical/geotechnical-site-investigations-onshore#tabbed5. [Accessed 28th May 2015].

Garrido, F., Ghodrati, M. & Chendorain, M. (1999) Small-scale measurement of soil water content using a fiber optic sensor. *Soil Science Society of America Journal*, 63(6), 1505–1512.

Gasmet, *in situ* (2015) *Gasmet™ in situ gas analyzer.* [Online] Available from: www.gasmet. com/products/stationary-analyzer-systems/gasmet-in-situ. [Accessed 14th August 2015].

Gasmet, portable (2105) *Gasmet™ DX4015 portable gas analyzer.* [Online] Available from: http://www.gasmet.com/products/portable-gas-analyzers/dx4015. [Accessed 14th August 2015].

GC-SAW (2015) *Portable gas-chromatograph with SAW detector* by Electronic Sensor Technology. [Online] Available from: www.estcal.com/products/model4600_portable_znose.html. [Accessed 24th August 2015].

Gee, G.W., Ward, A.L., Caldwell, T.G. & Ritter, J.C. (2002) A vadose zone water fluxmeter with divergence control. *Water Resources Research*, 38, 16-1–16-7.

Gee, G.W., Z.F. Zhang, & A.L. Ward (2003) A modified vadose-zone fluxmeter with solution collection capability, *Vadose Zone Journal*, 2, 627–632.

Gee, G.W., Zhang, Z.F. Ward, A.L. & Keller, J.M. (2004) Passive-wick water fluxmeters: theory and practice. In: *Supersoil 2004, Symposium 15: Water and solute transport in soil.* [Online] Available from: www.regional.org.au / au / asssi / supersoil2004 / s15 / oral / 1627_geeg.htm. [Accessed 14th August 2015].

Geoprobe (2015) *Direct push tools.* [Online] Available from: http://geoprobe.com/direct-push-tooling?gclid=COHJ4anSmcUCFYgewwodAAUA8Q. [Accessed 28th May 2015].

Geoprobe Anchor (2015) *Anchoring systems from Geoprobe.* [Online] Available from: http://geoprobe.com/anchoring-systems. [Accessed 28th May 2015].

Geoprobe dual tube (2015) *Dual tube soil sampling systems.* [Online] Available from: http://geoprobe.com/dual-tube-soil-sampling-systems. [Accessed 14th August 2015].

Geoprobe™ GC (2015) *Portable gas-chromatograph.* [Online] Available from: http://srigc.net/ home/product_detail/geoprobe-gc. [Accessed 14th August 2015].

Gerlach, J., Samways, M. & Pryke, J. (2013) Terrestrial invertebrates as bioindicators: an overview of available taxonomic groups. *Journal of Insect Conservation*, 17, 831–850. DOI: 10.1007/s10841-013-9565-9.

Ghodrati, M. (1999) Point measurement of solute transport processes in soil using fiber optic sensors. *Soil Science Society of America Journal*, 63, 471–479.

Ghodrati, M., Garrido, F., Campbell, C.G. & Chendorain, M. (2000) A multiplexed fiber optic miniprobe system for measuring solute transport in soil. *Journal of Environmental Quality*, 29(2), 540–550.

Ghodrati, M., Garrido, F. & Campbell, C.G. (2001) Detailed characterization of solute transport in a heterogeneous field soil. *Journal of Environmental Quality*, 30(2), 573–83. DOI: 10.2134/jeq2001.302573x.

Gimpel J., Zhang H., Davison W. & Edwards A.C. (2002) In situ trace metal speciation in lake surface waters using DGT, dialysis, and filtration. *Environmental Science & Technology*, 37(1), 138–146.

GLRL (2015) *Green Light Red Light* from Odyssey Data Recording. [Online] Available from: http://odysseydatarecording.com/download/GLRL_Handreader.pdf. [Accessed 28th May 2015].

Glud, R.N., Rysgaard, S. & Kühl, M. (2002) O₂ dynamics and photosynthesis in ice algal communities: quantification by microsensors, O₂ exchange rate, 14C-incubations and PAM-fluorometry. *Aquatic Microbial Ecology*, 27, 301–311.

GMR (2015) *Surface NMR-MRS technology* from Vista-Clara. [Online] Available from: http://www.vista-clara.com/instruments/gmr. [Accessed 28th May 2015].

Gopher (2015) *Gopher* from Odyssey Data Recording. [Online] Available from: http://odyssey datarecording.com/download/HBGOPH92.pdf. [Accessed 28th May 2015].

Gouda (2015) *Truck-Based CPT Penetrometer Rig.* [Online] Available from: http://www.gouda-geo.com/products/cpt-equipment/truck-based-cpt-rigs/truck-based-cpt-penetrometer-rig-max.-120-kn. [Accessed 28th September 2015].

Gouda Geo (2015) *Gouda Geo-Equipment.* [Online] Available from: http://www.gouda-geo.com/?gclid=COG4t7iUpcUCFcISwwod0aEAKQ. [Accessed 28th May 2015].

Gouda Geo VT (2015) *Electric vane tester* from Gouda Geo-Equipment. [Online] Available from: http://www.gouda-geo.com/products/cpt-equipment/special-modules-for-cpt/electric-vane-tester. [Accessed 28th May 2015].

Gruiz, K., Meggyes, T. & Fenyvesi, E. (eds.) (2015) *Engineering Tools for Environmental Management, Volume 2. Environmental Toxicology.* Boca Raton, Fl., CRC Press.

Gruiz, K., Molnár, M., Feigl, V., Hajdu, Cs., Nagy, Zs.M., Klebercz, O., Fekete-kertész, I., Ujaczki, É. & Tolner, M. (2015a) Terrestrial toxicology. In: Gruiz, K., Meggyes, T. & Fenyvesi, E. (eds.) (2015a) *Engineering Tools for Environmental Management, Volume 2. Environmental Toxicology.* Boca Raton, Fl., CRC Press. pp. 229–279.

Hach Immunoassay (2015) *Immunoassay test kits.* [Online] Available from: http://www.hach.com/test-kits/immunoassay-test-kits/family?productCategoryId=35547009711. [Accessed 14th August 2015].

Hajdu, Cs. & Gruiz, K. (2015) Bioaccessibility and bioavailability in risk assessment. In: Gruiz, K., Meggyes, T. & Fenyvesi, É. (eds.) *Engineering Tools in Environmental Risk Management. 2. Environmental Toxicology.* Boca Raton, London, New York, Leiden, CRC Press, Balkema.

Han, L., Shi, X., Wu, W., Kirk, F.L., Luo, J., Wang, L., Mott, D., Cousineau, L., Lim, SII., Lu, S. & Zhong, C-J. (2005) Nanoparticle-structured sensing array materials and pattern recognition for VOC detection. *Sensors and Actuators,* 106, 431–441.

Hapsite™ (2015) *Inficon Hapsite Portable Gas Chromatograph–Mass Spectrometer.* [Online] Available from: http://products.inficon.com/en-us/nav-products/Product/Detail/HAPSITE-ER-Identification-System?path=Products%2Fpg-ChemicalDetection. [Accessed 14th August 2015].

Hilker, T., Nesic, Z., Coops, N.C. & Lessard, D. (2010) A new, automated, multiangular radiometer instrument for tower-based observations of canopy reflectance (Amspec II). *Instrumentation Science & Technology*, 38, 319–340. DOI: 10.1080/10739149.2010.508357.

HNU (2015) *PID Models 102, 102+ and 112* by HNU. [Online] Available from: http://www.hnu.com/products.php. [Accessed 14th August 2015].

Ho, C.K., Robinson, A., Miller, D.R. & Davis, M.J. (2005) Overview of Sensors and Needs for Environmental Monitoring. *Sensors*, 5(2), 4–37. [Online] Available from: http://www.ncbi.nlm.nih.gov/pmc/articles/PMC3909362. [Accessed 14th August 2015].

Holder, M., Brown, K.W., Thomas, J.C., Zabcik, D. & Murray, H.E. (1991) Capillary-wick unsaturated zone soil pore water sampler. *Soil Science Society of America Journal*, 55, 1195–1202.

HydraSleeve (2015) *Direct interval, no-purge sampler.* [Online] Available from: http://www.hydrasleeve.com. [Accessed 14th August 2015].

HydroSCOUT (2015) *Groundwater sensor system* by Dexil. [Online] Available from: http://www.dexsil.com/products/detail.php?product_id=27. [Accessed 14th August 2015].

ICT (2015) *Soils Lysimeters Products* by ICT International. [Online] Available from: http://www.ictinternational.com/products/soils/lysimeters. [Accessed 14th August 2015].

Insam, H. & Seewald M.S.A. (2010) Volatile organic compounds (VOCs) in soils. *Biology and Fertility of Soils*, 46, 199–213. DOI: 10.1007/s00374-010-0442-3.

In-situ (2015) *TROLL® 9500 Multiparameter Instrument.* [Online] Available from: https://in-situ.com/wp-content/uploads/2014/11/TROLL-9500-Water-Quality-Instrument_Specs.pdf. [Accessed 28th May 2015].

Ismaël, A., Bousquet, B., Michel-Le Pierrès, K., Travaillé, G., Canioni, L. & Roy, S. (2011) In situ semi-quantitative analysis of polluted soils by laser-induced breakdown spectroscopy (LIBS). *Applied Spectroscopy*, 65(5), 467–73. DOI: 10.1366/10-06125.

ITRC (2005) *Technology Overview of Passive Sampler Technologies.* Interstate Technology & Regulatory Council, Authoring Team. Washington, D.C., ITRC. [Online] Available from: https://clu-in.org/download/char/passsamp/itrc-technology-overview-DSP-4.pdf. [Accessed 14th August 2015].

ITRC (2010) *Use and Measurement of Mass Flux and Mass Discharge.* Prepared by ITRC. MASSFLUX-1. Washington, D.C., ITRC, Integrated DNAPL Site Strategy Team. [Online] Available from: http://www.cluin.org/download/contaminantfocus/dnapl/Detection_and_Site_Characterization/DNAPL-Mass-flux-1.pdf. [Accessed 14th August 2015].

Ivask, A., Rolova, T. & Kahru, A. (2009) A suite of recombinant luminescent bacterial strains for the quantification of bioavailable heavy metals and toxicity testing. *BMC Biotechnology*, 9, 41–55.

Javelin (2015) *Soil moisture NMR for rapid soil and geotechnical application* from Vista-Clara. [Online] Available from: http://www.vista-clara.com/instruments/javelin. [Accessed 14th August 2015].

Jeffery, S., Gardi, C., Jones, A., Montanarella, L., Marmo, L., Miko, L., Ritz, K., Peres, G., Römbke, J. & van der Putten, W.H. (eds.) (2010) *European Atlas of Soil Biodiversity.* Luxembourg, Publications Office of the European Union.

Jones, M., Aptaker, P.S., Cox, J., Gardiner, B.A. & McDonald, P.J. (2012) A transportable magnetic resonance imaging system for *in situ* measurements of living trees: the Tree Hugger. *Journal of Magnetic Resonance*, 218, 133–140.

Kahru, A., Dubourguier, H-C., Blinova, I., Ivask A., & Kasemets, K. (2008) Biotests and biosensors for ecotoxicology of metal oxide nanoparticles: A minireview. *Sensors*, 8, 5153–5170. [Online] Available from: DOI: 10.3390/s8085153.

Kargas G. & Soulis K. (2012) Performance Analysis and Calibration of a New Low-Cost Capacitance Soil Moisture Sensor. *Journal of Irrigation and Drainage Engineering*, 138(7), 632–641.

Kaufmann, K., Chapman, S.J., Campbell, C.D., Harms, H. & Höhener, P. (2005) Miniaturized test system for soil respiration induced by volatile pollutants. *Environmental Pollution*, 140, 269–278.

Kaur, J., Adamchuk, V.I., Whalen, J.K. & Ismail, A.A. (2015) Development of an NDIR CO2 Sensor-Based System for Assessing Soil Toxicity Using Substrate-Induced Respiration. *Sensors*, 15, 4734–4748.

Kelly, M.J., Sweatt, W.C., Kemme, S.A., Kasunic, K.J., Blair, D.S., Zaidi, S.H., McNeil, J.R., Burgess, L.W., Brodsky, A.M. & Smith, S.A. (2000) *Grating light reflection spectroelectrochemistry* for detection of trace amounts of aromatic hydrocarbons in water. SAND 2000-1018, by Sandia National Labs., Albuquerque, NM (US); [Online] Available from: www.osti.gov/scitech/servlets/purl/754395. [Accessed 14th August 2015].

Kim, Y., Jabro, J.D. & Evans, R.G. (2010) Wireless lysimeters for real-time online soil water monitoring. *Irrigation Science*, 29(5), 423–430. DOI: 10.1007/s00271-010-0249-x.

Kimura, T., Geya, Y., Terada, Y., Kose, K., Haishi, T., Gemma, H. & Sekozawa, Y. (2011) Development of a mobile magnetic resonance imaging system for outdoor tree measurements. *Review of Scientific Instruments*, 82(5), 053704. DOI: 10.1063/1.3589854.

Knutson, J.H. & Selker, J.S. (1994) Unsaturated hydraulic conductivities of fiberglass wicks and designing capillary wick pore-water samplers. *Soil Science Society of America Journal*, 58, 721–729.

Koo, C., Malapi-Wight, M., Kim, H.S., Cifci, O.S., Vaughn-Diaz, V.L., Ma, B., Kim, S., Abdel-Raziq, H., Ong, K., Jo, Y-K., Gross, D.C., Shim, W-B. & Han, A. (2013) Development of a Real-Time Microchip PCR System for Portable Plant Disease Diagnosis. *PLOS ONE*, 8(12), e82704. DOI: 10.1371/journal.pone.0082704.

Kram, M.L. (1998) Use of SCAPS petroleum hydrocarbon sensor technology for real-time indirect DNAPL detection. *Journal of Soil Contamination*, 7(1), 73–85.

Kram, M., Lieberman, S. & Jacobs, J.A. (2000) *Direct sensing of soils and groundwater*. In: Lehr, J. (ed.) Standard Handbook of Environmental Science, Health, and Technology: New York, McGraw Hill. pp. 11.124–11.150.

Kühl, M. & Revsbech, N.P. (2001) Biogeochemical microsensors for boundary layer studies. In: Boudreau, B.P. & Jørgensen B.B. (eds.) *The Benthic Boundary Layer*. Oxford, Oxford University Press. pp. 180–210.

L2000DX (2015) *PCB/Chloride analyzer System* by Dexil. [Online] Available from: http://www.dexsil.com/products/detail.php?product_id=13. [Accessed 14th August 2015].

Lasat, M.M. (2002) Phytoextraction of toxic metals: A review of biological mechanisms. *Journal of Environmental Quality*, 31, 109–120.

Lawley, V., Lewis, M., Clarke, K. & Ostendorf, B. (2016) Site-based and remote sensing methods for monitoring indicators of vegetation condition: an Australian review. *Ecological Indicators*, 60, 1273–1283.

Lemos, S.G., Nogueira, A.R.A., Torre-Neto, A., Parra, A., Artigas, J. & Alonso, J. (2004) In-soil potassium sensor system. *Journal of Agricultural and Food Chemistry*, 52, 5810–5815.

Li, Y-F., Li, F-Y., Ho, Ch-L. & Liao, V.H-Ch. (2008) Construction and comparison of fluorescence and bioluminescence bacterial biosensors for the detection of bioavailable toluene and related compounds. *Environmental Pollution*, 152(1), 123–129. DOI: 10.1016/j.envpol.2007.05.002.

LIF, Clue-in (2015) *Laser-Induced Fluorescence* – Equipment. [Online] Available from: https://clu-in.org/characterization/technologies/lif.cfm. [Accessed 14th August 2015].

Liu, J. & Lu, Y. (2003) A colorimetric lead biosensor using DNAzyme-directed assembly of gold nanoparticles. *J. Am. Chem. Soc.* 125, 6642–6643.

Louie, M.J., Shelby, P.M., Smesrud, J.S., Gatcheil, L.O. & Selker, J.S. (2000) Field evaluation of passive capillary samplers for estimating groundwater recharge. *Water Resources Research*, 36(9), 2407–2416.

LUMIStox (1999) *LUMIStox 300 LPV321 User Manual*, Hach-Lange. [Online] Available from: http://www.manualsdir.com/manuals/333014/hach-lange-lumistox-300-lpv321-user-manual.html. [Accessed 24th March 2015].

LUMIStox (2010) Environmental Technology Verification Program. Joint Verification of the Hach-Lange GmbH LUMIStox 300 Bench Top Luminometer and ECLOX Handheld Luminometer Advanced Monitoring Systems Center. US EPA ETV and ETV Canada. [Online] Available from: http://nepis.epa.gov/Adobe/PDF/P100ELTP.pdf. [Accessed 24th March 2015].

McGeoch, M.A. (2007) Insects and bioindication: theory and practice. In: Stewart, A.J., New, T.R. & Lewis, O.T. (eds.) *Insect conservation biology*. Wallingford, CABI. pp. 144–174.

Meissner, R. & Seyfarth, M. (2004) Measuring water and solute balance with new lysimeter techniques. In: *Supersoil 2004, Symposium 15: Water and solute transport in soil*. [Online] Available from: www.regional.org.au/au/asssi/supersoil2004/s15/oral/1083_meissnerr.htm. [Accessed 28th May 2015].

Meroni, M., Barducci, A., Cogliati, S., Castagnoli, F., Rossini, M., Busetto, L., Migliavacca, M., Cremonese, E., Galvagno, M., Colombo, R. & Di Cella, U. M. (2011) The hyperspectral irradiometer, a new instrument for long-term and unattended field spectroscopy measurements. *Review of Scientific Instruments*, 82(4), 043106. DOI: 10.1063/1.3574360.

Mertens, J., Diels, J. & Vanderborght, J. (2007) Numerical Analysis of Passive Capillary Wick Samplers prior to Field Installation. *Soil Science Society of America Journal*, 71(1) 35–42. [Online] Available from: DOI: 10.2136/sssaj2006.010.

MicroBioTest (2015) *Miniaturized bioassays for toxicity screening*. [Online] Available from: http://www.microbiotests.be; http://www.microbiotests.be/rokdownloads/field%20tests%20product%20list.pdf. [Accessed 14th August 2015].

Micro Hound (2015) *Micro Hound Multi-Gas Analyzer* by Cerex. [Online] Available from: http://cerexms.com/micro-hound. [Accessed 14th August 2015].

Microtox® (2015) *Microtox toxicity testing*. [Online] Available from: http://www.leederconsulting.com/toxicology_services_microtox.html. [Accessed 14th August 2015].

Mini GC-FID (2015) *Portable GC* by Acquisition Solutions[k]. [Online] Available from: http://www.acqsol.com/GeoFox.htm and http://www.acqsol.com/FlyerGCDet1F.pdf. [Accessed 14th August 2015].

Model 8807 (2015) *MODel 8807 Portable Gas Chromatograph* by PCF Elettronica. [Online] Available from: http://www.pcfelettronica.it/en/products/portable-instruments/mod-8807-gc. [Accessed 14th August 2015].

ModernWater (2015) *Monitoring products, test kits*. [Online] Available from: http://www.modernwater.com/monitoring/environment/test-kits. [Accessed 14th August 2015].

Monoflex (2015) *Monoflex, porous cap lysimeter* by Baker Water Systems. [Online] Available from: www.bakerwatersystems.com/uploads/files/pdf/manuals/Lysimeter%20Manual%20updated%20092012.pdf. [Accessed 14th August 2015].

mPulse™ (2015) *mPulse Laser Induced Breakdown Spectroscopy* from Oxford Instruments. [Online] Available from: www.oxford-instruments.com/search-results?q=LIBS&t=pq-pr-pt. [Accessed 14th August 2015].

Multiparameter sensor (2015) *Groundwater sensor system* by INW. [Online] Available from: http://www.observatormeteohydro.com/products/hydrological-sensors/ground-water-quality/inw-adapter/208. [Accessed 14th August 2015].

NASA MicroFID (2015) *A Portable Gas Analyzer Using a Micro-FID/TCD* by Cbana Laboratories. [Online] Available from: https://technology.grc.nasa.gov/techdays2012/TechDays_TOPsheets/PS-00922-1112_B120_AS.pdf. [Accessed 14th August 2015].

Nielsen, M., Revsbech, N.P. & Kühl, M. (2015) Microsensor measurements of hydrogen gas dynamics in cyanobacterial microbial mats. *Frontiers in Microbiology*, 6, 726. DOI: 10.3389/fmicb.2015.00726.

Niton (2015) *Niton XL3t handheld XRF* by Thermo Scientific. [Online] Available from: http://www.niton.com/en/environmental-analysis/applications/contaminated-soil-testing & www.niton.com; www.thermoscientific.com & http://www.niton.com/en/niton-analyzers-products/which-xrf-analyzer-is-right-for-me. [Accessed 14th August 2015].

Nolan, A.L., Zhang, H. & McLaughlin, M.J. (2005) Prediction of zinc, cadmium, lead, and copper availability to wheat in contaminated soils using chemical speciation, diffusive gradients in thin films, extraction, and isotopic dilution techniques. *Journal of Environmental Quality*, 34, 469–476.

Ocean Optics, LIBS (2015) *Laser Induced Breakdown Spectroscopy* by Ocean Optics. [Online] Available from: http://oceanoptics.com/product/laser-induced-breakdown-spectroscopy-libs. [Accessed 14th August 2015].

Ocean Optics, plants (2015) *UAV microspectrometer for vegetation analysis*. [Online] Available from: http://oceanoptics.com/uav-capable-microspectrometers-analysis-vegetation-ground-cover & http://oceanoptics.com/applications/farm-to-table-technologies. [Accessed 14th August 2015].

Olaniran, A.O., Balgobind, A. & Pillay, B. (2013) Bioavailability of heavy metals in soil: impact on microbial biodegradation of organic compounds and possible improvement strategies. *International Journal of Molecular Sciences*, 14(5), 10197–10228. DOI: 10.3390/ijms140510197.

OxiTop (2015) *Respirometrics*. [Online] Available from: https://ecobiosoil.univ-rennes1.fr/ADEME-Bioindicateur/english/WS/WS5-Oxitop.pdf & http://www.wtw.de/en/products/lab/bodrespiration.html. [Accessed 14th August 2015].

Pagani, V.T. (2015) *Field Vane Test*, Pagani Geotechnical Equipment. [Online] Available from: http://www.pagani-geotechnical.com/index.php/en/ct-menu-item-13/ct-menu-item-43. [Accessed 28th May 2015].

Palintest (2015) *Complete soil kits*. [Online] Available from: www.palintest.com/products/category/soil-hydroponics-and-irrigation & http://www.palintest.com. [Accessed 14th August 2015].

Pawliszyn, J. (ed.) (2009) *Handbook of Solid Phase Microextraction*. Chemical Industry Press.

Peeper (2015) *Modified Hesslein In-Situ Pore Water Sampler* from Rickly. [Online] Available from: http://www.rickly.com/as/Hesslein.htm. [Accessed 14th August 2015].

Peranginangin, N.P., Richards, B.K. & Steenhuis, T.S. (2009) Assessment of vadose zone sampling methods for detection of preferential herbicide transport. *Hydrology and Earth System Sciences*, 6, 7247–7285. [Online] Available from: http://www.hydrol-earth-syst-sci-discuss.net/6/7247/2009/hessd-6-7247-2009.pdf. [Accessed 14th August 2015].

Perlo, J., Danieli, E., Perlo, J., Blümich, B. & Casanova, F. (2013) Optimized slim-line logging NMR tool to measure soil moisture in situ. *Journal of Magnetic Resonance*, 233, 74–79.

PetroFLAG (2015) *PetroFLAG high range extract* by Dexsil. [Online] Available from: http://www.dexsil.com/products/detail.php?product_id=25. [Accessed 14th August 2015].

Porcar-Castell, A., Mac Arthur, A., Rossini, M., Eklundh, L., Pacheco-Labrador, J., Anderson, K., Balzarolo, M., Martín, M.P., Jin, H., Tomelleri, E., Cerasoli, S., Sakowska, K., Hueni, A., Julitta, T., Nichol, C.J. & Vescovo, L. (2015) EUROSPEC: at the interface between remote sensing and ecosystem CO_2 flux measurements in Europe. Copernicus Publications on behalf of the European Geosciences Union. *Biogeosciences Discussions*, 12, 13069–13121.

Portable GC, Indonesia (2015) *Portable GC*, by PT Amerta Labindo Utama, Indonesia. [Online] Available from: http://amertalabindo.com/portable-gc. [Accessed 14th August 2015].

Portable GC-FID, Meta (2015) *Meta GC-FID/4HU* by Messtechnische Systeme, Dresden. [Online] Available from: http://www.meta-dresden.de/GC-FID_eng_v11.pdf. [Accessed 14th August 2015].

Potts, P.J. & West, M. (eds.) (2008) *Portable X-ray Fluorescence Spectrometry Capabilities for In Situ Analysis*. RSC Publishing. ISBN: 978-0-85404-552-5

PPO-LIBS (2015) *Laser Induced Breakdown Spectroscopy*, from PP Optica. [Online] Available from: www.ppo.ca. [Accessed 14th August 2015].

PSR+ and PSR-1100 (2015) *Field spectroradiometers* by Spectral Evolution. [Online] Available from: http://www.spectralevolution.com/files/PSR-1100_Series_42115_web.pdf & http://www.spectralevolution.com/portable_spectroradiometer_vegetation.html. [Accessed 8th August 2015].

Q-Box (2015) *Field portable soil respiration system.* [Online] Available from: http://www.eco search.info/sites/default/files/prodotto_scheda_tecnica/SR1LP%20combined.pdf. [Accessed 14th August 2015].

Quick Scan (2015) *Soil & compost analyzer.* [Online] Available from: http://www.challenge-sys. com/files/Quick_Scan_Soil_New_Brochure_2-12.pdf. [Accessed 14th August 2015].

Radcliff, D.E. & Simunek, J. (2012) Water flow in soils. In: Huang, P.M., Li, Y. & Sumner M.E. (eds.) *Handbook of Soil Sciences: Properties and Processes.* Second Edition. Boca Raton, CRC Press.

RAE (2015) *RAE equipment* by RAE Systems. [Online] Available from: http://www.raesystems. eu/products/ultrarae-3000 & http://www.geotechenv.com/photoionization_detectors.html. [Accessed 14th August 2015].

RAPID (2015) *Ruggedized Advanced Pathogen Identification Device, a biodetection system* by Biofire. [Online] Available from: http://biofiredefense.com/rapid. [Accessed 14th August 2015].

Reinke, K. & Jones, S. (2006) Integrating vegetation field surveys with remotely sensed data. *Ecological Management & Restoration*, 7, S18-S23.

Riether, K.B., Dollard, M.A. & Billard, P. (2001) Assessment of heavy metal bioavailability using *Escherichia coli* zntAp::lux and copAp::lux-based biosensors. *Applied Microbiology and Biotechnology*, 57(5–6), 712–716.

Rogers, H.H. & Bottomley, P.A. (1986) *In situ* nuclear magnetic resonance imaging of roots: influence of soil type, ferromagnetic particle content, and soil water. *Agronomy Journal*, 79(6), 957–965.

RoHS (2003) *Directive on the restriction of the use of certain hazardous substances in electrical and electronic equipment*, 2002/95/EC. [Online] Available from: http://eur-lex.europa. eu/legal-content/EN/TXT/?uri=CELEX:32002L0095. [Accessed 14th August 2015].

Rossini, M., Migliavacca, M., Galvagno, M., Meroni, M., Cogliati, S., Cremonese, E., Fava, F., Gitelson, A., Julitta, T., Morra di Cella, U., Siniscalco, C. & Colombo, R. (2014) Remote estimation of grassland gross primary production during extreme meteorological seasons. *International Journal of Applied Earth Observation*, 29, 1–10. DOI: 10.1016/j.jag.2013.12.008.

ROST (2015) *Rapid Optical Screening Tool* from Fugro Geoscience. [Online] Available from: www.fugroconsultants.com/downloads/CPT_ROST_Service_Sheet.pdf. [Accessed 14th August 2015].

Sampling plate, EK (2015) *Plastic soil water sampling plate* by Eijkelkamp. [Online] Available from: https://en.eijkelkamp.com/products/water-bottom-sampling-equipment/soil-water-sampling-plate-plastic.html. [Accessed 14th August 2015].

Sampling plate, SM (2015) *Soil water sampling plate* by Soilmoisture Co. [Online] Available from: www.soilmoisture.com/SOIL-WATER-SAMPLING-PLATE/. [Accessed 14th August 2015].

Samways, M.J., McGeoch, M.A. & New, T.R. (2010) *Insect conservation: A handbook of approaches and methods.* Oxford, Oxford University Press.

Sandia chemiresistor (2015) *Chemical microsensor for chemical warfare agent and toxic industrial chemical detection in air, soil, and water monitoring applications.* [Online] Available from: http://www.sandia.gov/mstc/_assets/documents/Fact_Sheets/sensors/2chemir esistor.pdf. [Accessed 28th May 2015].

Sandia LDRD (2015) *Micro-chemical sensors for in-situ monitoring and characterization of volatile contaminants* – The LDRD Project. [Online] Available from: http://www.sandia.gov/ sensor. [Accessed 28th May 2015].

Sandia micro (2015) μ*ChemLab* by Sandia National Laboratories. [Online] Available from: http://www.irn.sandia.gov/organization/mstc/organization/micro-analytical/chemlab.html. [Accessed 14th August 2015].

Sarkadi, A., Vaszita, E., Tolner, M. & Gruiz, K. (2009) *In situ* site assessment: short overview and description of the field-portable XRF and its application. *Land Contamination & Reclamation*, 17(3–4), 431–442.

SCAPS-LIF (2015) *Site Characterization and Analysis Penetrometer System.* Certification, U.S. Department of the Navy. [Online] Available from: https://www.dtsc.ca.gov/Technology Development/TechCert/navy-scaps-lif-techcert.cfm. [Accessed 14th August 2015].

Sentek (2015) *EnviroSCAN probe* from Sentek. [Online] Available from: http://www.sentek. com.au/products/enviro-scan-probe.asp. [Accessed 28th May 2015].

Shefsky, S. (1995) Lead in soil analysis using the NITON XL. In: *International Symposium on Field Screening Methods for Hazardous Wastes and Toxic Chemicals*, Las Vegas, Nevada, 22–24 February, A&WMA: Air & Waste Management Association. pp. 1106–1117.

Skye sensor (2015) *Low cost NDVI and PRI sensors* by Slye sensor. [Online] Available from: www.skyeinstruments.com/news-events/low-cost-ndvi-pri-sensors. [Accessed 14th August 2015].

SNAP Sampler (2015) *Snap Sampler for passive groundwater sampling* by ProHydro, Inc. [Online] Available from: www.SnapSampler.com. [Accessed 14th August 2015].

Soilmoisture (2015) *6050X3K5B – MiniTrase Kit* by Soilmoisture Equipment Corp. [Online] Available from: https://www.soilmoisture.com/RESOURCE_INSTRUCTIONS-all_products/Resource_Instructions_0898-6050_6050%20X3%20Mini%20Trase%20Kit%20Full%20Ops.Nov2012.pdf. [Accessed 14th October 2015].

SoilSpec (2015) *SOILSPEC tensiometer* by TK Systems. [Online] Available from: http://soilspec.com.au/system.html. [Accessed 28th May 2015].

Solinst (2015) *Multilevel groundwater monitoring.* [Online] Available from: www.solinst.com/ products/multilevel-systems-and-remediation/401-waterloo-multilevel-system. [Accessed 14th August 2015].

Søndergaard, J., Bach, L. & Gustavson, K. (2014) Measuring bioavailable metals using diffusive gradients in thin films (DGT) and transplanted seaweed (*Fucus vesiculosus*), blue mussels (*Mytilus edulis*) and sea snails (*Littorina saxatilis*) suspended from monitoring buoys near a former lead-zinc mine in West Greenland. *Marine Pollution Bulletin*, 78(1–2), 102–109.

Song, C., Dannenberg, M.P. & Hwang T. (2013) Optical remote sensing of terrestrial ecosystem primary productivity. *Progress in Physical Geography*, 37, 834–854.

SorbiCell (2015) *SorbiCell, for solute sampling from groundwater.* [Online] Available from: http://www.sorbisense.com. [Accessed 14th August 2015].

Sorbisense (2015) *Porous cartridge for measuring groundwater flux.* [Online] Available from: http://www.sorbisense.com. [Accessed 14th August 2015].

SPECTRO (2015) *SPECTRO xSORT family of XRF devices* by Spectro Metek. [Online] Available from: http://www.spectro.com/products/xrf-x-ray-fluorescence-spectrometer/handheld-xrf-xsort. [Accessed 14th August 2015].

Spittler, T.M. (1995) Assessment of lead in soil and housedust using portable XRF instruments. In: *Proceedings of International Symposium on Field Screening Methods for Hazardous Wastes and Toxic Chemicals.* Las Vegas, Nevada, 22–24 February, A&WMA: Air & Waste Management Association. pp. 1281–1290.

SRI instruments (2015) *Geoprobe Gas-chromatograph.* [Online] Available from: http://srigc. com/home/product_detail/geoprobe-gc. [Accessed 14th August 2015].

Stone, C., Carnegie, A., Melville, G., Smith, D. & Nagel M. (2013) Aerial mapping canopy damage by the aphid *Essigella californica* in a *Pinus radiata* plantation in southern New South Wales: what are the challenges? *Australian Forestry*, 76, 101–109.

Strosnider, H. (2003) *Whole-Cell Bacterial Biosensors and the Detection of Bioavailable Arsenic.* National Network of Environmental Management Studies Fellow, US EPA, Office of Solid Waste and Emergency Response Technology Innovation Office. [Online] Available from: https://clu-in.org/download/studentpapers/bacterial_biosensors.pdf. [Accessed 14th August 2015].

Supelco SPME (2015) *Gas and volatile samplers.* [Online] Available from: www.sigmaaldrich.com/analytical-chromatography/analytical-products.html?TablePage=9644862.   [Accessed 14th August 2015].

Swift, R.P. (1995) Evaluation of a field-portable X-ray fluorescence spectrometry method for use in remedial activities. *Spectroscopy,* 10(6), 31–35.

TarGOST (2015) *Tar Specific Green Optical Screening Tool* from Dakota technologies. [Online] Available from: http://www.dakotatechnologies.com/docs/default-source/Downloads/targost-brochure.pdf?sfvrsn=10. [Accessed 14th August 2015].

TempHion™ (2015) *Groundwater sensors,* INW. [Online] Available from: http://inwusa.com/products/smart-sensors/multi-parameter/multi-parameter & http://www.observatormeteohydro.com/products/hydrological-sensors/ground-water-quality/inw-temphion/207. [Accessed 14th August 2015].

Tibazarwa, C., Corbisier, P., Mench, M., Bossus, A., Solda, P., Mergeay, M., Wyns, L. & van der Lelie, D. (2001) A microbial biosensor to predict bioavailable nickel in soil and its transfer to plants. *Environmental Pollution,* 113(1), 19–26.

Titan (2015) *S1 TITAN XRF analyzer* by Bruker. [Online] Available from: www.bruker.com. [Accessed 14th August 2015].

Tolner, M., Nagy, G., Vaszita, E. & Gruiz, K. (2008) *In situ* delineation of point sources and high resolution mapping of polluted sites by X-ray fluorescence field-portable handheld device. In: Sarsby, R. (ed.) *Paper Abstracts, Green5 Conference 'Construction for a Sustainable Environment,* 1–4 July 2008, Vilnius, Lithuania. p. 35.

Tolner, M., Vaszita, E. & Gruiz, K. (2010) On-site screening and monitoring of pollution by a field-portable X-ray fluorescence measuring device. In: *ConSoil 2010 Conference Proceeding CD.* ISBN: 978-3-00-032099-6.

Toxi-ChromoPad™ (2015) *Rapid toxicity bioassay kit,* by ebpi. [Online] Available from: www.ebpi.ca/index.php?option=com_content&view=article&id=27&Itemid=54. [Accessed 14th August 2015].

ToxTrak (2015) *Colorimetric toxicity screening method* by Hach. [Online] Available from: http://hachhst.com/products/portable-rapid-response-testing-systems. [Accessed 14th August 2015].

TRIDION™ (2015) *Gas chromatograph with toroidal ion trap mass spectrometer* by Torion/Perkin Elmer. [Online] Available from: http://torion.com/products/tridion.html. [Accessed 14th August 2015].

Tuller, M. & Islam, M.R. (2005) Field methods for monitoring solute transport. In: Álvarez-Benedí J. & Muñoz-Carpena, R. (eds.) *Soil-Water-Solute Process Characterization: An Integrated Approach.* Boca Raton, FL, CRC Press. pp. 309–355.

Turner, W., Spector, S., Gardiner, N., Fladeland, M., Sterling, E. & Steininger, M. (2003) Remote sensing for biodiversity science and conservation. *Trends in Ecology & Evolution,* 18, 306–314.

Turpeinen, R. (2002) *Interactions between metals, anaerobes and plants – bioremediation of arsenic and lead contaminated soils.* Academic dissertation, Department of Ecological and Environmental Sciences, University of Helsinki. [Online] Available from: http://ethesis.helsinki.fi/julkaisut/mat/ekolo/vk/turpeinen/interact.pdf. [Accessed 14th August 2015].

TVA2020 (2015) *TVA2020 Toxic Vapor Analyzer* by Thermo Scientific. [Online] Available from: http://www.thermoscientific.com/en/product/tva2020-toxic-vapor-analyzer.html. [Accessed 14th August 2015].

UltraRAE (2015) *Ultra RAE Handheld Benzene and Compound-Specific VOC Monitor.* [Online] Available from: http://www.raesystems.eu/products/ultrarae-3000. [Accessed 14th August 2015].

UMS (2015a) *Lysimeters* from Umwelt Monitoring System. [Online] Available from: http://www.ums-muc.de/en/lysimeter_systems/lysimeter/meteo_lysimeter.html. [Accessed 14th August 2015].

UMS (2015b) *Tensiometer.* [Online] Available from: http://www.ums-muc.de/en/support/faq/tensiometer.html. [Accessed 28th May 2015].

UMS Lysimeter station (2015) *Research platform for interdisciplinary projects with 48 Lysimeters.* [Online] Available from: http://www.ums-muc.de/en/lysimeter_systems/reference_project/helmholtz_research_center_near_munich.html. [Accessed 14th August 2015].

US EPA Method 6200 (2007) *Field Portable X-ray Fluorescence spectrometry for the determination of elemental concentrations in soil and sediment.* [Online] Available from: http://www.epa.gov/osw/hazard/testmethods/sw846/pdfs/6200.pdf. [Accessed 14th March 2015].

Ustin, S.L. & Gamon J.A. (2010) Remote sensing of plant functional types. *New Phytologist*, 186, 795–816.

Ustin, S.L., Gitelson, A.A., Jacquemoud, S., Schaepman, M., Asner, G.P., Gamon, J.A. & Zarco-Tejada, P. (2009) Retrieval of foliar information about plant pigment systems from high resolution spectroscopy. *Remote Sensing of Environment*, 113, S67–S77.

UVOST (2015) *Ultra-violet Optical Screening Tool* from Dakota Technologies. *[Online] Available from: http://www.dakotatechnologies.com/docs/default-source/Downloads/uvost-brochure.pdf?sfvrsn=6.* [Accessed 14th August 2015].

Vaz, C., Jones, S., Meding, M. & Tuller M. (2013) Evaluation of Standard Calibration Functions for Eight Electromagnetic Soil Moisture Sensors. *Vadose Zone Journal*, 12(2). DOI: 10.2136/vzj2012.0160.

Verma, N. & Singh, M. (2005) Biosensors for heavy metals. *BioMetals*, 18, 121–129. DOI: 10.1007/s10534-004-5787-3.

Virtanen, S., Simojoki, A., Knuutila, O. & Yli-Halla, M. (2013) Monolithic lysimeters as tools to investigate processes in acid sulphate soil. *Agricultural Water Management*, 127, 48–58. DOI: 10.1016/j.agwat.2013.05.013.

Vista-Clara (2015) *NMR Geophysics.* [Online] Available from: http://www.vista-clara.com. [Accessed 28th May 2015].

Walsh, D., Turner, P., Grunewald, E., Zhang, H., Butler, J.J. Jr., Reboulet, E., Knobbe, S., Christy, T., Lane, J.W. Jr., Johnson, C.D., Munday, T. & Fitzpatrick, A. (2013) A small-diameter NMR logging tool for groundwater investigations. *Ground Water*, 51(6), 914–926. DOI: 10.1111/gwat.12024.

Wang, Q., Cameron, K., Buchan, G., Zhao, L., Zhang, E.H., Smith, N. & Carrick, S. (2012) Comparison of lysimeters and porous ceramic cups for measuring nitrate leaching in different soil types. *New Zealand Journal of Agricultural Research*, 55(4), 333–345. DOI: 10.1080/00288233.2012.706224.

Wang, K., Franklin, S.E., Guo, X. & Cattet, M. (2010) Remote sensing of ecology, biodiversity and conservation: a review from the perspective of remote sensing specialists. *Sensors*, 10, 9647–9667.

WET (2015) *Water content, electrical conductivity and temperature sensor* from Delta-T Devices. [Online] Available from: http://www.delta-t.co.uk/product-display.asp?id=WET-2%20Product&div=Soil%20Science. [Accessed 14th August 2015].

X-Met 8000 (2015) *X-MET8000 Series XRF analyzers* by Oxford Instruments. [Online] Available from: http://www.oxford-instruments.com/products/analysers/handheld-analysers/xrf-analyzer-x-met8000. [Accessed 14th August 2015].

Zhang, H., Davison, W. & Grime, G.W. (1995a) New *in situ* procedures for measuring trace metals in pore waters. *ASTM STP,* 1293, 170–181.

Zhang, H., Davison, W., Miller, S. & Tych, W. (1995b) *In situ* high resolution measurements of fluxes of Ni, Cu, Fe, and Mn and concentrations of Zn and Cd in pore waters by DGT. *Geochimica et Cosmochimica Acta,* 59, 4181–4192.

Zhang, H., Davison, W., Mortimer, R.J.G., Krom, M.D., Hayes, P.J. & Davies I.M. (2002) Localised remobilisation of metals in a marine sediment. *Science of the Total Environment,* 296, 175–187.

Zhang, H., Lombi, E., Smolders, E. & McGrath, S. (2004) Zn availability in field contaminated and spiked soils. *Environmental Science & Technology,* 38, 2608–3613.

Zhang, H., Zhao, F.J., Sun, B., Davison, W. & McGrath, S. (2001) A new method to measure effective soil solution concentration predicts copper availability to plants. *Environmental Science & Technology,* 35, 2602–2607.

Photos:

Delta (2015) *Handheld XRF device* by Olympus. [Online] Available from: www.olympus-ims.com/en/xrf-xrd/delta-handheld. [Accessed 14th October 2015].

Ecoprobe 5 (2015) *Portable Instruments for Soil* by RSdynamics. [Online] Available from: www.rsdynamics.com/main.php3?s1=products&s2=soilcont&s3=ecoprobe5 [Accessed 14th October 2015].

FAST (2015) *Fluorometric assay system.* [Online] Available from: www.resrchintl.com/Custom_Solutions.html. [Accessed 14th October 2015].

Hapsite™ (2015) *Hapsite* by Inficon. [Online] Available from: http://products.inficon.com/en-us/Product/Detail/HAPSITE-Smart-Plus-Chemical-Identification-System?path=Products%2Fpg-ChemicalDetection. [Accessed 14th October 2015].

L2000DX (2015) *PCB/Chloride Analyzer System* by Dexil. [Online] Available from: www.dexsil.com/products/detail.php?product_id=13 [Accessed 14th October 2015].

MiniTrase Kit (2015) [Online] Available from: http://www.soilmoisture.com/MINITRASE-KIT-WITHBLUETOOTH-AND-ANDROID. [Accessed 14th October 2015].

Niton XL3t (2015) *Handheld XRF devices* by Thermo Scientific. [Online] Available from: www.thermoscientific.com/en/search-results.html?keyword=Niton&matchDim=Y. [Accessed 14th October 2015].

S1 Titan (2015) *Handheld XRF device* by Bruker. [Online] Available from: www.bruker.com/products/x-ray-diffraction-and-elemental-analysis/handheld-xrf/s1-titan-series/overview.html. [Accessed 14th October 2015].

Sorbisense (2015) *Sorbicell, passive sampler.* [Online] Available from: www.sorbisense.dk/?action=text_pages_show&id=40&menu=23. [Accessed 14th October 2015].

Thermo Scientific (2015) *Niton handheld XRF device.* [Online] Available from: www.thermoscientific.com/en/search-results.html?keyword=Niton&matchDim=Y. [Accessed 14th October 2015].

TRIDION™ (2015) *Gas chromatograph with toroidal ion trap mass spectrometer.* The courtesy of Perkin Elmer. [Online] Available from: www.perkinelmer.com/PDFs/downloads/PRD-torion-t-9-gc-ms.pdf. [Accessed 14th October 2015].

TROLL® 9500 (2015) *Multiparameter sonde* by In-Situ Inc. [Online] https://in-situ.com/product-category/water-quality-testing-equipment. [Accessed 14th October 2015].

X-Met 8000 (2015) *Handheld XRF device* by Oxford Instruments. [Online] www.oxford-instruments.com/products/analysers/handheld-analysers/xrf-analyzer-x-met8000. [Accessed 14th August 2015].

Chapter 5

Dynamic site characterization for brownfield risk management

R.L. Nemeskeri[1,2], M. Neuhaus[3] & J. Pusztai[1,4]
[1]*Fugro Consult Kft., Hungary*
[2]*International Institute for Industrial Environmental Economics, Lund University, Sweden*
[3]*Fugro Consult GmbH, Germany*
[4]*Budapest University of Technology and Economics, Budapest, Hungary*

ABSTRACT

Common remediation and land management approaches often preclude economically attractive and low risk brownfield redevelopment. The authors argue that recently developed dynamic site characterization techniques can greatly improve sound decision making on land revitalization options. A number of these in-situ soil and groundwater test methods, such as MIP, ROST™/UVOST™, XRF and BAT, based on geotechnical cone pressure testing and developed by Fugro and its partners, are introduced.

I INTRODUCTION: BROWNFIELD DEVELOPMENT

Developers, regulators and other decision makers face a number of dilemmas when dealing with formerly used, and thus potentially contaminated, land. The issues that must be addressed include: the question of whether a specific site is indeed considerably polluted or not, and if polluted, what the nature and extent of that pollution is, and how that may influence the various risks involved in the development itself and in the subsequent commercial use of that redeveloped area in the longer term.

In urban areas which became industrialized long ago, for example in typical European and North American cities experiencing urban change and deindustrialization, many such formerly used industrial and commercial sites exist in the cadastres. In Budapest, Hungary, for instance, after the social, political and economic changes in the early 1990s, tens of thousands of former industrial sites changed ownership and became either vacant or derelict land, or were utilized by many newly formed small and medium-sized companies. Occasionally developers moved into these areas, cashing in on their prime inner-city locations and building either new commercial or residential building systems on the sites. Sometimes these sites were subject to proper environmental assessments, whereas at other times they were not. When the environmental liabilities of these sites were recognized, they were often hidden from investors throughout a frenetic privatization process. Nevertheless, at some sites where more serious contamination was detected, extensive efforts were made to remediate the land to some local or international standards, occasionally even with the intention of restoring the sites to become greenfields. At other contaminated sites, development has been pursued without any remediation and any due risk management procedure, as it would have significantly lowered the profit generation potential of the developer. Thus, the relevant risks might have even been increased by generating new sources of human exposure. As environmental legislation was harmonized with the more stringent European standards, many developers started to sense the risk of taking on environmental liabilities. Investors and industrialists as well as entrepreneurs started to be concerned about the risks of assuming potentially significant liabilities through ownership of polluted land.

This situation, in addition to other factors, has triggered a greenfield development era in which companies have rather opted for building new facilities outside the cities on either green or agricultural land, thus eliminating their potential for further alternative utilization. In the former Eastern bloc countries, environmental awareness and the entire conceptualization and the management of man-made and natural environments were so rudimentary that many people, including the developers, financiers, and even government decision makers regarded greenfield development as actually being 'green'! This false perception has not yet been completely reversed.

Meanwhile, in the past few decades a number of countries such as the US, Canada, UK, and countries in the Benelux, Scandinavian and German regions, have increasingly made attempts to revitalize the formerly industrially used, 'brown' land stock of major cities, to restore those sites to economically productive, environmentally healthy (or risk controlled) and socially vibrant use, and via considering them more as development opportunities, rather than planning problems (Adams *et al.*, 2010). Considerable efforts and initiatives have started, and international and national organizations have been formed to deal with the technical, economic, social and environmental aspects of redeveloping brownfields. A few of the most recognized efforts in this respect include the Contaminated Land Rehabilitation Network for Environmental Technologies (CLARINET, 2015) or the current Concerted Action on Brownfield and Economic Regeneration Network (CABERNET, 2015) and the Contaminated Land: Applications in Real Environments (CL:AIRE, 2015), plus such programs as the Regeneration of European Sites in Cities and Urban Environments (RESCUE, 2015), the National Round Table on the Environment and the Economy's National Brownfields Redevelopment Strategy (NTREE, 2003) in Canada, and the US EPA Action Plan by its Office of Brownfields and Land Revitalization (2009) (see also Chapters 3 and 7 in Volume 1).

A vast number of remediation companies have emerged in past decades, occasionally with environmentally and economically questionable approaches. Some of them, however, have understood that there is more to land management and planning than the routine implementation of the various expensive (and often futile) remediation measures.

The underlying logic behind this more economic approach to redeveloping brownfields is that once developers identify and grasp the environmental risks of their planned projects, they start to coordinate with the responsible regulators for zoning, construction, environmental and water protection, etc. The regulators, once they have recognized the respective (mainly environmental) liabilities, and thus risks of the sites to be redeveloped, engage in special agreements with the developers to execute shared responsibilities.

The majority of contaminated urban sites are remediated by the (very costly) removal and processing, or the safe disposal of, the pollution hot spots. The alternative option is *in-situ* remediation, which is not necessarily less expensive in the long run. When developers and regulators manage to strike sound deals through tactical negotiations, the involved challenges due to uncertainties can be overcome, saving an enormous amount of time, cost, and resources (Yousefi *et al.*, 2007). The regulator can also influence developers' excavations, for instance: driving them to the pollution hot spots where deep foundations and multi-story underground garages, etc. can be built. This is how an expensive remediation work component gets integrated in the typical construction process, while the excavated polluted earth or soil (unless it is highly toxic) can be carried away and deposited on safe landfills as covering materials. If necessary, the highly toxic parts can be transported to special treatment facilities or hazardous waste repositories or incinerators, as required by standards or considered economically reasonable. Clearly, the regulator and the sustainability-conscious municipality can clean up many of their liabilities much more efficiently this way.

What is, in turn, the benefit to the developer of this approach? The regulator can provide them various benefits, from easier permission processes to tax cuts, while actually lowering the shared risks. Even more interesting is the controlled move to prime urban sites where complete infrastructure from roads and easy public access, energy, water and wastewater utilities already exist, and thus their development costs can be minimized, in comparison to greenfield developments outside of or near city limits.

A number of criteria have to be satisfied, or conditions have to prevail in order to achieve this state of coordinated efforts. The most important of these might be that this approach can only work when there is a strong driver behind it, this driver being the strict management and enforcement of environmental and public health policies. Where this condition is not met, developers can easily find alternative 'economic' (read 'profit maximization') solutions. It is easy to ascertain in which countries and regions prudent and strategic land development is currently being practiced.

Other preconditions to this economic approach to brownfield development include: the lack of available land for development, for example in areas of high population density, and the relatively high level of national or regional competitiveness (Oliver *et al.*, 2010), which manifests itself in all kinds of strategic approaches to sound and sustainable socio-economic progress.

This economic approach to brownfield development creates a logical conceptual and management framework for the revitalization of former industrial land, mainly in urbanized areas. Nevertheless, it is important to note that this framework has to be adapted to the larger frame of planned future land use of sustainable communities and regional development (Schadler *et al.*, 2010).

Logically, all actors in this process should aim at creating a brownfields cadastre which provides information on the specifics of the impacted fields, their attractive properties, and risks to be considered. A number of expert geo-engineering and geo-environmental companies such as, for instance, Fugro (2015) or Golder (2015), are greatly assisting this process as they attempt to map a number of cities and other (formerly) commercially utilized land, as terms of their suitability for planning and redevelopment. Clearly, all private and public sector developers can gain a great deal from the availability of these brownfield cadastres, but of course the generation of these cadastres comes at a cost. Therefore, it is vital to aim at advanced technologies capable of providing the necessary quality information at reasonable price. The authors suggest that dynamic site characterization, through real-time measurement techniques, is a fast and safe process to quickly assess the environmental risks involved in brownfields, considering that this assessment phase of brownfield development is typically characterized by limited data availability, but also by flexibility in land-use planning and development.

2 DYNAMIC SITE CHARACTERIZATION

Dynamic site characterization refers to the process of making *in-situ* measurements on the brownfields, and evaluating their results on the spot, which enables reaching rapid decisions on where to conduct further tests if required (see Chapter 1 for more details). This technology is built on cone penetrometer testing (CPT), which is a worldwide-known geotechnical investigation method, to determine soil and groundwater characteristics for understanding local soil bearing capacity, and for other geotechnical engineering purposes, including soil classification. The tool is pushed into the soil layers, and piezometric CPTu cones measure cone tip resistance and sleeve friction, plus pore water pressure, providing a continuous profile of soil parameters in their real environments, from which geotechnical engineers and designers can determine their necessary design parameters for foundation work. The CPT-based soil investigation method has gained global recognition for its high accuracy, speed and flexibility of deployment, and continuous soil conditions profiling. Thus it has reduced overall cost for engineering, design, and analytical use (Mayne, 1995), compared to other, more traditional soil testing methods, for instance: drilling, sampling, laboratory testing, and analysis or standard penetration testing (SPT).

Fugro has developed a variety of penetrometers, probes and samplers which are hydraulically pushed into the subsurface and attached next to the CPT or CPTu cone, to obtain physical and chemical data from soil and groundwater layers. Lightweight detachable CPT units are offered for difficult access sites such as basements or tunnels, while large trucks and all-terrain vehicles, with weights in the range of 15 to 30 tons to provide penetration reaction, are typically deployed on most test sites.

Several screening tools are available that are able to detect, characterize and delineate soil and groundwater contamination in both the unsaturated (vadose) and saturated zones. These probes, being attached to the CPT cone, reveal the vertical contaminant distribution at each test location via the simultaneous logging of tip resistance and sleeve friction (lithology).

This information can be processed in real time, as the data from the CPT system is digitally recorded on the on-board computer, from where analysis, including the 2 or 3D visualization of the sensed pollution and conceptual site modeling (see Figure 5.1 as an example) can be performed. This enables on-site decision making as to whether further site investigations are required, and provides the basis for remediation intervention design. Moreover, this information is an important input to land-use planning, as it can influence higher-level decision making on the further socioeconomic use of the subject area. For instance, it helps the specific spatial design of civil engineering and architectural systems to be built on the site, or even determine (i.e. via urban zoning) for what kind of uses the site might be considered suitable.

In the following, the authors introduce a number of these CPT-based *in-situ* measurement techniques that Fugro has developed in cooperation with its partners.

Due to their highly mobile and toxic nature, chlorinated and aromatic hydrocarbon molecule groups, commonly called volatile organic compounds (VOC) and/or

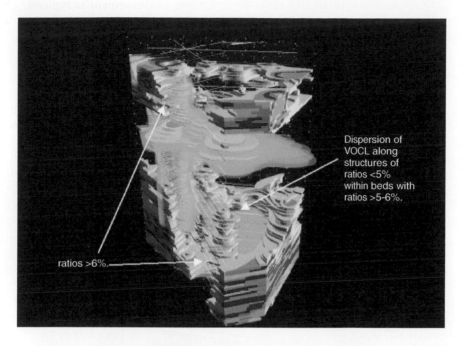

Dispersion of
VOCL along
structures of
ratios <5%
within beds with
ratios >5-6%.

ratios >6%.

Figure 5.1   An example of 3D conceptual site modeling.

persistent organic pollutants (POP), pose high risks to exposed populations and to various environmental and ecological media (Upton, 1990; IPCS, 1995; EEA, 2005; Baan *et al.*, 2009; US EPA, 2015). This is why these types of chemical pollution have become the primary target of engineering testing and abatement technologies.

3   THE MEMBRANE INTERFACE PROBE (MIP) SYSTEM

The MIP system is used to screen chlorinated hydrocarbons (CHCs), aromatic hydrocarbons of benzene-toluene-ethylbenzene-xylenes (BTEX), perchloroethylene (PCE), trichloroethylene (TCE) and their biodegradation products, and other volatile organic compounds (VOCs), in both the saturated and unsaturated zones. The heating element of the MIP cone, a heated membrane on the cone's sleeve (see Figure 5.2), mobilizes the VOCs. When heated, these compounds diffuse through the membrane and are then transported by a carrier gas stream up to the truck, where they are detected with a gas chromatograph equipped with a photoionization detector (PID), flame ionization detector (FID), and dry electrolytic conductivity detector (DELCD).

This detector combination allows for selective specification of the contaminant type as shown in Table 5.1. The PID is equipped with a 10.6 eV UV lamp and detects unsaturated chemical compounds such as ethylenes and aromatics with lower ionization energy. The FID detects organic carbon, while the DELCD is able to detect organic bound chlorine.

The MIP-CPT log in Figure 5.3 shows a PCE plume migrating in groundwater at two different depth levels, at around 6–7 and 23–24 meters respectively below the

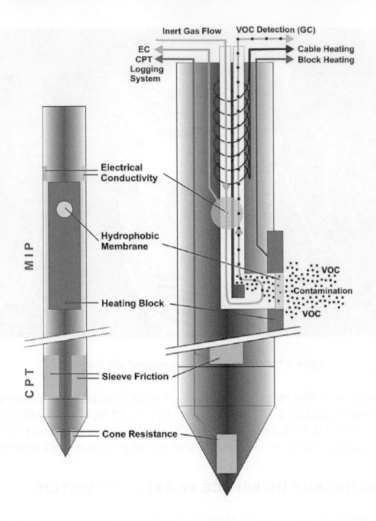

Figure 5.2 The MIP cone.

Table 5.1 Detector Sensitivities to CHCs and BTEX.

COMPOUND	PID	FID	DELCD
PCE	+++	+	+++
TCE	+++	+	++
cDCE	++	+	+
tDCE	++	+	+
VC	+	+	+
TCA	–	+	++
Benzene	++	++	–
Toluene	+++	++	–
Xylene	+++	++	–

PCE: tetrachloroethylene, TCE: trichloroethylene, cDCE: cis-1,2-Dichloroethylene,
tDCE: trans-1,2-Dichloroethylene, VC: vinylchloride, TCA: trichloroethane

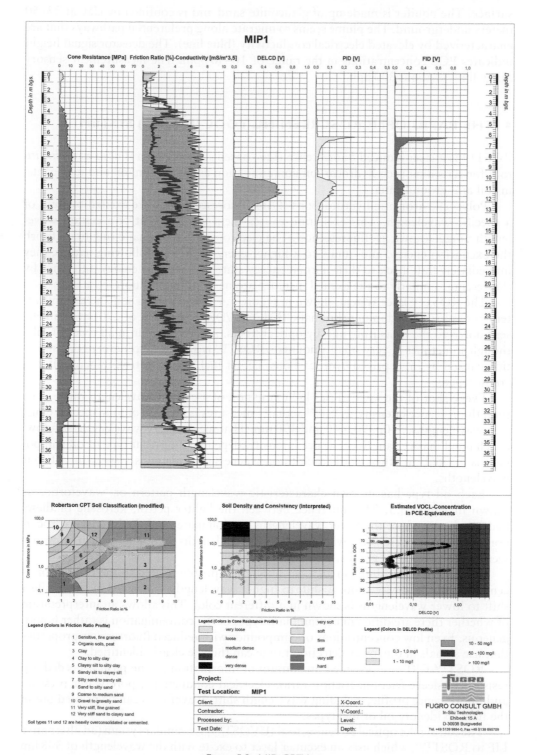

Figure 5.3 MIP-CPT log.

surface. The aquifer is made up of glauconite sand and is confined by clay at 33–50 meters underground. The plume seems to migrate along preferential pathways that are characterized by elevated electrical conductivity (blue line). The detector signal height indicates PCE concentrations in the range of 10–50 ppm. The measurement sensory range limit is about 300 ppb.

4 THE RAPID OPTICAL SCREENING TOOL (ROST™) SYSTEM

ROST™ is a laser-induced fluorescence (LIF) based sensor, which is used as an *in situ* tool for screening such hydrocarbon contamination as petroleum, oil and lubricants, in soil and groundwater.

Polycyclic aromatic hydrocarbons (PAH) fluoresce if they are excited by light of a specific wavelength. This excitation causes them to emit light of a certain wavelength range, known as 'fluorescence'. The electrons of the aromatic ring first absorb the excitation energy, and 'jump' to a higher energy state. By following nature, they immediately 'fall' back to their normal energy state and emit the energy difference by fluorescing, similarly to the excited mercury atoms in common fluorescent lamps.

The ROST™ laser system, shown in Figure 5.4 below, excites soil and groundwater contaminants with monochromatic (Nd-YAG/Dye) laser light of 290 nm at 50 Hz, sent down to the measuring unit by optical fibers. As PAHs occur in all types of oil – though sometimes in very small amounts – ROST™ is able to detect every contamination caused by oil-deprived hydrocarbons, i.e. jet fuel, diesel, petrol, mineral oil, tar, creosote, etc.

Every shot of the laser causes light emission of a certain wavelength range. The total fluorescence intensity ROST™ actually measures is the sum of four specific emission wavelengths: 340, 390, 440 and 490 nm. That means every wavelength 'window' or 'channel' has certain fluorescence intensity, depending on what type of oil component is excited. In other words each fluorescence signal contains a spectrum of four wavelengths.

Each oil type has its own characteristic wavelength pattern or 'waveform', as shown in Figure 5.5. The difference between the lighter petrol, kerosene or diesel hydrocarbons, where the lower wavelengths predominate, and the heavier hydrocarbons such as tar or creosote, where the higher wavelengths predominate, can be clearly seen in the two sets of graphs.

This wavelength shift can also be shown in a profile related to the total fluorescence at a given test location. A shift to higher wavelengths is marked in red, while a shift to lower wavelengths is shown in a different color. This allows an interpretation of whether there are different contaminant types or the contamination is rather homogeneous. Even low concentrations or compounds with reduced fluorescence properties that use signals in the lower detection range can thus be clearly identified.

ROST™-CPT can provide a maximum of information in one push if needed: tip resistance, sleeve friction, friction ratio, electrical conductivity, pore water pressure and fluorescence profile, including wavelength shift. Waveforms can be printed out in the field to identify oil or fuel types.

The ultraviolet optical screening tool (UVOST™) is a similar system (an alternative LIF to ROST™), which uses an excimer laser to excite with the wavelength of 308 nm

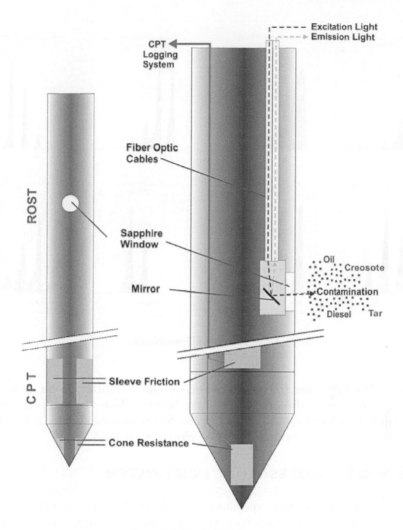

Figure 5.4 The ROST™ cone.

and emit at the wavelengths of 350, 400, 450, and 500 nm. Just as in ROST™, the UVOST™ waveforms serve to identify the product type, while fluorescence intensity indicates the concentration levels in the Non-Aqueous Phase Liquid (NAPL) zones.

The ROST™-CPT log shown in Figure 5.6 illustrates a test in a kerosene spill area on an underground storage tank site. As can be seen, kerosene NAPL was present at different depths and could be clearly identified by its typical waveform. A layer of 200 percent fluorescence intensity with sharp boundaries at 9.20 meters below ground was detected, indicating a floating light NAPL on the groundwater table. Significant amounts of kerosene deposit were found in several other layers between 3.5 and 9.5 meters, differentiated from the background clearly by the blue color in the graph in

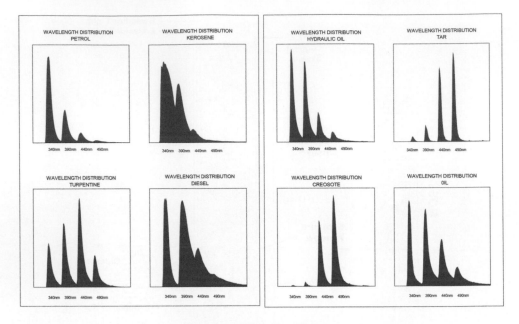

Figure 5.5 ROST™ waveforms.

Figure 6.6. The light blue fraction between 7.6 and 9.0 meters point to weathered or degraded kerosene, while the green layer between 10.6 and 12.6 meters below indicates dissolved fractions. All these different fractions are characterized and can be easily discerned on the computer generated plots, see Figure 5.7.

5 THE X-RAY FLUORESCENCE (XRF) SYSTEM

There is considerable need and expectation from the land use and remediation markets for maps of inorganic compounds such as toxic metals. Most of the metals in the deep subsurface may be less problematic than highly mobile dissolved or NAPL organic contaminants because of their typically lower mobility through environmental fluids. However, the attention paid by regulators to mercury, lead and cadmium in particular, seems warranted due to their proven toxic nature and the associated health impacts by cancer, cardiovascular disease, reproductive dysfunction, developmental and nervous system disorders (Upton, 1990; EEA, 2005; US EPA, 2015).

Originally developed by the US Naval Research Laboratory (2015), the XRF-Detector, made suitable for CPT application by a cooperation of Austin Automation and Instrumentation (2015), GreenLab Europe (2015) and Fugro, provides an elegant option for rapid metal testing on suspected or known contaminated sites. The fully-fledged XRF-CPT technique which utilizes advanced Energy Dispersive Spectrometry was developed.

The XRF-CPT cone, shown in Figure 5.8, works somewhat similarly to the ROST™ system, as its physical principle is the excitement of various chemicals or

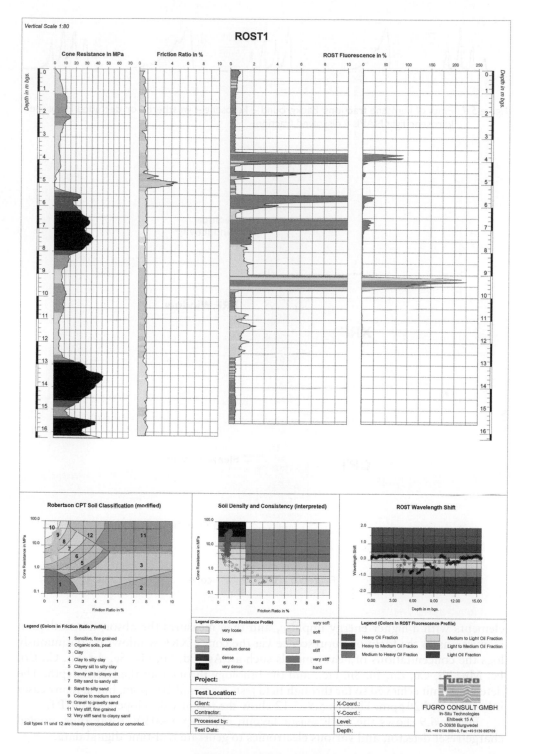

Figure 5.6 ROST™-CPT log.

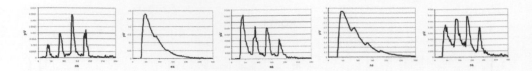

Figure 5.7 Plots of kerosene fractions: Background – kerosene – weathered or degraded kerosene – kerosene light NAPL – dissolved fraction.

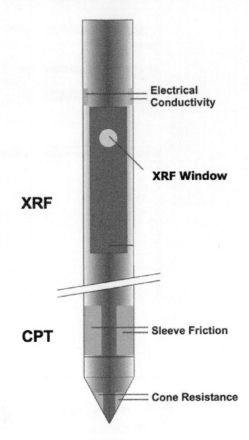

Figure 5.8 The XRF-CPT cone.

elements with a higher energy impulse, and then measures the absorption radiation reemissions in the form of fluorescence via the cone's XRF window. It continuously logs the concentrations of heavy metals such as: As, Cd, Pb, Hg, Se, Ag, Sb, Cr, Co, Fe, Mn, Ni, Cu, Zn, In, Sn, etc., and the lithology throughout the tested depth. The detection limit achieved so far through field tests is around 100 ppm, but in many cases it has reached 20 ppm. Figures 5.9–5.12 show the recently developed XRF-CPT.

Figure 6.9 shows the XRF sensor layout, applying a 40 kV X-ray tube, cooled charge-reset detector, and various electronics to generate real-time digital data.

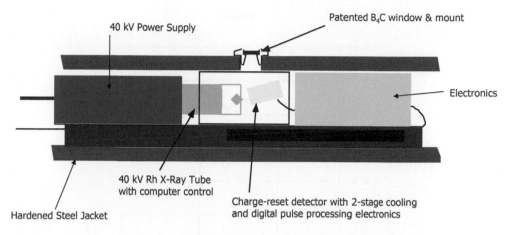

40 kV Power Supply

Patented B₄C window & mount

Electronics

40 kV Rh X-Ray Tube
with computer control

Charge-reset detector with 2-stage cooling
and digital pulse processing electronics

Hardened Steel Jacket

Real-time data reduction software with GUI interface

Figure 5.9 The XRF sensor layout.

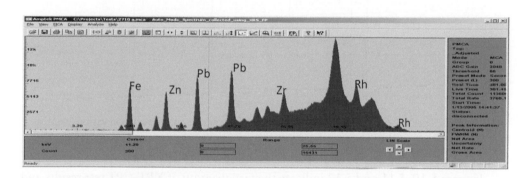

Figure 5.10 The XRF-CPT test result: visualization of heavy metal elements and concentrations.

```
SRM 2710 a.txt - Notepad                                    _|□|x|
File  Edit  Format  View  Help
XRF Summary Report  --  Date: 01-23-05,  Time: 11:39:39
Sample ID: C:\Projects\Tests\SRM 2710 a
Depth (mm):
Results:---
   Cr  =   86 ppm
   Mn  = 9191 ppm
   Fe  =  3.82 wt.%
   Ni  =   36 ppm
   Cu  = 2316 ppm
   Zn  = 6372 ppm
   As  =  716 ppm
   Rb  =  158 ppm
   Sr  =  411 ppm
   Zr  =   78 ppm
   Cd  =   17 ppm
   Hg  =   39 ppm
   Pb  = 4807 ppm
   Sio2 = 93.76 wt.%

Total: 100.0 wt.% (Normalized)
```

Figure 5.11 The XRF CPT typical sample read out report.

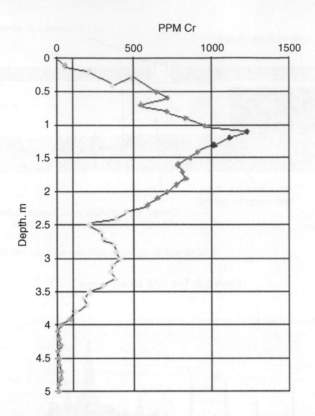

Figure 5.12 The XRF-CPT test result: visualization of elements concentration in various depths.

Figures 5.10, 5.11 and 5.12 indicate the visualization of XRF-CPT test results and sample reports.

The markets for XRF-CPT testing include mining and metal smelter sites, arsenic warfare sites, harbor mud/silt testing via near-shore CPT gear, hazardous waste deposits, etc.

6   THE BAT *IN-SITU* GROUNDWATER SAMPLER

In conclusion, it is worth noting that there is opportunity to take real-time and real-space discrete groundwater samples for traditional lab analysis, commonly requested by environmental and water regulators, with the help of the CPT technique. The BAT groundwater sampler (developed by BAT Geosystem, 2015), is ideal for quick, discrete sampling of both liquid and gaseous phases at CPT penetrated depths. It allows for exact placement of a filter cone in the contaminated layer, and collection of hermetically sealed, encapsulated water samples, thus preventing the loss of volatile components.

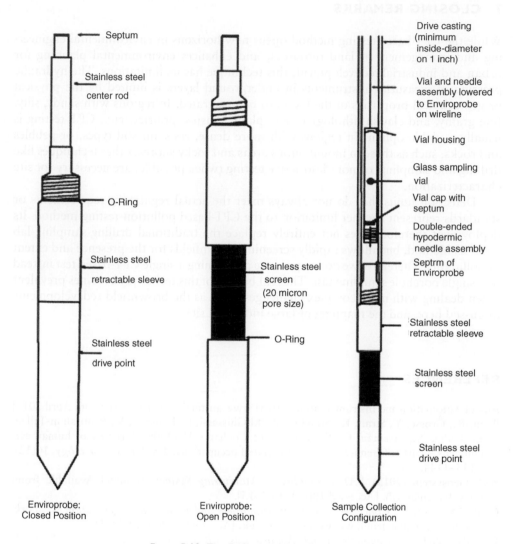

Septum

Stainless steel
center rod

O-Ring

Stainless steel
retractable sleeve

Stainless steel
drive point

Enviroprobe:
Closed Position

Stainless steel
screen
(20 micron
pore size)

O-Ring

Enviroprobe:
Open Position

Drive casting
(minimum
inside-diameter
on 1 inch)

Vial and needle
assembly lowered
to Enviroprobe
on wireline

Vial housing

Glass sampling
vial

Vial cap with
septum

Double-ended
hypodermic
needle assembly

Septrm of
Enviroprobe

Stainless steel
retractable sleeve

Stainless steel
screen

Stainless steel
drive point

Sample Collection
Configuration

Figure 5.13 The BAT sampler (not to scale).

The probe is pushed in a closed position as can be seen in Figure 5.13. At the required depth, a stainless steel filter screen is opened, and exposed to the groundwater by retracting the case about 0.3 meters. The groundwater sample is then collected under *in situ* chemical and physical conditions (pressure, redox potential, etc.) by lowering an evacuated sampling vial onto a double-ended needle connecting the filter chamber to the vial. Both are sealed by a septum, allowing for chemically undisturbed sampling, which is a significant advantage over alternative, exhaustive, high quality or low-tech and quality sampling methods.

7   CLOSING REMARKS

While the CPT-based testing method opens new horizons in environmental engineering and management of land resources, and enhances environmental planning for urban and industrial redevelopment, this technique has its limitations. The hydraulic pressing of measuring instruments in underground layers is limited by the physical or geotechnical properties of the layers to be penetrated. In regions with sandy, silty, fine grainy, and clayish lithology of e.g. plains, basins, prairies, etc., CPT testing is usually a viable option. In regions with more dense, resistant soil types, or pebbles and rocks, such as those in mountainous areas and rocky shores, other techniques like drilling and sampling, or non-destructive testing (when possible) are necessary for site characterization.

Detection limits that do not always meet the actual regulatory requirements or standards represent another limitation to the CPT-based pollution-testing method. Its deployment therefore does not entirely replace the traditional drilling-sampling-lab analysis method, but allows rapidly screening brownfields for the presence and extent of pollutants. Therefore, we do not propose performing a single CPT-based test instead of a single borehole sampling test. The real benefit of this technique becomes prevalent when dealing with more complex challenges, such as the brownfield redevelopments discussed here, and the mapping of large industrial sites.

REFERENCES

Austin Automation and Instrumentation (2015) www.austinai.com. [Accessed 10th April 2015]

Baan, R., Grosse, Y., Straif, K., Secretan, B., El Ghissassi, F., Bouvard, V., Benbrahim-Tallaa, L., Guha, N., Freeman, C., Galichet, L. & Cogliano, V. (2009) A review of human carcinogens – Part F: Chemical agents and related occupations. *The Lancet Oncology*, 10(12), 1143–1144.

BAT Geosystem (2015) *BAT Groundwater Monitoring System*. [Online] Available from: www.bat.eu/en-GB. [Accessed 10th April 2015].

CABERNET (2015) *Concerted Action on Brownfield and Economic Regeneration Network*. [Online] Available from: www.cabernet.org.uk. [Accessed 10th April 2015].

CL:AIRE (2015) *Contaminated Land: Applications in Real Environments*. [Online] Available from: www.claire.co.uk. [Accessed 10th April 2015].

CLARINET (2015) *Contaminated Land Rehabilitation Network for Environmental Technologies*. [Online] Available from: www.clarinet.at. [Accessed 10th April 2015].

EEA (2005) *Environmental and Health, EEA Report No 10/2005*. The European Environment Agency, Copenhagen, Denmark.

Fugro (2015) [Online] Available from: www.fugro.com. [Accessed 10th April 2015].

Golder (2015) [Online] Available from: www.golder.com. [Accessed 10th April 2015].

GreenLab Europe (2015) [Online] Available from: www.greenlab.hu. [Accessed 10th April 2015].

IPCS (1995) *A Review of Selected Persistent Organic Pollutants*. The International Programme on Chemical Safety, within the framework of the Inter-Organisation Programme for the Sound Management of Chemicals (IOMC), established by UNEP, ILO, FAO, WHO, UNIDO, and OECD. Document PCS/95.39. [Online] Available from: www.who.int/ipcs/assessment/en/pcs_95_39_2004_05_13.pdf. [Accessed 10th April 2015].

Mayne, P.W., Auxt, J.A., Mitchell, J.K. & Yilmaz, R. (1995) US National Report on CPT. In: *Proceedings of the International Symposium on Cone Penetration Testing, Vol. 1 (CPT'95).* Report 3:95. Linköping, Swedish Geotechnical Society. pp. 263–276.

NTREE (2003) *Cleaning up the Past, Building the Future: A National Brownfield Redevelopment Strategy for Canada.* National Round Table on the Environment and the Economy. Ottawa, ON, Renouf Publishing Co.

Oliver, L., Ferber, U., Grimski, D., Millar, K. & Nathanail, P. (2010) *The Scale and Nature of European Brownfields.* CABERNET document. [Online] Available from: http://www.cabernet.org.uk/resourcefs/417.pdf. [Accessed 10th April 2015].

RESCUE (2014) *Regeneration of European Sites in Cities and Urban Environments.* [Online] Available from: www.rescue-europe.com. [Accessed 16th November 2014].

Schadler, S., Morio, M., Bartke, S., Rohr-Zanker, R. & Finkel, M. (2010) Designing sustainable and economically attractive brownfield revitalization options using an integrated assessment model. *Journal of Environmental Management*, 92, 827–837.

Upton, A.C. (1990) Environmental Medicine: introduction and overview. *The Medical Clinics of North America*, 74(2), 235–244.

US EPA (2009) *US EPA Action Plan on Brownfields and Land Revitalization.* [Online] Available from: http://www.epa.gov/brownfields/overview/09brochure.pdf. [Accessed 10th April 2015].

US EPA (2015) *Contaminants focus – clean-up information* by the US Environmental Protection Agency. [Online] Available from: https://clu-in.org/contaminantfocus/. [Accessed 16th November 2014].

US Naval Research Laboratory (2015) [Online] Available from: www.nrl.navy.mil. [Accessed 10th April 2015].

Yousefi, S., Hipel, K.W., Hegazy, T., Witmer, J.A. & Gray, P. (2007) Negotiation Characteristics in Brownfield Redevelopment Projects. In: *Proceedings of the IEEE International Conference on Systems, Man and Cybernetics. 7–10 October, 2007, Montreal, QUE.*

Chapter 6

Environmental geochemistry modeling: Methods and applications

G. Jordan & K. Z. Szabó

Department of Chemistry, Institute of Environmental Science,
Szent István University, Gödöllő, Hungary

ABSTRACT

As a result of technological development, the interaction between human society and the environment has become extremely complex, and environmental problems have become a major concern. Environmental problems such as pollution emission, waste disposal and treatment, soil and water quality degradation, and health disorders, are being caused by natural and anthropogenic compounds. These are related to the distribution and behavior of chemical elements, and require the understanding of geochemical processes in earth systems. State-of-the-art spatio-temporal modeling is needed to describe and predict toxic compound behavior for the prevention, mitigation and management of the associated risks. Environmental geochemical processes are dynamic, and their behavior must be understood in order to control these processes, and prepare for the inherent risks to society, such as contamination dispersion induced by climate change. Section 1 describes a systematic time series analysis procedure, and provides an example for the analysis of measurement series for high-frequency soil gas radon activity concentration (^{222}Rn) in terms of seasonal climatic effects, long-term trends, persistence (auto-correlation), and sudden events. From the spatial perspective, representative sampling, spatial data structure recognition and description, and spatial processes modeling, including interpolation, are the key issues for the management of contamination risk. This chapter also presents a systematic spatial data analysis procedure, and an example is provided for spatially continuous parameter distribution in an urban attic dust investigation on airborne contamination assessment in Section 3. This local-scale study also shows the methods for stochastic geochemical process modeling. Next, we take a step forward and describe methods for advanced spatial structure analysis, such as trend and anisotropy analysis at the regional scale. A detailed spatial data analysis procedure is described, and an example is provided for the spatial characterization of regional geogenic radon potential in Section 4. In this case, modeling and mapping of geogenic radon potential enable development of a radon potential map, which can help to reduce the cumulative radiation risk. The next section goes beyond the previous sections' spatial description of geochemical processes, and applies multivariate statistical analysis methods for contamination assessment in a mining catchment. Moreover, a spatially distributed sediment transport model is described and applied, to estimate the sediment-bound toxic element liberation, transport and deposition in the catchment, and to calculate the total sedimentary contamination export from the catchment to downstream agricultural lands. Finally, geochemical modeling is concluded with a numerical study of contamination risk assessment in Section 6. A semi-quantitative risk-based ranking method of mine waste sites (EU Mining Waste Directive Pre-selection Protocol) is presented, compared to other recognized contamination risk assessment methods, adopted to local conditions and tested for parameter uncertainty in a GIS environment.

I INTRODUCTION

Environmental geochemistry is a scientific discipline, dealing with the relative abundance, distribution, and migration of the Earth's chemical elements and their isotopes. Its scope includes the definition of elemental abundances in minerals and rocks, cycling of the Earth's constituent materials through geologic processes, and the cyclic flow of individual elements (and their compounds) between living and non-living systems (Plant & Raiswell, 1983; Thornton, 1983). Numerous sampling methodologies, using various media, have been developed (Garrett, 1983; Garrett & Sinding-Larsen, 1984; Ottesen et al., 1989; Plant & Ridgway, 1990), and sample analysis techniques have been improved in concert with modern methods of instrumental chemical analysis, sensitive extraction procedures (Tessier et al., 1979; Kheboian & Bauer, 1987), and methods of controlled analytical systems (Howarth & Thompson, 1976; Ramsey et al., 1987). The availability of large geochemical data sets enabled the development of spatial modeling of geochemical gradients, and application of advanced multivariate statistical analysis (Howarth & Sinding-Larsen, 1983; Davis, 1986; Gaál, 1988). Quality control and decision analysis tools have also been developed and applied (Plant et al., 1975; Garrett & Goss, 1979; Ramsey, 1993) that have been used, for example, in the multi-element, multi-media environmental geochemical atlas of Europe (Salminen et al., 2005; Reimann et al., 2014). Sections 2 and 3 of this chapter present these spatial mapping techniques and their applications using GIS technology. These sections also provide step-by-step case studies for the univariate and bivariate statistical analysis of geochemical data. Section 4 provides guidelines for multivariate data analysis in geochemistry.

Environmental geochemical time series, however, have quite rarely been studied, most probably due to the lack of appropriate monitoring data (WFD, Directive 2006/21/EC). Section 2 develops a detailed time series analysis procedure, and shows a case study for soil radon concentration analysis. Description of the migration and flow of geochemical compounds in the environment, in terms of reaction and transport models, is a recent development of environmental sciences and geochemistry. These techniques are briefly demonstrated through case studies on mining-related toxic element contamination assessment in Section 4.

The objective of environmental geochemistry is to support decisions on the improvement of environmental quality, economic development, and quality of life, by identifying relevant geochemical processes and associated risks. Still, it remains a challenge how scientific knowledge can be used for decision support in the most efficient way. Section 5 presents a recent attempt to develop, use and test a contamination risk assessment method for mine site evaluation, in order to support the implementation of EU environmental legislation.

2 TIME SERIES ANALYSIS AND MODELING OF GEOCHEMICAL PROCESSES: AN EXAMPLE FOR SOIL RADON DYNAMICS

Environmental geochemical processes are dynamic phenomena, and their behavior must be understood, in order to control these processes and prepare for the inherent risks to society, such as climate change-induced contamination dispersion. High quality monitoring data of climatic and geochemical parameters is needed, in addition to the appropriate modeling methods, to analyze these data series. This section describes a systematic time series analysis procedure, and an example is provided for the analysis of measurement series of soil gas radon activity concentration (^{222}Rn), in terms of seasonal climatic effects, long-term trends, persistence (auto-correlation), and sudden events (Szabó et al., 2013). Eleven 'monthly week' datasets were collected, which contain measurements at 15-minute equidistant intervals for the duration of about one week (3–10 days) every month, between August 2010 and July 2011. The 15-minute sampling time during the observation weeks in each month ('monthly weeks') enabled the capture of high-frequency radon activity concentration changes on the one hand, whereas the one-year observation period enabled the capture of seasonal changes and long-term trends on the other. The 11 monthly week datasets altogether represent a year, with missing periods between them. In this case, the missing values in the unobserved periods were omitted, and each monthly week dataset was analyzed separately for high temporal resolution dynamics (e.g. diurnal). Low-resolution temporal features (e.g. seasonality) were analyzed using the median central values of the 11 monthly week data series. Soil gas radon activity concentration was measured in situ using a RAD7 Electronic Radon Detector (Durridge Company Inc., 2000), coupled with a soil probe.

2.1 Data processing and data analysis

Data analysis of monitoring time series starts with the statistical description of the data (see Volume 2, p. 484–513). First, summary statistics measuring central tendency and variability are calculated. Since geochemical data series, such as soil gas radon concentration measurements, are often characterized by non-normality, heterogeneity and outliers (Jordan et al., 1997; Kurzl, 1988; Reimann et al., 2008), robust statistics like the median for location (central tendency), median averaged deviation (MAD), and the inter-quartile range (IQR) for measure of scale (variability), are used (Hoaglin et al., 1983). The robust statistics to be calculated are the five-letter display variables: minimum, lower quartile, median, upper quartile and maximum, in addition to average, mode, and standard deviation, coefficient of variation, MAD, range and inter-quartile range. Variability parameters play an important role, since they are particularly suitable for the characterization of changes in a time series (altering variation, called heteroscedasticity, due to seasonal effects of soil radon in our case). For example, radon exhalation to the open air might be more variable in summer than in winter, indicating less topsoil sealing, and more dynamic response to surface temperature conditions. In this study, the major yearly seasonal period (change in central tendency or location), and the seasonal alteration of variation (change in variability), was described by using the 11 medians and MAD values of the monthly week data series, respectively, and visualized by box-and-whisker plots (see Figure 6.2). Extreme variability, i.e. outlying values, was also captured in the plots. An interesting parameter is range/median indicative of total variability containing the outliers, too. In order to account for seasonal differences among the monthly week measurement series, robust variability measures were normalized to the monthly week central values, and the monthly week MAD/median relative variability parameters were used for comparison. In this study, IQR/median and MAD/median values were found to be very similar, however, IQR/median was systematically higher. Variability of the original monthly week series contains not only the random variations, but also the seasonally dependent amplitude of the diurnal periods, in addition to cycle and trend components. For the pure random component (noise) characterizing system stability, the cycle, trend, periodicity and auto-correlation components have to be removed from the series. Various measures for location and scale were compared using simple least-squares regression analysis. Also, the location dependency of scale was assessed by regression analysis between median and MAD. Sub-population identification followed the 'natural break' method, i.e. a data series was separated where the cumulative distribution function (CDF) had an inflection point (natural break) identified visually on the cumulative distribution plot. This point corresponds to a local minimum in the frequency histogram. A homogeneity test between these sub-populations can reveal similarity, if any, between seasons. Separation of sub-populations was confirmed at the 95% confidence level by the Mann-Whitney homogeneity test, based on the comparison of medians. Outlying values represent sudden and unusual events, essential for identifying very fast processes, such as soil gas radon activity concentration changes, due to heavy torrential summer rainfall or gust. Tukey's (1977) inner-fence criteria were used for outlier definition. All discussed statistical tests, including trend and auto-correlation analyses and homogeneity tests, are significant at the 95% confidence level. Summary statistics were calculated for original data series, and for the identified sub-populations, separately.

2.2   Time series analysis and signal processing

A time series consists of a set of sequential numeric data taken at equally spaced intervals, usually over a period of time or space. Time series analysis (TSA) defines pattern, according to an additive decomposition of the radon measurement series, into trend $(T(t))$, cycle $(C(t))$, periodicity $(P(t))$, auto-correlation $(A(t))$, white noise residuals $(\varepsilon(t))$, and events (outliers or transients) $(E(t) = E_O(t) + E_T(t))$ components (Equation 1) (Szucs & Jordan, 1994):

$$c(t) = T(t) + C(t) + P(t) + A(t) + E(t) + \varepsilon(t) \tag{6.1}$$

Note that this study applies exploratory time series analysis that accommodates the identification and characterization of transients and outlying singular events (Figure 6.1). Unlike classical time series decomposition, this approach captures sudden time series features that are essential for the understanding of e.g. climate change-induced catastrophic events.

In the demonstrative study, the additive decomposition was carried out separately on the 11 equidistant monthly week soil gas radon time series $(c(t))$. First, a 5RSSH type non-linear moving median smoother algorithm was used. This algorithm starts with a 5-point window (i.e. 5×15 minutes $= 75$ minutes) moving median calculation, then the Re-smooth and Split algorithm developed by Tukey (1977) is applied. Finally,

a

TSA- Classical Statistics

$$Y_{(t)} = T_{(t)} + P_{(t)} + A_{(t)} + \varepsilon_{(t)} \quad \textit{(additive model)}$$

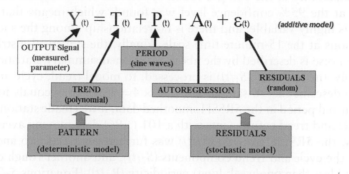

OUTPUT Signal
(measured
parameter)

PERIOD
(sine waves)

RESIDUALS
(random)

TREND
(polynomial)

AUTOREGRESSION

PATTERN
(deterministic model)

RESIDUALS
(stochastic model)

b

TSA- Exploratory Data Analysis

$$Y = SMOOTH + ROUGH$$

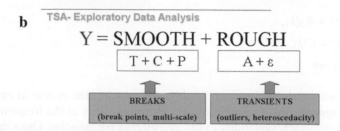

$T + C + P$

$A + \varepsilon$

BREAKS
(break points, multi-scale)

TRANSIENTS
(outliers, heteroscedacity)

Figure 6.1 a. Classical time series decomposition scheme (additive model). b. Exploratory time series analysis scheme. Note that this approach accommodates the analysis of transient events in the rough. See text for details.

it calculates Hanning-type 3 point average (Velleman & Hoaglin, 1981). This process separates the series into 'smooth' $(S_1(t))$ carrying pattern (cycle, trend, periodicity), and 'rough' or 'residual' $(R_1(t))$ containing auto-correlation, noise and outliers, according to Tukey (1977) (Equations 2–4).

$$c(t) = S_1(t) + R_1(t), \tag{6.2}$$

$$S_1(t) = T(t) + C(t) + P(t), \tag{6.3}$$

$$R_1(t) = A(t) + E(t) + \varepsilon(t). \tag{6.4}$$

All features or time periods shorter than 75 minutes (very fast component), join the rough (residuals), eliminating random noise and the effect of outliers. The residuals are stationary (constant in the mean), and represent the natural variability of soil gas radon, in addition to the stochastic and the sampling uncertainties.

Firstly, the above obtained 'rough' $(R_1(t))$ is processed, and outliers are defined by the inner-fence criteria, and subsequently removed. The outlier-free series is then subject to tests for randomness of median, sign and Box-Pierce tests, to check that no pattern remains in the noise as trend, periodicity and auto-correlation, respectively. In this study, all residual series were found random at the 95% confidence level.

A detailed auto-correlation analysis is performed to identify the autogressive property in the outlier free residual series ($R_1(t)$), and to describe the 'memory effect', inertia, or the predictability of the soil gas radon system. In this study, no auto-correlation significant at the 95% confidence level was found, which means that the observed soil radon is highly variable, and there is no relationship among the successive radon concentrations at the 15-minute time scale. Finally, the statistical distribution of the outlier-free noise is described by the above mentioned summary statistics.

Secondly, the 'smooth' ($S_1(t)$) is processed, to model trend, cycle and periodicity. In order to describe the 96-sample long ($24 \times 4 = 96$; 1 hour equals four 15 minute periods) diurnal period in the 5RSSH smoothed data, it was made stationary by removing the cycle and trend components with a 101 (>96) data moving average smoother. In this way, the 5RSSH 'smooth' ($S_1(t)$) was further separated into another smooth, containing the cycle and trend components ($S_2(t)$), and another rough containing the diurnal (and less than one week-long) periodicity ($R_2(t)$) (Equations 5–7).

$$S_1(t) = S_2(t) + R_2(t), \tag{6.5}$$

$$S_2(t) = T(t) + C(t), \tag{6.6}$$

$$R_2(t) = P(t). \tag{6.7}$$

Periodicity was analyzed by the periodogram showing the power at each Fourier frequency (see Figure 6.3b). The periodogram shows the data in the frequency domain by considering how much variability exists at different frequencies. Once the frequencies in the data were identified, periodicity was modelled with sine waves fit to each monthly week data series using the least-squares method. The best fit was indicated by the smallest root-mean-square error (RMSE) value. The amplitude of the calculated sine waves may reveal seasonal differences. From the 101 moving average smoothed data, the trend component was modelled by a simple linear least-squares regression line to $S_2(t)$ (see Figure 6.3a). After subtracting the trend line from the smoothed series, the pure cycle component ($C(t)$) is obtained (see Figure 6.3a).

2.3 Interpretation of data series features: an example for soil gas radon concentration

2.3.1 Long-term change: variability in different seasons

According to the relative MAD/median measure of variability of the monthly week datasets, the highest data scatter occurs in August and September in 2010 and in May, June, July in 2011 (Figure 6.2). These are the summer months, while the overall relative variability is low in the winter months (except for December 2010). Maximum scatter in the data based on range/median measure of variability, which is sensitive to the effect of extreme values, also occurs in the summer period (except for June 2010). Again, the weather conditions impacting soil characteristics, such as soil temperature, wetness and pore gas pressure, are more variable in summer, often associated with extreme events of sudden temperature variations (Sundal et al., 2008; Smetanová et al., 2010). The average relative variability (median of the monthly weeks' MAD/median values) is 13% for the summer, and 6% for the winter period. It is interesting to note that there is no obvious transition between the

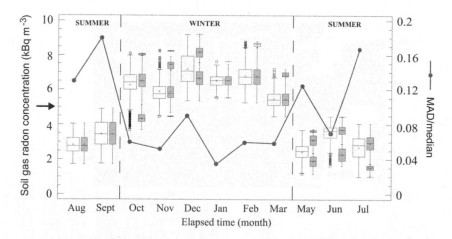

Figure 6.2 Box-and-whisker plot of the original monthly week soil gas radon activity concentration data series (empty box), and the identified 28 sub-populations (gray plots). Main population is the population with the most observations of a monthly week. Minor population is separated from the main population, and it can be high or low according to the values. Vertical dashed lines separate the seasons. The black arrow on the ordinate indicates the natural break (inflection point) at 5 kBq m^{-3} soil gas radon activity concentration, separating summer and winter seasons on a statistical basis. Points connected by the solid line show the MAD/median monthly week relative variations. Data series for the observed week in April 2011 was lost due to unfortunate field conditions (after Szabó *et al.*, 2013).

two seasons, and the soil radon activity concentration changes from one state to the other, both in terms of seasonal level (median value) and variability (MAD/median) (Figure 6.2). This indicates that soil radon is controlled by a factor(s) with definite threshold(s).

2.3.2 Long-term change: trend and cycle

In order to study long-term trend in the soil gas radon activity concentration data, trend analysis was carried out for the whole one year data represented by the 11 weekly median values. Similarly, trend was fitted separately to the seasons. With respect to temporal scales beyond the seasonal (half-year) period, there is no obvious long-term pattern according to trend analysis with a 95% confidence level. The 11 monthly week data series were also studied separately, and a simple linear regression line was fitted to the data series denoised with a 5RSSH smooth, and subsequently treated with a 101 moving average smooth to remove diurnal (96-sample-long and shorter) periodicity ($S_2(t)$). The success of the smooth was confirmed by the observed lack of any periodicity in the smoothed series by Fourier and auto-correlation analysis. The removal of trend from the treated series reveals the cycle component ($C(t)$). All the monthly week data series cycles have two local minima and maxima located at equal distances, indicating a half-week periodicity (Figure 6.3a). This was confirmed by the periodograms showing significant 2.5–3 days periodicity in the cycle.

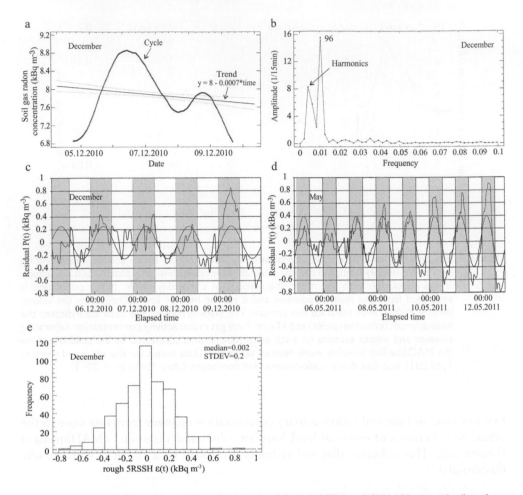

Figure 6.3 a. Trend and cycle modeling in smooth of $S_2(t)$ (5RSSH and SMA101 smoothed) soil gas radon activity concentration dataset for December 2010. b. Periodogram of the rough of $S_2(t)$ (5RSSH and SMA101 smoothed) soil gas radon activity concentration dataset for December 2010. It shows the period of time (96) equal to one day and a harmonics. c. Fitted diurnal sine wave to the $P(t)$ for December 2010. Transparent black rectangles show nights from 22:00 to 10:00. d. Fitted diurnal sine wave to the $P(t)$ for May 2011. Transparent grey rectangles show nights from 22:00 to 10:00. e. $R_1(t)$ (rough of the 5RSSH smooth) of December 2010 dataset shows noise (after Szabó *et al.*, 2013).

2.3.3 Short-term change: diurnal periodicity

In the second 'rough' ($R_2(t)$), remaining after the 5RSSH and 101 moving average smooths, significant diurnal periodicity was found in all the 11 monthly week data series, with an average 93 data point lengths (0.97-day frequency) (Figure 6.3b). In order to numerically model diurnal periodicity and to check for possible seasonal variations, the obtained diurnal sine waves were fitted with the least-squares method (Makridakis *et al.*, 1998) (Figures 6.3c and 6.3d). The best fit to the data was defined by the minimum value of the root mean square error (RMSE). In summer, wave length is more constant, but it is highly variable in winter (Figures 6.3c and 6.3d). However, the average amplitude was found at $0.3\,kBq\,kg^{-1}$, and it was twice as high in summer $(0.4\,kBq\,kg^{-1})$ than in winter $(0.2\,kBq\,kg^{-1})$ (Figures 6.3c and 6.3d). Again, besides the random variability, this shows a higher and more regular systematic (diurnal) variation in summer than in winter, and is most likely driven by the climatic and soil conditions (Baykut et al., 2010).

2.3.4   Short-term change: outliers and transients

No pattern (trend, periodicity, auto-correlation) was found among the outliers in the first 'rough' $(R_1(t))$ (residuals after the 5RSSH smooth, the ε random noise component). Outliers occurred at any time of day. Sudden events in soil gas radon are most probably associated with climatic events of torrential rainfall or wind storms. An interesting transient phenomenon is that there are amplitude and frequency differences between the soil gas radon activity concentration and the sine wave fitted to the smoothed monthly week series. Figures 6.3c and 6.3d show the $P(t)$ and the fitted sine wave for December 2010 and May 2011. There are high deviations both in the frequency and amplitude during winter months; however, soil gas radon time series fit much better the regular sine wave during the summer months. No interpretable auto-correlation was found in the outlier-free 11 monthly weeks noise (ε) (5RSSH rough) data series $(R_1(t))$. This indicates the high variability of soil gas radon and the lack of relationship between successive measurements taken at the 15 minutes intervals. Finally, the outlier-free first rough $(R_1(t))$ for the 11 monthly week time series, the noise component, had a zero average and median (at the 95% confidence level), with homogeneous and symmetric distribution (Figure 6.3e). Summary statistics clearly show that the stochastic variability of the soil gas radon activity concentration is significantly higher in winter than in summer, as confirmed by the F Test with the 95% confidence level.

3   SPATIAL ANALYSIS AND MODELING OF GEOCHEMICAL PROCESSES. STATISTICAL ANALYSIS AND INTERPOLATION AT THE LOCAL SCALE: ATTIC DUST URBAN GEOCHEMICAL CONTAMINATION

Environmental geochemical processes define the liberation, transport and distribution of contaminants in the environment. Representative sampling, spatial data structure recognition and description, and spatial processes modeling including interpolation, are key issues for the management of contamination risk. In this section, a systematic spatial data analysis procedure is described, and an example is provided for spatially continuous parameter distribution in an urban attic dust investigation on airborne contamination assessment (Völgyesi et al., 2014). This local-scale study also shows the methods for stochastic geochemical process modeling.

Ajka, a town in Hungary, has a total area of $95\,km^2$ and more than 29,000 inhabitants. Multi-source contamination originates from lignite mines, with several large mine waste heaps located in the southeastern part of town (see Figure 6.4b). In the center of the town is the lignite power plant, which has been in operation since the

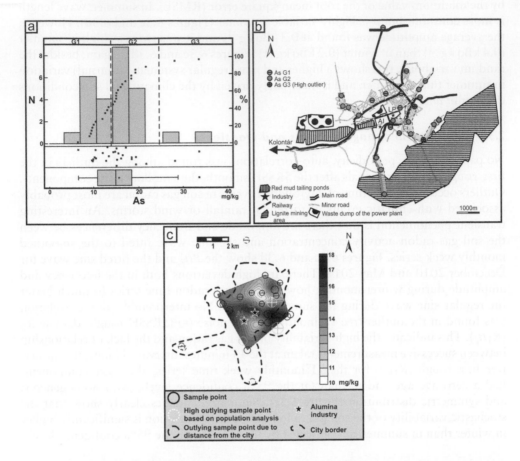

Figure 6.4 Univariate exploratory data analysis and As distribution map in Ajka attic dust. a (upper part). Frequency histogram of As concentration and its cumulative distribution function. Vertical dashed lines: separation of sub-populations, based on histogram analysis and homogeneity test. 'G1 to G3' correspond to group of samples. a (lower part). Scatter plot and box-and-whiskers plot of original As data. b. Spatial distribution of the As sub-groups in the attic samples. Different colours, circles, and circles with minus and with plus signs: sample populations (G1, G2 and G3). Light grey: samples belonging to a homogeneous group showing spatial pattern. c: Arsenic contour map at Ajka. Circles and circles with minus and plus signs show sample populations (G1, G2 and G3) based on the statistical analysis (Figure 6.4a). The circle diameter is proportional to element concentrations. Circles in dashed lines: samples that were not used to create the contour map due to the large distance from other points. Circles in dotted lines were also removed from the contour map due to low or high concentrations, disturbing the spatial trend of element concentrations in the attic samples, based on the method of Reimann *et al.* (2008) (after Völgyesi *et al.*, 2014).

1940s, and produces the large fly ash and slag ponds nearby. Bauxite mining in the region supported the large-scale alumina factory and an aluminum smelter since the 1940s. Twenty-seven houses in Ajka and its vicinity, including industrial areas, urban dwelling locations and background locations were sampled (Figure 6.4b). The sampling strategy followed a grid-based, stratified random sampling design. The 64 km² project area was covered by a 1×1 km grid. In each cell, one house, located closest to a randomly generated point in the cell, was selected for attic dust collection. A further selection criterion, superimposed on the grid design, was related to the position of the alumina and power plant industrial areas, of the lignite mines, and of the red mud tailings ponds (Figure 6.4b). To obtain representative geochemical background samples, material was collected from the upwind direction (NW). Four ash samples were collected from the waste dump of the power plant, three samples from the waste heaps in the lignite mining area, and ten red mud samples close to the tailings pond two days after the catastrophe on October 6, 2010, to study the possible effect of the main contamination sources of the lignite mines, the lignite-fired power plant, and the alumina industry. As, Cd, Cu, Ni, Pb, Zn and Hg total concentrations were defined in the samples (Völgyesi et al., 2014).

3.1 Data processing and data analysis

Robust statistics were used in this study since geochemical data series often display non-normality, heterogeneity and outliers. Tukey's (1977) resistant five-letter summary statistics, containing the minimum, lower quartile, median, upper quartile and the maximum values, were calculated and displayed using box-and-whiskers plots. To compare the overall variability of trace elements in the samples, their relative variability, represented by values of robust inter-quartile range per median (IQR/median), were compared. The range/median measure was used to characterize extreme concentrations related to point-source contamination. Enrichment factors (EF) were computed by dividing the element concentration in the dust sample by the regional geochemical background level (Ódor et al., 1996), and also with the more expanded European mean topsoil levels (FOREGS atlas, Salminen et al., 2005) (EF, $median_{element}$/regional background $value_{element}$). There are no environmental standards for airborne dust. Therefore, this study used the national pollution limit values for earth materials (soil and sediments; Government Decree 6/2009). Data were also analyzed for heterogeneity. Sub-populations were identified using the natural brake method based on cumulative histogram analysis according to Reimann et al. (2008). The natural brake is an inflection point in the cumulative distribution function (CDF), which corresponds to the local minimum in the frequency histogram (multi-modal histogram; e.g. Figure 6.4a). Assuming that a homogeneous distribution represents a single stochastic process, such as anthropogenic trace metal contamination from a single source, or the natural geochemical background, each identified sub-population may reveal a geochemical process in the univariate data space. Separation of sub-populations was confirmed at the 95% confidence level by the Mann-Whitney (Wilcoxon) homogeneity test (Mann and Whitney, 1947), based on the comparison of medians. Outlying values represent sudden and unusual events, essential to identify elements as a contaminant at point sources. Tukey's (1977) inner-fence criterion was used to identify outliers. Summary statistics were calculated separately for the original data series and the identified sub-populations.

Relationships between measured parameters were studied with the Pearson's correlation coefficient (Reimann et al., 2008), using the robust interactive outlier rejection regression method. No more than 10% of bivariate outliers were rejected in all cases. All correlations were checked visually using the regression bivariate scatter plots, and all correlations discussed are significant at the 95% confidence level. Correlations possibly induced by the action of a background variable were checked by means of partial correlations. Attic dust concentration groups were plotted in maps, to assess the spatial distribution of contaminated dust and its possible association with industrial sources.

A further spatial characterization of the trace elements, using contour maps, was generated with the linear and accurate Triangular Irregular Network (TIN) interpolation (Guibas and Stolfi, 1985). A grid size of 25 m was used, based on the shortest distance between the closest two samples. Successive moving average smoothing was applied to generalize the TIN model, and to capture the major spatial trends of attic dust contamination distribution. Single high values were removed from the interpolation procedure. Local maxima and minima in the interpolated surfaces were accepted only if they were detected by at least two samples. Thus, local extremes, represented by a single outlying value, do not affect the smoothed interpolated contamination trend surface maps.

3.2   Interpretation of features in the geometric space and the variable space: airborne contamination in attic dust

3.2.1   Statistical analysis

Enrichment factors calculated as the ratio of median element concentrations in attic dust, and the Hungarian regional geochemical background (HRGB) element concentration in stream sediments, follow the order of Pb >> Zn > Hg >> Cu >= Cd > As > Ni. At least 50% of the samples are contaminated with As (median $= 15$ mg kg^{-1}), Cd (median $= 1.33$ mg kg^{-1}) and Zn (median $= 276$ mg kg^{-1}) as compared to the national pollution limit (hereinafter referred to as 'pollution limit') value for earth materials. The relative variability of Zn, Pb and Hg, expressed as the IQR/median statistics, standing out with the highest values (0.9–1.39), suggests that they have one or more sources which produce a highly diverse distribution of airborne particulate concentrations. Anthropogenic emissions of these elements, due to industrial activity such as coal and lignite-fired power plants and mining (Glodek & Pacyna, 2009; Gosar & Miler, 2011; Sajn, 2002), and fuel combustion in heavy traffic (Davis & Gulson, 2005), may explain the observed high variability. The analysis of univariate summary statistics implies that Pb, Cd, Hg and Zn, characterized by high concentrations and enrichment factors, high overall variability, and high extreme variability, are most likely dominated by anthropogenic, mainly point-source contamination (Cizdziel & Hodge, 2000; Davis & Gulson, 2005; Gosar et al., 2006; Hlawiczka et al., 2003; Sajn, 2005). As and Cu, and particularly Ni, have low concentrations at the natural background level, low enrichment and low overall variability, with few outliers. Therefore, these elements are probably associated with wind-blown soil particles, and their spatial distribution is not driven by prominent point sources (Sajn, 2005).

3.2.2   Distribution analysis and spatial mapping

The simple, yet thorough, univariate data analysis clearly delineates areas with houses where lignite mining, the lignite-fired power plant, the waste dumps and traffic have deposited airborne dust. As coal can contain appreciable amounts of As (Fordyce et al., 2005; Goodarzi, 2009; Salminen et al., 2005), the effect of the lignite mining area and the waste dump of the power plant is obvious, as shown by the statistical analysis and population spatial pattern of As concentrations (Figure 6.4a). The two high outlier samples (with concentrations of 28.9 mg kg^{-1} and 34.5 mg kg^{-1}), and the higher sub-population members of the two major As populations, are located near the lignite mines and next to the waste dump of the power plant (Figure 6.4b). 75% of these samples exceed the pollution limit value, indicating that areas are uniformly As-contaminated. A value of 14.3 mg kg^{-1} As can be used as a tentative threshold to delineate areas impacted by the lignite mining. The As-contamination nature of the higher sub-group is supported by its extreme variability, higher than that of the lower significant sub-group (Figure 6.4a). Thus, the samples from the lower group are affected by one source, most probably the natural background. The maps show the transport of As-bearing lignite from mines to the central industrial area and the waste dump of the lignite-fired power plant (Figure 6.4c), corroborating the lignite-bound As content of the attic dust samples (Yudovich & Ketris, 2005).

The higher sub-group of Hg, where eleven concentration values lie above the pollution limit, and the outlier sample (1.97 mg kg^{-1}), depict a well-defined area around the power plant (Figure 6.5b). The lower group of samples (Hg 6 0.335 mg kg^{-1} with 14 samples) shows a higher relative overall variability than the extreme values (Figure 6.5a). This agrees with the observation that the extreme Hg values derive from a single source. A mixture of non-contaminating geochemical processes, such as natural background and the potential contribution of an additional anthropogenic source, defines Hg concentrations in the lower group. The trend surface contour map indicates that the Hg concentrations steadily decrease away from the pollution source in a well-defined concentric manner, showing the fast and short-distance deposition of larger particles (Gosar et al., 2006; Shah et al., 2008; Huggins & Goodarzi, 2009; Figure 6.5c).

While Hg is most probably bound to finer particles in the background attic dust (Balabanova et al., 2011; Gosar et al., 2006; Huggins & Goodarzi, 2009), Hg speciation also influences atmospheric transport and residence time (Gosar et al., 2006; Shah et al., 2008). An N-S oriented trend can be seen in the centre of the study area (Figure 6.5c). This is sub-parallel to the prevailing NW-SE wind direction, suggesting that the atmospheric pathway is the main Hg transport route. A spatial pattern analysis of soil samples showed similar results, indicating a strong association between soil and attic dust in Ajka (Zacháry et al., 2012).

Four out of the five Cu-outlier samples are situated in the south-eastern region at the lignite mines and the industrial area (Figure 6.6b). Only two of them exceed the value of the pollution limit (75 mg/kg). The higher sub-population shows a larger relative variability than the lower sub-group of 9 samples. An obvious spatial correspondence can be observed between locations of samples of the higher concentration group and lines of the main roads and railways (Figures 6.6a and b). In contrast, this linear spatial Cu pattern is invisible in the contour map (Figure 6.6c). Hence, the grid-based trend surface map (Figure 6.6c) is not sufficient to identify the linear pattern and the effect of transport in Ajka. This underlines the relevance of detailed statistical and data analysis, prior to geochemical map generation (Reimann et al., 2008). The simple and detailed statistical analysis of data revealed an important transportation-related contamination Cu source in Ajka, similar to observations in Birmingham, Coventry (UK, Charlesworth et al., 2003) and in Ulsan (South Korea, Duong & Lee, 2011).

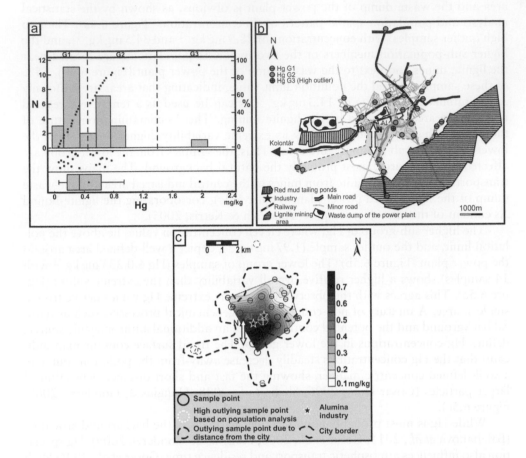

Figure 6.5 Univariate exploratory data analysis and Hg distribution map in Ajka attic dust. a (upper part). Frequency histogram of Hg concentration and its cumulative distribution function. Vertical dashed lines: separation of sub-populations, based on the histogram analysis and homogeneity test. 'G1 to G3' corresponds to group of samples. a (lower part). Scatter plot and box-and-whiskers plot of original Hg data. b. Spatial distribution of the mercury sub-groups in the attic samples. Different colors, circles and circles with minus and with plus signs: sample populations (G1, G2 and G3). Light grey area: samples belonging to a homogeneous group showing spatial pattern. White arrow: N-S trend of the Hg concentration sub-parallel to the prevailing wind direction in Ajka. c. Mercury contour map at Ajka. Circles and circles with minus and plus signs show sample population (G1, G2 and G3) based on the statistical analysis (Figure 6.5a). The circle diameter is proportional to element concentrations. Circles in dashed lines: samples that were not used to create the contour map due to the large distance from other points. Circles in dotted lines were also removed from the contour map due to low or high concentrations, disturbing the spatial trend of element concentrations in the attic dust samples based on the method of Reimann *et al.* (2008). The white arrow indicates the N-S trend of the Hg concentration sub-parallel to the prevailing wind direction in Ajka (after Völgyesi *et al.*, 2014).

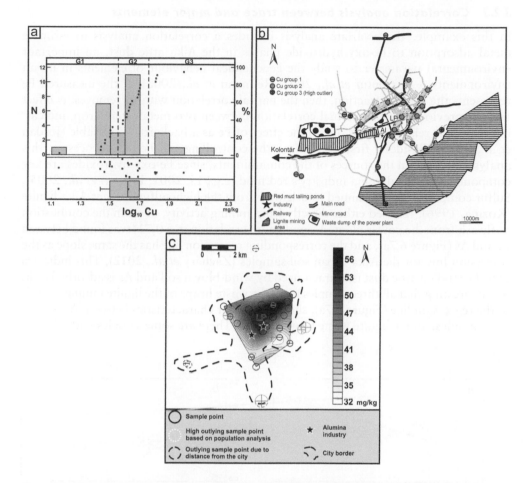

Figure 6.6 Univariate exploratory data analysis and Cu distribution map in Ajka attic dust. a (upper part). Frequency histogram of Cu concentration and its cumulative distribution function. Vertical dashed lines: separation of sub-populations based on histogram analysis and homogeneity test. 'G1 to G3' corresponds to group of samples. a (lower part). Scatter plot and box-and-whiskers plot of original Cu data. b. Spatial distribution of the copper sub-groups in the attic dust samples. Different colors, circles and circles with minus and with plus signs: sample populations (G1, G2 and G3). Light gray area: samples belonging to a homogeneous group showing spatial pattern. c: Copper contour map at Ajka. Circles and circles with minus and plus signs show sample population (G1, G2 and G3) based on the statistical analysis (Figure 6.6a). The circle diameter is proportional to element concentrations. Circles in dashed lines: samples that were not used to create the contour map due to the large distance from other points. Circles in dotted lines were also removed from the contour map due to low or high concentrations disturbing the spatial trend of element concentration in the attic dust samples, based on the method of Reimann *et al.* (2008) (after Völgyesi *et al.*, 2014).

3.2.3 Correlation analysis between trace and major elements

In this example, the bivariate analysis includes a correlation analysis to estimate metal adsorption to Fe-oxy-hydroxide phases in the Ajka attic dust, an important environmental parameter to study the geochemical behaviour of elements in urban environments (e.g. Contin *et al.*, 2007; Reimann *et al.*, 2008). If the measured Fe represents the sorbing fraction, then the partial correlation with Fe removes its effect and the previously strong virtual correlation between two metals may drop, indicating that their relationship is due to the effect of Fe as a background variable (Jordan *et al.*, 1997). Besides a first insight into the controlling geochemical processes, this analysis may reveal the sources of the contaminants, since Fe-oxy-hydroxides are key components of the alumina industry's red mud (e.g., Li, 2001; Brunori *et al.*, 2005). Sulfur components are assumed to originate from the high-sulfur content Ajka lignite (Kozma, 1996), released either directly from mining activity, or from the combustion as fly ash from the power plant. A significant correlation ($r = 0.71$) was found between Fe and As (Figure 6.7a), and the corresponding regression line has the same slope as the regression line for the Ajka urban soil samples (Zacháry *et al.*, 2012). This indicates that the studied attic dust is characterized by wind-blown soil and As is adsorbed to it. It is interesting that all three samples from the waste heaps of the lignite mining area fit to the regression line (Figure 6.7a), showing similar characteristics between As in attic dust samples, and in lignite waste heap samples. There are some samples with high As

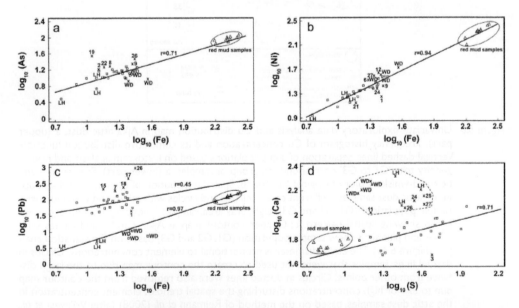

Figure 6.7 Pearson's correlation coefficients of As, Ni, Pb and Ca with Fe and S. The sampling points fitting on the regression line are indicated as empty boxes, and bivariate outliers are indicated with 'x'. 'LH' corresponds to samples from lignite waste heap, and 'WD' corresponds to samples from waste dump of the lignite-fired power plant. Triangles show red mud samples. Concentrations of Fe, Ca and S are given in $g\,kg^{-1}$, whereas the other elements are given in $mg\,kg^{-1}$ (after Völgyesi *et al.*, 2014).

concentrations, falling above the Fe-As regression line, corresponding to samples with excessive As that are not controlled by Fe compound sorption (Figure 6.7a). These points are found mostly in the lignite mining area and close to the lignite-fired power plant (Figures 6.4 and 6.7a), suggesting that As is present in other chemical forms as well, most likely associated with sulfur compounds including As sulfides in the original Ajka lignite. Goodarzi (2009) and Shah *et al.* (2008), studying coal-fired power plants in Canada and in Australia, reported similar findings. The strong correlation (r = 0.94) between Fe and Ni for most of the samples, captures a similar geochemical behavior of Fe and Ni in natural earth materials (e.g., Kabata-Pendias, 2000; Figure 6.7b).

There is an overall moderate correlation (r = 0.45) between Fe and Pb, explaining only 20% of the Pb variability in the majority of the dust samples (Figure 6.7c), and showing the general sorption mechanism between Fe oxides and Pb (Banerjee, 2003). Bivariate outliers, falling above the Fe-Pb regression line (7, 15, 17, 18, 26), are randomly situated samples, probably showing local Pb sources within the attic. This slight correlation reveals that processes other than sorption by Fe compounds primarily control the Pb distribution in the airborne attic dust. The Pb supply most likely originates from traffic, as was observed in attic dust samples from Las Vegas, Toquerville, Washington City, Dover and New Jersey (US), by Ilacqua *et al.* (2003) and Cizdziel and Hodge (2000), respectively. All attic samples lie far above the seven samples from the waste heaps in the mining area and the waste dump of the power plant, verifying that excessive Pb content of the dust samples originates from another source, most probably from the traffic in the study urban area.

The significant correlation (r = 0.71) between Ca and S, shows (Figure 6.7d) the presence of gypsum in the majority of attic dust samples. Electron microscopy and XRD analysis of the collected attic dust confirm that most samples contain significant amounts of gypsum, most likely as a secondary mineral phase (Völgyesi *et al.*, 2013). The samples display a total S content in the same range as the seven samples from the lignite waste heap and the waste dump of the power plant (Figure 6.7d).

Significant and mutual correlations (r = 0.5–0.79) exist within the Pb-Zn-Cd-Ni-Cu group. This is particularly strong (r > 0.73) for Pb-Cd-Ni. The correlations remain unchanged when the possible effect of Fe-oxy-hydroxide sorption is taken into account, by applying a partial correlation with Fe for these relationships. Our study confirms that the strong association of these typical metal industry-related elements relates to their common source(s) (Davis and Gulson, 2005; Sajn, 2005), and it is not driven by their common sorption to Fe compounds in the airborne soil particles. The lack of Fe effect on metal correlations also confirms that the Fe oxide-rich red mud dust is not a major component of the attic dust. It is most interesting that neither Hg nor As displays any notable correlation with the other elements, clearly showing that their origin and geochemistry is different from the other potential contaminants.

4 SPATIAL ANALYSIS AND MODELING OF GEOCHEMICAL PROCESSES. ADVANCED PROCEDURES AT THE REGIONAL SCALE: RADON RISK ASSESSMENT

This section makes a step forward and describes methods for advanced spatial structure analysis, such as trend and anisotropy analysis at the regional scale. A detailed spatial data analysis procedure is described, and an example is provided for the spatial characterization of regional geogenic radon potential (Szabó *et al.*, 2014). It is noted that the applied 2D method is very similar to the 1D time series analysis procedure, presented in the first section of this chapter.

Soil gas radon activity concentration and soil gas permeability were measured in situ at 192 measurement sites in Pest County and surrounding areas in Hungary, along a 10×10 km grid (Dubois *et al.*, 2010; Tollefsen *et al.*, 2011) over the 80×90 km target area. This resulted in about three measurement sites per cell with a 3.2 km average nearest-neighbor distance between the sites. The three dominating geological formations were selected in the cell and a measurement in each formation was made. Thus, the sampling scheme is similar to a stratified (grid-based) random sampling. Measured rock formations include: volcanic rocks, limestone, marl and sandstone, clay, sand, gravel, Holocene alluvial mud and sand, in addition to sand, drift sand, gravel, loess, marl, silt, clay or limestone (Gyalog, 1996). Geogenic radon potential provides information about the potential risk from radon. Its calculation takes into account the equilibrium soil gas radon activity concentration, with saturation at an infinite depth. Modeling and mapping of geogenic radon potential (GRP) provide an opportunity to identify radon-prone areas (Dubois *et al.*, 2010). Therefore, a radon potential map helps to reduce cumulative radiation risk.

4.1   Statistical analysis

Summary statistics used in this study include measures of central tendency and variability. These statistics are the minimum, lower quartile, median, upper quartile, maximum and average (arithmetic mean), mode, standard deviation, median absolute deviation (MAD), range and inter-quartile range. Tukey's (1977) inner-fence criteria were used for outlier definition, and the robust MAD/median measure was employed for comparison of variability of parameters. Spatial radon measurements empirical distribution often has heavy-tail property, and outliers indicate real anomalies in many cases. This has been confirmed by several authors (Appleton *et al.*, 2011; Bertolo & Verdi, 2001; Bossew *et al.*, 2008; Kemski *et al.*, 2001; Tóth *et al.*, 2006). The Kruskal-Wallis and Mann-Whitney Tests (Kruskal & Wallis, 1952; Mann & Whitney, 1947) were applied under the null hypothesis that the medians of measured parameters (soil gas radon activity concentration, soil gas permeability, GRP) within each of the geological formations are the same. Levene's Test (Levene, 1960) was applied to test the null hypothesis if the standard deviations within each of the geological formations are the same. Kolmogorov-Smirnov Test (Kolmogorov, 1933; Smirnov, 1948) was applied to test the null hypothesis, if the distributions of two datasets are homogeneous. Simple least-squares linear regression analysis with a constant additive was performed to explore the linear relationship between the meteorological parameters (atmospheric temperature, humidity and absolute pressure) and soil gas radon concentration, soil permeability and geogenic radon potential. Strength of relationship is expressed by the Pearson's linear correlation coefficient (r) (Rodgers & Nicewander, 1988). All of the statistical tests applied in this study were at a 95% confidence level.

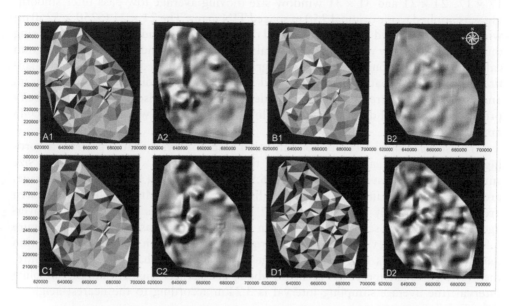

Figure 6.8 Results of the smoothing procedure on theTIN maps of soil gas radon activity concentration (A), soil gas permeability (B) (also in logarithmic scale (D)) and geogenic radon potential (C). The first maps (A1, B1, C1 and D1) areTIN maps made from original data. The second maps (A2, B2, C2 and D2) are the 21 × 21 (5250 × 5250 m) window size smoothed TIN maps, which revealed the spatial trends and pattern without losing much detail. They were used for spatial autocorrelation and directional variogram calculations. Coordinates are in meters (after Szabó *et al.*, 2014).

4.2 Mapping and spatial analysis

The smallest grid cell that resolves all measurement points for the interpolated parameter surfaces is defined by the two closest points located 9.5 m apart from each other found in one garden. This would imply an unmanageable high grid density in the 80 × 90 km study area. Based on a trial-and-error approach, a 250 m grid size proved to be the optimal compromise between the loss of information (8 data pairs fall in shared grid cells, i.e. 4.2% of all data) and digital data processing efficiency. Figure 6.8 shows the original 250 m spaced TIN map and the 21 × 21 (5250 × 5250 m) window size moving average filter smoothed map of the soil gas radon activity concentration, soil gas permeability and geogenic radon potential. In order to study the large-scale spatial trend in the measured parameters, a smoothing procedure was applied to filter the small scale 'noise' from the data. This spatial characterization is based on a contour map generated with the linear and accurate (fitting to the original measured values at sample locations) Triangular Irregular Network (TIN) interpolation (Guibas & Stolfi, 1985). A grid size of 250 m was used based on the shortest distance between the closest two measurement sites. Successive moving average smoothing was applied to generalize the TIN model, and to capture the major spatial trends of radon and other parameter distribution. Firstly, a series of 3 × 3, 9 × 9, 13 × 13,

17×17, 21×21 and 31×31 window size moving average low-pass filter smoothing was applied to the original 250 m spaced TIN maps. The 21×21 (5250×5250 m) window size smoothed TIN maps revealed the spatial trends and pattern without losing much detail, and they were used for spatial autocorrelation and directional variogram calculations. Directional empirical variograms were calculated to capture and quantitatively describe spatial anisotropy and periodicity. In addition to the variogram analysis, 2D autocorrelograms were made also to reveal anisotropy and periodicity present in the spatial data. Spatial modeling was performed with Surfer and ArcGIS applications.

The soil gas radon activity concentration map displays a spatial pattern, according to the contour map generated from the 21×21 (5250×5250 m) window size smoothed TIN trend surface map (Figure 6.9a). Higher values ($30\,\mathrm{kBq\,m^{-3}}$ on average) are in the hilly areas in the Buda Mts., Pilis Mts., Visegrád Mts. And in the northern areas in the Börzsöny Mts. And Cserhát Mts. Lower values ($9.5\,\mathrm{kBq\,m^{-3}}$ on average) characterize the southern and eastern plane areas, such as the Pest Plane (Figure 6.9a). The empirical variogram of the original soil gas radon activity concentration data without outliers has a strong nugget effect showing that, besides a measurement error, this parameter has great variability at distances smaller than the sampling interval (the average sampling interval is 3.2 km according to the nearest-neighbor distance).

4.3   Interpretation of features in the regional radon concentration maps: advanced spatial analysis

Since the GRP spatial pattern was less variable than the soil gas radon concentration, it shows the major spatial structures, whereas the soil gas radon was more sensitive to local geological mechanisms (e.g. fault line effects). Fault lines are located in the mountains within the study area (Gyalog and Síkhegyi, 2010), with the appearance of higher soil gas radon activity concentration values. GRP involves the probability of radon escape from the geological formations and soils as described earlier. Permeability is spatially highly heterogeneous in the study area. The smoothing procedure on TIN maps sheds light on the fact that GRP has the same pattern as soil gas radon at a larger spatial scale, but permeability becomes dominant at lower scale (Figure 6.8). In addition, contour maps were prepared from the selected 21×21 window size (5250×5250 m) of the smoothed TIN trend surface maps and their median and upper quartile were indicated. The upper quartiles are continuous in both the soil gas radon concentration (Figure 6.9a) and GRP contour maps (see Szabó et al., 2014), and follow the mountainous topography well. This means indeed that the soil gas radon concentration and GRP follow the different geologies associated to hills and mountains on the one hand, and to plains on the other.

This pattern cannot be related to soil gas permeability, because it does not display same spatial pattern. In fact, soil gas permeability data have no discernible spatial pattern at all. There was a strong nugget effect in the empirical variogram of the soil gas radon activity concentration and the soil gas permeability without outliers, whereas GRP had a lower nugget effect, confirming the applicability of GRP. Autocorrelation and directional variograms calculated for the smoothed TIN trend surface maps showed periodicity in the NE-SW direction (azimuth: 20° with

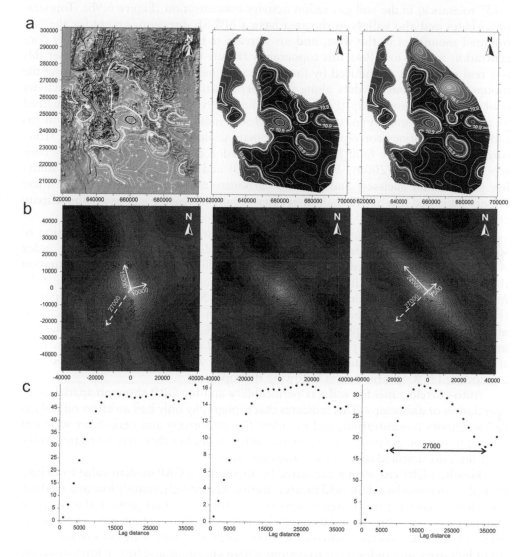

Figure 6.9 a. Shaded relief map and the superimposed contour lines of the soil gas radon activity concentration (kBq m^{-3}), also showing the measurement sites (white crosses). The bold white lines indicate the median, thick black lines correspond to the upper quartile of the soil gas radon activity concentration. The middle map shows the contour lines of the soil gas radon activity concentration when the mountain areas are removed. The map on the right shows the contour lines of the soil gas radon activity concentration when only the N-S mountain areas are removed. Mountain areas were removed along the upper quartile of the soil gas radon activity concentration (first map of this figure). b. Auto-correlograms for the soil gas radon activity concentration (kBq m^{-3}) calculated from the smoothed TIN trend surface map. White arrows indicate the direction of anisotropy, white dashed arrows correspond to periodicity of soil gas radon activity concentration. The first map comprises all data. The middle map shows the data without the mountain areas (see corresponding maps in Figure 6.9a), the right one displays data when only the N-S mountains are removed. Note the strong NW-SE anisotropy and NE-SW periodicity soil gas radon activity concentration. c. Directional variograms of the maps corresponding to the soil gas radon activity concentration maps in Figure 6.9a above. Variograms are in SW-NE direction (azimuth = 20°, tolerance = 15°). Note the ca. 27 km NE-SW periodicity. Coordinates are in meters (after Szabó *et al.*, 2014).

±15° tolerance) in the soil gas radon activity concentration (Figure 6.9b). Topography, ridges and also valleys in the area have a NW-SE direction, except for the N-S oriented mountains in the west, and an increase in the soil gas radon activity concentration seemed to follow this topography (Figure 6.9a). We verified whether this is a real periodicity or induced by the dominating mountain features. Removing the mountain areas from the data along the upper quartile of the soil gas radon activity concentration contour map (Figure 6.9a), the observed 27 km periodicity remained as a persistent feature (Figure 6.9c). This clearly indicates, in turn, that the soil gas radon concentrations corresponded to the main geological, basically tectonic, structures. Influence of the fault lines on soil gas radon concentration has already been reported by Barnet (2008), Papp *et al.* (2010) and Swakon *et al.* (2005), studying granites and gneisses, andesite and carbonate rocks, respectively. All of these papers suggested that active fault lines cause locally increasing soil gas radon activity concentration.

A two-dimensional auto-correlogram showed strong spatial autocorrelation of soil gas radon activity concentration in the N-S direction. The anisotropy index is 0.6 (Figure 6.9b) according to the anisotropy ellipse. We tested the assumption that this anisotropy was due to the N-S running mountain range in the west. If the mountain areas were eliminated from the map, the N-S autocorrelation disappeared (Figure 6.9b). However, the strong NW-SE direction anisotropy emerges, which was also captured by the directional variograms (Figure 6.9b). This confirms that the N-S anisotropy was induced by the N-S mountains, and does not characterize the entire study area. The soil gas radon concentration of the whole area is characterized by a pronounced NW-SE orientation following the topography.

Auto-correlograms for soil gas permeability and for GRP show no spatial autocorrelation or anisotropy. This indicates that topography only had an effect on soil gas radon activity concentration, and the identified anisotropy and periodicity were not related to the soil gas permeability spatial distribution, but they may be related to the fault lines documented only in the mountain areas.

Finally, a GRP risk map was created by attributing a GRP median value to all geological formations based on field measurements. Accordingly, mainly low and medium risk characterizes the study area (Szabó *et al.*, 2014). Only two geological formations (fluvial-proluvial sediment and deluvial clay and sand) showed high risk. These partly affect 18 settlements in the northern part of the study area, which are close to hills made up of limestone and andesite, or to regions where sandstone and lignite formations are predominant.

In summary, it can be concluded that the systematic spatial analysis of geochemical data applied to regional soil gas radon concentration measurements was successful, revealing a clear spatial structure for soil gas radon activity concentration and GRP. Exploratory data analysis revealed that soil gas permeability data had no discernible spatial pattern at the available spatial resolution of the data, whereas the soil gas radon activity concentration and GRP did have such features. The latter two had almost the same pattern; however, GRP was less variable. The pattern of both of them follow the topography of the area. They have the same 27-km-long NE-SW periodicity as the topography represented by the digital elevation model. Moreover, a persistent NW-SE spatial anisotropy was shown in the soil gas radon activity concentration.

5 GEOCHEMICAL TRANSPORT MODELING: TOXIC ELEMENT CONTAMINATION TRANSPORT IN A MINING CATCHMENT

This section goes beyond the previous sections' spatial description of geochemical processes, and applies multivariate statistical analysis methods for contamination assessment in a mining catchment (Jordan et al., 2009). Moreover, a spatially distributed sediment transport model is described and applied to estimate the sediment-bound toxic element liberation, transport and deposition in the catchment, and to calculate the total sedimentary contamination export from the catchment to downstream agricultural lands.

Seventy-nine stream water and sediment samples were collected at ca. 250-m intervals along the stream courses within the Recsk Copper Mines, Hungary, study area. The catchment area of the Recsk mining area is 87.2 km^2, and it is characterized by numerous small streams originating from the mountain slopes (Somody, 2005). Stream water temperature, pH and electrical conductivity (EC) were measured in the field. The dissolved concentrations of the elements As, Cd, Cu, Fe, Mn, Ni, Pb, Sb and Zn, in filtered (45 μm) and acidified stream water samples were determined using an ICP-AES analyzer. The sulfate concentrations in the filtered and unacidified water samples were determined using spectrophotometry. Stream-sediment samples sieved through a 2-mm mesh were digested using hot aqua regia and were analyzed for total metal contents using ICP-MS.

5.1 Multivariate methods, reaction and sediment transport modeling

Detailed quantitative geochemical modeling involving statistical methods and thermodynamic reaction models was used to determine the fate of the contamination in the studied mining catchment. In this study, geochemical modeling utilized univariate exploratory data analysis (EDA) techniques similar to the previous sections, followed by multivariate statistical analyses, including cluster analysis (CA) and principal component analysis (PCA), to investigate element distribution patterns and gradients in stream water and sediments (Jordan et al., 2003; Somody & Jordan, 2005). The measured concentration parameters were normalized to the 0–1 scale for multivariate analysis, in order to ensure equal weightings for the variables in the multivariate analyses. The concentration parameters were log-transformed to stabilize variation for bivariate and multivariate procedures. The cluster analysis used the Euclidian distance similarity measure and the nearest-neighbor linkage method.

Chemical reaction modeling was then performed using the thermodynamic reaction model PHREEQC (Parkhurst & Appelo, 1999) to describe the prevailing processes controlling contamination in environments affected by mining. In this study, the WATEM/SEDEM distribution model (Van Oost et al., 2000) was used to study the fate of metals (in this study copper) bound to soils and sediments, and to assess the mean annual export of heavy metals from the studied catchment area. The spatial pattern of metal concentrations and the topological relationships between the sediment sources and sinks (Van Rompaey et al., 2005) were taken into account. Each grid cell in the catchment area is connected with a flow path to a permanent river channel. Depending on the transport capacity along the flow path, soil erosion, sediment transport or sediment deposition will occur (Van Rompaey et al., 2001). In the case of erosion, the metal concentration of the transported sediment can increase or decrease, depending on the metal concentration of the locally eroded soil material. When considering sediment deposition, a metal enrichment factor (Steegen et al., 2001) is used in order to take into account metal adsorption onto the soil and sediment particles. Metal concentrations in point sources of waste dumps and non-point sources of soils in the studied catchment area were taken from previous studies (Farsang, 1996; Bats, 2006). Sediment deposition data collected from retention ponds at the outlet of the catchment area were used to calibrate the model for erosion and sediment transport (Bats, 2006). Archive data (VITUKI Consult Rt., 1996) on measured suspended sediment and metal concentrations in this sediment were used to verify the model. The impact of a set of possible land-use scenarios was evaluated based on the calibrated and verified model.

5.2   Multivariate data modeling for geochemical inference

The statistical distribution analysis of the studied metals showed a polymodal behavior, with fewer than 20% of outliers being found in the stream water samples. Outliers (forming the 'anomalous population'), were located either at mine sites around Lahóca Hill, or within the hydrothermal alteration area on the upper reaches of the Ilona Creek (Figure 6.10). However, based on concentration transects taken along the stream courses, the metal contamination is attenuated within a short distance (200–250 m). Apart from Fe and Zn, all dissolved metals in the stream water samples were under the detection limits in samples from other locations (which formed the 'background population'). A strong Spearman correlation ($r > 0.7$) of pH with the SO_4 content, the EC and metals content in the anomalous sample group at the Lahóca mines is characteristic of AMD. Correlations in the 'background population' in stream water showed a strong relationship ($r > 0.7$) between the SO_4 content, the EC, the Mn content and, to a lesser extent, the Fe content. In the background population, the lack of correlation with pH indicates that the pH is most probably determined by the abundant bicarbonates in the related upstream areas. A strong mutual correlation ($r > 0.8$) for As, Cu, and Zn in the sediments, and to a lesser extent Pb in the 'anomalous' samples, is typical of the mineralization in the area. All metals in the sediments have a strong correlation with the Fe content, in particular close to waste dumps where 'yellow boy' (the yellowish-orange precipitate that results from acid mine drainage) sediments were observed. This indicates that the dissolved metal concentrations are attenuated by adsorption and co-precipitation as iron oxyhydroxides along the stream course.

Cluster analysis and principal component analysis were able to distribute the 'background population' into further groups, corresponding to well-defined geochemical regimes in the stream waters of the study area. In Figure 6.10a, the CA dendogram makes an obvious distinction between the 'anomalous' (Group A) and 'background' populations, but it also reveals that the geochemical background (Group B) constitutes a strong group along the Baláta Creek draining carbonate-rich Tertiary sedimentary rocks. The group of samples in the upper reaches of Ilona Creek (Group C) is the most distinct from the other 'background' samples, and it corresponds to a surface hydrothermal alteration zone (Figure 6.10b). Further downstream, in the lower reaches of the Ilona Creek, the samples represent stream water draining Eocene strato-volcanic

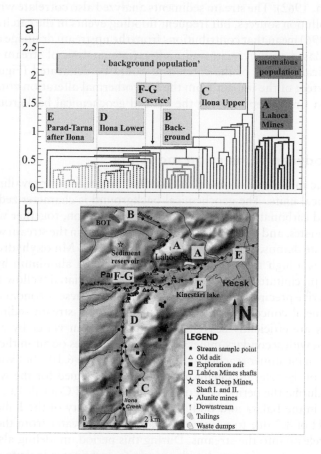

Figure 6.10 a. Cluster analysis (CA) dendogram showing the association of stream water samples. The vertical axis indicates the Euclidean distance for variables normalized to the 0–1 scale. CA used the nearest-neighbor linkage method. b. Map of the study area with the location of stream-water sample groups identified by CA. Stream water and sediment sample locations (black dots) are also shown. A shaded relief model shows the topography in the background, and the black lines are drainage lines in the study area (after Jordan *et al.*, 2009).

andesite (Group D). After its confluence with Parádi-Tarna Creek, a stream that has a more alkaline character, Ilona Creek shows mixed characteristics (Group E) that dominate even after the confluence with Bikk Creek which drains the Lahóca and Recsk mines (Figure 6.10b). The samples in groups F-G are found in areas where gas (primarily CO_2) exhalation is known to occur, often manifested as bubbling water in the streams. Finally, the stream sediment samples fell into the same groups as the stream water samples, but with less well-defined boundaries, indicating mixing due to sediment transport. The above results show that drainage water has dynamic contact with its surroundings in the study area, and thus the water chemistry accurately reflects the locations of AMD pollution. This is consistent with the findings of previous studies (Gedeon, 1962). The stream sediments analyzed also correlate with the ambient geology and pollution sources, but frequent flooding events in the catchment (VITUKI Consult Rt., 1996) mean that contributions from the upstream drainage area are mixed in with them. Based on univariate and multivariate analysis of stream chemistry, the main geochemical regimes of the catchment could be delineated (Figure 6.10). The location and extent of the impact from the hydrothermal alteration zone in the upper reaches of Ilona Creek, representing the natural geochemical background, were also identified.

5.3   Thermodynamic reaction modeling

Thermodynamic modeling was used to describe the processes prevailing in the identified geochemical units. The geochemical background is characterized by carbonate equilibrium and carbonate (calcite, aragonite) precipitation, together with precipitating metal carbonates, and it controls metal concentrations in the stream water upstream of the mine waste dumps. Where AMD discharges Fe and Mn oxyhydroxides into the streams, there is a high thermodynamic probability that aluminum hydroxides and jarosite will be precipitated. Field observation of the associated 'yellow boy' sediments confirms the active precipitation of these components. These secondary minerals lead to significant metal concentrations in the corresponding stream sediments, suggesting that metals are efficiently scavenged from the stream water by adsorption and co-precipitation with secondary minerals. Similar processes occur in the hydrothermal alteration zone in the upper reaches of Ilona Creek (Figure 6.10), but with lower intensity. Modeling of the relevant chemical reactions, performed for the water chemistry data collected during the period of active mining (VITUKI Consult Rt., 1996) showed that the water in the Baláta and Bikk creeks in the vicinity of the Lahóca and Recsk mines had a pH of 8.2 due to pumping of saline groundwater from the underground mines and its release into the streams. During this period, modeling also showed that water samples were oversaturated with respect to carbonates (calcite and dolomite), and that metals could co-precipitate with carbonates. Carbonate-rich sediments found at these sites at that time had significant metal contents (primarily As, Pb and Zn) (VITUKI Consult Rt., 1996), thus verifying the co-precipitation of metals with carbonates. A comparison of the hydrochemistry observed during the two periods shows that the chemical processes controlling the impact of mine effluents can differ markedly during and after active mining.

5.4   Soil erosion and contaminated sediment transport modeling

Modeling of soil erosion and sediment transport, using a land-use map for the year 2003, made it possible to produce a map of these processes, as well as another map showing metal transportation (Figure 6.11). The model also predicted that on average about 9,000 t/year of sediment, and about 1.3 t/year of particulate copper, were exported from the catchment. The amount of copper calculated to be exported from the catchment area shows good agreement (within 15% difference) with measured suspended sediment and particulate Cu data, taken from a previous study (VITUKI Consult Rt., 1996). The impact of land use on the export of metals was evaluated for land use scenarios, taken from aerial photographs in 1987 and a field survey

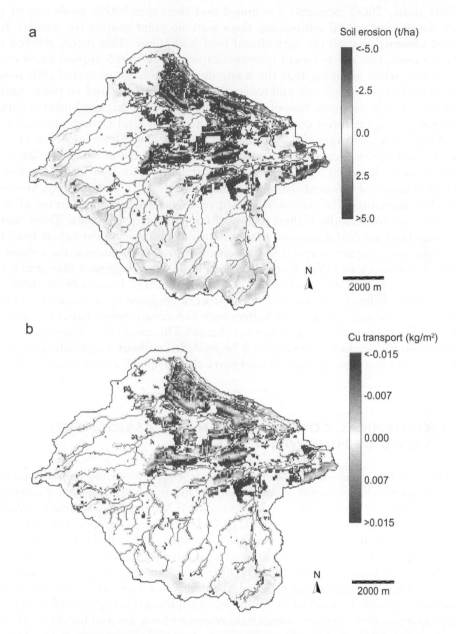

Figure 6.11 Erosion and metal (Cu) transport maps generated from distributed modeling of the Recsk copper mines catchment. a. Map of soil erosion and sediment deposition (t/ha). Negative values indicate net erosion; positive values represent net deposition. b. Map of metal (Cu) erosion and deposition (kg/m²). Negative values represent metal erosion; positive values represent net metal deposition. Fainter lines indicate streams. See text for details (after Jordan *et al.*, 2009).

in 2003 (Bats, 2006). Scenario 3 assumed that there was 100% protection of the waste dumps from metal release (i.e. there were no point sources for metals). Scenario 4 assumed that all the agricultural land was fallow. This model showed the highest erosion rate and sediment transport capacity. Scenario 5 applied forest cover everywhere, while assuming that the waste dumps would be protected (this model showed the lowest erosion rate and sediment transport capacity and no point sources of metals). While the protection of waste dumps obviously left catchment sediment loss intact, complete forest cover resulted in the lowest total erosion and sediment loss. Assuming that all agricultural areas were fallow also had little impact on erosion and sediment release, because the agricultural land is located in low-lying flat accumulative areas. Scenarios 3 and 5 were the most effective for minimizing the loss of metal-contaminated sediments from the mining catchment. It is interesting, however, that while Scenario 4 ('all fallow') did not result in a significant increase of sediment loss, it produced the highest export of contaminated sediments. These results show that land use configurations, that minimize the total sediment export from the catchment area, are not necessarily the same as those that minimize the volume of exported polluted waste (Van Rompaey *et al.*, 2005). It is suggested that land cover interferes with metal (Cu) export from: (1) non-point sources, because of the volume of total exported sediment and of enrichment processes induced by sediment deposition; (2) from point sources, because of buffers such as forested stream banks on the flow path between the point source and the river channel. This means that sedimentary metal export from the catchment area cannot be modelled without a spatially distributed approach, and that spatially explicit land cover change models are necessary for future estimations.

6 GEOCHEMICAL CONTAMINATION RISK ASSESSMENT: RANKING OF MINE WASTE SITES

This final section concludes this chapter on geochemical modeling with a numerical contamination risk assessment (Abdaal *et al.*, 2013). In this study a semi-quantitative risk-based ranking method of mine waste sites (EU Mining Waste Directive (MWD) Pre-selection Protocol, Directive 2006/21/EC) is presented, compared to other recognized contamination risk assessment (RA) methods (EEA, 2005), adopted to local conditions, and finally tested for parameter uncertainty. Altogether, 145 ore mine waste sites were tested using the EU MWD Pre-selection Protocol as a case study from Hungary. Then, by running the protocol, the number of YES, NO and UNKNOWN responses to questions (marked as 'Q') on input parameters is registered for each site. The proportion of the certain to uncertain responses for a site and for the total number of sites may give an insight to specific and overall uncertainty in the data we use (Table 6.1). The distance from mine waste sites to the nearest receptors such as human settlements (Q15) is measured using proximity analysis tools (Point Distance and Generate Near Table) in ArcINFO® 10. Statistical analysis is carried out using STATGRAPHICS Centurion XV.II® software, such as the topographic slope (Q10), and the measured distance to the nearest surface water courses (Q11), settlements (Q15), groundwater bodies (poor status, Q16), protected areas (Natura 2000 sites, Q17), and agricultural areas (Q18). The objective of this section is to demonstrate

Table 6.1   Summary statistics of the EU Pre-Selection Protocol responses to questions Q1–18, showing the number of YES and NO responses based on the EU Pre-Selection Protocol thresholds, the local median-based thresholds and on the local highest group-based thresholds (after Abdaal et al., 2013).

Preselection	Protocol	Number of sites	EU thresholds		Local thresholds (median based)		Local thresholds (highest group)		U	U%
			Yes	No	Yes	No	Yes	No		
Impact	Q1	145	19	126	19	126	19	126	0	0
Source	Q2	145	101	40	101	40	101	40	4	3
	Q3	145	126	15	126	15	126	15	4	3
	Q4	145	7	138	7	138	7	138	0	0
	Q5	145	9	136	9	136	9	136	0	0
	Q6	9	9	0	9	0	9	0	0	0
	Q7	9	4	2	4	2	4	2	3	33
	Q8	136	34	92	34	92	34	92	10	7
	Q9	136	9	115	9	115	9	115	12	9
	Q10	136	**110**	26	**74**	62	**2**	134	0	0
Pathway	**Q11**	145	**64**	81	**73**	72	**144**	1	0	0
	Q12	145	120	25	120	25	120	25	0	0
	Q13	145	17	128	17	128	17	128	0	0
	Q14	145	17	128	17	128	17	128	0	0
	Q15	145	**45**	100	**73**	72	**141**	4	0	0
	Q16	145	**28**	117	**73**	72	**142**	3	0	0
	Q17	145	**131**	14	**112**	33	**142**	3	0	0
	Q18	145	**84**	61	**73**	72	**142**	3	0	0

The number (*U*) and percentage of uncertain responses (U%) for each question, based on the number of UNKNOWN responses. Bold indicates questions and statistics depending on thresholds.

a contamination risk assessment procedure, based on the geochemical knowledge generated by modeling described in the previous sections.

Two types of data are used in this study. Waste site data include: (1) location of mine waste sites, (2) composition of mine waste including sulfides, toxic metals and dangerous processing substances (Q2–Q4), (3) geometry of the waste heap (height and area) and slope of foundation (Q6–Q10), and (4) other data such as presence of impermeable layer beneath the waste site, and if the facility is uncovered and the waste is thus exposed to wind or direct contact (Q13, 14). Information on the mine waste facility engineering design was obtained from mine archives, aerial photos and field studies. Spatial and census data include topographic data of location of settlements as polygons, surface water courses and slope data, calculated from 50 m digital elevation model using the ILWIS® 3.7 open source raster GIS software and census data from 2009. Data on the national protected areas (Natura 2000, etc.) and the location and status classification of groundwater bodies in Hungary under the Water Framework Directive (WFD, Directive 2006/21/EC) were also obtained. Land use/land cover data was obtained from the European CORINE Land Cover website. Presence of permeable layer beneath the mine waste site (Q12) was derived from the 1:100,000 national surface geological map. Polygons of the mine waste sites were overlaid by Google Earth® aerial photographs (2010–2011) in order to identify whether or not the material within the mine waste sites is exposed to wind (Q13), or covered or not (Q14). A detailed review and comparison of available contamination risk assessment methods for mining is presented in the review paper by Jordan and Abdaal (2013).

6.1   The EU pre-selection mine waste contamination risk assessment method

The EU MWD Pre-selection Protocol (Stanley *et al.*, 2011) is based on a 'YES-or-NO' questionnaire, and consists of 18 questions using simple criteria available in existing databases, readily enabling the preliminary screening of mine waste sites for environmental risk. In the case of lack of knowledge or information, i.e. in the presence of uncertainty, an 'UNKNOWN' response is entered for the particular parameter, which is the same as a YES response, and the site is selected for further examination which is a precautionary position. This screening should result in the elimination of those sites which do not cause, or have the potential to cause, a serious threat to human health and the environment from the inventory of closed waste sites. Note that even if a waste facility passes the pre-selection protocol, and it is classified as EXAMINE FURTHER, it does not mean that the closed waste facility will necessarily be included in the final inventory. The EU MWD Pre-selection Protocol 18 risk parameters or 'questions' (marked with 'Q') are arranged in four sections as follows: (1) Known serious impact (Q1); (2) Source: waste contains sulfide minerals (Q2), heavy metals (Q3) or the mine uses dangerous chemicals (Q4), type of the facility (tailings pond or waste heap) (Q5), area of the tailings pond ($>10.000\,m^2$) (Q6) and its height ($>4\,m$) (Q7), area of the waste heap ($>10.000\,m^2$) (Q8), its height is ($>20\,m$) (Q9) and the topographic slope under the waste heap ($\geq 5°$) (Q10); (3) Pathway: surface water course distance from mine waste site ($\geq 1\,km$) (Q11), high permeability layer beneath the mine waste site is present (Q12), waste material is exposed to air (Q13) and waste site is uncovered allowing direct contact (Q14); (4) Receptors: human settlement is present (>100 people, $\geq 1\,km$) (Q15), distance of groundwater body in 'poor status' ($\geq 1\,km$) (Q16), distance to Natura 2000 site ($\geq 1\,km$) (Q17), and distance of the nearest agricultural area ($\geq 1\,km$) (Q18).

The possible responses to each question are YES, NO or UNKNOWN. A YES answer means the presence of a risk factor, such as a toxic metal in the waste, the potential of transport by groundwater, or a nearby located settlement as a receptor. An UNKNOWN response indicates uncertainty in information, and uncertainty implies risk. Thus, UNKNOWN follows the same route as the YES response, pointing towards further examination according to the precautionary principle. If there is at least one YES or UNKNOWN response in each of the three sections of source, pathway and receptor, the assessor is directed to the EXAMINE FURTHER endpoint. This case means that there possibly exists a contamination source, at least one possible pathway, and a sensitive receptor. If the answers to all questions in at least one section are NO, then the source-pathway-receptor chain is broken, no risk exists for the site and the assessor is directed to NO NEED TO EXAMINE FURTHER end point. Threshold values, such as distance to pathways or sensitive receptors, topographic slope and census data, are defined for some of the key parameters in the Protocol. For example, if there is a stream or protected ecosystem within 1 km of the site, or there is a nearby settlement with more than 100 inhabitants, the site potentially bears high risk. The Protocol thresholds are based on the Irish regulation for the operation of ponds with respect to quarries (Safe Quarry, 2008).

6.2 Application of the EU Pre-Selection Protocol

In this example, a detailed statistical analysis is carried out using the 145 ore mine test cases, and the original 1 km threshold value is modified to the values identified as natural breaks in the distance histograms (see Figure 6.12). The lowermost break in the histogram identifies sites that are located within the closest distance, and

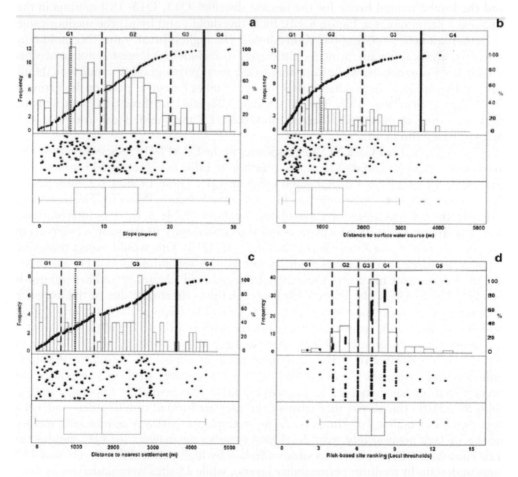

Figure 6.12 One-variable analysis (including histogram, scatterplot, box and whisker and quantile plot) was performed on the slope and distance measurements for the Pre-Selection Protocol a. for the slope (Q10), b. for the distance to the nearest surface water course (Q11), c. for the distance to the nearest settlement (Q15), d. for the total site ranking classes based on the number of YES responses and using local (median-based) thresholds. Solid line indicates the highest group of the local thresholds (after Abdaal et al., 2013).

therefore, these have the highest risk. In this way, the distance threshold is adopted to, for example, the settlement and stream course density conditions in Hungary. Also, the median of the 145 distances is calculated for all threshold-limited parameters, allowing a threshold estimation representing a 50% probability of the site falling within the risk-limiting distance (median-based threshold). The same calculations are performed for the census and slope data. Therefore, each Member State can choose a different threshold, which can meet their particular topographic and census conditions.

The contamination RA, according to the EU MWD Pre-Selection Protocol, is carried out in two runs. The first run uses the original EU thresholds (slope $\leq 5°$, 1 km distance and number of people in the nearest settlement ≥ 100). The second run uses local thresholds defined by (1) the highest natural break in the parameter [slope (Q10) and the lowest natural break for the nearest distance (Q11, Q15–18)] minima in the frequency histogram, see Figure 6.12); local threshold, and by (2) the median value of these parameters (median-based threshold). The highest break value threshold represents the precautionary principle, and tries to include the largest number of sites for further examination, while adjusting to the local physiographic conditions (Hungary in this study). The median-based threshold takes a neutral position by giving a 50% chance of relative risk. This test results altogether in three final selections of sites according to the three different thresholds (EU threshold, local threshold and median-based threshold).

The YES, NO and UNKNOWN responses of the EU MWD Pre-Selection Protocol are registered and calculated for each question in Table 6.1. Out of 145 mine waste sites, only 19 sites have a documented incident (Q1). These 19 sites are immediately directed to further examination in the EU MWD Pre-Selection Protocol. The remaining risk parameters (questions) are calculated similarly (Table 6.1). It is interesting to note that there is lack of information and thus uncertainty in the simple engineering properties of abandoned mine waste facilities (Q5–Q10). One would expect that mine archives of former active mines shall contain readily this information. For example, in Q9, 9 waste rock heap sites are >20 m in height, and 12 sites (9%) have unknown heights. The height of the waste rock heap is hard to determine due to the irregular geometry of the rock mass over a sloping terrain. The slope of the foundation upon which the waste rock heap rests is of concern with respect to stability. The greater the slope angle, the greater the risk of waste heap failure. The EU threshold chosen is 1:12 which equates to 8.3% or a slope angle of almost 5°. Based on the slope values derived from the 50-m DEM, 110 waste heap sites with YES responses are greater than or equal to 1:12 (5°) in slope and 26 sites with NO responses are less than 5° (Q10). This shows that most of the sites are located in hilly areas under the topographic conditions of Hungary. As an example for pathway parameters, the use of the surface permeability map developed to generate answers for Q12 resulted in 120 sites with YES responses (3 sites underlain by high-permeability layers and 117 sites underlain by medium-permeability layers), while 25 sites were underlain by low-permeability layers. Under the local physiographic conditions of high-density drainage network, for Q11, 64 sites are within 1 km distance to the nearest surface water bodies (streams or lakes) (Table 6.1). A preliminary risk-based site ranking is possible based on the EU thresholds (slope of almost 5° and 1 km distance) by counting and ranking the YES responses of the Pre-Selection Protocol, and ranging in scores from 3 to 12

Table 6.2 Site ranking classification based on the number of YES responses of the EU Pre-Selection Protocol using the original EU thresholds and the local median-based thresholds with risk classes according to Figure 6.12d. The number of waste sites in each class is also shown (after Abdaal et al., 2013).

Class	EU thresholds	Number of sites	Local thresholds	Number of sites
5	3–4	13	2–3	9
4	5	41	4–5	25
3	6–7	48	6	35
2	8–9	28	7–8	62
1	10–12	15	9–13	20
No pathway	18		16	
Examine further	127		129	

in each site (Table 6.2). Obviously, the site has a higher risk if there is more than one dangerous substance at the source or there are multiple contamination pathways and receptors.

6.3   EU MWD Pre-Selection Protocol with local thresholds

A distribution analysis has identified various sub-groups in the studied parameter thresholds (topographic slope, distance and census data; Figure 6.12). For example, in Q10 (Figure 6.12a), 3 sites have a topographic slope greater than 25°, 8 sites a slope of 20–25°, 64 sites a slope of 9–20°, and 70 sites a slope of less than 9°. This result suggests the 9° slope as a natural threshold reflects the local (Hungarian) conditions, instead of the original 5° slope threshold. Also, there are 11 (8 + 3) sites located on very steep slopes above 20°, which may single out these sites for specific attention in terms of slope movement and facility stability. According to Figure 6.12b (Q11), 57 sites are within a distance less than 500 m to the nearest surface water bodies, 66 sites are within a distance of 531–1,997 m, 19 sites within 2,029–3,014 m, and 3 sites are within a distance of 3,014–4,021 m. This shows that almost half of the mine waste sites are significantly (at the 90% confidence level) closer ($\leq$ 500 m) to receiving streams than the other sites, specifying these sites for a more detailed surface transport modeling if identified for 'further examination' in the EU MWD Pre-selection Protocol. Moreover, the second group of 531–1,997 m distance contains the initial 1 km threshold and thus the 2 km (1,997 m) threshold may better reflect the local topographic conditions for this question. In Q15 (Figure 6.12c), 33 sites with a population of more than 820 inhabitants are within a distance less than 680 m to the nearest settlement, indicating that these sites require prime attention if settlement protection is the concern. It is interesting that 25 sites lie directly above the groundwater bodies with 'poor status' (Q16), and 91 sites are located inside the protected Natura 2000 sites (Q17). The surprisingly high portion (63%) of mine waste sites lying directly in protected ecosystems calls for immediate special attention if landscape protection is a priority, while 81 sites are within a distance of less than or equal to 861 m to the nearest agricultural areas in Q18.

The neutral local thresholds based on median values (median-based threshold), selecting half of the sites for YES response, yield 10° for the slope below the waste site (Q10), 760 m for the distance to surface water bodies (Q11), and 1,722 m for the distance to settlements with 820 inhabitants (median-based; Q15). This is all consistent with the fact that mining areas lie in forested hilly areas with high-density drainage networks and sparse population. Sites are located on steep 5°–10° slopes, close (760 m < 1 km) to an abundant stream network, and with settlements remote (1,722 m >> 1 km) from mine sites. The settlement population cut-off value is much higher than the initial EU value (820 >> 100 inhabitants) since people live in villages in Hungary unlike farm areas in Ireland. This calls for a stringent catastrophe response in the case of civil protection and rescue. The 6,044 m distance to the nearest groundwater bodies with 'poor status' (Q16) however is reassuring, unlike the median distances of 470 m to Natura 2000 sites (Q17), and 612 m agricultural areas (Q18). The distribution analysis was performed on the population census data of Hungary (census 2009), to develop a population threshold number for Q15 of the EU MWD Pre-Selection Protocol, resulting in 53 classes ranging from <45 to >45,000 persons of the two extreme groups. The analysis indicates that 1,670 of the total 3,157 settlements with less than or equal to 820 persons represent 53% of the total number of settlements in Hungary. Therefore, this number, 820 persons, is a reasonably representative choice as a local (median-based) threshold for the population in Q15. By running the EU MWD Pre-Selection Protocol using these local (median-based) thresholds, the YES, NO and UNKNOWN responses were compared to those of EU thresholds as depicted in Table 6.1. Table 6.1 shows that the number of waste sites with YES responses of the EU MWD Pre-Selection Protocol varies from using the EU thresholds to local (median-based) thresholds. For example, sites on an underlying terrain slope with YES responses (Q10) decreased from 110 (EU thresholds) to 74 (local (median-based) thresholds), and to two sites with the highest threshold group, while in Q11 on the distance to the nearest surface water course, the sites with YES responses increased from 64 (EU thresholds) to 73 (local median-based threshold), and 144 (the highest group). The local threshold of the highest distance group boundary represents the worst-case scenario by selecting the possible largest number of sites for a YES response, and therefore for further examination based on a reasonable level of risk, indicated by solid lines in Figure 6.12a–c. Thus, this threshold selection follows the precautionary principle. Summing up, according to the existing pre-screening risk assessment of the mine waste sites in Hungary, 127 mine waste sites are directed toward EXAMINE FURTHER based on the EU thresholds (Table 6.2), 18 sites have no risk (these sites have no pathway), while 129 sites are directed to EXAMINE FURTHER based on the local (median-based) thresholds, 16 mine waste sites have no risk (these sites have no pathway). In the case of using the local threshold (lowest group boundary) in Q10 (5°), Q11 (270 m), Q15 (319 m), Q16 (0 m), Q17 (0 m) and Q18 (167 m), 118 sites are directed toward EXAMINE FURTHER, and 27 sites have no risk (19 sites with no pathway and 8 sites with no receptor). Using the local (highest group boundary) threshold, all 145 mine waste sites are directed toward EXAMINE FURTHER in Q10 (29°), Q11 (3,643 m), Q15 (4,083 m), Q16 (13,635 m), Q17 (2,732 m) and Q18 (3,956 m). It is obvious that this threshold selection represents the worst-case scenario and follows the precautionary principle.

6.4 Sensitivity and uncertainty analysis of the EU MWD Pre-Selection Protocol

Uncertainty analysis is indispensable in the assessment of environmental hazard, exposure and the consequent risks to human health, and it occurs at every stage in these assessments (Ramsey 2009). It causes an increased risk of incorrect decisions being made in the assessment, particularly if uncertainty is ignored in a deterministic approach, or merely underestimated in a probabilistic one. In this example, uncertainty assessment is limited to the UNKNOWN responses (U) in each question of the EU MWD Pre-Selection Protocol, due to missing site-specific data. The number of uncertain responses is simply counted for each site. The higher the number of uncertain responses for a site, the higher the risk the site bears due to lack of information, and it requires a more detailed further examination in the follow-up Tier 1 RA. Similarly, the number of uncertain responses can be lumped for each question, which provides an overall indicator of parameter uncertainty. For example, if a question receives the response UNKNOWN for ten sites, it represents a more uncertain parameter than a question for which all sites have reliable data available for a solid YES/NO answer. According to a preliminary site ranking based on the number of UNKNOWN responses (U), which ranges from 0 to 2 U responses in the sites, and results in 125 sites which have no uncertain responses (U = 0), 7 sites have one (U = 1) and 13 sites have two (U = 2), using the EU threshold and local median-based threshold within the EU Pre-Selection Protocol. Table 6.1 indicates that UNKNOWN (U) responses are located only in the source questions in the EU MWD Pre-selection Protocol, ranging from 3% in Q2 (presence of sulfide minerals in waste), and Q3 (toxic element potential in waste), and 7% in Q8 (size of the waste heap), to 33% in Q7 (height of dam of the tailings pond). Thus, relaxing the source questions, the percentage of uncertain responses (U%) is reduced to zero. This is the most unexpected outcome of this study because high certainty about the source, i.e. the mine waste facilities, was expected due to the assumed mining engineering archive documentation. One explanation is that mining flourished in the period of centrally directed economy in the 1950s to 1980s, when waste treatment and environmental issues were not among the priorities, leading to poor documentation of related facilities. This is confirmed by the surprising fact that the overwhelming majority of mine sites have no environmental monitoring data whatsoever available. In order to identify the key parameters and to check the sensitivity (in terms of final selection for further examination), by removal of parameters (questions of the MWD Pre-Selection Protocol) from Q2 to Q18, the number of YES responses is recalculated in the other questions for all sites using the EU and local median-based thresholds. By removal of question Q1 (whether the site has a known impact with documented incident), there is no change to the total source-pathway-receptor site ranking, because the 19 sites with known impact are directed to 'Examine Further' in one step. For the Source Q2 to Q10, the removal of Q2 directs 125 sites to 'Examine Further' using EU thresholds, while 141 sites to 'Examine Further' using local median-based thresholds. In Q3, 126 sites were directed to 'Examine Further' using EU thresholds, while 136 sites to 'Examine Further' using local thresholds. In Q4, Q5, Q6, Q7 and Q9, 126 sites were directed to 'Examine Further' using EU thresholds, while 142 sites with 'Examine Further' using local thresholds. In Q8, 125 sites with 'Examine Further' using EU thresholds, while 141 sites with 'Examine Further' using local thresholds.

In Q10, 120 sites went to 'Examine Further' using EU thresholds, while 139 sites to 'Examine Further' using local thresholds. For the Pathway Q11 to Q14, by removal of Q11, 127 sites were directed to 'Examine Further' using EU thresholds, while 139 sites to 'Examine Further' using local thresholds. In Q12, 69 sites went to 'Examine Further' using EU thresholds, while 92 sites to 'Examine Further' using local thresholds. In Q13 and Q14, 127 sites were directed to 'Examine Further' using EU thresholds, while 142 sites to 'Examine Further' using local thresholds. For the Receptor Q15–Q18, the removal of Q15 and Q16 directed 127 sites to 'Examine Further' using EU thresholds, and 142 sites to 'Examine Further' using local thresholds. In Q17, 74 sites went to 'Examine Further' using EU thresholds, while 140 sites to 'Examine Further' using local thresholds. In Q18, 124 sites were directed to 'Examine Further' using EU thresholds, while 128 sites to 'Examine Further' using local median-based thresholds. The key parameters as depicted above are Q3 (if sites are producing minerals with toxic heavy metals) and Q10 (slope) for source questions, Q12 (presence of higher-permeability layer beneath the waste site) for pathway, and Q17 (distance to the nearest surface water course) and Q18 (distance to the nearest agricultural areas) for receptor questions. The final selection of sites for further examination will be sensitive to and depends most strongly on these parameters.

In order to estimate the spatial uncertainty in distance measurements, buffer distances 100, 200, 500, 1,000, 1,500 and 2,000 m were delineated to pathway Q11, and the receptor Q15–18, and the number of waste sites were counted within each buffer distance. In terms of the distance to the nearest surface water course (Q11), 6 sites are within 100 m, 9 sites within 200 m, 20 sites within 500 m, and 30 sites within 1,000 m. In terms of the distance to the nearest settlements (Q15), 4 sites are within 100–200 m, 10 sites within 500 m, 18 sites within 1,000 m, 34 sites within 1,500 m, and 42 sites within 2,000 m. In terms of the distance to the nearest groundwater bodies (Q16), 24 sites are within 100–200 m, 25 sites within 500 m, 26 sites within 1,000 m, and 30 sites within 1,500–2,000 m. As far as the distance to the nearest Natura 2000 sites is concerned (Q17), 9 sites are within 100 m, 12 sites within 200 m, 20 sites within 500–1,000 m, 24 sites within 1,500 m, and 26 sites within 2,000 m. Concerning the distance to the nearest agricultural areas (Q18), 22 sites are within 100–200 m, 37 sites within 500 m, 47 sites within 1,000 m, 55 sites within 1,500 m, and 63 sites within 2,000 m. It is obvious from the above that there is no change in the number of sites from 100 to 200 m buffer distance. There is no big change in Q16 from 100 to 2,000 m. For Q11, the number of sites has increased from 100 to 1,000 m while only four sites increased up to 2,000 m. In Q18, there is a continuous increase in the number of sites from 200 to 2,000 m. Moreover, most of the digital topographic maps used in this study have a 1:100,000 scale; therefore, ±100 m will be reasonably accepted as spatial uncertainty in the distance measurements. For the topographic slope (Q10), an increase of the slope from 1° to 5° (EU threshold), decreased the number of sites from 138 to 111. 78 sites are risky at 9°, 74 sites at 10° and 69 sites at 11°. The trend continues and the number of risky sites decreases to 39 at 15°, to 11 sites at 20° and to 3 sites at 25°.

ACKNOWLEDGEMENTS

The authors are grateful to the colleagues contributing to the research work summarized in this chapter: Ahmed Abdaal, Andras Bartha, Andrea Farsang, Ubul Fügedi,

Ákos Horváth, Jorg Matschullat, Aniko Somody, Csaba Szabo, Péter Szilassi, Andrea Szücs, Anton Van Rompaey, Peter Völgyesi and Dora Zachary. The support of the Hungarian National Scientific Research Foundation (OTKA, Grant No. PD115810) is acknowledged. This paper is an original summary of the authors' previous and on-going research work partly described in Journal of Environmental Radioactivity, Applied Geochemistry, Water Air Soil Pollution and Journal of Land Contamination and Reclamation.

REFERENCES

Abdaal, A., Jordan, G. & Szilassi, P. (2013) Testing Contamination Risk Assessment Methods for Mine Waste Sites. *Water Air Soil Pollution*, 224, 1416.

Appleton, J.D., Miles, J.C.H. & Young, M. (2011) Comparison of Northern Ireland radon maps based on indoor radon measurements and geology with maps derived by predictive modeling of airborne radiometric and ground permeability data. *Science of the Total Environment*, 409, 1572–1583.

Balabanova, B., Stafilov, T., Šajn, R. & Baceva, K. (2011) Distribution of chemical elements in attic dust as reflection of their geogenic and anthropogenic sources in the vicinity of the copper mine and flotation plant. *Archives of Environmental Contamination and Toxicology*, 61, 173–184.

Banerjee, A.D.K. (2003) Heavy metal levels and solid phase speciation in street dusts of Delhi. India. *Environmental Pollution*, 123, 95–105.

Barnet, I. (2008) *Radon in Geological Environment: Czech Experience.* Czech Geological Survey, Prague.

Bats, M. (2006) *Optimal Land Use for Sediment-bound Heavy Metal Export in a Historical Mining Area, Recsk, Hungary.* M.Sc. Thesis, Catholic University of Leuven (in Dutch).

Baykut, S., Akgül, T., Inan, S. & Seyis, C. (2010) Observation and removal of daily quasi-periodic components in soil radon data. *Radiation Measurements*, 45, 872–879.

Bertolo, A. & Verdi, L. (2001) Validation of a geographic information system for the evaluation of the soil radon exhalation potential in South-Tyrol and Veneto (Italy). *Radiation Protection Dosimetry*, 97, 321–324.

Bossew, P., Dubios, G. & Tollefsen, T. (2008) Investigations on indoor Radon in Austria, part 2: geological classes as categorical external drift for spatial modeling of the Radon potential. *Journal of Environmental Radioactivity*, 99, 81–97.

Brunori, C., Cremisini, C., Massanisso, P., Pinto, V. & Torricelli, L. (2005) Reuse of a treated red mud bauxite waste: studies on environmental compatibility. *Journal of Hazardous Materials*, 117, 55–63.

Charlesworth, S., Everett, M., McCarthy, R., Ordonez, A. & De Miguel, E. (2003) A comparative study of heavy metal concentration and distribution in deposited street dusts in a large and a small urban area: Birmingham and Coventry, West Midlands, UK. *Environment International*, 29, 563–573.

Cizdziel, J.V. & Hodge, V.F. (2000) Attics as archives for house infiltrating pollutants: trace elements and pesticides in attic dust and soil from southern Nevada and Utah. *Microchemical Journal*, 64, 85–92.

Contin, M., Mondini, C., Leita, L. & De Nobili, M. (2007) Enhanced soil toxic metal fixation in iron (hydr)oxides by redox cycles. *Geoderma*, 140, 164–175.

Davis, J.E. (1986) Statistics and data analysis in geology. Chichester, Wiley.

Davis, J.J. & Gulson, B.L. (2005) Ceiling (attic) dust: a "museum" of contamination and potential hazard. *Environmental Research*, 99, 177–194.

Directive 2006/21/EC the European Parliament and of the Council on the management of waste from extractive industries and amending Directive 2004/35/EC. Commission of the European Communities, Brussels. [Online] http://eur-lex.europa.eu/LexUriServ/LexUriServ.do?uri= CONSLEG:2006L0021:20090807:EN:PDF. [Accessed 13rd April 2015].

Dubois, G., Bossew, P., Tollefsen, T. & De Cort, M. (2010) First steps towards a European atlas of natural radiation: status of the European indoor radon map. *Journal of Environmental Radioactivity*, 101, 786–798.

Duong, T.T.T. & Lee, B.-K. (2011) Determining contamination level of heavy metals in road dust from busy traffic areas with different characteristics. *Journal of Environmental Management*, 92, 554–562.

Durridge Company Inc. (2000) *RAD7 Electronic Radon Detector, User Manual.* [Online] www.durridge.com/documentation/RAD7Manual.pdf. [Accessed 13rd April 2015].

EEA, European Environment Agency (2005) *Towards an EEA Europe-wide assessment of areas under risk for soil contamination Vol. 3. PRA.MS: scoring model and algorithm.* [Online] http://sia.eionet.europa.eu/activities/reportste/PRAMS3. [Accessed 13rd April 2015].

Farsang, A. (1996) *Heavy Metal Distribution in SoUs in a Study Area in the Mátra Mts. with Special Emphasis on Anthropogenic Sources.* Ph.D. Thesis, Szeged University, Szeged (in Hungarian).

Fordyce, F.M., Brown, S.E., Ander, E.L., Rawlins, B.G., O'Donnell, K.E., Lister, T.R., Breward, N. & Johnson, C.C. (2005) GSUE: urban geochemical mapping in Great Britain. *Geochemistry: Exploration, Environment, Analysis*, 4, 325–336.

Gaál, G. (ed.) (1988) *Exploration target selection by integration of geodata using statistical and image processing techniques: An example from Central Finland.* Geochemical Survey of Finland, Report of Investigation 80(1).

Garrett, R.G. & Goss, TI. (1979) The evaluation of sampling and analytical variations in regional geochemical surveys. In Wattenson, J.R. & Theobald, P.K. (eds.) *Geochemical Exploration 1978: 7th International Geochemical Exploration Symposium. Special Publications*, 7, 374–384. Toronto, Association of Exploration Geochemists.

Garrett, R.G. (1983) Sampling methodology. In: Howarth, R.J. (ed.) *Handbook of exploration geochemistry, Vol. 2, Statistics and data analysis in geochemical prospecting.* Amsterdam, Elsevier.

Garrett, R.G. & Sinding-Larsen, R. (1984) Optimal composite sample size selection: Applications in geochemistry and remote sensing. In: Björklund, A.J. (ed.) Geochemical Exploration 1983: 10th International Geochemical Exploration Symposium. *Journal of Geochemical Exploration*, 21, 421–435.

Gedeon, A. (1962) *Geochemical survey in the Mátra Mts. Annual Report of the Geological Institute of Hungary.* Budapest, Geological Institute of Hungary (in Hungarian with English abstract). pp. 337–348.

Glodek, A. & Pacyna, J.M. (2009) Mercury emission from coal-fired power plants in Poland. *Atmospheric Environment*, 43, 5668–5673.

Goodarzi, F. (2009) Environmental assessment of bottom ash from Canadian coal fired power plants. *Open Environmental & Biological Monitoring Journal*, 2, 1–10.

Gosar, M. & Miler, M. (2011) Anthropogenic metal loads and their sources in stream sediments of the Meza River catchment area (NE Slovenia). *Applied Geochemistry*, 26, 1855–1866.

Gosar, M., Šajn, R. & Biester, H. (2006) Binding of mercury in soils and attic dust in the Idrija mercury mine area (Slovenia). *Science of the Total Environment*, 369, 150–162.

Government Decree No 6/2009. Joint Government Decree on the environmental standards for Earth materials, B pollution threshold. [Online] http://www.complex.hu/jr/gen/hjegy_doc. cgi?docid=A0900006.KVV. [Accessed 13rd April 2015].

Guibas, L. & Stolfi, J. (1985) Primitives for the manipulation of general subdivisions and the computation of Voronoi diagrams. *ACM Transactions on Graphics*, 4, 74–123.

Gyalog, L. (ed.) (1996) *Signal Code of the Geological Maps and Short Description of the Stratigraphical Units I* (in Hungarian). Budapest, Geological Institute of Hungary. Special Paper 187.

Gyalog, L. & Síkhegyi, F. (eds.) (2010) *1:100.000 Geological Map of Hungary*. Budapest, Geological Institute of Hungary. (CD-ROM).

Hlawiczka, S., Kubica, K. & Zielonka, U. (2003) Partitioning factor of mercury during coal combustion in low capacity domestic heating units. *Science of the Total Environment*, 312, 261–265.

Hoaglin, D.C., Mosteller, F. & Tukey, J.W. (1983) *Understanding Robust and Exploratory Data Analysis*. New York, John Wiley and Sons Inc.

Howarth, R.J. & Sinding-Larsen, R. (1983) Multivariate analysis. In: Howarth, R.J. (ed.) *Handbook of exploration geochemistry, Vol. 2, Statistics and data analysis in geochemical prospecting*. Amsterdam, Elsevier.

Howarth, R.J. & Thompson, M. (1976) Duplicate analysis in geochemical practice: Part II. Examination of proposed method and examples of its use. *Analyst*, 101, 699–709.

Huggins, F. & Goodarzi, F. (2009). Environmental assessment of elements and polyaromatic hydrocarbons emitted from a Canadian coal-fired power plant. *International Journal of Coal Geology*, 77, 282–288.

Ilacqua, V., Freeman, N.C.J., Fagliano, J. & Lioy, P.J. (2003) The historical record of air pollution as defined by attic dust. *Atmospheric Environment*, 37, 2379–2389.

Jordan, G. & Abdaal, A. (2013) Decision support methods for the environmental assessment of contamination at mining sites. *Environmental Monitoring and Assessment*, 1859, 7809–7832.

Jordan, G., van Rompaey, A., Somody, A., Fügedi, U., Bats, M. & Farsang, A. (2009) Spatial Modeling of Contamination in a Catchment Area Impacted by Mining: a Case Study for the Recsk Copper Mines, Hungary. *Journal of Land Contamination and Reclamation*, 17, 415–424.

Jordan, G., Rukezo, G., Fügedi, U., Carranza, E.J.M., Somody, A., Vekerdy, Z., Szebenyi, G. & Lois, L. (2003) Environmental impact of metal mining on catchment drainage in the historic mining area of Recsk-Lahóca mines, Hungary. *Proceedings, 4th European Congress on Regional Geoscientific Cartography and Information Systems, 2003, Bologna, Italy*. Regione Emilia-Romagna, Servizio Geologico, Italy. pp. 713–715.

Jordan, G., Szucs, A., Qvarfort, U. & Szekely, B. (1997) Evaluation of metal retention in a wetland receiving acid mine drainage. In: Xuejin, Xie (ed.) *Proceedings of IGC 30, Geochemistry* 19. pp. 189–206.

Kabata-Pendias, A. (2000) *Trace Elements in Soils and Plants*. Boca Raton, CRC Press. p. 432.

Kemski, J., Siehl, A., Stegemann, R. & Valdivia-Manchego, M. (2001) Mapping the geogenic radon potential in Germany. *Science of the Total Environment*, 272, 217–230.

Kheboian, C. & Bauer, C.F. (1987) Accuracy of selective extraction procedures for metal speciation in model aquatic sediments. *Analytical Chemistry*, 59, 1417–1423.

Kolmogorov, A. (1933) Sulla determinazione empirica di una legge di distribuzione. *Giornale dell'Istituto Italiano degli Attuari*, 4, 83.

Kozma, K. (1996) *The history of the power plant in Ajka* (in Hungarian). Ajka, Pri-Comp Kft., 399.

Kruskal, W.H. & Wallis, W.A. (1952) Use of ranks in one-criterion variance analysis. *Journal of the American Statistical Association*, 47, 583–621.

Kurzl, H. (1988) Exploratory data analysis: recent advances for the interpretation of geochemical data. *Journal of Geochemical Exploration*, 30, 309–322.

Levene, H. (1960) Robust tests for equality of variances. In: Olkin, I., Hotelling, H. *et al.* (eds.) *Contributions to Probability and Statistics: Essays in Honor of Harold Hotelling*. Stanford University Press. pp. 278–292.

Li, L.Y. (2001) A study of iron mineral transformation to reduce red mud tailings. *Waste Management*, 21, 525–534.

Makridakis, S.G., Wheelwright, S.C. & Hyndman, R.J. (1998) *Forecasting: Methods and Applications*. 3rd edition. New York, Whiley Publisher.

Mann, H.B. & Whitney, D.R. (1947) On a test of whether one of two random variables is stochastically larger than the other. *Annals of Mathematical Statistics*, 18, 50–60.

Ódor, L., Horváth, I & Fügedi, U. (1996) Low-density geochemical survey of Hungary. In: Volume of abstracts, *Environmental Geochemical Baseline Mapping in Europe Conference, May 21–24, 1996, Spisska Nova Ves, Slovakia*. pp. 53–57.

Ottesen, RT, Bogen, J., Bölviken, B. & Volden, T. (1989) Overbank sediments: a representative sample medium for regional geochemical mapping. *Journal of Geochemical Exploration*, 32, 257–277.

Papp, B., Szakács, A., Néda, T., Papp, S. & Cosma, C. (2010) Soil radon and thoron studies near the mofettes at Harghita Bai (Romania) and their relation to the field location of fault zones. *Geofluids*, 10, 586–593.

Parkhurst, D.L. & Appelo, C.A.J. (1999) *User's Guide to PHREEQC – a Computer Program for Speciation, Batch-Reaction, One-Dimensional Transport, and Inverse Geochemical Calculations*. Water-Resources Investigations Report, 99-4259. Denver, USA, US Geological Survey.

Plant, I.A. & Raiswell, R. (1983) Principles of environmental geochemistry. In: Thornton, I. (ed.) *Applied Environmental Geochemistry*, 1–40. London, Academic Press.

Plant, J.A. & Ridgway, J. (1990) Inventory of geochemical surveys in Western Europe. In: Demetriades, A., Locutura, J. & Ottesen, R.T. (eds.) *Geochemical mapping of Western Europe towards the year 2000*. Pilot project report. Appendix 10, NGU Report: 90–115. Geological Survey of Norway.

Plant, J.A., Jeffrey, K., Gill, E. & Fage, C. (1975) The systematic determination of accuracy and precision in geochemical exploration data. *Journal of Geochemical Exploration*, 4, 467–486.

Ramsey, M.H. (1993) Sampling and analytical quality control (SAX) for improved error estimation in the measurement of Pb in the environment using robust analysis of variance. *Applied Geochemistry*, Suppl. Issue 2, 149–153.

Ramsey, M.H. (2009) Uncertainty in the assessment of hazard, exposure and risk. *Environmental Geochemistry and Health*, 31, 205–217.

Ramsey, M.H., Thompson, M. & Banerjee, E.K. (1987) Realistic assessment of analytical data quality from inductively coupled plasma atomic emission spectrometry. *Analytical Proceedings*, 24, 260–265.

Reimann, C., Birke, M., Demetriades, A., Filzmoser, P. & O'Connor, P. (eds.) (2014) The GEMAS Project Team. Chemistry of Europe's agricultural soils – Part A: Methodology and interpretation of the GEMAS data set. *Geologisches Jahrbuch* (Reihe B). Hannover, Schweizerbarth pp. 528.

Reimann, C., Filzmoser, P., Garrett, R.G. & Dutter, R. (2008) *Statistical Data Analysis Explained. Applied Environmental Statistics with….* Chichester, Wiley.

Rodgers, J.L. & Nicewander, W.A. (1988) Thirteen ways to look at the correlation coefficient. *American Statistician*, 42, 59–66.

Safe Quarry: Guidelines to the Safety, Health and Welfare at Work (Quarries) Regulations (2008) (S.I. No. 28 of 2008). Published by the Irish Health and Safety Authority.

Sajn, R. (2002) Influence of mining and metallurgy on chemical composition of soil and attic dust in Meza valley. Slovenia. *Geologija*, 45(2), 547–552. DOI: http://dx.doi.org/10.5474/geologija.2002.063. [Accessed 13rd April 2015].

Sajn, R. (2005) Using attic dust and soil for the separation of anthropogenic and geogenic elemental distributions in an old metallurgic area (Celje, Slovenia). *Geochemistry: Exploration, Environment, Analysis*, 5, 59–67.

Salminen, R., (Chief-editor), Batista, M.J., Bidovec, M., Demetriades, A., De Vivo, B., De Vos, W., Duris, M., Gilucis, A., Gregorauskiene, V., Halamic, J., Heitzmann, P., Lima, A., Jordan, G., Klaver, G., Klein, P., Lis, J., Locutura, J., Marsina, K., Mazreku, A., O'Connor, P.J., Olsson, S.A., Ottesen, R.T., Petersell, V., Plant, J.A., Reeder, S., Salpeteur, I., Sandstrom, H., Siewers, U., Steenfelt, A. & Tarvainen, T. (2005) *FOREGS Global Geochemical Baselines Programme. Geochemical Atlas of Europe.* [Online] http://weppi. gtk.fi/publ/foregsatlas. [Accessed 13rd April 2015].

Shah, P., Streezov, V., Prince, K. & Nelson, P.F. (2008) Speciation of As, Cr, Se, and Hg under coal fired power station conditions. *Fuel*, 87, 1859–1869.

Smetanová, I., Holy, K., Müllerová, M. & Poláskova, A. (2010) The effect of meteorological parameters on radon concentration in borehole air and water. *Journal of Radioanalytical and Nuclear Chemistry*, 283, 101–109.

Smirnov, N.V. (1948) Tables for estimating the goodness of fit of empirical distributions. *Annals of Mathematical Statistics*, 19, 279–281.

Somody, A. (2005) *Hydrogeological Analysis for the Long-term Suspension of the Recsk Deep Mines by Flooding.* Ph.D. Thesis, University of Miskolc (in Hungarian with English summary).

Somody, A. & Jordan, G. (2005) Mining impact or neutral environment? In: *Proceedings, MicroCAD2005 Conference, 2005, Miskolc, Hungary.* pp. 136–140.

Stanley, G., Jordan, G., Hamor, T. & Sponar, M. (2011) *Guidance document for a risk-based selection protocol for the inventory of closed waste facilities as required by Article 20 of Directive 2006/21/EC.* February, 2011. [Online] http://ec.europa.eu/environment/ waste/mining/pdf/Pre_selection_GUIDANCE_FINAL.pdf [Accessed 13rd August 2015].

Steegen, A., Govers, G.,Takken, I., Nachtergaele, J., Poesen, J. & Merckx, R. (2001) Factors controlling sediment and phosphorus export from two Belgian agricultural catchments. *Journal of Environmental Quality*, 30, 1249–1258.

Sundal, A.V., Valen, V., Soldal, O. & Strand, T. (2008) The influence of meteorological parameters on soil radon levels in permeable glacial sediments. *Science of the Total Environment*, 389, 418–428.

Swakon, J., Kozak, K., Paszkowski, M., Gradzinski, R., Loskiewicz, J., Mazur, J., Janik, M., Bogacz, J., Horwacik, T. & Olko, P. (2005) Radon concentration in soil gas around local disjunctive tectonic zones in the Krakow area. *Journal of Environmental Radioactivity*, 78, 137–149.

Szabó K.Z., Jordan G., Horváth Á. & Szabó Cs. (2014) Mapping the geogenic radon potential: methodology and spatial analysis for Central Hungary. *Journal of Environmental Radioactivity*, 129, 107–120.

Szabó, K.Z., Jordan, G., Horváth, Á. & Szabó, Cs. (2013) Dynamics of soil gas radon concentration in a highly permeable soil based on a long-term high temporal resolution observation series. *Journal of Environmental Radioactivity*, 214, 74–83.

Szucs, A. & Jordan, G. (1994) Analysis of sampling frequency in groundwater quality monitoring systems: a case study. *Water Science and Technology*, 30, 73–78.

Tessier, A., Campbell, PG.C. & Bisson, M. (1979) Sequential extraction procedure for the speciation of particulate trace metals. *Analytical Chemistry*, 51, 844–851.

Thornton, I. (ed.) (1983) *Applied Environmental Geochemistry.* London, Academic Press.

Tollefsen, T., Gruber, V., Bossew, P. & De Cort, M. (2011) Status of the European indoor radon map. *Radiation Protection Dosimetry*, 145, 110–116.

Tóth, E., Hámori, K. & Minda, M. (2006) Indoor radon in Hungary (lognormal mysticism). In: Barnet, I., Neznal, M., Pacherová, P. (eds.) *Radon Investigations in the Czech Republic XI and the 8th International Workshop on the Geological Aspects of Radon Risk Mapping.* Czech Geological Survey/Radon v.o.s./JRC-IESREM. Prague, Ispra. pp. 252–257.

Tukey, J.W. (1977) *Exploratory Data Analysis.* Addison-Wesley.

Van Oost, K., Govers, G. & Desmet, P. (2000) Evaluating the effects of changes in landscape structure on soil erosion by water and tillage. *Landscape Ecology*, 15, 577–589.

Van Rompaey, A.J.J., Notebaert, B., Bats, M., Jordan, G., Somody, A & Van Dessei, W. (2005) Optimal land use scenarios for the minimalization of polluted mining waste export: a case study in the uplands of the Tisza River (Hungary). In: *International Conference on European Union Expansion: Land Use Change and Environmental Effects in Rural Areas, 4–7 September 2005, Luxemburg*. Abstracts, p. 61.

Van Rompaey, A.J.J., Verstraeten, G., Van Oost, K., Govers, G. & Poesen, J. (2001) Modeling mean annual sediment yield using a distributed approach. *Earth Surface Processes and Landforms*, 26, 1221–1236.

Velleman, P.F. & Hoaglin, D.C. (1981) *Applications, Basics and Computing of Exploratory Data Analysis*. Boston, Duxbury Press.

VITUKI Consult Rt. (1996) *Environmental Re-examination of Recsk Deep-level Mine*. Report for the Recski Ércbányák Ltd., Miskolc (manuscript) (in Hungarian).

Völgyesi, P., Jordan, G., Zacháry, D., Szabó, C., Bartha, A. & Matschullat, J. (2014) Attic dust reflects long-term airborne contamination of an industrial area: A case study from Ajka, Hungary. *Applied Geochemistry*, 46, 19–29.

Völgyesi, P., Jordan, G., Zacháry, D. & Szabó, Cs. (2013) Complex urban geochemical analysis of attic dust samples in an industrial area, Ajka, Hungary. In: Goldschmidt Conference 2013, August 25–30, 2013, Florence (Italy), Abstract, USB Flash Drive, pp. 2426. [Online] http://goldschmidtabstracts.info/2013/2426.pdf. [Accessed 13rd April 2015].

Yudovich, Y.E. & Ketris, M.P. (2005) Arsenic in coal: a review. *International Journal of Coal Geology*, 61, 141–196.

Zacháry, D., Jordan, G. & Szabó, Cs. (2012) Geochemical study of urban soils in public areas of an industrialized town (Ajka, Western Hungary). In: *European Geosciences Union General Assembly April 22–27, 2012, Vienna (Austria), Geophysical Research Abstracts*, vol. 14, EGU2012-4762.

Chapter 7

Potential of cyclodextrins in risk assessment and monitoring of organic contaminants

É. Fenyvesi[1], Cs. Hajdu[2], K. Gruiz[2]

[1]CycloLab Cyclodextrin Research & Development Laboratory Ltd., Budapest, Hungary
[2]Department of Applied Biotechnology and Food Science, Budapest University of Technology and Economics, Budapest, Hungary

ABSTRACT

This chapter provides a brief overview of the literature on sensors and bioassays using cyclodextrins, the special carbohydrate molecules which form inclusion complexes with various organic chemicals. They can be applied in risk assessment and monitoring in various ways:

i)   Cyclodextrin derivatives bearing a chromophore/fluorophore moiety are especially selective and sensitive as sensors due to their changed fluorescence spectrum in the presence of an appropriate competitive contaminant.

ii) The samplers containing cyclodextrin collect and concentrate the target contaminants from the contaminated phases of chemical analyses and bioassays.

iii) Soil extraction with aqueous cyclodextrin solution has become an everyday method to predict the microbial bioavailability of organic contaminants.

iv) In toxicity assays or bioassays cyclodextrins can sensitize or attenuate the effects depending on the conditions of the tests.

1 INTRODUCTION

Cyclodextrins (CDs) are cyclic oligosaccharides produced by enzymatic degradation of starch. Although there are a lot of variants, those consisting of six, seven or eight glucopyranose units, called alpha-, beta- and gamma-CDs respectively, are those which are most frequently applied. Their structure is unique: all the hydroxyl groups are located on the rims. The primary and the secondary hydroxyls are situated at different rims giving the shape of truncated cones to the molecules (Figure 7.1). Therefore the cavity is rather hydrophobic, while both rims are hydrophilic (Szejtli, 1998). This special structure explains the inclusion complex-forming ability: hydrophobic guest molecules or moieties enter the cavity of 0.5–0.8 nm in diameter. The included guest of low aqueous solubility becomes water-soluble due to the hydrophilic surface of the CDs. Not only does the solubility of the included molecules change but so do a lot of other properties (chemical stability, volatility, spectral characteristics, etc.) as a consequence of wrapping them individually into the CDs (molecular encapsulation).

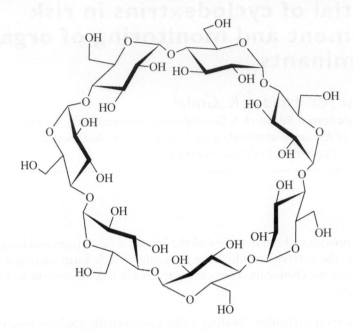

Figure 7.1 The structure of beta-cyclodextrin (BCD).

The high number of hydroxyl groups makes various chemical modifications possible. Single isomers can be prepared with special care but most of the CD derivatives contain a wide variety of randomly substituted isomers. Such products are characterized by the degree and pattern of substitution (Jicsinszky *et al.*, 1996). The degree of substitution (DS) indicates the average number of substituents in a CD molecule. The distribution pattern is typically a fingerprint chromatogram obtained by HPLC or capillary electrophoresis showing the ratio of the fractions of various DS. Depending on the type, number and pattern of the substituents, the properties of the CDs can be tailor-made: derivatives which are highly water-soluble and water-insoluble, but soluble in organic solvents, can be prepared. Various derivatives having fluorescent moieties have become important in risk assessment and monitoring methods. CDs immobilized either by crosslinking or by coupling them to a macromolecular surface are used as sorbents (samplers) and as membranes (films) in specific sensors.

The CD derivatives with high aqueous solubility show high solubilizing effect, too. Hydroxypropyl, randomly methylated and carboxymethyl beta-CDs (HPBCD, RAMEB and CMBCD respectively) are the most thoroughly studied solubilizing agents, of which the first two are industrially produced and widely used in household products, cosmetics, pharmaceutical formulations, etc. Their environmental application has steadily increased for approximately 20 years. Gruiz *et al.* (2011) recently reviewed their principles of application in the technologies of environmental risk management showing examples of utilization of CDs for intensification of soil and groundwater remediation and of application of CD-containing sorbents as samplers for contaminated soil gas and groundwater. The versatile benefits of CDs in analytical chemistry have been also reviewed (Szente & Szemán, 2013).

In this chapter we give an overview of how CDs can be applied to risk assessment and monitoring. All these applications are based on the complex-forming ability of CDs with high-K_{ow} organic compounds such as typical hydrocarbon contaminants, chlorinated organic solvents and pesticides (Szaniszló et al., 2005; Shirin et al., 2004; Villaverde et al., 2007). Chemical sensors with enhanced sensitivity can be obtained due to the spectral changes of special CD derivatives as a result of complex formation; the sensitivity of bioassays is also usually enhanced because of the increased solubility and bioavailability of the contaminants. The immobilized CDs can capture the organic molecules from gas or liquid phase in the samplers. The non-exhaustive extraction of soils with CD solutions gives information on the easy-to-desorb, bioavailable fraction of the contaminants in the soil and enhances the sensitivity of the bioassays. Application of CDs as a bioavailability-enhancing additive in bioassays represents a risk-based pessimistic approach.

2 CYCLODEXTRIN-SENSITIZED CHEMICAL SENSORS FOR ASSESSMENT AND MONITORING

The selectivity and sensitivity of sensors in detecting environmental contaminants in gas and water can be enhanced by applying CDs (Ogoshi & Harada, 2008). Detection based on various physico-chemical phenomena (absorption, fluorescence spectra, acoustic wave, surface plasmon resonance, electrical resistance and conductivity, etc.) can be sensitized by the inclusion complex formation, which alters the microenvironment of the analyte. CDs are derivatized either to enhance their adhesion to the sensor (e.g. electrode) surface or to improve the specificity of the detection. For the latter purpose, moieties emitting signals in the given detection system and forming self-inclusion complexes are attached to the CD cavity. In the presence of competitive environmental pollutants, which show higher affinity toward the CD cavity than the substituents, the signal will be changed.

Such sensors can be applied in early warning systems working near the source, at the hot spots along the transport route or at the receptors enabling the decision makers to respond quickly and effectively and protect public health in the event of contamination of the environment (Gruiz, 2009).

2.1 Enhancement of chemical detection sensitivity

CDs can enhance the sensitivity of fluorescent, piezo-electric, electrochemical or surface acoustic detection (Fenyvesi & Jicsinszky, 2009). Among others, pyrene (Alarie & Vo-Dinh, 1991), bromonaphthalene (Bissel & de Silva, 1991) and aromatic acids (Jyisy et al., 2005) were detected at much lower concentrations due to the enhanced fluorescence of the encapsulated fluorophores.

Piezo-electric sensors were fabricated for detecting p-nitrophenol by using sulfonated CD (Kanclerz et al., 1995), organic vapors such as benzene and cyclacene derivatives, nitrobenzene and other aromatic compounds by using (tert-butyldimethylsilyl)-alpha-CD (Thomas, 1990), and automotive gasoline in air by using beta-CD and TiO_2 together (Si *et al.*, 2005). The combined application of CDs and TiO_2 proved to be beneficial also in surface polarization impedance spectroscopy for detection of nitroaromatic compounds (Ju *et al.*, 2007).

CD-modified electrodes can be prepared either by incorporating CD into carbon paste or preparing membrane with crosslinking or immobilizing the proper CD derivative. Such sensors can sensitively detect *p*-nitrophenolate (Komiyama, 1988), carcinogenic polycyclic aromatic amines, aminopyrenes and hydroxypyrenes (limit of detection is in the range of 10^{-8} to 10^{-7} mol dm^{-3}) (Ferancová *et al.*, 2005), etc. Recently, further enhancement in selectivity and sensitivity has been achieved by combined application of CDs with carbon nanotubes for the analysis of benzene isomers, pentachlorophenol (PCP), organophosphate pesticides and herbicides (Yu *et al.*, 2010; Xu *et al.*, 2010; Du *et al.*, 2010; Rahemi *et al.*, 2012, 2013). Electrodes modified with gold nanoparticles bearing CDs attached to the surface of the nanoparticles *via* thiol groups provide enhanced sensitivity (1 nM) in detection of 2,4,6-trichlorophenol (Zheng *et al.*, 2015). Amino-CDs attached to the surface of graphene sheets are sensors of high sensitivity for persistent organic pollutants (POPs) with a detection limit of 1 nM for 1-aminonaphthalene as model POP (Zhu *et al.*, 2012). CD conjugated to fullerene was found to be highly sensitive to electrochemical detection of *p*-nitrophenol (Rather *et al.*, 2013). The outstanding molecular recognition of CDs can be further improved by molecular imprinting (by polymerizing the CD rings in the presence of the guest molecules to be determined and thus ensuring the best fit for the guest). Polymer layers of high sensitivity can be obtained in this way for selective detection of xylene isomers and polycyclic aromatic hydrocarbons (PAHs) in drinking water (Dickert *et al.*, 1999).

Species-selective, thin-film sensors based on surface-acoustic-wave transducers using trimethyl-CD were developed to monitor nitrotoluene (Swanson *et al.*, 1996).

Immobilized CD enhanced the sensitivity of surface plasmon resonance ensuring minimum detectable concentrations of estrogen and environmental chemicals at ppb level (Hattori *et al.*, 2007). Recently, silver nanoparticles embedded in CD-silicate composite were developed for sensing Hg(II) ions and nitrobenzene in the presence of other environmentally relevant metal ions (Manivannan & Ramaraj, 2013). Silver and gold were consecutively deposited on silica nanospheres; this surface was then modified with CD substituted with mercapto moieties to be used for selective detection of endocrine disrupting chemicals (EDCs) from river water, applying surface-enhanced Raman scattering (SERS). EDC model compounds, including 3-amino-2-naphthoic acid, potassium hydrogen phthalate and the EDC beta-estradiol, were captured by the CD-decorated surface and detected at a micromolar concentration (Fang *et al.*, 2013).

2.2   Sensors based on competitive complex formation

The fluorescent, UV and circular dichroism spectra of CD derivatives functionalized by chromophore or fluorophore groups, and of inclusion complexes with chromophore or fluorophore compounds, can change in the presence of guests with good complex-forming ability (Ueno, 1993; Ikeda *et al.*, 2007). The chromophore or fluorophore group is included in the cavity and will be displaced upon the effect of another guest molecule (Figure 7.2). These guest-induced changes in the spectra can be applied to the development of selective sensors with enhanced sensitivity.

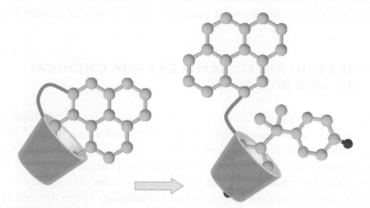

Figure 7.2 Schematic of the mechanism of sensing based on competitive complex formation (pyrene appended to the CD is complexed until another contaminant such as bisphenol A displaces it from the cavity, thus changing the fluorescence).

The types of fluorophore CD derivatives and their application have recently been reviewed (Malanga, 2011).

If pyrene/dimethyl beta-CD complex is embedded in plasticized poly(vinyl) alcohol film in the sensor, the high fluorescence of pyrene will be quenched in the presence of a guest, such as bisphenol A, competing for the CD cavities (Wang *et al.*, 2006). This sensor is suitable for continuous monitoring of compounds with a complex-forming ability better than that of pyrene. Bisphenol A can be detected at concentrations as low as 7×10^{-8} mol/L in surface water and soil leachates.

The fluorophore moiety can be bound covalently to the CD ring as shown in Figure 7.2. The pyrene, dansyl or disulfonyl dibenzosulfolane-diphenyl groups attached to CD enable the sensitive detection of polychlorinated biphenyls (PCBs), dioxin analogs, alkylphenols, 2,4-dichlorophenoxyacetic acid (2,4-D), bisphenol A, diethyl phthalate, etc. (Hamada *et al.*, 2001). Carbon tetrachloride and other alkyl chlorides, as well as the odorous geosmin and methyl isoborneol, can be sensed in surface waters by naphthyl beta-CD (Wang & Ueno, 2000). Indolizine beta-CD is useful for the detection of aromatic compounds such as benzene, toluene and other VOCs such as phenols and cresols (Fourmentin *et al.*, 2006). A fluorescein-modified methylated CD (FITC-RAMEB) was found to be highly sensitive to PCP (Fenyvesi & Jicsinszky, 2009).

2.3 Chiral sensors

A lot of contaminants, especially pesticides, are chiral. Since the two enantiomers might have different toxicity and might be biodegraded differently, it may be important to detect chirality. CDs are chiral compounds with a remarkable chirality-recognizing ability (the stability of the complexes formed with the two enantiomers is usually different); therefore a great number of chiral sensors recognizing chiral compounds have been developed (Szemán, 2007). These sensors are used primarily in pharmaceutical analysis, but they can play a role in environmental analysis too, especially in monitoring the enantioselectivity of biodegradation, which is of special importance when the bioaccumulation of the two enantiomers of pesticides and of their degradation products is different.

3   CYCLODEXTRIN-BASED SAMPLERS FOR CHEMICAL ANALYSIS AND BIOASSAYS

CD polymers and CDs immobilized on polymer surfaces are widely used as sorbents for the removal of contaminants from wastewater and contaminated groundwater (Fenyvesi & Balogh, 2009). Such sorbents are able to collect and concentrate the specific contaminants and can be used for further purification of purified wastewater, e.g. for the removal of pharmaceutical residues, pesticides, etc. (Nagy *et al.*, 2014). Their efficiency depends on the CD content and availability in the sorbent as well as on the type and concentration of the target contaminant. The application of such sorbents as samplers has not been very widely published.

Solid CD was applied to ambient air sampling based on complexation of PAHs in the gas phase (Butterfield *et al.*, 1996). CD as an additive to a polyurethane foam tube was used for sample collection in a modified high-volume air sampling method to monitor PAHs (Trevino *et al.*, 2005). Various CD-containing sorbents, such as CD-modified textiles (cotton and polyethylene terephthalate), CD-containing cellulose acetate films and epichlorohydrin-crosslinked CD polymer beads, as passive samplers showed enhanced sorption capacity in capturing volatile aromatic and chlorinated hydrocarbons from air (Gruiz *et al.*, 2011). The latter polymer beads proved to be useful also for sampling contaminated water and selectively binding certain contaminants such as bisphenol A. The captured contaminants can be extracted from the sampler by proper chemical extractant or by heating prior to analysis. Another possibility is to use the sampler as it is for ecotoxicological tests, but in this case the CD-entrapped contaminant might behave differently in the bioassays (see later in this chapter).

The CD-based samplers can be used for enrichment of the components to be measured by various analytical techniques. Especially in the case of chromatography, homogeneous samples free of interfering compounds are needed. The contaminants to be measured are selectively bound by the CD-containing sorbents filling a cartridge and can be eluted by an appropriate solvent. The eluted sample contains the contaminant at a higher concentration than the original sample, without the undesired impurities (Figure 7.3). Another possibility is to capture the interfering components, e.g. organochlorine pesticides DDT, DDE and DDD, from a sample containing PCBs prior to analysis by gas chromatography (Zhang *et al.*, 2014). This technique is known as solid-phase extraction (SPE), used frequently for sample preparation prior to chromatography. Its miniaturized implementation is solid-phase microextraction (SPME). In SPME the sorbent is applied on the surface of a syringe. The syringe is immersed in a liquid or is kept in the headspace above a solid or liquid sample to collect the contaminants from environmental samples in an amount sufficient to be detected. The syringe content saturated with the contaminant is then injected into the gas chromatograph (GC). The sorptive layer can be applied on a stir bar, making it possible to collect the target analyte compounds from large volume samples such as contaminated surface and groundwater by simple magnetic stirring (Figure 7.4). Some CD-based sorbents are listed in Table 7.1.

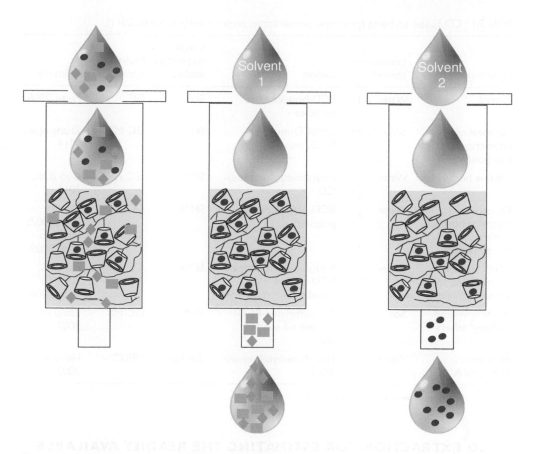

Figure 7.3 Solid-phase extraction (SPE) for cleaning and concentrating a sample by using a CD polymer: the solution containing various components is eluted through the cartridge, then the components non-interacting with CD are washed down by using solvent 1, and then the component(s) interacting with CD are eluted with solvent 2.

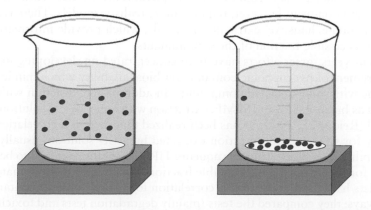

Figure 7.4 Concentrating the contaminants from a large-volume sample by using a stir bar coated with a CD polymer.

Table 7.1 CD-based sorbents for sample preparations prior to analysis (Oláh, 2011).

Contaminant	Environmental element	Sorbent	Sample preparation method	Analytical method	Reference
Phenol	Water	BCD-bonded silica particles	SPE	GC	Faraji, 2005
Removal of interfering components	Soil extract	CMBCD-modified Fe_3O_4 nanoparticles	SPE	GC-MS*	Zhang et al., 2014
Triazine herbicides	Water	Poly(dimethylsiloxane)-CD	SPME	GC, HPLC	Guo et al., 2011
Geosmin, methylisoborneol N-nitroso-dimethylamine	Water	BCD-polyurethane polymers	SPME	GC-MS	Mamba et al., 2007; Mhlongo et al., 2009
PAHs, phenolic compounds, amines	Water	Poly(dimethylsiloxane)/BCD-coated membrane	SPME	GC-MS	Hu et al., 2005; Fu et al., 2006
Polybrominated diphenyl ethers	Soil	Permethylated-BCD/silicone oil coated fiber	SPME	GC-MS	Zhou et al., 2007
Estrogens, bisphenol A	Water	Poly(dimethylsiloxane)/BCD	Stir bar	HPLC	Hu et al., 2007

*Gas chromatography—mass spectrophotometry

4   CD EXTRACTION FOR ESTIMATING THE READILY AVAILABLE (BIOAVAILABLE) FRACTION OF ORGANIC CONTAMINANTS

The total contaminant content of the soil is traditionally determined by exhaustive extraction of the soil sample prior to chemical analyses. Solvents which are able to extract as much and as many contaminants as possible are selected. Risk-based assessment of the environment, however, requires more realistic models. There is a growing demand for 'non-exhaustive' extraction methods, which provide information on the available or accessible fraction of the contaminants.

In recent years, great efforts have been concentrated on developing methods for the measurement of hydrophobic contaminant bioavailability which mimic microbial interactions with soil-associated compounds. In addition to extraction with mild solvents, such as butanol, SPE and SPME, extraction with aqueous CD solutions was also introduced (Reid *et al.*, 2000). It has been realized that there is a correlation between availability for CD (the concentration extracted by a CD solution, usually HPBCD) and availability for a biota. Thus the aqueous HPBCD solution seemed to be a suitable extractant for predicting the bioavailable fraction of an organic contaminant. Numerous scientists have tried to prove this correlation for various organic contaminants in different ways: they compared the tests (mainly degradation tests and toxicity tests) to CD's availability (Table 7.2).

Table 7.2 Comparison of the results of biodegradation tests and HPBCD extractability from soil and sediment (Hajdu et al., 2009a).

Contaminant	Environmental element	Bioassay	CD extraction	Result	Reference
Phenanthrene	Spiked soil	Biodegradation assay	HPBCD	Good correlation	Swindell & Reid, 2007
Phenanthrene, pyrene, benzo[a]pyrene	Spiked soil	Microbial mineralization	HPBCD	Good correlation	Papadopoulos et al., 2007
Phenanthrene, cresol	Silt, loamy sand, peat, sandy clay	Mineralization test with adapted Pseudomonas sp.	HPBCD	Good correlation	Allan et al., 2006
Naphthalene	Spiked soil	Biodegradation assay	HPBCD	Good correlation	Paton et al., 2009
PAHs	Contaminated sediment	Biodegradation assay (84 days)	HPBCD	Good correlation for low molecular weight PAHs only	Cuypers et al., 2002
PAHs	Spiked soil	Mineralization test with adapted Pseudomonas sp. (123 days)	HPBCD	Good correlation	Doick et al., 2005
PAHs	Spiked soil	Biodegradation assay (6 weeks)	HPBCD	Good correlation	Stokes et al., 2005
PAHs	Motorway site	Biodegradation assay	HPBCD	Good correlation	Johnsen et al., 2006
PAHs	Spiked soil	Biodegradation assay	HPBCD	Good correlation	Rhodes et al., 2010
PAHs	Spiked soil	Biodegradation assay	HPBCD	Good correlation for 3- & 4-ring PAHs only	Juhász et al., 2005
PAHs	Soil from gas plant, wood preservation site, etc.	Biodegradation assay	HPBCD	Good correlation for total PAHs	Rostami & Juhász, 2013
PAHs	Spiked sediment	Biodegradation assay	MeBCD*	Good correlation for phenanthrene	Spasojević et al., 2014
PAHs, cyanide, phenol compounds	Contaminated soil	Biodegradation assay (6 weeks)	HPBCD	Good correlation	Hickman et al., 2008
Creosote	Contaminated soil	Different soil microcosms tests (200 days)	HPBCD	Poor correlation	Sabaté et al., 2006
Diesel oil, black oil	Spiked soil	Closed bottle test	HPBCD	Good correlation	Fenyvesi et al., 2008
Diesel oil, transformer oil	Spiked soil	4 bioassays	HPBCD and RAMEB*	Good correlation	Molnár et al., 2009; Gruiz et al., 2009
Hexadecane	Spiked soil	Biodegradation assay	HPACD**	Good correlation	Stroud et al., 2009

*Methyl-beta-cyclodextrin **Hydroxypropyl-alpha-cyclodextrin

HPBCD was mostly used because it showed better correlation with the microbial decomposition than RAMEB. For the contaminants of linear hydrocarbons such as hexadecane, however, the smaller sized alpha-CD gives a better fit than beta-CD. Indeed, there was good correlation between hydroxypropyl-alpha-cyclodextrin (HPACD) extraction and the biodegradation tests for soils containing no other contaminants (Stroud *et al.*, 2009). In the case of mixed contaminants, the correlation is usually poorer because the other contaminants may enhance bioavailability but not extractability. In most cases, however, the availability for HPBCD extraction may mimic the availability for the microbes. With multi-component contaminated soils, Sabaté *et al.* (2006) examined the selectivity of HPBCD, the extent to which selected organic contaminants would be degraded with differently enhanced biodegradation microcosms (mixed and amended with N, P and K), and found that, with the exception of benz[*a*]anthracene and chrysene, there is a good correlation between biodegradable and HPBCD-extracted amounts of PAHs. Sabaté *et al.* (2006) explained this phenomenon with the recalcitrant nature of these compounds, but also because their diameter seems to be wider (diameter data: anthracene 0.5 nm, pyrene 0.71 nm, fluoranthene 0.71 nm, phenanthrene 0.58 nm) (Cuypers *et al.*, 2002, Wang & Brusseau, 1995) than the other compounds, which may be the reason for the difference in the complex inclusion ratio. These data were supported by molecular modelling studies (Morillo *et al.*, 2012). The geometrical fitting into the HPBCD cavity (diameter: 0.75 nm) seems to be important. Cuypers *et al.* (2002) went on to show that HPBCD primarily extracted the readily bioavailable PAHs from aged sediments without increasing the total fraction size and demonstrated that, based on HPBCD extraction, the biodegradability for low molecular weight PAHs (pyrene, fluoranthene, chrysene, benz[*a*]anthracene, etc.), but not for high molecular weight ones, could be predicted when compared to biodegradation studies. HPBCD extraction of creosote-contaminated soil was able to predict microbial degradability of the three- and four-ring PAHs; the biodegradability of five-ring PAHs was, however, overestimated (Juhasz *et al.*, 2005). Similarly, in the case of petroleum hydrocarbon contamination (diesel oil and mazut), the fractions extracted by HPBCD contained the low molecular weight components that are easy to desorb and taken up by the microbes (Molnár *et al.*, 2009).

It is well known that the size and shape of the target molecule and the HPBCD cavity need to be complementary for the formation of the inclusion complexes and this structure-selectivity can be a limiting factor. A comparison of HPBCD with RAMEB failed to show any significant difference in the extraction of two- and three-ring PAHs, while RAMEB was a bit more effective in extracting four- and five-ring compounds (Figure 7.5) (Sánchez-Trujillo *et al.*, 2013). The efficiency of the extraction with hydroxypropyl-gamma-cyclodextrin (HPGCD), containing larger rings (diameter: 0.75–0.83 nm), was comparable with those of beta-CD derivatives in extracting four- and five-ring PAHs.

The efficiency of the extraction depends not only on the chemical properties of the contaminants but also on soil properties such as clay content and porosity (Duan *et al.*, 2015).

Sequential extraction with HPBCD solutions gives information on the residual contaminant concentration after biodegradation (Sabaté *et al.*, 2006).

PAH biodegradation data from microcosm studies was compared to PAH bioaccessibility data determined by HPBCD extraction in order to develop PAH

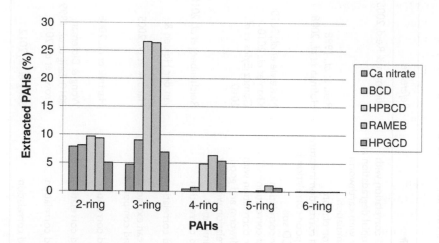

Figure 7.5 Extraction of sixteen PAHs from an aged, contaminated soil collected at a chemical plant in Spain using aqueous solutions of various CDs and calcium nitrate as control (Sánchez-Trujillo *et al.*, 2013).

bioaccessibility – biodegradability predictive models (using linear regression) (Juhasz *et al.*, 2014). The fifteen soils with a total PAH content in the range of 40–400 mg/kg involved in this study were collected from various locations. Another set of soils (ten soils with 60–2800 mg/kg) was used for validation of the models. Validated linear regression models were developed for eight of the twelve PAHs assessed (acenaphthylene, phenanthrene, anthracene, fluoranthene, pyrene, chrysene, dibenz[*a*,*h*]anthracene and indeno[1,2,3-*cd*]pyrene) and are useful for predicting the final PAH concentration after enhanced natural attenuation. Lower correlation was obtained for the four remaining PAHs (benz[*a*]anthracene, benzo[*a*]pyrene, benzo[*b*,*p*,*k*]fluoranthene and benzo[*ghi*]perylene).

There are a few studies on higher organisms such as earthworms, Collembola (Hartnik *et al.*, 2005) and algae (Bi Fai *et al.*, 2008), as well as plants (Gomez-Eyles *et al.*, 2010) (Table 7.3). The results are controversial: Hartnik *et al.* (2005, 2008) observed a good correlation on earthworm bioaccumulation for cypermethrin and chlorfenvinphos. Hickman and Reid (2005), however, found poor correlation for phenanthrene. The reason for the poor correlation may be that earthworms can access compounds not only from the aqueous phase but also the solid phase (Gomez-Eyles *et al.*, 2010). It was also shown that the profile of PAHs extracted by various techniques and accumulated in plants or earthworms was significantly different.

The results of Molnár *et al.* (2009) draw attention to the fact that correlation with bioavailability does not refer to the correlation with biodegradation in soil because the latter depends on the biodegradation potential of the metagenome of the microbial society. On the other hand, only the available compounds can be biodegraded. As microbial availability is the prerequisite of biodegradation, according to Gruiz *et al.* (2009) good prediction from the amount of CD-extracts is expected only for the biodegradation limited by bioavailability.

Table 7.3 Comparison of the results of toxicity/bioaccumulation assays and HPBCD extraction to mimic bioavailability of contaminants in soil (Hajdu et al., 2009a).

Contaminant	Environmental element	Bioassay	Result	Reference
Phenanthrene	soil (sand, silt, clay, peat)	earthworms (Lumbricus rubellus) phenanthrene degrading microorganisms	good correlation with microbial degradation and poor with earthworm accumulation	Hickman & Reid, 2005
Phenanthrene	sandy loam	uptake of earthworms	no correlation	Reid et al., 1998
Phenanthrene	spiked OECD and natural soil	uptake of earthworms (Enchytraeus albidus)	good correlation in natural soil, poor correlation in OECD soil	Hofman et al., 2008
Phenanthrene	spiked peat soil	earthworm uptake (Eisenia foetida)	poor correlation	McKelvie et al., 2010
Pyrene	spiked, aged soil	earthworm uptake	good correlation	Khan et al., 2010
PAHs	spiked, aged soil	earthworm (Eisenia foetida) and rye grass root (Lolium multiflorum) accumulation bioassays	poor correlation with earthworm and good correlation with plant accumulation	Gomez-Eyles et al., 2010
PAHs	manufactured gas plant soil	chronic toxicity earthworms (Lumbricus rubellus)	good correlation	Reichenberg et al., 2010
PAHs	sediment	bioaccumulation of aquatic worms (Lumbriculus variegatus)	good correlation	van der Heiden & Jonker, 2009
Alpha-cypermethrin	2 different types of soils	Collembola (Folsomia candida), earthworm (Eisenia foetida) acute and subacute tests	CD can extract the weakly bound compound	Hartnik et al., 2005
Cypermethrin, chlorfenvinphos DDT and PCBs	agricultural and forest soil	uptake of earthworms: bioaccumulation	good correlation	Hartnik et al., 2008
	spiked soil	earthworm uptake	good correlation	Wong & Bidleman, 2010; Tang et al., 1999
Phenanthrene	spiked soil	earthworm (Eisenia foetida) NMR metabolomic response	good correlation	Brown et al., 2010
Hexachlorobenzene	biochar-amended soil	earthworm (Eisenia foetida) uptake	good correlation	Song et al., 2012

Limited bioavailability was the reason for the lack of biodegradation at a PAH-contaminated site where the bioavailable fraction of PAHs measured by HPBCD extraction was <10% (Mahmoudi *et al.*, 2013).

On the other hand, the HPBCD extraction methods were found to be in correlation with the results of various toxicity tests (Molnár *et al.*, 2009). The adverse effects such as toxicity also depend on bioavailability since the first step is always the binding or taking up of the contaminant by the test organism.

5 CYCLODEXTRINS IN BIOASSAYS

CDs as bioavailability-enhancing agents influence the results of the biotests used for the risk assessment of a contaminant or contaminated soil and water (Hajdu *et al.*, 2009a). Aqueous CD solutions can be used for selective extraction of organic contaminants from soil for the purposes of biotesting, given that they are non-toxic and they form water-soluble inclusion complexes with the contaminants. Increasing bioavailability, bioaccessibility and uptake from solid matrix CDs enhance sensitivity of the direct-contact bioassays and results in hazardous-effect-based methods, which may be of importance in modern environmental risk assessment and risk management. This is a pessimistic approach that takes into account not only the actual bioavailable fraction but also the easily desorbable, bioaccessible fraction of the contaminant.

Approaches based on total extraction of contaminants do not take into account the importance of bioavailability and ageing processes, thus leading to possible overestimation of risk. Application of CDs may be a good tool to create optimally conservative estimates (moderate overestimation) by creating a realistic environmental scenario with maximal occurring availability and effect in biotests.

5.1 Effect of cyclodextrin solution

Compared to the organic solvents, the usage of CD solution as an extractant is closer to the 'biological extraction' process when microorganisms in soil or sediment contact (bind or take up) contaminants. Extraction of soil with organic solvents results in overestimation of the risk of the contaminants, but water-based extraction leads to underestimation. CD extraction offers a good compromise: a more realistic, but still pessimistic, result. Some examples of the application of CD solution in bioassays are collected in Table 7.4.

Reid *et al.* (1998) tested the effect of phenanthrene, pyrene, and benzo[*a*]pyrene in a 50 mM HPBCD solution using lux assay and ATP assay and concluded that HPBCD decreased the toxic effect of the tested PAHs because the effective part of the complexed PAH molecules was most likely covered by HPBCD. Decreased toxicity of the studied compounds was observed in the algae test (Bi Fai *et al.*, 2008) and in the white-rot fungi test for PCP (Boyle, 2006). As a consequence of the enhanced leaching, the toxicity of lichen extracts, however, increased in relation to water (Kristmundsdóttir *et al.*, 2005). The HPBCD extract resulted in toxicity in between that of extracts obtained by water and by organic solvent for dioxins,

Table 7.4 Effect of CD on the results of toxicity/bioaccumulation assays (Hajdu *et al.*, 2009a).

Contaminant	Bioassay	CD extraction	Result	Reference
Benzo[a]pyrene, pyrene, phenanthrene	lux assay, ATP assay	HPBCD (~5%)	decreased toxicity in HPBCD solution compared to water solution	Reid *et al.*, 1998
2,4-dichlorophenol, diuron, isoproturon, ZnSO$_4$	algae toxicity test	HPBCD (~2.5%)	decreased toxicity compared to water solution	Bi Fai *et al.*, 2008
Pentachlorophenol	white-rot fungi growth and respiration test	gamma-CD and MeBCD (2.5%)	decreased toxicity	Boyle, 2006
Atranorin, fumarprotocetraric acid, dibenzofuran derivative, (+)-usnic acid	cell proliferation assay	HPBCD, HPGCD (10–40%)	enhanced toxicity	Kristmundsdóttir *et al.*, 2005
Dioxins, PAHs, PCBs	dioxin-responsive, chemically-activated luciferase expression (DR-CALUX) bioassay	HPBCD	enhanced toxicity compared to water and decreased compared to solvent extract	Puglisi *et al.*, 2007
PCP	Ames mutagenicity test	RAMEB	enhanced mutagenicity	Hajdu *et al.*, 2009b

PAHs and PCBs (Puglisi *et al.*, 2007). The mutagenic effect of PCP that was not detectable in water or ethanol/water was detected in RAMEB solution (Hajdu *et al.*, 2009b). Two effects of complexation explain these controversial observations: the protective effect (the contact/uptake of the included compound by the microbes is hindered by the CD ring) and the solubilizing effect (the enhanced concentration of the contaminant in the aqueous phase results in an enhanced biological effect) (Figure 7.6).

A method proposed by Ashworth and Bullecer (2012) avoids the effect of CDs by using CD to extract the easy-to-desorb (bioavailable) fraction of petroleum hydrocarbons (PHC) from soil, then decomposing the CD with amylase. The toxicity is correlated to the available PHC content by subjecting the CD-free extract to a standard Microtox(R) bioassay.

In zebrafish toxicity bioassays, the small molecules are delivered to zebrafish embryos and larvae by aqueous exposure. As many compounds exhibit limited solubility in aqueous solution, CD complexation can help to avoid the use of organic solvents. HPBCD in 1% concentration was well-tolerated by both embryos and larvae, indicating the utility of this carrier for compound screening in zebrafish toxicity tests (Maes *et al.*, 2012).

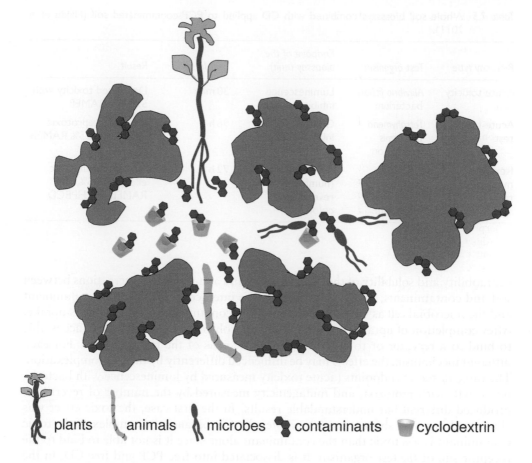

plants   animals   microbes   contaminants   cyclodextrin

Figure 7.6 CDs either enhance the bioavailability of contaminants sorbed on soil particles by increasing their solubility or protect them through inclusion.

5.2 Cyclodextrins as an additive in direct toxicity assessment of soil

Direct-contact or whole-soil bioassays do not extract the contaminant, but the test organisms are in contact with the soil surface. When the mutual interactions between the contaminant, the soil and the test organism are integrated, the results have higher environmental relevance in characterization of the risk.

PCP-contaminated soil was mixed with solid RAMEB 24h before various bioassays (Hajdu *et al.*, 2009b). In this way, the advantages of a direct-contact soil test were combined with those of CD application, assuming that the hydrophobic PCP ($logK_{ow} = 5.12$) is solubilized and hence can be integrated into the biofilm of the soil (Table 7.5).

The RAMEB application brought inconsistent results in all of the soil toxicity tests, too, because the formation of an inclusion complex with CD changes not only

Table 7.5 Whole soil bioassays combined with CD applied to PCP-contaminated soil (Hajdu *et al.*, 2011).

Bioassay type	Test organism	Endpoint of the bioassay (unit)	Test duration	Result
Acute toxicity test	*Aliivibrio fisheri* bacterium	Luminescence inhibition (%)	30 min	Decreased toxicity with 5%, 10% RAMEB
Acute toxicity test	*Tetrahymena pyriformis* protozoon	Growth inhibition (%)	96 h	Increased reprotoxic effect with 2.5% RAMEB
Reverse mutation test	*Salmonella typhimurium* TA 1538 bacterium	Reverse mutation (number of revertant colonies)	72 h	Increased mutagenic effect with 5%, 10% RAMEB and HPBCD

the mobility and solubility of the contaminant, and as a result the interactions between soil and contaminants, but has an effect on the interaction between the contaminant and the microbial cell as well. Absorption of the contaminant is limited by its uptake. After completion of uptake, it is still a crucial problem whether the substance is able to bind to a receptor or to an active site of the cells of the test organism. For each different mechanism, the effects may be influenced differently by the CD complexation. The three different endpoints (acute toxicity measured by luminescence with bacteria, by growth with protozoa, and mutagenicity measured by the number of revertants) produced different but understandable results. In the first case, the toxic effect was eliminated in a very short term (30 minutes) by CD because the complex of the toxic contaminant is less toxic than the contaminant alone since it is not able to bind to the receptor site of the test organism. It is dissociated into free PCP and free CD. In the second and third cases, the toxic effect in the growth-inhibition and mutagenicity assays did appear and was greater than with the non-complexed substance. In these tests, the test organism has the time to take up more toxic agent as it is delivered continuously to its body. The increase in the effect in both cases is due to the increased uptake.

This review shows that for the purposes of risk assessment CDs usually enhance the sensitivity of both chemical and biological methods. CD-based sensors applicable in early warning systems have been developed. Samplers containing immobilized CD can collect the target contaminants from water and wet soil. CD application can solve the major problems of sample extraction and limited bioavailability in biological testing because CDs can transfer hydrophobic molecules into a hydrophilic phase. Their usage in non-exhaustive extraction showed that CDs may help the readily bioavailable contaminants to desorb from the soil. Mixing CDs directly to soil ensures a biological model scenario with enhanced bioavailability. Although bioavailability is not the only parameter which influences the effect of a hazardous chemical substance, in environmental bioassays the bioavailability can be the limiting factor of the process, and accessibility and uptake of contaminants can be either increased or decreased by CDs under proper conditions.

REFERENCES

Alarie, J.P. & Vo-Dinh, T. (1991) A fiber-optic cyclodextrin-based sensor. *Talanta*, 38, 529–534.

Allan, I.J., Semple, K.T., Hare, R. & Reid, B.J. (2006) Prediction of mono- and polycyclic aromatic hydrocarbon degradation in spiked soils using cyclodextrin extraction. *Environmental Pollution*, 144, 562–571.

Ashworth, J. & Bullecer, I. (2012) A rapid bioassay to screen soils for toxicity of residual petroleum hydrocarbons. *Canadian Journal of Soil Science*, 92, 901–904.

Bi Fai, P., Grant, A. & Reid, B.J. (2008) Compatibility of hydroxypropyl-β-cyclodextrin with algal toxicity bioassays. *Environmental Pollution*, 157, 135–140.

Bissell, R.A. & de Silva, A.P. (1991) Phosphorescent PET (photoinduced electron transfer) sensors: Prototypical examples for proton monitoring and a 'message in a bottle' enhancement strategy with cyclodextrins. *Journal of the Chemical Society, Chemical Communications*, 17, 1148–1150.

Boyle, D. (2006) Effects of pH and cyclodextrins on pentachlorophenol degradation (mineralization) by white-rot fungi. *Journal of Environmental Management*, 80, 380–386.

Broan, C.J. (1995) *A method and material for the detection of ionizing radiation*. PCT International Application WO 1995016210 A1.

Brown, S.A.E., McKelvie, J.R., Simpson, A. & Simpson, M.J. (2010) 1H NMR metabolomics of earthworm exposure to sub-lethal concentrations of phenanthrene in soil. *Environmental Pollution*, 158, 2117–2123.

Butterfield, M.T., Agbaria, R.A. & Warner, I.M. (1996) Extraction of volatile PAHs from air by use of solid cyclodextrin. *Analytical Chemistry*, 68, 1187–1190.

Cuypers, C.P., Grotenhuis, T. & Rulkens, W. (2002) The estimation of PAH bioavailability in contaminated sediments using hydroxypropyl-β-cyclodextrin and Triton X-100 extraction techniques. *Chemosphere*, 46, 1235–1245.

Dickert, F.L., Greibl, W., Hayden, O., Licberzeit, P., Sikorski, R., Tortschanoff, M. & Weber, K. (1999) Development of materials for chemical sensors – from molecular cavities to imprinting techniques. *Advances in Science and Technology (Faenza, Italy)*, 25(Smart Materials Systems), 175–182.

Doick, K.J., Dew, N.M. & Semple, K. T. (2005) Linking catabolism to cyclodextrin extractability: Determination of the microbial availability of PAHs in soil. *Environmental Science & Technology*, 39, 8858–8864.

Du, D., Wang, M., Cai, J. & Zhang, A. (2010) Sensitive acetylcholinesterase biosensor based on assembly of beta-cyclodextrins onto multiwall carbon nanotubes for detection of organophosphates pesticide. *Sensors and Actuators B: Chemical*, 146(1), 337–341.

Duan, L., Naidu, R., Liu, Y., Palanisami, T., Dong, Z., Mallavarapu, M. & Semple, K.T. (2015) Effect of ageing on benzo[a]pyrene extractability in contrasting soils. *Journal of Hazardous Materials*, 296, 175–184. DOI: 10.1016/j.jhazmat.2015.04.050

Fang, C., Bandaru, N.M., Ellis, A.V. & Voelcker, N.H. (2013) Beta-cyclodextrin decorated nanostructured SERS substrates facilitate selective detection of endocrine disruptor chemicals. *Biosensors and Bioelectronics*, 42, 632–639.

Faraji, H. (2005) Beta-cyclodextrin-bonded silica particles as the solid-phase extraction medium for the determination of phenol compounds in water samples followed by gas chromatography with flame ionization and mass spectrometry detection. *Journal of Chromatography*, 1087(1–2), 283–288.

Fenyvesi, É. & Balogh, K. (2009) CD-containing sorbents for removal of toxic contaminants from waste water. *Cyclodextrin News*, 23(4), 1–8.

Fenyvesi, É. & Jicsinszky, L. (2009) Cyclodextrin-containing sensors to provide an early warning of contamination. *Land Contamination & Reclamation*, 17, 405–412.

Fenyvesi, É., Molnár, M., Kánnai, P., Balogh, K., Illés, G. & Gruiz, K. (2008) Cyclodextrin extraction of soils for modelling bioavailability of contaminants. In: *Proceedings of the 14th Intern Cyclodextrins Symp – Kyoto, Japan, May 8–11, 2008*. pp. 238–241.

Ferancová, A., Korgová, E., Labuda, J., Zima, J. & Barek, J. (2002) Cyclodextrin modified carbon paste based electrodes as sensors for the determination of carcinogenic polycyclic aromatic amines. *Electroanalysis*, 14, 1668–1673.

Ferancová, A., Buckova, M., Korgova, E., Korbut, O., Gruendler, P., Waernmark, I., Stepan, R., Barek, J., Zima, J. & Labuda, J. (2005) Association interaction and voltammetric determination of 1-aminopyrene and 1-hydroxypyrene at cyclodextrin and DNA based electrochemical sensors. *Bioelectrochemistry*, 67, 191–197.

Fourmentin, S., Surpateanu, G.G., Blach, P., Landy, D., Decock, P. & Surpateanu, G. (2006) Experimental and theoretical study on the inclusion capability of a fluorescent indolizine beta-cyclodextrin sensor towards volatile and semi-volatile organic guest. *Journal of Inclusion Phenomena and Macrocyclic Chemistry*, 55, 263–269.

Fu, Y., Hu, Y., Zheng, Y. & Li, G. (2006) Preparation and application of poly(dimethylsiloxane)/beta-cyclodextrin solid-phase microextraction fibers. *Journal of Separation Science*, 29(17), 2684–2691.

Gomez-Eyles, J.L., Collins, C.D. & Hodson, M.E. (2010) Assessing the capability of butanol, cyclodextrin and Tenax extractions to predict polycyclic aromatic hydrocarbon bioavailability: Are accumulation bioassays valid reference systems? *Environmental Pollution*, 158, 278–284.

Gruiz, K. (2009) Early warning and monitoring in efficient environmental management. *Land Contamination & Reclamation*, 17, 385–404.

Gruiz, K., Molnár, M., Fenyvesi, É., Nagy, Zs., Illés, G. & Kánnai, P. (2009) Comparative evaluation of biological and chemical tools for characterization of biodegradation in soil. In: *14th Intern Symp Toxicity Assessment – Metz, France, August 30-September 4, 2009, Program and Abstract Book*. p. 117.

Gruiz, K., Molnár, M., Fenyvesi, É., Hajdu, Cs., Atkári, Á. & Barkács, K. (2011) Cyclodextrins in innovative engineering tools for risk-based environmental management. *Journal of Inclusion Phenomena and Macrocyclic Chemistry*, 70, 299–306.

Guo, W., Hu, X., Xu, S. & Feng, L. (2011) *Solid phase extraction tube of triazine herbicide and its preparation*, CN 101972637. (Chemical Abstract: 154:287342).

Hajdu, Cs., Fenyvesi, É. & Gruiz, K. (2009a) Cyclodextrins in environmental bioassays. *Cyclodextrin News*, 23(10), 1–10.

Hajdu, Cs., Gruiz, K. & Fenyvesi, É. (2009b) Bioavailability- and bioaccessibility-dependent mutagenicity of pentachlorophenol (PCP). *Land Contamination & Reclamation*, 17, 473–481.

Hajdu, Cs., Gruiz, K., Fenyvesi, É. & Nagy, Zs. (2011) Applications of cyclodextrins in environmental bioassays for soils. *Journal of Inclusion Phenomena and Macrocyclic Chemistry*, 70, 307–313.

Hamada, F., Ogawa, N. & Narita, Y. (2001) *Environmental hormones testing reagent by using fluorescent molecular sensory system*. Kokai Tokkyo Koho, JP 2001281153. (Chemical Abstract: 135:307859).

Hartnik, T., Styrishave, B., Busser, F., Hermens, J. & Jensen, J. (2005) Dose response relationships in soil ecotoxicology: Relating toxicity to bioavailable concentrations instead of total concentrations. *SETAC Europe: 15th annual meeting – Lille, France, 22–26 May, 2005*.

Hartnik, T., Jensen, J. & Hermens, J.L.M. (2008) Nonexhaustive beta-cyclodextrin extraction as a chemical tool to estimate bioavailability of hydrophobic pesticides for earthworms. *Environmental Science & Technology*, 42, 8419–8425.

Hattori, K., Takeuchi, T., Ogata, M., Takanohashi, A., Mikuni, K., Nakanishi, K. & Imata, I. (2007) Detection of environmental chemicals by SPR assay using branched cyclodextrin as sensor ligand. *Journal of Inclusion Phenomena and Macrocyclic Chemistry*, 57, 339–342.

Hickman, Z.A. & Reid, B.J. (2005) Towards a more appropriate water based extraction for the assessment of organic contaminant availability. *Environmental Pollution*, 138, 299–306.

Hickman, Z.A., Swindell, A.L., Allan, I.J., Rhodes, A.H., Hare, R., Semple, K.T. & Reid, B.J. (2008) Assessing biodegradation potential of PAHs in complex multi-contaminant matrices. *Environmental Pollution*, 156, 1041–1045.

Hofman, J., Rhodes, A. & Semple, K.T. (2008) Fate and behaviour of phenanthrene in the natural and artificial soils. *Environmental Pollution*, 152, 468–475.

Hu, Y., Yang, Y., Huang, J. & Li, G. (2005) Preparation and application of poly (dimethyl-siloxane)/ beta-cyclodextrin solid-phase microextraction membrane. *Analytica Chimica Acta*, 543(1–2), 17–24.

Hu, Y., Zheng, Y., Zhu, F. & Li, G. (2007) Sol-gel coated polydimethylsiloxane/beta-cyclodextrin as novel stationary phase for stir bar sorptive extraction and its application to analysis of estrogens and bisphenol A. *Journal of Chromatography*, 1148(1), 16–22.

Ikeda, H., Sugiyama, T. & Ueno, A. (2007) New chemosensor for larger guests based on modified cyclodextrin bearing seven hydrophobic chains each with a hydrophilic end group. *Journal of Inclusion Phenomena and Macrocyclic Chemistry*, 57, 83–87.

Jicsinszky, L., Fenyvesi, É., Hashimoto, H. & Ueno, A. (1996) Cyclodextrin derivatives. In: Szejtli, J. & Osa, T. (eds.) *Comprehensive Supramolecular Chemistry*. Volume 3. Oxford, Elsevier. pp. 57–188.

Johnsen, A.R., De Lipthay, J.R., Reichenberg, F., Sorensen, S.J., Andersen, O., Christensen, P., Binderup, M.L. & Jacobsen, C.S. (2006) Biodegradation, bioaccessibility, and genotoxicity of diffuse polycyclic aromatic hydrocarbon (PAH) pollution at a motorway site. *Environmental Science & Technology*, 40, 3293–3298.

Ju, M.J., Yang, D.H., Lee, S.W., Kunitake, T., Hayashi, K. & Toko, K. (2007) Fabrication of TiO2/gamma-CD films for nitro aromatic compounds and its sensing application via cyclic surface-polarization impedance (cSPI) spectroscopy. *Sensors and Actuators B: Chemical*, 123, 359–367.

Juhasz, A.L., Waller, N. & Stewart, R. (2005) Predicting the efficacy of polycyclic aromatic hydrocarbon bioremediation in creosote-contaminated soil using bioavailability assays. *Bioremediation Journal*, 9, 99–114.

Juhasz, A.L., Aleer, S. & Adetutu, E.M. (2014) Predicting PAH bioremediation efficacy using bioaccessibility assessment tools: Validation of PAH biodegradation-bioaccessibility correlations. *International Biodeterioration & Biodegradation*, 95, 320–329.

Jyisy, Y., Lin, H.J. & Huang, H.Y. (2005) Characterization of cyclodextrin modified infrared chemical sensors. Part II. Selective and quantitative determination of aromatic acids. *Analytica Chimica Acta*, 530, 213–220.

Kanclerz, K., Koziel, K., Suchecka, A., Strojek, J.W., Lapkowski, M., Niedzielski, C. & Szeja, W. (1995) A modification of Pt minielectrode surface by polyaniline film with cyclodextrin. *Polish Journal of Chemistry*, 69, 316–319. (Chemical Abstract: 123:211258).

Khan, M.I., Cheema, S.A., Shen, C., Zhang, C., Tang, X., Malik, Z., Chen, X. & Chen, Y. (2011) Assessment of pyrene bioavailability in soil by mild hydroxypropyl-β-cyclodextrin extraction. *Archives of Environmental Contamination and Toxicology*, 60, 107–115.

Komiyama, M. (1988) Preparation of cyclodextrin membrane-modified electrodes as sensor materials. *Angewandte Makromolekulare Chemie*, 163, 205–207.

Kristmundsdóttir, T., Jónsdóttir, E., Ógmundsdóttir, H.M. & Ingólfsdóttir, K. (2005) Solubilization of poorly soluble lichen metabolites for biological testing on cell lines. *European Journal of Pharmaceutical Science*, 24, 539–543.

McKelvie, J.R., Wolfe, D.M., Celejewski, M., Simpson, A.J. & Simpson, M.J. (2010) Correlations of Eisenia fetida metabolic responses to extractable phenanthrene concentrations through time. *Environmental Pollution*, 158(6), 2150–2157.

Maes, J., Verlooy, L., Buenafe, O.E., de Witte, P.A.M., Esguerra, C.V. & Crawford, A.D. (2012) Evaluation of 14 organic solvents and carriers for screening applications in zebrafish embryos and larvae. *PLoS ONE*, 7(10): e43850. DOI: 10.1371/journal.pone.0043850

Mahmoudi, N., Slater, G.F. & Juhasz, A.L. (2013) Assessing limitations for PAH biodegradation in long-term contaminated soils using bioaccessibility assays. *Water, Air, and Soil Pollution*, 224(2), 1–11.

Malanga, M. (2011) Fluorescent cyclodextrin derivatives: Major recent advances. *Cyclodextrin News*, 25(2), 1–9.

Mamba, B.B., Krause, R.W., Malefetse, T.J., Mhlanga, S.D., Sithole, S.P., Salipira, K.L. & Nxumalo, E.N. (2007) Removal of geosmin and 2-methylisoborneol (2-MIB) in water from Zuikerbosch Treatment Plant (Rand Water) using beta-cyclodextrin polyurethanes. *Water SA*, 33(2), 223–228.

Manivannan, S. & Ramaraj, R. (2013) Silver nanoparticles embedded in cyclodextrin-silicate composite and their applications in Hg(II) ion and nitrobenzene sensing. *Analyst*, 138, 1733–1739.

Mhlongo, S.H., Mamba, B.B. & Krause, R.W. (2009) Monitoring the prevalence of nitrosamines in South African waters and their removal using cyclodextrin polyurethanes. *Physics and Chemistry of the Earth*, 34(13–16), 819–824.

Molnár, M., Fenyvesi, É., Gruiz, K., Illés, G., Hajdu, Cs. & Kánnai, P. (2009) Laboratory testing of biodegradation in soil: A comparative study on five methods. *Land Contamination & Reclamation*, 17(3–4), 495–508.

Morillo, E., Sánchez-Trujillo, M.A., Moyano, J.R., Villaverde, J., Gómez-Pantoja, M.E. & Pérez-Martínez, J.I. (2012) Enhanced solubilisation of six PAHs by three synthetic cyclodextrins for remediation applications: Molecular modelling of the inclusion complexes. *PLoS ONE*, 7(9): e44137. DOI: 10.1371/journal.pone.0044137

Nagy, Zs.M., Molnár, M., Fekete-Kertész, I., Molnár-Perl, I., Fenyvesi, É. & Gruiz, K. (2014) Removal of emerging micropollutants from water using cyclodextrin. *Science of the Total Environment*, 485–486, 711–719.

Ogoshi, T. & Harada, A. (2008) Chemical sensors based on cyclodextrin derivatives. *Sensors*, 8(8), 4961–4982.

Oláh, E. (2011) Cyclodextrins in analytical sample preparation. *Cyclodextrin News*, 25(6), 1–9.

Papadopoulos, A., Reid, B.J. & Semple, K.T. (2007) Prediction of microbial accessibility of carbon-14-phenanthrene in soil in the presence of pyrene or benzo[a]pyrene using an aqueous cyclodextrin extraction technique. *Journal of Environmental Quality*, 36, 1385–1391.

Paton, G.I., Reid, B.J. & Semple, K.T. (2009) Application of a luminescence-based biosensor for assessing naphthalene biodegradation in soils from a manufactured gas plant. *Environmental Pollution*, 157, 1643–1648.

Puglisi, E., Murk, A.J., van den Berg, H.J. & Grotenhuis, T. (2007) Extraction and bioanalysis of the ecotoxicologically relevant fraction of contaminants in sediments. *Environmental Toxicology and Chemistry*, 26, 2122–2128.

Rahemi, V., Vandamme, J.J., Garrido, J.M.P.J., Borges, F., Brett, C.M.A. & Garrido, E.M.P.J. (2012) Enhanced host-guest electrochemical recognition of herbicide MCPA using a beta-cyclodextrin carbon nanotube sensor. *Talanta*, 99, 288–293.

Rahemi, V., Garrido, J.M.P.J., Borges, F., Brett, C.M.A.E. & Garrido M.P.J. (2013) Electrochemical determination of the herbicide bentazone using a carbon nanotube β-cyclodextrin modified electrode. *Electroanalysis*, 25, 2360–2366.

Rather, J.A., Debnath, P. & De Wael, K. (2013) Fullerene-β-cyclodextrin conjugate based electrochemical sensing device for ultrasensitive detection of p-nitrophenol. *Electroanalysis*, 25, 2145–2150.

Reichenberg, F., Karlson, U.G., Gustafsson, O., Long, S.M., Pritchard, P.H. & Mayer, P. (2010) Low accessibility and chemical activity of PAHs restrict bioremediation and risk of exposure in a manufactured gas plant soil. *Environmental Pollution*, 158, 1214–1220.

Reid, B.J., Semple, K.T., Macleod, C.J., Weitz, H.J. & Paton, G.I. (1998) Feasibility of using prokaryote biosensors to assess acute toxicity of polycyclic aromatic hydrocarbons. *FEMS Microbiology Letters*, 169, 227–233.

Reid, B.J., Jones, K.C. & Semple, K.T. (2000) Bioavailability of persistent organic pollutants in soils and sediments, a perspective on mechanisms, consequences and assessment. *Environmental Pollution*, 108, 103–112.

Rhodes, A.H., McAllister, L.E. & Semple, K.T. (2010) Linking desorption kinetics to phenanthrene biodegradation in soil. *Environmental Pollution*, 158, 1348–1353.

Rostami, I. & Juhasz, A.L. (2013) Bioaccessibility-based predictions for estimating PAH biodegradation efficacy – Comparison of model predictions and measured endpoints. *International Biodeterioration & Biodegradation*, 85, 323–330.

Sabaté, J., Vinas, M. & Solanas, A.M. (2006) Bioavailability assessment and environmental fate of polycyclic aromatic hydrocarbons in biostimulated creosote-contaminated soil. *Chemosphere*, 63, 1648–1659.

Sánchez-Trujillo, M.A., Morillo, E., Villaverde, J. & Lacorte, S. (2013) Comparative effects of several cyclodextrins on the extraction of PAHs from an aged contaminated soil. *Environmental Pollution*, 178, 52–58.

Shirin, S., Buncel, E. & vanLoon, G.W. (2004) Effect of cyclodextrins on iron-mediated dechlorination of trichloroethylene – a proposed new mechanism. *Canadian Journal of Chemistry*, 82, 1674–1685.

Si, S.H., Fung, Y.S. & Zhu, D.R. (2005) Improvement of piezoelectric crystal sensor for the detection of organic vapors using nanocrystalline TiO_2 films. *Sensors and Actuators B: Chemical*, 108, 165–171.

Song, Y., Wang, F., Bian, Y., Kengara, F.O., Jia, M., Xie, Z. & Jiang, X. (2012) Bioavailability assessment of hexachlorobenzene in soil as affected by wheat straw biochar. *Journal of Hazardous Materials*, 217–218, 391–397.

Spasojević, J.M., Maletić, S.P., Rončević, S.D., Radnović, D.V., Čučak, D.I., Tričković, J.S. & Dalmacij, B.D. (2014) Using chemical desorption of PAHs from sediment to model biodegradation during bioavailability assessment. *Journal of Hazardous Materials*, 283, 60–69.

Stokes, J.D., Wilkinson, A., Reid, B.J., Jones, K.C. & Semple, K.T. (2005) Prediction of PAH biodegradation in contaminated soils using an aqueous hydroxypropyl-β-cyclodextrin. *Environmental Toxicology and Chemistry*, 24, 1325–1330.

Stroud, J.L., Paton, G.I. & Semple, K.T. (2009) Linking chemical extraction to microbial degradation of 14C-hexadecane in soil. *Environmental Pollution*, 156, 474–481.

Swanson, B.I., Li, D.Q., Shi, J.X., Johnson, S. & Yang, X. (1996) Smart-film sensors hydrocarbons for halogenated and VOCs. In: *Proceedings of the International Conference on Incineration Thermal Treatment Technologies*. Irvine, California, University of California. pp. 393-397. (Chemical Abstract: 128:183849).

Swindell, A.L. & Reid, B.J. (2007) The influence of a NAPL on the loss and biodegradation of C-14-phenanthrene residues in two dissimilar soils. *Chemosphere*, 66, 332–339.

Szaniszló, N., Fenyvesi, É. & Balla, J. (2005) Structure-stability study of cyclodextrin complexes with selected volatile hydrocarbon contaminants of soils. *Journal of Inclusion Phenomena and Macrocyclic Chemistry*, 53, 241–248.

Szejtli, J. (1998) Introduction and general overview of cyclodextrin chemistry. *Chemical Reviews*, 98, 1743–1753.

Szemán, J. (2007) Cyclodextrin-based chiral sensors. *Cyclodextrin News*, 21(4), 1–4.

Szente, L. & Szemán, J. (2013) Cyclodextrins in analytical chemistry: Host-guest type molecular recognition. *Analytical Chemistry*, 85, 8024–8030.

Tang, J., Robertson, B.K. & Alexander, M. (1999) Chemical-extraction methods to estimate bioavailability of DDT, DDE, and DDD in soil. *Environmental Science & Technology*, 33(23), 4346–4351.

Thomas, J.D.R. (1990) Membrane systems for piezoelectric and electrochemical sensing in environmental chemistry. *International Journal of Environmental Analytical Chemistry*, 38, 157–169. (Chemical Abstract: 112:190760).

Trevino, A.L., Watanabe, Y., Billiot, F. & Billiot, E.J. (2005) Development of a modified air sampling technique for the measurement of PAHs using a PUF/cyclodextrin sorbent. *National Meeting of American Chemical Society, Division of Environmental Chemistry*, 45, 218–222.

Ueno, A. (1993) Modified cyclodextrins as supramolecular sensors of molecular recognition. In: Tsuruta, T. (ed.) *New Functional Materials*. Volume C. Amsterdam, Elsevier. pp. 521–526.

van der Heiden, S.A. & Jonker, M.T.O. (2009) PAH bioavailability in field sediments: Comparing different methods for predicting in situ bioaccumulation. *Environmental Science & Technology*, 43, 3757–3763.

Villaverde, J., Maqueda, C., Undabeytia, T. & Morillo, E. (2007) Effect of various cyclodextrins on photodegradation of a hydrophobic herbicide in aqueous suspensions of different soil colloidal components. *Chemosphere*, 69, 575–584.

Wang, J. & Ueno, A. (2000) Naphthol-modified β-cyclodextrins as fluorescent sensors for detecting contaminants in drinking water. *Macromolecular Rapid Communications*, 21, 887–890.

Wang, X. & Brusseau, M.L. (1995) Cyclopentanol-enhanced solubilization of polycyclic aromatic hydrocarbons by cyclodextrins. *Environmental Science & Technology*, 29, 2346–2351.

Wang, X., Zeng, H., Zhao, L. & Lin, J.M. (2006) Selective determination of bisphenol A (BPA) in water by a reversible fluorescence sensor using pyrene/dimethyl beta-cyclodextrin complex. *Analytica Chimica Acta*, 556, 313–318.

Wong, F. & Bidleman, T.F. (2010) Hydroxypropyl-beta-cyclodextrin as non-exhaustive extractant for organochlorine pesticides and polychlorinated biphenyls in muck soil. *Environmental Pollution*, 158, 1303–1310.

Xu, H., Zhang, X. & Zhan, J. (2010) Determination of pentachlorophenol at carbon nanotubes modified electrode incorporated with beta-cyclodextrin. *Journal of Nanoscience and Nanotechnology*, 10(11), 7654–7657.

Yu, Q., Liu, Y., Liu, X., Zeng, X., Luo, S. & Wei, W. (2010) Simultaneous determination of dihydroxybenzene isomers at MWCNTs/beta-cyclodextrin modified carbon ionic liquid electrode in the presence of cetylpyridinium bromide. *Electroanalysis*, 22, 1012–1018.

Zhang, J., Pan, M., Gan, N., Cao, Y. & Wu, D. (2014) Employment of a novel magnetically multifunctional purifying material for determination of toxic highly chlorinated polychlorinated biphenyls at trace levels in soil samples. *Journal of Chromatography A*, 1364, 36–44.

Zheng, X., Liu, S., Hua, X., Xia, F., Tian, D. & Zhou, C. (2015) Highly sensitive detection of 2,4,6-trichlorophenol based on HS-beta-cyclodextrin/gold nanoparticles composites modified indium tin oxide electrode. *Electrochimica Acta*, 167, 372–378.

Zhou, J., Yang, F., Cha, D., Zeng, Z. & Xu, Y. (2007) Headspace solid-phase microextraction with novel sol-gel permethylated-beta-cyclodextrin/hydroxyl-termination silicone oil fiber for determination of polybrominated diphenyl ethers by gas chromatography-mass spectrometry in soil. *Talanta*, 73(5), 870–877.

Zhu, G.B., Zhang, X.H., Gai, P.B. & Chen, J.H. (2012) Enhanced electrochemical sensing for persistent organic pollutants by nanohybrids of graphene nanosheets that are noncovalently functionalized with cyclodextrin. *ChemPlusChem*, 77(9), 844–849.

Subject index

Engineering Tools for Environmental Risk Management

Editors: Katalin Gruiz, Tamás Meggyes & Éva Fenyvesi

Engineering Tools for Environmental Risk Management: 1
Environmental Deterioration and Contamination – Problems and their Management
©2014
Editors: Katalin Gruiz, Tamás Meggyes & Éva Fenyvesi
ISBN: 9781138001541 (Hardback)
e-book ISBN: 9781315778785
Cat# K22815

Engineering Tools for Environmental Risk Management: 2
Environmental Toxicology
©2015
Editors: Katalin Gruiz, Tamás Meggyes & Éva Fenyvesi
ISBN: 9781138001558 (Hardback)
e-book ISBN: 9781315778778
Cat# K22816

Engineering Tools for Environmental Risk Management: 3
Site Assessment and Monitoring Tools
Editors: Katalin Gruiz, Tamás Meggyes & Éva Fenyvesi
ISBN: 9781138001565 (Hardback)
e-book ISBN: 9781315778761
Cat# K22817

Forthcoming:

Engineering Tools for Environmental Risk Management: 4
Risk Reduction Technologies
Editors: Katalin Gruiz, Tamás Meggyes & Éva Fenyvesi
ISBN: 9781138001572 (Hardback)
e-book ISBN: 9781315778754
Cat# K22818

Engineering Tools for Environmental Risk Management: 5
Integrated Environmental Risk Management – Case Studies
Editors: Katalin Gruiz, Tamás Meggyes & Éva Fenyvesi
ISBN: 9781138001589 (Hardback)
e-book ISBN: 9781315778747
Cat# K22819

Printed and bound by CPI Group (UK) Ltd, Croydon, CR0 4YY

24/10/2024

01778285-0002